HANDBOOK FOR
CHEMICAL
PROCESS
INDUSTRIES

Himanshu J. Patel

Department of Applied Science & Humanities
Pacific School of Engineering, Gujarat, India

CRC CRC Press
Taylor & Francis Group
Boca Raton London New York

CRC Press is an imprint of the
Taylor & Francis Group, an **informa** business

A SCIENCE PUBLISHERS BOOK

First edition published 2024
by CRC Press
6000 Broken Sound Parkway NW, Suite 300, Boca Raton, FL 33487-2742

and by CRC Press
4 Park Square, Milton Park, Abingdon, Oxon, OX14 4RN

CRC Press is an imprint of Taylor & Francis Group, LLC

ISBN: 978-1-032-53482-4 (hbk)
ISBN: 978-1-032-53486-2 (pbk)
ISBN: 978-1-003-41228-1 (ebk)

DOI: 10.1201/9781003412281

Typeset in Times New Roman by
Radiant Productions

Preface

The Chemical Process Industries are play pivotal role for each and every country, as economy of country is depends upon its chemical process industries enhancement. It is also anticipating the job opportunities to all over the World. It was estimated the chemical industry was contributed US US$171 billion in 1970, growth to $4.12 trillion in 2010. And now-a-day, it shares USD 5.7 trillion to gross domestic product (GDP), an equivalent of 7% of the World's GDP. It also gives 120 million jobs. According to OECD (Organization for Economic Co-operation and Development), chemical industries would grow to 6.0% in 2021 and 4.9% in 2022 in developing countries such as the United States, Brazil, Russia, China, India, Japan, etc. Beside economic growth and livelihood, chemical becomes essential part of daily life. We cannot imagine our lives without chemical. We all are surrounded by different chemicals. From the walk-up in morning, tooth-past and tooth-brush and after end of our day, bed-sheet and pillow, all the important materials are made from different natural as well as man-made chemicals. Although these chemicals are increasing potential risk to human life and eco-system, but new chemicals are continuously manufactured for smoothing and comfortable our lives. More than 144,000 chemicals are existed and about 2000 new chemicals are being introduced in every year.

Present book emphasized on the chemical processes of several industries. This book contains different processes of directly used chemicals from natural materials like petroleum, coal and ores from the Earth. It included chemical's general knowledge, properties, manufacturing process (such as raw materials, chemical reactions, quantitative requirement, flow sheet diagram, procedure) and its uses. The book is specially prepared for quantitative requirement, flow sheet and its procedure, which are directly collected from industrial personnel and author industrial experiences. Other materials like general knowledge, properties, raw materials and uses are taken from different internet as well as industrial resources.

The book contain fifteen chapters, which covers entire range of organic and inorganic industries. The components of chemical process industries such as concept, raw materials, unit process, unit operation, flow sheet diagram, parts of diagram, etc. are explained briefly in first chapter. Second chapter comprises introduction, extraction, component, classification, purification and convert into important chemicals of three most important fossil fuels, i.e., crude oil, coal and metal. Chapter three to five enclosed wide extent of organic chemical containing one, two, three and four carbons, while chapter six contains process of aromatic chemicals. Remaining chapters cover the chemical process of several industries like dye, pigment, drug, pharmaceutical, fermentation, agrochemical, explosive, polymer, Period – II chemicals, Period – III chemicals, sugar, coating, starch, soap, detergent, paper, pulp, glass and cement industries. These chapters consist of brief introduction, classification, manufacturing process of particle chemical and their usages.

Contents

CHAPTER 1

Introduction

1.1 Basic Concepts of Chemical Industries

Generally, students conduct experiments with a few grams of raw material, use a Bunsen burner for heating, and prepare a few grams of final product. However, such a practice is not used by chemical industries. Chemical Industries play a vital role in the development of each and every country. These industries depend on various aspects, such as raw materials, unit processes, unit operations, number of chemical reactions involved, types of processes, by-products, quantities and quality of final products, different utilities, details of reaction vessels and stirrers, emission of gases during chemical reactions, discharge of effluents and solid wastes during chemical processes and operations. All these features need to be maintained by chemical industries. Some of these features are discussed in this chapter. Also, we are discussing all the requirements step-by-step.

1.2 Raw Material

The starting materials of chemical industries are called raw materials. Also, some chemicals required for purification, precipitation, and solubilization are also referred as raw materials. The quantity and quality of raw materials depends upon the required final product. They are come in un-pure or pure form as per requirement. They may be natural, semi-synthetic or synthetic. They may be in solid, liquid or gaseous state (Englezos and Kalogerakis, 2001; Spero et al., 2000).

1.3 Unit Processes and Unit Operations

There are various unit processes and unit operations involved in chemical industries. There is a difference between a unit process and unit operation. Chemical conversion of raw materials takes place in the unit process, while physical or mechanical conversion is conducted in the unit operation. The unit process is defined as a method in which a reaction between two or more chemical reactants takes place and gives one or more chemical products. The examples of various unit processes are sulfonation, nitration, polymerization, oxidation, reduction, halogenation and alkylation. The reactor names are sometimes derived from the name of the unit process, i.e., nitration, sulfonation autoclave, and more. The unit operation is defined as a method which involves only physical or mechanical changes during the process such as filtration, evaporation, crystallization, is omerization, separation, size reduction, drying, distillation and liquid-liquid extraction.

There are single or multipleunit processes (chemical reactions) involved in the conversion of raw materials to the final product. Some chemical industries are manufacturing different intermediates from raw materials and selling them in the market. Mostly dye, drug, pigment and metal intermediates are being prepared by companies.

There are three different processes by which raw materials are converted into products namely batch, continuous and semi-batch processes, but industries prefer batch or continuous processes. In a batch process, an arrangement of steps followed in a specific order are involved. Batch processes are preferred for the manufacture of drugs, dyes, foods, beverages, pharmaceuticals, paints, fertilizers, and cement. The inflow of raw material inputs and the outflow of product outputs is constant in a continuous process. Some industries, i.e., chemical and petrochemical industries conduct continuous processes. In a semi-batch process, reactants are added and products are removed gradually over a certain time period. Some drug manufacturing industries prefer semi-batch processes (UMPRC, 1995; Galbraith, 2014).

1.4 Utilities

Chemical industries require different utilities, such as electricity, water (such as raw water, purified water, cool water, hot water brine, water for injection and ice), air (compressed air, nitrogen, oxygen and another medical gases), steam, fuel (natural gas, diesel and coal), steam, refrigeration, furnaces and insulations extensively. These utilities are providing to chemical industries for maintaining suitable conditions in a manufacturing unit.

1.5 Process Flow Diagram

There are many types of flowcharts used for more than one purpose. Only material and energy flows and operating conditions are mentioned in a simple block flow-chart. However, in chemical engineering flowcharts require certain more parameters than in a simple block flow-chart. These parameters include the point of entrance for raw materials, the point of exit of impurities, by-product and product. It also indicates the fine-line, instrument and process control systems, air lines, drains, energy flows, sequences of unit processes and unit operations. Using a Process Flow Diagram (PFD), engineers can understand the common scheme of operations and its relative equipments. It reveals the real system of chemical processes, associated instruments/equipment, structure of pipe-lines between equipment, support requirements of pumps and other apparatus. We are discussing some major components of a flow sheet diagram.

Mechanical apparatus such as storage tanks, heaters, furnaces, chemical reactors, autoclaves, receiver tanks, heat exchangers, autoclaves, distillation columns, scrubbers, filter presses and different attachments such as stirrers, motors, shafts, jackets, materials of construction (MOC) and pipelines play a vital role in chemical industries. Each mechanical component has individual applications. Materials of construction of each component depend upon the properties of its raw materials, intermediates and final products. Mild steel, different grades of stainless steel (302, 304 & 316L), cast iron, carbon steel, glass and bronze are more usable construction materials in chemical industries (Snyder Jr., 1983).

1.5.1 Storage tank

Hazardous and flammable liquids (like LPG, benzene, petrol, acids, oleum, diesel, toluene, acetone, cyclo-hexane, aniline, chloroform, diethyl ether, pentane, ligroin, petroleum ether, kerosene, methanol, ethanol and xylenes) and gases (natural gas, nitrogen, hydrogen, oxygen, methyl isocyanate, phosgene, chlorine, ethylene oxide and nitrogen) are stored in separate storage tanks with proper MOC. These storage tanks are kept far from the production unit to avoid being part of process mishaps (e.g., electric shock, falling) and process safety disasters (e.g., suffocation, poisoning fire, explosion) and noxious activity. The storage tank is a closed system, so, proper safety measures like safety values, vents, high level interlocks, alarms and flare systems are vital. Also, these gases and liquids have corrosive properties, so, selection of MOC for storage tanks is quite challenging (Carson and Mumford, 2002).

1.5.2 Heater

Some raw materials and intermediates are heated to high temperatures ranging from 50 to 500°C during chemical reactions. These temperatures are achieved using direct heaters or direct fired heaters. Moderate temperatures of 50–80°C are achieved using direct heaters, in which steam is used. Heat is transferred from the steam to the raw materials. Higher temperatures of 80 to 500°C are achieved in a fired heater, in which liquid or gaseous fuel undergoes a combustion reaction in the presence of excess air (oxygen) to give energy (heat). This heat is transferred into materials like water, oil and other compounds. MOC of heaters is either nickel and chromium alloy (Nichrome) or ceramic (alumina-silica) (Towler and Sinnott, 2022).

1.5.3 Reactor and autoclave

Different types of chemical reactions (unit processes) occur in the chemical reactor and autoclave. The name of the reactor depends on the unit process conducted in it. If nitration occurs in the reactor, then its name is "Nitration

Reactor". The chemical reactor is an open system that is facilitated with a stirrer and jacket, while the autoclave is closed system, in which the chemical reaction is conducted at a higher pressure, temperature with more vigorous stirring. An autoclave is facilitated with a proper safety valve, stirrer (agitator) and jacket. Raw materials are charged from the main-hole in the upper part in the beginning of the reaction and the final reaction mass is collected from the bottom after the completion of the chemical reaction in both (chemical reactor and autoclave) components, if the reaction occurs in the liquid phase. Vapor phase reactions occur in column type reactors, in which raw materials are inserted from the bottom and the reacted mass is collected from the top of the reactor. In the liquid phase reaction a stirrer is not required. The unit processes are performed at higher or lower temperatures sometime seven at room temperature depending upon the nature of the chemical reaction (exothermic or endothermic). In addition the temperature of these two components is controlled by passing steam, hot water, cool water or brine into the jacket. There are mainly three types of jackets, i.e., plain, piped and dimple. The main function of the agitator is to make the reaction mixture homogenous thus, increasing the reaction rate. Also, there are different types of agitators such as paddle, anchor, radial propeller, propeller, turbine, and helical. The materials of construction for the reactor; its jacket and stirrer are the same in most of the chemical reactions. Some special types of reactions require different MOC for the reactor and jacket. Cast iron or mild steel reactors having jackets made of abrasive materials such as silica-alumina brick, glass, stainless steel or carbon steel are used in drugs, dyes, paints and agro-based industries. If the materials of construction (MOC) of the reactor and jacket are cast iron and glass respectively, then this type of reactor is known as "Glass Lined Cast Iron Reactor" (Brodzki et al., 1994).

1.5.4 Receiver tank and filter press

After completion of the reaction, the reaction mass is transferred into the receiver tank, where some unit operations like cooling, heating, neutralization and precipitation are performed by chemical or other means. The MOC of this tank is most probably cast iron or mild steel having agitation and sometimes jacketed facilities. After completion of reaction and precipitation, the final product is collected by filtration using a filter press. The main function of the filter press is to separate the solid product from mother liquor. In chemical industries, there are mainly two types of filtration: (1) Gravity filtration and (2) Pressure or Vacuum filtration, in which gravity filtration called cloth filtration or filter press is preferred more for cost reduction. The filter press is sub-divided to into the plate and frame and overhead filter press. The plate and frame filter press contains a number of plates with different frames and arranged in an alternative way. In filtration, solid material is kept in the frame, while liquor is separated out from the bottom. Dye, drug, pharmaceutical, petroleum, polymer, explosive and agriculture industries use this type of filter press. In the overhead filter press, the reaction mass is retained for a certain time period. The suspension is released in the upper area and solid material is collected from the bottom. It has a drawback that it takes long time. It is especially used for ceramic and sand washing industries. A recent development is to use a membrane filter press to separate out solid materials from liquor.

In the chemical industry flow-sheet charts and some another mechanical apparatus such as heat exchanger, distillation column, activated charcoal column, evaporator, crystallizer, centrifuge and dryer are mentioned. These apparatuses are shown in-between the raw materials and the final product, which have different applications. No chemical reaction takes place in each apparatus and they only conduct several unit operations.

1.5.5 Heat exchanger

Heat exchangers are used to transfer heat between two fluids. They are used for providing thermal energy needed for cooling and heating the material or reaction mass. The concept of heat exchanges is based on the simple principle of the Second Law of Thermodynamics. The heating or cooling effect of a body on another occurs with respect to the difference in temperature of the two fluids and thus, heat transfer occurs from or to a fluid resulting in cooling or heating. There are different types of heat exchangers: shell and tube type, heat generated, plate type, direct contact type, wet surface type with rotary, stem injection type and reciprocating regenerative heat exchanger. Out these types of heat exchanges, there are only three types mostly used in industries: (1) Shell and tube, (2) Plate type and (3) Regenerative heat exchanger. The shell and tube heat exchanger is the most versatile and usable. In this a big shell contains a large number of stacked tubes, so, it is also called a tubular heat exchanger. Now, material to be heated or cooled, is passed through the tubes, and already heated or cooled fluid is passed through the shell, thus energy transfer is achieved. In a plate type heat exchanger, stacked and bonded corrugated plates instead of tubes are kept and responsible for heating or cooling effects due to heat transfer. Regenerative heat exchange is also called energy efficient exchange, because, hot or cool fluid is stored in interim thermal insulated storage; required heat is transferred to or from the fluid (Ng, 2021; Shah and Sekulic, 2003).

1.5.6 Distillation column

The distillation column is used to separate individual liquids from liquid mixtures. The concept of distillation in a column is based upon the difference in boiling points of each liquid. In this column, the boiled liquid mixture is inserted into the bottom or middle of distillation column. The liquid travels upstream and cools down to its condensation point. Thereafter, it is condensed and collected. There are two types of distillation columns: (1) batch column and (2) continuous column. In a batch column, the liquid mixture is inserted batch-wise and individual components are collected. Thereafter, a new batch of liquid mixture is added. In a continuous column the liquid mixture is continuously inserted into the column and components are collected simultaneously. This process is constantly carried out until liquid mixture is consumed or a problem occurs in the column. A batch column is less efficient, highly flexible, controllable and can be used for smaller quantities while a continuous column is highly efficient, less flexible, un-controllable and can be used for larger quantities. Batch distillation is used in drug, dye, food processing, rubber, textile, paper, agricultural and polymer industries, while continuous distillation is widely used in petroleum and chemical industries. Also, it is also classified according to the internal material of the column, used for distillation: (1) tray column and (2) packed column. In a tray column, stacks of various trays are kept in the column to provide better contact between vapor and liquid. In a packed column a proper packing of ceramic, metal or plastic material, instead of trays, is used to increase the contact between vapor and liquid. Tray columns and packed columns are more suitable for high and low capacity operations respectively. There are five parts of in a distillation column: (1) Vertical narrow closed vessel having higher length, where distillation (separation of liquid) occurs. (2) Internal material: The vessel either has internal trays or packed material for improving the separation process. (3) Reboiler (heat exchanger): Material (most preferable liquid or semi-liquid) is collected from the bottom and heated in the reboiler to convert it to vapor. The vapor is further transferred into the column, while liquid materials left behind in the reboiler are collected as the bottom product. Different reboilers such as kettle, thermosyphon, fired and forced circulation reboiler are being used in distillation columns. (4) Condenser: Vapors, collected from the bottom, are cooled down to liquid formin the condenser. (5) Reflux drum: It is used to hold the condensed vapor and thereafter, liquid is sent back to the distillation column (Beneke et al., 2013; Boozarjomehry et al., 2012).

1.5.7 Evaporator

Another component of the flow-sheet chart is the evaporator. Some reactions are conducted to produce vapor forms of C1 to C4 chemicals, so, raw materials (solid or liquid) are converted to gaseous form by heating them in an evaporator. The required heat is taken from steam or other heat utilities in the evaporator. There are five sections of the evaporator column: (1) Vertical narrow vessel having lower length, where evaporation occurs. (2) Feed section, where the starting solution is fed into the vessel. (3) Heating section, where steam is fed from the middle of the vessel. By means of steam, water or solvent is evaporated from the initial solution to increase its concentration. (4) Separating section, which separates water or solvent vapor and increases the concentration of the solution. Vapor (condensate) is collected from the middle of the vessel, while the concentrated solution is collected from the bottom of the vessel. The concentrated solution is further fed into another evaporator to get the desired concentration. (5) Condenser: the condensate is cooled down to separated the vapor and the water/solvent. The vapor is further circulated in the evaporator. Sometimes, it is also kept as an intermediate in the flow chart for different applications, such as desired product concentration, volume reduction, water/solvent recovery, dryer feed pre-concentration, re-vaporization of liquefied gases, refrigeration applications (cooling or chilling) and crystallization. In the sugar industry, multi-effect evaporators are used to increase the juice concentration. Solvents are recovered and re-cycled in the pharmaceutical and chemical industry using evaporators. The food and fermentation industries are use evaporators for crystallization. The dairy industry uses evaporators in dryer feed pre-concentration. About 50% volume reduction is conducted using evaporators in food, beverage and milk products. There are several types of evaporators in the market with four main types : (1) Falling film evaporator, in which steam and feed solution are circulated by gravity and no additional pump or other control is required. This type of evaporator is used in dairy, sugar, paper and fertilizer industries. (2) Forced circulation evaporators, in which a pump is used to circulate steam and the feed contains some solids or crystalline molecules. The manufacture of most sodium, magnesium and nitrogen based inorganic chemical sutilizes this type of evaporator. (3) Long tube vertical (LTV) evaporator, in which a long tube is used as the main evaporating vessel instead of a small tube. This type of evaporator is used in the sugar, paper and some inorganic industries. (4) Plate Evaporators, in which the plates keep the thin layer of liquid flowing. They are mainly used in the polymer and food industries. (5) Mechanical vapor recompression (MVR) evaporator, in which evaporation is conducted by compressing the liquid. Dairy, brewing, sugar, saline, pulp, chemical and alcohol industries use this type of evaporator (Lage and Campos, 2004; Kalogirou, 2001).

1.5.8 Crystallizer

This technique is used for the separation and purification of compounds by means of heat transfer. This process is complex and stimulating, in which several variables like purity, crystal size, shape and the solid structure are important issues. It is conducted by dissolving the feed material to form a saturated solution in the proper solvent, such as water, methanol and acetone; thereafter, heating/cooling the saturated solution or adding an anti-solvent to the saturated solution to form solid crystals. If the feed material contains impurities, then they remain intact in solution and pure solid crystals are separated. Mostly, a saturated solution is heated above the melting point of the final product by means of steam. Here, supersaturation is developed, where the solvent is transformed into vapor and the final product is transformed into liquid and thus, liquid-solid separation occurs.

Basically, there are two consecutive processes in the crystallizer: (1) Nucleation, in which pure molecules get together in the form of clusters in a particular style. (2) Crystal formation and growth, in which the size and number of clusters are increases to finally form crystals. Crystallization is conducted in the crystallization chamber, which is divided into three parts: (1) Lower part, where the feed solution is collected and recycled into the chamber. Also, pure solid crystalline products are collected using fluid beds or simple stirring. (2) Middle part, where fresh and recycled steam is charged into the chamber. Also, fresh and recycled saturated solution is fed into the chamber (3) Upper part, where, supersaturation is conducted for liquid-vapor seperation. The vapor component of the pure product is collected from the top of the chamber and recycled into it via a heat-exchanger, while the liquid component is transferred to the bottom by gravity.

By elimination of the solvent, more preferably water, the upper part contains a liquid-vapor system where supersaturation is involved. Liquor which is slightly supersaturated will flow down through an underground pipe and relieved of its supersaturation by contact with the fluidized bed of crystals. The supersaturation is relieved gradually as the circulating mother liquor charges upwards through the classifying bed before being collected in the top part of a chamber. Then it leaves via a conduit and is sent to an upper chamber where additional heat make-up is provided. Afterwards, it passes through a heat exchanger where additional heat is removed (Grosch et al., 2008; Power et al., 2015).

1.5.9 Centrifugation

It is mainly used to separate solids from liquids, liquid-liquid separation, and liquid-liquid-solid separation by means of mechanical centrifugal force. It is also used to exact or wash materials from liquid. This process is utilized when the solid particles' density is suspected to be the same as that of the liquid preventing them from settling down to the bottom. A centrifugal force is applied to this solution, and the denser solid particles are separated. This force is several hundred or thousand times the earth's gravitational force. After the application of centrifugal force, solid particles according to their size, shape, density and viscosity descend to the bottom, while the remaining liquid is separated as an upper layer called supernatant. In this process, the solid-liquid mixture is kept in the container which is rotated at a high speed to separate solid particles as pellets from the liquid. It has uses in several industries such as dairy, pharmaceutical, wine and chalks. It is also used for wastewater treatment, air pollution control and in blood laboratories (Regel et al., 2001).

1.5.10 Dryer

It is used to eliminate moisture/water from intermediates or final products by the application of heat. There are different types of dryers depending upon the desired final product and its quantity. Some of the dryers are fluidized bed, rotary, rolling bed, spray, conduction and convection suspension/paste dryers. A wide range of industries like petroleum, organic, inorganic, agricultural, food, ceramic, detergent, cement, dye, polymer, pigment, pharmaceutical and fermentation industries use dryers (Chou et al., 200).

Now, we will discuss chemical process industries one-by-one. This book includes, a brief introduction (like inventor name, year of invention, history, etc.), properties (such as chemical name, IUPAC name, appearance, solubility, boiling point, melting point, etc.), manufacturing process, which includes chemical reactions, names of raw materials, quantities of raw materials required, chemical reactions, flow sheet diagrams and uses of each chemical. This book is clearly focussed on the quantity requirements of raw materials and flow sheet diagrams, because this information plays a vital role for any industry.

References

Beneke, D., Peters, M., Glasser, D. and Hildebrandt, D. 2013. Understanding Distillation Using Column Profile Maps. John Willey & Sons. DOI:10.1002/9781118477304

Boozarjomehry, R.B., Laleh, A.P. and Svrcek, W.Y. 2012. Evolutionary design of optimum distillation column sequence. The Canadian Journal of Chemical Engineering, 90(4): 956–972. https://doi.org/10.1002/cjce.20589

Brodzki, D., Djega-Mariadassou, G., Li, C. and Kandiyoti, R. 1994. Comparison of product distributions from the thermal reactions of tetralin in a stirred autoclave and a flowing-solvent reactor. Fuel 73(6): 789–794. https://doi.org/10.1016/0016-2361(94)90270-4

Carson, P. and Mumford, C. 2002. Hazardous Chemicals Handbook, Second Edition, Butterworth-Heinemann. Elsevier Science.https://doi.org/10.1016/B978-0-7506-4888-2. X5000-2

Chou, S.K., Hawlader, M.N.A., Ho, J.C. and Chua, K.J. 1999. The contact factor for dryer performance and design. International Journal of Energy Research, 23(14): 1277–1291.https://doi.org/10.1002/(SICI)1099-114X(199911)23:14<1277::AID-ER556>3.0.CO;2-U

Englezos, P. and Kalogerakis, N. 2001. Applied parameter estimation for chemical engineers. Chemical Industries, 1st Edition, Marcel Dekker.

Galbraith, J.R. 2014. Designing Organizations: Strategy, Structure, and Process at the Business Unit and Enterprise Levels, Jossey-Bass.

Grosch, R., Monnigmann, M. and Marquardt, W. 2008. Integrated design and control for robust performance: Application to an MSMPR crystallizer, Journal of Process Control. 18(2): 173–188. https://doi.org/10.1016/j.jprocont.2007.07.002

Kalogirou, S.A. 2001. Design of a new spray-type seawater evaporator. Desalination, 139(1–3): 345–352. https://doi.org/10.1016/S0011-9164(01)00329-0

Lage, P.L.C. and Campos, F.B. 2004. Advances in Direct Contact Evaporator Design. Chemical Engineering & Technology 27(1): 91–96. DOI: 10.1002/ceat.200401760

Ng, X.W. 2021. Concise Guide to Heat Exchanger Network Design: A Problem-based Test Prep for Students, Springer International Publishing. https://doi.org/10.1007/978-3-030-53498-1

Power, G., Hou, G., Krishna, V., Morris, G., Zhao, Y. et al. 2015. Design and optimization of a multistage continuous cooling mixed suspension, mixed product removal crystallizer. Chemical Engineering Science 133: 125–139. https://doi.org/10.1016/j.ces.2015.02.014

Regel, L.L., Wilcox, W.R. and Derebail, R. 2001. Processing by Centrifugation, Springer US.

Shah, R.K. and Sekulic, D.P. 2003. Fundamentals of Heat Exchanger Design. John Willey & Sons. DOI:10.1002/9780470172605

Snyder Jr., O.P. 1983. A Computerized Flow Chart System for Food Production, Foodservice Research International 2(4): 211–228.

Spero, J.M., DeVito, B. and Theodore, L. 2000. Regulatory Chemicals Handbook, Chemical Industries, CRC Press. https://doi.org/10.1201/9781482270389

Towler, G. and Sinnott, R. 2022. Chemical Engineering Design, Third Edition, Principles, Practice and Economics of Plant and Process Design, Chapter 19—Heat transfer equipment 823–951.

UMPRC 1995. Unit Manufacturing Process Research Committee, Commission on Engineering and Technical Systems, National Research Council, Unit Manufacturing Processes: Issues and Opportunities in Research, National Academies Press.

CHAPTER 2

Basic Chemicals from Natural Materials

2.1 Introduction

We all know that the environment is divided into four spheres: hydrosphere, lithosphere, biosphere and atmosphere, in which the lithosphere covers a small portion of the Earth. Actually, the Earth has two layers: (1) Mantle and (2) Core. The mantle is divided into three sub-parts: (1) Lithosphere, (2) As the nosphere (outer mantle) and (3) mesosphere (inner mantle). Further, the core is divided into two sub-parts: (1) Outer core and (2) Inner core. Each zone has a different temperature range and depth. Off these parts, the lithosphere is the rocky rigid outer surface of the earth and its coolest zone. It covers only 1% of the Earth by volume Lithosphere enlarged up to 70–100 km in depth. This depth is differs on the basis of location; the oceanic lithosphere occupies only 5–15 km, while the continental lithosphere occupies only 34–40 km. Some mountainous lithospheres extend upto 80–100 km. Further, the temperature of the lithosphere varies from 0 to 500°C depending upon th geographical conditions. Eight most abundant elements, i.e., oxygen, silicon, aluminum and iron are found in lithosphere. Other metals such as silica, calcium, sodium, potassium, magnesium, copper, gold, platinum, silver, zinc and chromium are also available in the lithosphere. Some non-metallic minerals (sand, stone, limestone, gravel, mica, gypsum halite, uranium) and fossil fuels (coal and crude oil) are also extracted from the lithosphere. Crude oil is available at a range of 2–10 km depending upon the terrestrial conditions. Gulf countries, USA, Russia, Europe and some parts of China and India are the main parts of the World, where oil reservoirs are available. Availability of coal is in the range of 30 to 1500 km, so, it exists in the lithosphere and the as the nosphere. It is present all over the World in varying quantities and qualities, but the United States, Russia, China, Australia, and India have the largest coal reservoirs (Collins et al., 2209). In this chapter, we will discuss three natural materials, i.e., petroleum (crude oil), coal and metals in details.

2.2 Petroleum

Crude petroleum is typically thick, luminous, dark brown in color, and has a distinctly unpleasant odor.

2.2.1 Concert of petroleum

It is generally trapped deep below the layers of the earth and often floats on salt water. The oil producing countries include The United States of America which is the world's biggest country having crude oil; the others are Russia, Venezuela, Iran, Rumania, Iraq, Gulf countries, Burma, India and Pakistan. All alkanes (C_1 to C_{40}), cycloalkanes or naphthenes and aromatic hydrocarbons are the main components of crude oil. Considering both its production value and its significance to the global economy, crude oil is without a doubt the king of all commodities. Today, crude oil is a non-financial commodity that is traded the most on a global scale. It meets 40% of the world's total energy needs, making it the second most traded non-financial commodity. The fractional distillation of crude oil yields a variety of products, some of which are utilized as fuel for both industrial and residential purposes, including gasoline (44%), heating oil and diesel fuel (19%), jet fuel (8%), residual fuel oil (5%), asphalt (3%), and miscellaneous products (21%). These fractioned materials are also used to create a variety of petrochemicals. Petrochemicals, broadly defined, are compounds and polymers utilized in the chemical industry that are either directly or indirectly produced from petroleum. Plastics, synthetic fabrics, rubber, detergents, and nitrogen fertilizers are some examples of the principal applications for petrochemical goods. Other significant chemical industries include paints, adhesives, aerosols, insecticides, and pharmaceuticals, all of which may use one or more petrochemical compounds during the course of their production. The first raw resources utilized in the production of petrochemicals are gas and petroleum. However, other carbonaceous resources like coal, oil shale, and tar sand are commonly handled in order to produce these compounds (at a cost). They also produce useful byproducts.

Crude oil is currently the world's most important energy source and is likely to remain so for many decades to come, even under the most optimistic estimates for the development of alternative energy sources. The majority of nations are considerably harmed by changes in the oil market, whether they are producers, consumers, or both. About 38 percent of the world's energy requirements were met by oil in 2014, and oil is anticipated to be the primary energy source for the foreseeable future. Oil is now regarded as the most important raw commodity. We utilize a lot of products produced from oil or gas every day. Refining was first practiced more over 5,000 years ago. In the Middle East, boats were known to be made waterproof by oil seeping up from the bottom. This oil was also used to treat illnesses and paint other objects. Since there was a far greater demand for oil than it could really supply, the idea of setting up boring businesses-collectively known as the refining industry-was developed. The value of the oil determines a lot of the refining industry's operations, and it has been noted that as oil prices rise, the value of many different products rises as well. Through boring, the refining industry is also responsible for a disproportionate amount of energy use. The countries in Europe finish with the lowest consumption since the Middle East is in first place in this issue. About 30 billion oil barrels are consumed by the world, off which 7.5 billion oil barrels are utilized by the United States of America only according to the statistics. Upstream and downstream are the two segments of the oil industry. Oil-based products go through a long process before being noticed, but once they are, they become one of the most important things in people's life. Additionally, there is a rapid rise in oil use in Non Asian nations. Around 40% of the world's oil is used in Asia outside of the Organization for Economic Co-operation and Development (OECD) (Including both India and China). The consumption of oil is expected to increase from 80 million barrels per day in 2003 to 118 million barrels per day in 2030. Today, oil plays a significant role in our daily lives and the global economy for a number of reasons.

Because oil is a widely traded and fiercely competitive commodity, basic economics dictates that profit margins will decline as far as they can without companies leaving the market. The smallest risk premium you can provide a company in this situation to persuade it to keep doing business is:

1. A market where your product can almost entirely be replaced by a competing one a cyclical sector with 4–5x price swings for finished goods and constantly rising raw costs.
2. A sector of the economy where each $100 million exploratory well has a 50–90% failure rate.
3. A company where a mistake would result in $40 billion in fines and damages.
4. A market dominated by government-run businesses that are subject to laxer legal and environmental regulations.
5. Nations with a record of forcibly taking over oil infrastructure (Islam et al., 2010; Zhijun, 2004).

2.2.2 Crude oil: occurrence and origin

Petroleum usually crude oil is viscous, dark brown colored and fluorescent having a definite offensive odor. The oil producing countries include The United States of America which is world biggest country having crude oil, others are Russia, Venezuela, Iran, Rumania, Iraq, Gulf countries, Burma, India and Pakistan. It is generally trapped deep below layers of the earth and often floats on salt water. A hundred years ago, plants and animals from the Paleolithic era gave rise to crude oil. It is thought that oil and gas-containing hydrocarbons were created due to the thermal maturation of organic materials exposed to high pressure and heat while being buried deep in the soil over an extended period of time. In nature, petroleum occurs in three states: (1) Solid, (2) Liquid and (3) Gas. Solid or semi-solid petroleum is known as pitch, which is usually black in color. Gas deposits in the form of natural gas. As per postulated theory, petroleum was formed several million years ago in anaerobic conditions. It is believed that hydrocarbons were formed by organic matter in marine deposits. Extreme conditions of high temperature and pressure with selective bacterial attacks destroyed protein and carbohydrates and left over the fats which accumulated as crude oil or fossil fuel. The rate of formation compared to present consumption is nil. In general, it's generally trapped below layers of the earth and often floats on salt water. There are two origin theories: organic and inorganic (abionic) (bionic).

2.2.2.1 Inorganic theory

When it became clear that there had been widespread petroleum deposits on the earth, early hypotheses proposed an inorganic origin. Different scientists had given their hypothesis on the formation of crude based on inorganic materials. Acording to Dmitri Mendelev (1877) a Russian scientist and the father of the periodic table of elements, heavy hydrocarbons were created when acetylene (C_2H_2) condensed from the reaction of metallic carbides with water at high temperatures. In a lab, this reaction is simple to reproduce. Two scientists, Berthelot in 1860 and Mendelev in 1902, offered additional hypotheses that modified the acetylene theory. They postulated that iron carbide, which would combine with percolating water to produce methane, was present in the mantle (blanket) beneath the surface of the Earth. The reaction was,

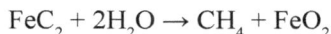

$$FeC_2 + 2H_2O \rightarrow CH_4 + FeO_2$$

The issue was, and still is, the dearth of evidence indicating cementite's presence in the mantle. The ingrained terrestrial theory prompts the mention of these theories. Another inorganic theory put out by Sokoloff indicated a cosmic origin. According to his idea, hydrocarbons from the system's initial protoplanetary matter were first expelled from the interior of the earth. and then precipitated as rain on rocks on the surface. The extraterrestrial hypothesis is the reason why this theory and others like it are mentioned. Two discoveries in the 20th century led to variations and a resurgence of interest in the inorganic mode of origin among others: the existence of the carbon-containing chondrites (meteorites) consequently verified that methane-containing atmospheres are present in a few celestial bodies like Saturn, Titan, and Jupiter. Methane is only known to come from inorganic processes. Methane, ammonia, hydrogen, and water vapor were allegedly present in the earth's original atmosphere. This idea argues that the outcome is the production of an oily, sticky external layer that acts as the host to a range of prebiotic molecules, including the precursors of life. Photochemical reactions (induced by UV radiation) increase. According to Muueller, the discovery of a group of meteorites known as carbonaceous chondrites in 1963 also triggered a renaissance in interest in an inorganic approach for producing organic compounds. Chondritic meteorites include evidence of several hydrocarbons, including amino acids, and organic stuff older than 6 June 1944 (but not graphite). The main argument in favor of an inorganic origin is the frequent inorganic composition of the hydrocarbons methane, ethane, acetylene, and benzene. For instance, congealed magma containing gaseous and liquid hydrocarbons (mostly methane with traces of ethane, propane, and isobutane) has been discovered in the Kola Peninsula in Russia. Other igneous rocks have also been found to contain paraffinic hydrocarbons.

The inorganic hypothesis has the following issues.

1. There is no data to suggest that the chondritic meteorites' organic content was formed, or that it was in a clever parent material that was organically generated or was a byproduct of a relatively inorganic origin. Other celestial bodies follow a similar logic.

2. Despite the expanding body of data supporting an organic beginning, no scientific information has ever been found to support the existence of inorganic processes in the environment.

3. If the major mechanism for the formation of hydrocarbons is of inorganic origin, then significant quantities of hydrocarbons ought to be released from volcanoes, congealed magma, and other igneous rocks. White and Waring noted the presence of gaseous hydrocarbons, the most frequent of which was methane (CH_4), coming from volcanoes. The majority of the time, volumes are less than 1%, but they have occasionally reached 15%. However, igneous rocks lack enormous pools. When significant deposition is present, they are typically discovered in volcanic rocks that have invaded or are stacked atop sedimentation materials, suggesting that the hydrocarbons likely formed within the sedimentary rock and moved into the volcanic materials (Bluemle, 2000; Carter, 2004; Krauskopf and Bird, 1995).

2.2.2.2 Organic hypothesis

There are many strong arguments in favor of the organic development theory. The percentages of carbon and hydrogen in crude oil are higher, and there are also traces of oxygen, nitrogen, and sulfur. A variety of arguments are used to justify the organic theory. The first and most well-known argument is the link between carbon, hydrogen, and organic matter. The initial components of organic material, both plant and animal, are carbon and hydrogen. Furthermore, the life processes of plants and animals repeatedly made certain elements and compounds such as carbon, hydrogen, and hydrocarbons. These life processes break when these elements and compounds are left unchanged or barely altered within the deposits. Organic matter contains nitrogen and porphyrins, which are products of blood and chlorophyll in plants and animals, respectively. The existence of porphyrins also suggests that anaerobic conditions must have evolved early in the formation process because porphyrins are easily and swiftly oxidized and destroyed under aerobic settings. Because a low oxygen content also indicates a dropping environment, there is a considerable possibility that petroleum originates in an anaerobic and lowering atmosphere. Thirdly, there were physical distinctive observations. Petroleum is almost always found in sediments that are predominantly marine in origin.

Petroleum that was present in sediments that weren't marine most likely flowed into these regions from neighboring marine sources. Additionally, temperatures in the deeper petroleum reservoirs hardly ever rise above 141°C. However, because porphyrins are destroyed above this degree, temperatures there have never gone above 200°C. Thus, a low-temperature event is probably where petroleum first formed. Finally, 1 MM years could be needed; recent oil findings in Pliocene sediments frequently confirm this. The development of liquid petroleum, however, would have taken much longer because the physical conditions of the world may have been different in the geologic past. As a result, the organic idea about the turn of the century was eventually accepted since the oil and gas industry was starting to fully grow and geologists were looking for brand-new resources. According to the organic explanation, marine plankton and other early marine life forms that were present on Earth during the geologic past provided the carbon and hydrogen needed for the production of oil and gas. Despite being minuscule, the ocean is so full of them that they account for almost 95% of all living things there. All living creatures, including plankton and other types of marine life, receive their energy from the sun. As these young forms passed away, abrasion and sedimentation processes gathered their remains. Over time, layers on the ocean floor were generated that were rich in the fossilized remains of former life as successive layers of mud and silt buried earlier layers of organically rich sediments. Thermal maturation procedures including heat, pressure, and decay gradually transformed organic materials into oil and gas. The organically rich sediments were transformed into layers of rocks as more geologic time (millions of years) was added to the process. The layers were distorted, buckled, broken, and raised as more geologic time passed; the liquid petroleum flowed upward through porous rock until it became trapped and could not flow any further, generating the oil and gas that we search for at present. However, the hydrocarbons present in the final product (oil, gas) have a somewhat different chemistry than those found in living things. Between the dumping of the organic waste and the creation of the finished product, various adjustments and transformations occurred. This chapter discusses several types of crude, its makeup, how to evaluate oil using different criteria, and the end products of refineries (Walters, 2007; Abbas, 1997; Tissot and Welte, 1984).

2.2.3 Exploration of crude oil

Petroleum geologists and geophysicists use a wide range of methodologies when it comes to the exploration of petroleum resources. These consist of:

2.2.3.1 Gravity survey

To determine what might be below the surface, a gravity reader (gravimeter) detects the gravitational attraction in specific rock formations. "MilliGal" units are used to measure gravitational attraction. The usual measurement of an un-mineralized rock formation on the surface of the planet is 980,000 milliGal. However, a petroleum reserve in a rock formation would hardly ever exceed 300–400 milliGal. This kind of survey is simple, reasonably affordable, and non-intrusive, making it helpful in densely inhabited areas. For readings in challenging underwater environments, it is normally and most correctly done on the bottom, but it can also be done from the air or on a ship.

2.2.3.2 Magnetic survey

In order to determine whether specific rock formations have magnetic anomalies, which can indicate an oil reserve at certain levels of a magnetic reading, a magnetic reader (magnetometer) is typically flown on the back of an airplane. Particularly in difficult-to-examine locations, such as hilly terrain, this form of survey is appropriate.

2.2.3.3 Seismic reflection surveys

Seismographs are used beneath, although they are controlled from above. Three to five meters is the shallowest depth at which a seismograph is frequently employed. There are a few ways for geologists and geophysicists to utilize a seismograph to find petroleum, but a geophone, a device that transforms ground movement into voltage in the form of waves that a seismograph can read, is the most successful one. Seismic reflection surveys are more intrusive into the ground, more expensive, and more complicated than gravity and magnetic choices, hence they are less useful.

2.2.3.4 Exploration drilling

There are a number of exploratory drilling techniques, but they all include drilling up to three miles out from known reserves to see whether there are any undiscovered reserves. This approach often carries a high level of risk because it is more expensive and relies less on data gathering up front. Since this does not require drilling for data, it is typically more beneficial in unpopulated areas. There are a number of exploratory drilling techniques, but they all include drilling up to three miles out from known reserves to see whether there are any undiscovered reserves. This approach often carries a high level of risk because it is more expensive and relies less on data gathering up front. Since this does not require drilling for data, it is typically more beneficial in unpopulated areas.

2.2.3.5 Visual oil seeps

It is exactly what it says, it is-visual proof of an oil reserve. This can take many different forms, such as geographical images, bubbling seas, and many others. Both a negative and positive value could apply to it. It is inexpensive because it requires no equipment, but depending on where these seeps are found, it may signify harm to animals or water supplies (Reddy et al., 2019; Zhang et al., 2019).

2.2.4 Composition of crude oil

The origin of formation has a significant impact on the composition of crude oil. It is a uniform blend of different saturates and ring-structured hydrocarbons. Methane (C1) and asphalt are only a few of the thousands of distinct chemical compounds that make up crude petroleum (C70). Carbon (83–87%) and hydrogen (12–14%) make up the majority of petroleum (also known as crude oil), which also contains complex hydrocarbon mixtures such paraffins, naphthenes, aromatic hydrocarbons, and gaseous hydrocarbons. In addition, crude oil also contains trace amounts of non-hydrocarbons, such as nitrogen, sulfur, and oxygen compounds (0.1 to 1.5%). Inorganic crude oil that contains different minerals (up to 0.5%) is heavier and has more sulfur. Due to their intrinsic characteristics like odor, color, corrosiveness, and ash production, these non-hydrocarbons are typically treated as contaminants. Petroleum is categorized as either paraffin base, intermediate base, or naphthenic base depending on the amount of hydrocarbons it contains. (Abdel-Raouf, 2012).

2.2.5 *Classification of crude oil*

Crude oil is classified into the following four types according to their chemical composition.

2.2.5.1 *Paraffin base crude*

This has paraffins with a higher molecular weight. Hydrocarbon and branched chains contain more than 50 carbon atoms, saturated ones having C 1–4: gas, C5–15: liquid, C15 and above: solid. All are mixed as a complex in the form of light paraffinic oil or light oil. Crude oils on distillation yield residues containing gasoline, paraffin waxes and high-grade lubricating oils.

2.2.5.2 *Intermediate base crude*

Crude oils are categorized as asphaltic base if the atmospheric distillation residue contains asphaltic components. Many crudes can be atmospherically distilled to produce paraffin waxes and asphaltic compounds.

2.2.5.3 *Naphthenic base or hybrid base crudes*

These are crudes that, after atmospheric distillation, leave behind asphaltic material and a negligible quantity of paraffin wax and contain a significant proportion of saturated cycloalkane derivatives.

2.2.5.4 *Asphaltic base crude oil*

This oil has little to no paraffin and high quantities of asphaltic particles. Some crudes that produce lubricating oil are more susceptible to temperature variations than paraffin-base crudes because they are mostly composed of naphthenes.

It is categorized based on the flash point value. Any petroleum's flash point is defined as the lowest temperature at which it produces a vapor that, when ignited, will cause a brief flash. Categories include:

1. Petroleum with a flash point lower than 230°C is classified as Class "A" petroleum.
2. Petroleum with a flash point of 230°C or greater but less than 650°C is classified as petroleum class "B".
3. Petroleum with a flash point of 650°C or more but lower than 930°C is classified as petroleum class "C".

Another classification of petroleum is as per physical characteristics like API or specific gravity. The primary US trade association for the oil and gas industry is called API, or the American Petroleum Institute. The ratio of one substance's density to that of a reference substance, typically water, is known as relative density. Oils are categorized as light, medium, heavy, or exceptionally heavy using the API gravity. Since an oil's weight is the main factor influencing its market value, API gravity is extremely significant. A light petroleum has an API value greater than 31.1, a medium petroleum has an API value between 22.3 and 31.1, a heavy petroleum has an API value less than 22.3, and an extra heavy petroleum has an API value less than 10.0.

Other physical features, such as the Correlation Index (CI), Viscosity, Carbon Distribution, Viscosity-Gravity Constant, UOP Characterization Factor and Pour Point, are also used to categorize it.

2.2.6 Distillation of crude oil

Pretreatment and pre-heating are required prior to distillation.

2.2.6.1 Pretreatment of crude oil

Water, inorganic salts, suspended particles, and water-soluble trace metals are typically found in crude oil. Desalting is required to get rid of these impurities as the first step in the refining procedure in order to lessen corrosion, clogging, and fouling of machinery and to avoid poisoning the catalysts in processing units (dehydration). Since heat from atmospheric distillation is used to heat the crude throughout the desalting process, desalting is typically thought of as a component of the crude distillation unit. The most prevalent salts in crude oil are sodium, calcium, and magnesium chlorides, which are either crystallized or ionized in the water present in the crude (NaCl: 70 to 80 weight percent, $CaCl_2$: 10 weight percent, and $MgCl_2$: 10 to 20 weight percent). In the event that a salt is not eliminated, the high temperatures experienced during petroleum refining may result in water hydrolysis, which subsequently permits the creation of acid (HCl), posing major corrosion issues within the apparatus. Unremoved salt particles may also contribute to fouling issues in furnaces, heat-transfer devices, and pipes. The metals in salts, especially sodium, may also improve the deactivation of catalysts (such as the zeolite-type catalysts used in fluid catalytic cracking). The maximum amount of salt typically permitted in the feed to crude distillation units is 50 PTB (pounds of salt per thousand barrels of crude oil). The two main methods for desalting crude oil are chemical and electrostatic separation, both of which use hot water as the extraction agent.

Chemical desalting promotes the coalescence of water drops by adding water and chemical surfactants (de-emulsifiers) to the crude. Chemicals like soda ash, sodium hydroxide, fatty acid salts, and petroleum sulfonates are employed as de-emulsifiers because they speed up the agglomeration of water droplets. Thereafter, it is heated upto 75–80°C at a pressure of 15 atmospheres. Pressure ensures the retention of volatile matters in crude. After adding the chemicals, crude oil is allowed to stand at a temperature of 75–80°C for 48 hours, such that

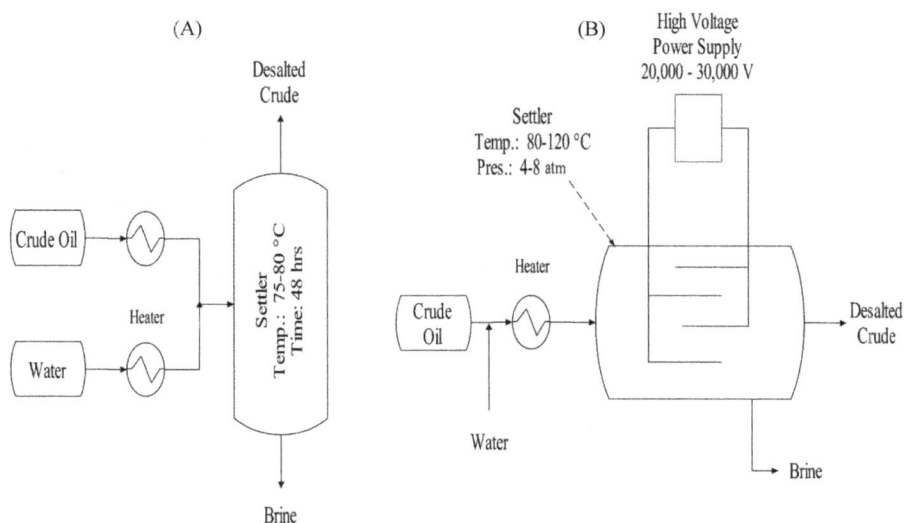

Figure 2.1 (A) Desalting by settling process, (B) Desalting by electrostatic process

salts and other contaminants attach to the water or dissolve in it, and then the water is kept in a tank where they settle out as per Fig. 2.1(A). Desalted oil is collected from the top of the settler, while effluent (brine) is collected from its bottom. The height of the settler tank is 16 meters with a 3000 kl capacity.

Oil and water are fed to the settler during the electric desalting process, and high-voltage electrical desalting is used. The voltage used to transmit crude between electrodes ranges from 20,000 to 30,000 volts. The pertinent pressure and temperature ranges are 4–6 atmospheres and 40–120°C. In the stream, emulsion water also congeals and clumps together, encasing all the salts in the process. Brine collects at the bottom of the desalter, while crude floats above and forms a different stream. This quick, straightforward, compact, and simple technique takes less than 30 minutes to remove about 90% of the salt. Per barrel, about 0.01 kW of power is used. Surfactants are only introduced when the crude has a large number of suspended particles. Diatomaceous earth is used to filter hot crude in a third and less common step. Other methods for dehydration and desalting are centrifugal separation and gravity settling treatment. Centrifugal separation allows efficient separation, but requires high energy (Cronquist, 2001).

2.2.6.2 Heating of crude

Following the desalting procedure, the crude oil passes via a furnace, heater, or heat exchanger where it is heated by the lengthy residual from the atmospheric distillation unit or the bottom product that is being sent out. A direct fired heater or industrial furnace is a piece of machinery that generates heat for a process or can act as a reactor to generate heat for a reaction. The purpose, degree of heating, kind of fuel, and technique for delivering combustion air differ among furnace designs. However, the majority of process furnaces share a few characteristics. Air from an air blower is used to burn the fuel as it enters the burner. In a particular furnace, there are frequently only one or two burners that are grouped in cells to heat a particular set of tubes. Depending on the design, burners can also be put on the floor, a wall, or a ceiling. The fluid inside the first section of the furnace, known as the radiant section or firebox, is heated as a result of the flames heating up the tubes. The warmth is primarily delivered to the tubes around the fire inside the combustion chamber by radiation. Through the tubes, the heating fluid is heated to the desired temperature. Flue gas is the name for the combustion byproducts.

Most furnace designs contain a convection section where more heat is collected before venting to the atmosphere through the flue gas stack after the flue gas leaves the firebox. Depending on the geometrical layout of the radiating section, fired furnaces can be either vertical cylindrical or box-type heaters. Box-type furnaces are typically employed in situations that call for huge capacities and heavy heat loads. The radiation portion of this furnace typically has a square or oblong cross section. The burners are situated on the ground or on the bottom portion of the longest side wall, where there are no tubes, because the tubes in the radiation section might potentially be oriented horizontally or vertically along the heater walls. Pipe still furnaces are another type of furnace; they may hold 25,000 bbl or more of crude oil per day and vary widely in size, shape, and interior layout. Gas or oil burners are installed through one or more walls, and the walls and ceiling are insulated with firebrick. The furnace's interior is split into two sections: a smaller convection area where oil is introduced first and a larger area into which burner exhaust is directed and where oil is heated to its peak temperature. These kinds are widely used in thermal cracking, atmospheric distillation, and vacuum distillation. Th temperature of the oil at the furnace exit is around 350°C.

Additionally, the feedstock is preheated through heat exchangers before entering the furnace. These exchangers are collections of tubes that have been organized in a shell so that a stream can pass through the tubes while another

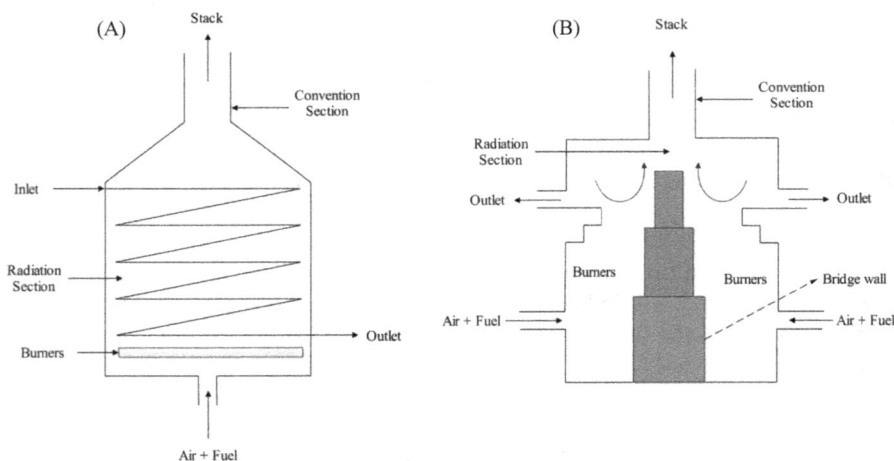

Figure 2.2 (A) Diagram of box-type fire furnace, (B) Diagram of pipe still furnace

stream flows through the shell. Therefore, cold petroleum may be a key component in the efficient operation of refineries by passing through a series of heat exchangers where hot products obtained from the distillation tower are cooled before entering the furnace (Babalola and Susu, 2019).

2.2.6.3 Atmospheric distillation

By atmospheric distillation, which involves preheating the desalted crude feedstock using a heater, crude oil is first separated into several fractions or straight-run cuts (such as fuel gases, LPG, naphtha, kerosene, diesel, and fuel oil). In order to further separate the hydrocarbons under lower pressure, the heavy hydrocarbon residue that was left at the bottom of the atmospheric distillation column is transferred to the vacuum distillation column. The feedstock is then injected into the vertical atmospheric distillation unit (ADU) slightly above the bottom after passing through a direct-fired crude charge heater. The horizontal steel trays in the fractionating distillation column, a steel cylinder that is about 120 feet high, are used to separate and collect the liquids. To create a column, trays are placed one on top of the other and encased in a cylinder. For a majority of the time, a packing is sandwiched between the hold-down and support plates on the inside of a cylindrical shell. Depending on a number of factors, including scale, flexibility, and the solid content of the feed, the column may also be operated continuously or in batch mode. One or more locations along the column shell introduce the feed material.

There are 15 to 25 trays in all, with an average ratio of 7 plates for light to heavy naphtha and 6 plates for heavy to light distillates, depending on how many products are extracted. The average number of plates is five for light distillation to middle distillates and four for middle distillates to diesel. There are three trays in all for the flash zone to be drawn first and the bottom. Vapors enter holes and bubble caps at each tray. At the temperature of that tray, they allow some condensation to occur when the vapors bubble through the liquid. Each tray's condensed liquids are drained back into the tray below via an overflow pipe, where the higher temperature prompts re-evaporation. Until the desired level of product purity is attained, the evaporation, condensing, and cleaning processes are repeatedly carried out. The furnace's entryway, where the crude is introduced, has a temperature of 200–280°C. It is then heated within the furnace to an additional 330–370°C. About one bar of pressure is maintained. The pressure profile in the atmospheric distillation unit is on the edge of the air pressure, with the highest pressure at the bottom of the column and a progressive decline in pressure until the top stage, where the heaviest fractions suddenly turn to vapor. The temperature of the heated vapor falls as it ascends in the tower. The asphalt or heavy fuel oil residue is gathered at the bottom. The desired fractions are then obtained by removing side streams from specific trays. Products from a fractionating tower are frequently extracted constantly, ranging from uncondensed fixed gases at the top to heavy fuel oils at the bottom. In towers, steam is typically used to create a partial vacuum and lowers the vapor pressure. Gas + naphtha, kerosene, light fuel oil, heavy fuel oil, and atmospheric residue are the typical five products produced. The heavy (atmospheric) residue, also known as reduced oil, is often further distilled in a vacuum distillation column at a high vacuum.

2.2.6.4 Vacuum distillation

Unrecovered gas oils can be found in the atmospheric tower's residue. Because of worries about thermal cracking, the outlet temperature of the crude unit heater is limited. In order to collect extra heavy distillates, the residue from the bottoms of the atmospheric distillation unit is frequently piped to a different distillation column that is run under vacuum (0.4 PSI), known as a Vacuum Distillation unit (VDU). Decreased pressure is necessary to prevent thermal cracking and further distil the residuum or reduced oil from the atmospheric tower at higher temperatures. The tower's operation is more expensive than an ADU. Absolute pressures of 25 to 40 mm Hg are used for distillation in the tower flash zone area (Fig. 2.4). The addition of steam to the furnace inlet and the base of the vacuum tower lowers the effective pressure even more (to 10 mm Hg or less) in order to increase vaporization. The addition of steam to the furnace's entrance enhances the furnace's tube velocity, reduces coke formation inside the furnace, and lowers the vacuum tower's overall hydrocarbon partial pressure. The amount of stripping steam used is a function of the boiling range of the feed and the fraction

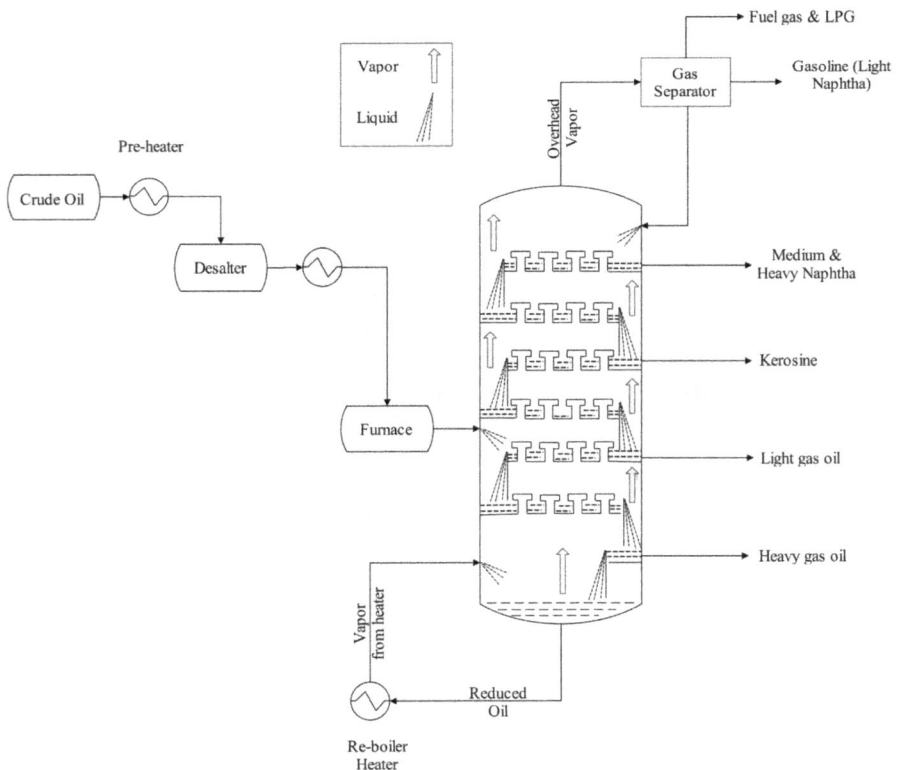

Figure 2.3 Atmospheric distillation unit of crude oil

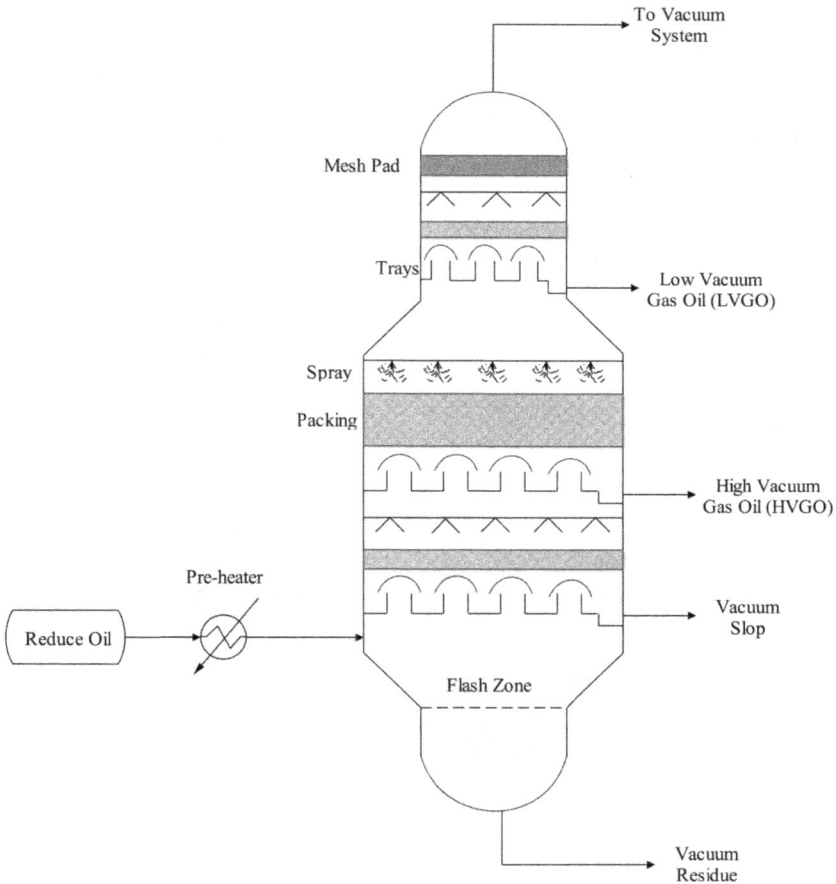

Figure 2.4 Vacuum distillation unit of crude oil

vaporized, but generally ranges from 10 to 50 lb/bbl feed. The temperature is kept at around 380–420°C. The process takes place in one or more vacuum distillation tower. These towers are called vacuum distillation units (VDU). The vacuum pressure is typically maintained by a three-stage steam ejector system. The boiling point decreases under vacuum and this causes a big increase in the volume of vapor. Therefore, the VDU is meant to have large diameters to take care of the comparable vapor velocities. The principles of vacuum distillation are identical to those of fractionation, and the equipment is also similar, with the exception that larger-diameter columns tend to retain equivalent vapor velocities at the lower pressures. Some vacuum columns have different interior layouts than atmospheric towers; instead of using trays, they use ad hoc packing and demister pads. There is mesh pad at the uppermost layer and thereafter, it is packed with three simultaneous spray, packing and tray layers. And finally, there is a flash zone at the bottom. Gas oils, lubricating-oil base stocks, and heavy residuals for propane deasphalting may be produced using a standard first-phase vacuum tower. Both surplus residue from the atmospheric tower, which is not used for processing lube stock, and surplus residue from the first vacuum tower, which is not used for deasphalting, may be distillated in a second-phase tower running at a lower vacuum. Vacuum towers are typically used to extract surplus residue from catalytic cracking material. Vacuum towers are only intended for a few goods. A tower's suitability for one product may not apply to others (Serge-Bertrand, 2020). The four primary commercial products are bitumen, heating oil, lubricating oil, and cracking stocks.

The products of ADU and VDU are mentioned as per Table 2.1.

Table 2.1 Various products of atmospheric and vacuum distillation

Column	Name of fraction	Temperature range	Carbon range	Uses
Atmospheric column	Fuel Gases	> 40°C	C1–C2	Fuel
	LPG		C3–C4	Domestic fuel
	Gasoline	20–90°C	C6–C10	As motor fuel, solvent and in dry cleaning
	Naphtha (Medium and heavy)	130–180°C	C6–C10	Catalytic reforming and aromatic plant feed stock, Steam cracker, synthesis gas manufacture
	Kerosene	150–270°C	C11–C12	Aviation turbine fuel, Domestic fuel, Lab feed stock (paraffin source)
	Light gas oil	230–320°C	C13–C17	High speed diesel component
	Heavy gas oil	320–380°C	C18–C25	High speed diesel component
Vacuum column	Light vacuum gas oil	370–425°C	C18–C25	Feed to Fluid Catalytical Cracking (FCC)/Hydro Cracking Unit (HCU)
	Heavy vacuum gas oil	425–550°C	C26–C38	
	Vacuum slop	550–560°C	> C38	Resid Fluid Catalytical Cracking unit (RFCCU) feed
	Vacuum Residue	> 560°C		Bitumen/ Visbreaker feed

2.2.7 Treatment of refinery products

From crude oil, a range of intermediate and final products are produced, such as middle distillates, gasoline, kerosene, jet fuel, and heavy oil. Impurities are to be removed from it. Also, some treatments are to be given to petroleum to produce more valuable products. These treatments differ greatly amongst fractions, and occasionally a series of treatments is unavoidable. Corrosiveness and treatment rigor are inversely related. Of course, such extreme treatments are applied to naphthenic crude fractions. All fractions have impurities, which fall into two categories: physical and chemical. Particles, muck, moisture, and other physical impurities including solvent droplets and catalyst dust occasionally appear in crude. In the initial step, these impurities must be separated from crude. Dust particles, moisture droplets and oily mater are removed by solvents like water, washing and thereafter, a drying process. If oily matter is high, then it can be dissolved by passing it through gas. Chemical impurities exist in all fractions to varying extents and create several problems while it undergoes distillation. Sulfur (0–3%), nitrogen (0–0.6%) and oxygen (0.5–6%) are chemical impurities found in petroleum, in which a major part is covered by sulfur [As discuss in previous chapter-1 (Composition of crude)]. Free sulfur, hydrogen sulfide, polysulfides, mercaptans, sulfonates, sulfates, and thiohenes are just a few examples of the forms it can take. The petroleum fractions' thiol content contributes to corrosion issues, catalyst poisoning, and unpleasant lab and plant odors. Thiols are frequently utilized as odorants in Liquefied Petroleum Gases (LPG) for safety reasons even though they are not particularly dangerous in low concentrations. Thiols are stronger acids than alcohols and use is made of this property to remove low molecular weight thiols from light gasolines with caustic soda solution. Hydroprocessing techniques are employed to desulfurize other oil fractions and here combined sulfur is eliminated as hydrogen sulfide from all types of compounds containing this heteroatom. Crude oils may contain organic nitrogen molecules that are simply heterocyclic in nature, such as pyridine (C_5H_5N) and pyrrole (C_4H_5N), or complicated in structure, such as porphyrin. These compounds can lead to a loss in catalyst activity color instability of the product and formation of gum. Numerous oxygen compounds, such as carboxylic acids, cresylic acids, phenols, and naphthenic acids, are present in sort of weak acids in unrefined crude oil. Compared to sulfur compounds, oxygen molecules in crude oils are more complicated. However, processing catalysts are not poisoned by their presence in petroleum streams. The oxygen concentration of oxygen increases from lighter fractions to heavier ones.

2.2.7.1 Dehydration process

Dehydration is the technique used to remove the water content found in natural gas vapor. Additionally, it is essential to ensure the efficient operation of gas transmission lines, avoiding the formation of gas hydrates, and lessening corrosion. Glycols such as triethylene glycol (TEG), diethylene glycol (DEG), glycol (MEG), and tetraethylene glycol (TREG) are utilized throughout this procedure. The glycol that is most frequently used in industry is TEG. According to Fig. 2.5, a glycol contactor, also known as an absorber, is fed with lean, water-free glycol (purity > 99%). It comes into contact with the stream of wet natural gas here. The natural gas is inserted from the bottom of the contactor. Glycol contactors can be either tray columns or packed columns. Here, glycol physically absorbs water from the natural gas to eliminate it. From the top of the column, dry gas is collected and cooled in a heat exchanger. Rich glycol is fed with water into a flash vessel, where liquid hydrocarbons are skimmed off and hydrocarbon fumes are removed. The pressure must be decreased before the regeneration process since the absorber is often operated at high pressure, making this step crucial. When the pressure is reduced, a vapor phase with a high hydrocarbon content develops. Rich glycol is heated in a cross-exchanger after exiting the flash vessel before being delivered to the glycol regenerator, a stripper. Column, overhead condenser, and reboiler are components of the glycol stripper. Thermal regeneration is used to remove extra water from the glycol and restore its high

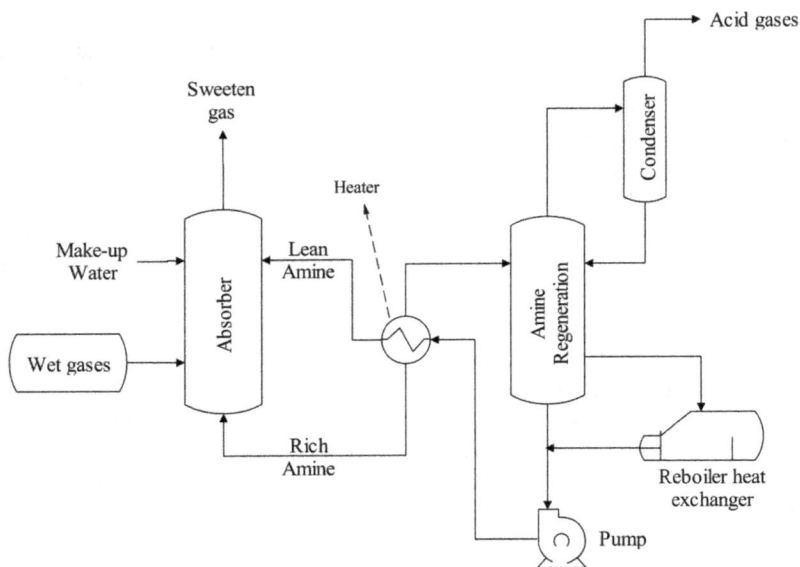

Figure 2.5 Dehydrogenation of gases and regeneration unit of amine

purity. Cross-exchange cooling between the hot lean glycol and the rich glycol coming from the stripper. The pressure is then increased so that it is applied to the glycol absorber through a lean pump. Before being sent back to the absorber, the lean solvent is cooled once more with a trim cooler. This trim cooler can either be an air-cooled exchanger or a cross-exchanger with the dry gas leaving the absorber.

2.2.7.2 Sweetening process

The process of removing hydrogen sulfide (H_2S) and carbon dioxide (CO_2) from gases is called the sweetening process. Amines are utilized for the sweeetening process, so this process is additionally referred as amine gas treatment or amine scrubbing or acid gas removal. Diethanolamine (DEA), Monoethanolamine (MEA), Methyldiethanolamine (MDEA), Diisopropanolamine (DIPA), and Diglycolamine (DGA) are only a few of the amines used in the treatment of gas. The alkanolamines DEA, MEA, and MDEA are the amines that are utilized in industrial facilities most frequently. In many oil refineries, these amines are also used to remove sour gases from liquid hydrocarbons like liquefied petroleum gas (LPG). Figure 2.6 depicts a situation where a sour gas stream and lean amine (purity > 99%) come into contact at the top of an absorber, also known as a "glycol contactor." The natural gas is fed into the absorber at its base. Absorbers can either be tray columns or packed columns. A temperature range of 35 to 50°C and a pressure range of 5 to 250 atmospheres are maintained in the absorber. Here, hydrogen sulfide (H_2S) and carbon dioxide (CO_2) are absorbed from the gas and collected as "sweetened gases" from the top of the adsorber. In a heat exchanger, "rich amine" is cooled before being given to the stripper known as an amine regenerator. The overhead condenser, reboiler, and column make up

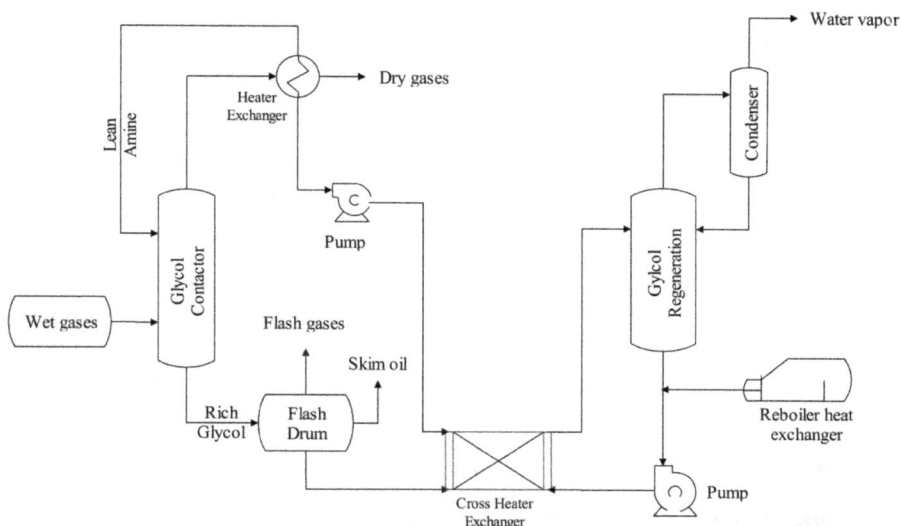

Figure 2.6 Sweeting of gases and regeneration unit of glycol

the glycol stripper. To remove extra water and restore the high amine purity, the amine is thermally regenerated. Heat exchangers are used to cool the hot, lean amine before it is pumped back into the absorber.

2.2.7.3 Catalytic desulfurization

The catalytic chemical process known as catalytic desulfurization (HDS) is widely used to remove sulfur (S) from fuels such as gasoline, natural gas, jet fuel, kerosene, diesel fuel, and fuel oils. This process is conducted with the help of hydrogen, so, this process is known as 'Catalytic Hydro desulfurization (HDS)'. Also, this process improves the color and odor of gasoline. Sulfur is removed from gasoline due to control of pollution by the SO_2 produced in the combustion of gasoline and prevention of poisoning of the metal catalysts by it. Through a heat exchanger and furnace, the gasoline oil feed is combined with makeup and reclaimed hydrogen before being injected into catalytic fixed bed reactors. The feed that enters the reactor is kept between 290 and 425°C. In this reactor, sulfur and nitrogen molecules are transformed into hydrogen sulfide (H_2S) and ammonia. In a high-pressure separator (flash drum), the reactor product is cooled before the hydrogen-rich stream is flashed away from it and recycled. The final product is collected from the bottom of the separator. Only sulfur atoms are removed by this process.

2.2.7.4 Solvent extraction

Solvent treatment plays an important role in refinery operations, as it has major applications, like viscosity index improvement, extraction, deasphalting and dewaxing. By eliminating undesired components (such as aromatics, naphthenes, and unsaturated compounds) from the charge material, these solvent procedures produce products that meet the required criteria. Desirable constituents

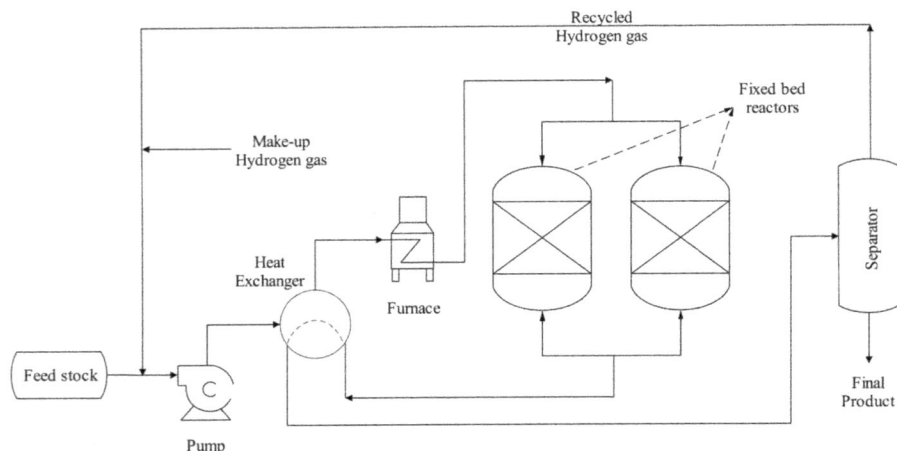

Figure 2.7 Catalytic desulfurization process

(aromatics and other chemicals) are extracted from the petroleum fraction using solvent refining techniques. The choice of solvent depends on a number of factors, including the hetero atom's high electronegativity, high dipole moment, the presence of O, N, and S atoms with the ability to form hydrogen bonds, the molecule's free rotatability, the absence of additional substituent screening, and finally the accumulation of selectively acting groups. Numerous solvents are employed in this process; in this area, phenol and furfural are advantageous. In Fig. 2.7, the phenol extraction procedure is described.

2.2.7.5 Solvent dewaxing

The lube oil contains at least some amount of waxy material. Wax affects the pour point of the oil. So, it is removed to attain a desirable pour point and the process is called dewaxing. It is done mainly by two different processes: (1) solvent dewaxing and (2) catalyst dewaxing. Solvent dewaxing is performed by utilizing a proper solvent like propane which can successfully remove wax from the lube oil cut. In this process, solvent is added with the feedstock. The mixture is heated to dissolve the wax properly. Thereafter, it is fed into the chiller, where wax components are solidified. Solid wax is subjected to the filter to eliminate it from oil. Thereafter, flash distillation and stripping are used in the solvent recovery stage to recover the solvent. Then propane is recycled. Toluene, which dissolved the oil and kept fluidity at low temperatures, and Methyl ethyl ketone (MEK), which dissolved little wax at low temperatures and served as a wax precipitating agent, are the other two solvents utilized in this method. MEK dewaxing is the main solvent dewaxing procedure used in the refinery. Additional dewaxing solvents include benzene, propane, methyl isobutyl ketone, petroleum naphtha, ethylene dichloride, methylene chloride, sulfur dioxide, and others. A chemical shift known as catalytic dewaxing is inherently a conversion process. It uses n-paraffin catalytic cracking, however as the goal is to remove wax, it is categorized as a separation process. The current technique is not very detailed, although it does involve hydrogen addition to prevent coking. Given that it produces a product with lower pour points, a high yield, and a high degree of stability, it is undoubtedly the easiest method of dewaxing the feedstock. Since the n-paraffins undergo splitting during catalytic dewaxing, it is also possible to assemble sunlight distillates like gasoline.

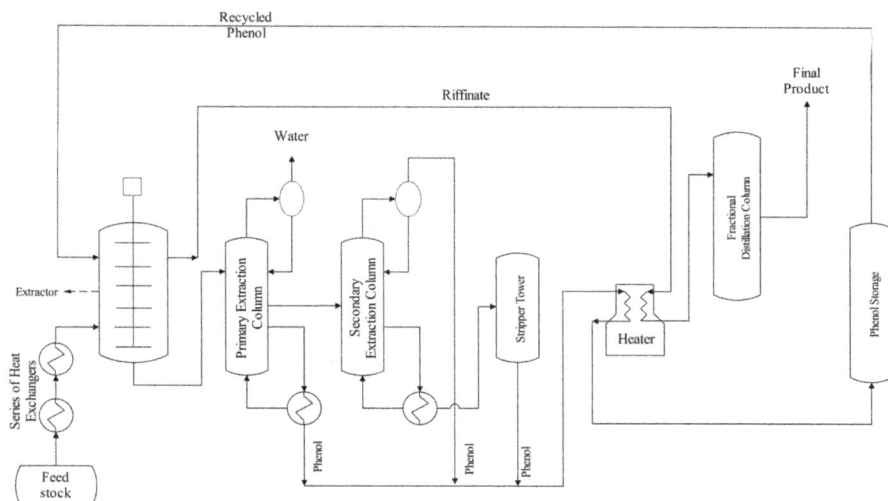

Figure 2.8 Phenol extraction process

2.2.7.6 Solvent deasphalting [Propane deasphalting]

Solvent deasphalting (SDP) or solvent extraction is the process in which the asphalt content of lube (or similar materials) is removed or reduced, especially by solvent extraction. It is a method used in refineries to remove asphaltenes and resins from heavy vacuum fuel oil, atmospheric residue, vacuum residue, or other petroleum-based products in order to supply valuable deasphalted oil that can't otherwise be recovered from the residue by other refinery processes. Asphaltenes are molecules with a high relative molecular mass that contain carbon, hydrogen, nitrogen, oxygen, sulfur, and minute amounts of nickel and vanadium. These are high carbon: hydrogen ratio molecules frequently condensed ring hydrocarbons and hetero-atom compounds. The hydrocarbons are precursors to coke formation in some refinery processes, heteroatom compounds are liable for poisoning certain secondary processing catalysts and metals are particularly detrimental to cracking catalysts. Solvent deasphalting is essential in the production of high-quality conversion unit (FCC and hydrocracking) feedstock. Here, the solvent deasphalting process uses light paraffinic solvents like propane, isobutene, butane, and pentane. Olefin solvents may also be used. Most preferred solvent is propane in the deasphalting process. The differential in solubility between asphaltene and deasphalted oil (DAO) in a specific solvent determines how easily asphaltene can be removed from oil. Except this, solvent composition, temperature of operation and solvent to oil ratio are important parameters for

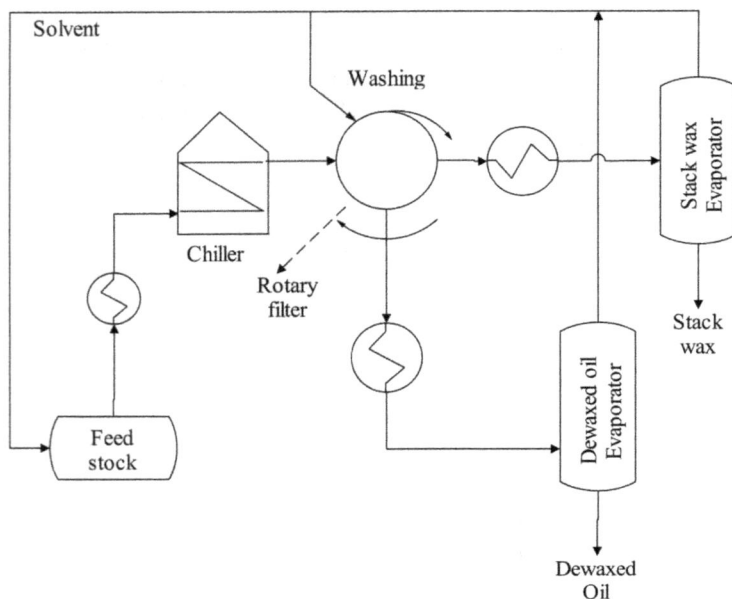

Figure 2.9 Solvent dewaxing process

separation. Typical uses of DAO are lube oil base stock, hydrocracker feed and FCC feed.

In the solvent deasphalting process, there are five main parts: (1) atmospheric and vacuum distillation (pretreatment of deasphalting), (2) mixer, in which feed is mixed with solvent, (3) extractor, where separation is done, (4) deasphalted oil (DAO) separator and (5) pitch and DAO stripper. First, the feed for deasphalting undergoes atmospheric and vacuum distillation thereafter. Its residue is collected from the bottom and further used. Following distillation, the residue and solvent are combined in the feed-solvent mixer before being fed into the top of the extractor. Now, more solvent is counter currently added into the separator through the bottom distributor for efficient separation. Here, the solvent does not dissolve asphaltene, but dissolves the oil. Hence, asphaltene is dropped out of the solution and comes through the bottom of the extractor. The DAO/solvent mixture comes out from the top. Here, DAO yield and quality are primarily hooked into the operating temperature. Higher temperatures yield smaller quantities of DAO, and, lower temperatures produce higher quantities of reduced quality. The solvent cooler governs the operating temperature of the separator, and in turn controls the yield and quality of DAO. DAO/solvent solution, rich in solvent, is fed to a DAO separator to separate the solvent from DAO by heating above the critical temperature of the pure solvent. This is a supercritical phase separation step. After separation, the DAO containing some amount of solvent (in one volume of DAO, slightly but one volume of solvent) is shipped to a DAO stripper to separate the remaining solvent. Asphaltene obtained from the asphaltene separator is additionally related to some solvent and is again introduced into an asphaltene stripper to recover the trace solvent. Superheated steam is used to strip off the solvent from both the components. The stripped solvent vapor with steam is received from the top of the stripper and thereafter condensed. Water and solvent are separated and the solvent is recycled to the process (Jones, 2015).

2.2.7.7 Cracking

Considering the vast usage of gasoline, diesel, natural gases (low molecular weight compounds), it is feared that the world's supplies of natural crude oil might get exhausted over time. Also, distillation of crude oil produces large amounts of high molecular weight compounds, which are less useful. So, a cracking process has to be developed that converts high molecular weight compounds into low molecular weight compounds. By using heat or a catalyst, high molecular weight chemicals are broken down into lower molecular weight ones during the cracking process. This process helps to improve the quality of petroleum products with respect to their costs and decreased residues. In order to transform heavy hydrocarbon feedstock into lighter fractions like kerosene, gasoline, liquefied petroleum gas (LPG), heating oil, and petrochemical feedstock, this process reorganizes the molecular structure of hydrocarbon molecules. The process mechanism, may be divided into three major steps the first one being initiation, in which a carbon free radical or carbonium ion is formed by carbon-carbon cleavage. The second and third steps are propagation and termination respectively.

(A) Thermal cracking

Thermal cracking was discovred by Viadmir Shukov in 1891. It was modified by William Burton in 1908. In Baku, USSR, a factory for the Shukhov cracking method was created in 1934. In Baton Rouge, Standard Jersey created the first steam cracker in history. One of the most crucial processes in the refinery is thermal cracking, which transforms crude oil into lower-molecular-weight and more valuable petroleum products like gasoline (or petrol), diesel fuel, residual heavy oil, coke, and numerous gases like C1, C2, C3, and C4 (methane, ethane, propane, and butane) from higher-molecular products. The huge hydrocarbons are often broken down into smaller ones by thermal cracking, which uses high pressures and temperatures (typically between 450 and 750°C). Alkenes, a kind of double-bonded hydrocarbons, are produced in high concentrations in mixes as a result of thermal cracking.

Mechanism of thermal cracking

The C-C and C-H bonds are broken during thermal cracking. In the course of thermal cracking, acid is used to break the C-C bonds. This cracking takes place without a catalyst and proceeds by a free radical mechanism. Free radicals are electrically neutral, highly reactive particles formed by the cleavage of C-C and C-H bonds. The following steps are involved in the thermal cracking mechanism.

1. Initiation:

$$C_4H_{10} \rightarrow C_4H_9 \cdot + \quad H\cdot$$
$$C_6H_{14} \cdot \rightarrow C_2H_5 \cdot + C_4H_9 \cdot$$

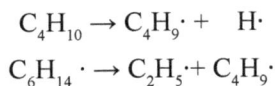

2. Propagation:

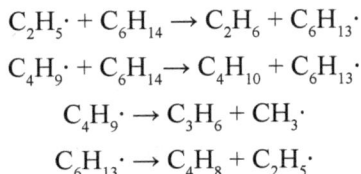

$$C_2H_5 \cdot + C_6H_{14} \rightarrow C_2H_6 + C_6H_{13} \cdot$$
$$C_4H_9 \cdot + C_6H_{14} \rightarrow C_4H_{10} + C_6H_{13} \cdot$$
$$C_4H_9 \cdot \rightarrow C_3H_6 + CH_3 \cdot$$
$$C_6H_{13} \cdot \rightarrow C_4H_8 + C_2H_5 \cdot$$

3. Termination:

$$C_2H_5 \cdot + CH_3 \cdot \rightarrow C_3H_8$$

Typically, the cracking process results in some coke being created. Coke is a byproduct of the polymerization synthesis, which involves the joining of two big olefin molecules to create a much larger olefinic molecule.

Properties of cracked material

The process of thermal cracking is quite endothermic. The production of smaller fragments causes a considerably bigger entropy change, which consumes the majority of the energy. This cracking primarily involves five reactions: (1) cracking of the side chains to free aromatic groups, (2) dehydrogenation of naphthenes to generate aromatics, (3) condensation of aliphatics to form aromatics, (4) condensation of aromatics to form higher aromatics, and (5) dimerization or oligomerization. This procedure involves coking, steam cracking, and visbreaking. The cracking circumstances have a significant impact on the product's qualities. The following characteristics alter as a result of cracking:

a) Characterizing element (decreases due to high aromatic content)
b) Decreases in pour point, viscosity, and boiling point
c) Increases in unsaturation and aromatization
d) Gasoline's octane rating rises
e) Sulfur concentration in fractured products generally rises, however only in greater amounts in heavier portions.

Various parameters such as cracking temperature, pressure, residence time, feed stock properties, affect various products of thermal cracking. Under really harsh circumstances, there is a tendency for coke production in this cracking. Hydrocarbons with two bonds are also produced during cracking (olefins). Additionally, a variety of side reactions, such as condensation and polymerization, result in the creation of gum and tar-like polymerization products (To prevent this type of development, thermally cracked gasoline or diesel blends are hydro-treated to create stable, useable products). Due to their extremely low stability, thermal cracking products need additional treatment. So, parameters are strictly maintained in thermal cracking. Increases in cracking time and temperature increases the rates of all the reactions. So, an increase in temperature enhances the ultimate yield of lighter volatile (required) products like C1 to C6. It also minimizes the formation of heavy polymers and coke. Theoretically, pressure retards cracking reactions. Pressure has a negligible effect on the reaction velocity. A slight pressure of 10 to 15 atmospheres, minimizes coke formation. At higher pressures, lighter fracton yields decrease, but, they are favorable for the production of diesel or circulating oil in the early stages. Other factors considered for thermal cracking are rate of reaction, depth of cracking, soaking factor and heat of decomposition.

This process includes visbreaking, Dubb's Two Coil, Delayed Coking and Naphtha Cracking. The visreaking flow sheet diagram is given below.

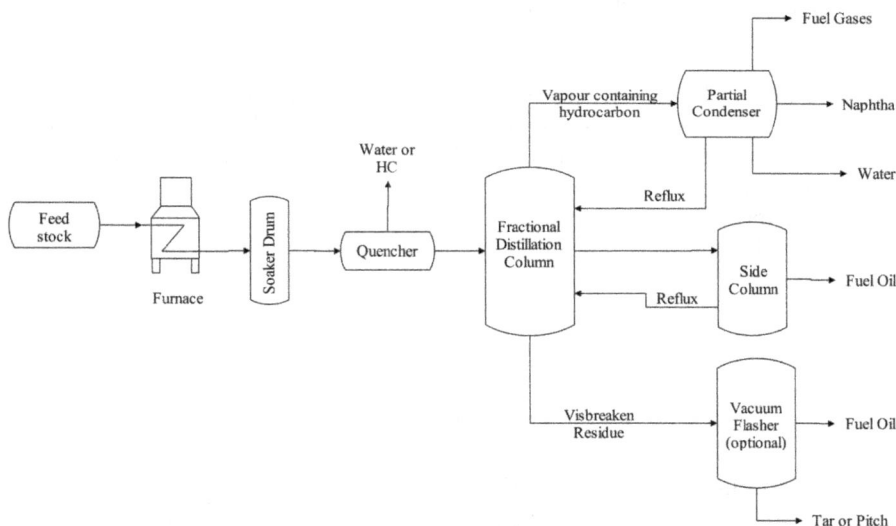

Figure 2.10 Visbreaking thermal cracking process

(B) Catalytic cracking

Eugene Houdry created the catalytic cracking process in 1920 for upgrading residue, and it was later commercialized in 1930. Today, catalytic cracking units process around 20% of all distilled crude oil. A wide range of feedstock is used because of this method's flexibility, including molecules with carbon counts ranging from simple C7 to C8 molecules to complicated structures with 100 or more carbon atoms. Naphthalenes, aromatics, and paraffins make up the majority of the feedstock. Fuel gases (H2, C1, C2), LPG (C3, C4), gasoline range naphtha (C5 to C10), diesel range light cycle oil (C11 to C18), heavy cycle oil (C19+), and coke are among the byproducts of this process. The primary catalytic cracking reactions include

1. Conversion of high Paraffins into lower paraffins and olefins.
2. Olefins are converted into LPG olefins, Naphthalene, coke and branched paraffins.

3. Naphthalene is converted into lower molecular weight olefins and aromatic compounds.
4. Aromatic compounds are converted into different alkyl aromatic compounds, un-substituted aromatics and coke.

With the exception of simultaneous cracking, several processes include isomerization, dehydrogenization, hydrogen transfer, cyclization, condensation, alkylation, and dealkylation lead to the formation of these diverse types of compounds.

The presence of acid catalysts in powder, bead, or pellet form-typically solid acids such as silica-alumina and zeolites, aluminum hydrosilicate, treated bentonite clay, fuller's earth, or silica-alumina is necessary for the catalytic cracking process. This catalyst encourages heterolytic (asymmetric) bond breaks that produce pairs of ions with opposing charges, typically as the first step in the formation of carbon free radicals. Carbon-localized free radicals are extremely unstable and go through processes including intra- and intermolecular hydrogen transfer, C-C scission at position beta as in cracking, and chain rearrangement. Both reactions continue through a self-propagating chain mechanism because the associated reactive intermediates (ions, radicals) are continuously renewed. Radical or ion recombination eventually puts an end to the series of reactions. The mechanism of catalytic cracking is as follows.

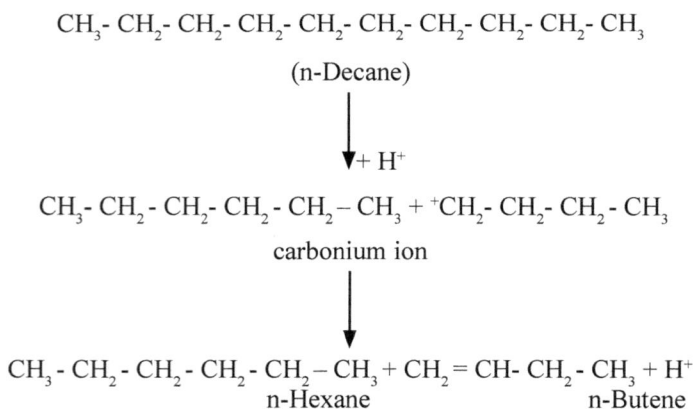

$$CH_3- CH_2- CH_2- CH_2- CH_2- CH_2- CH_2- CH_2- CH_2- CH_3$$

(n-Decane)

$$\downarrow + H^+$$

$$CH_3- CH_2- CH_2- CH_2- CH_2- CH_3 + {}^+CH_2- CH_2- CH_2- CH_3$$

carbonium ion

$$\downarrow$$

$$CH_3- CH_2- CH_2- CH_2- CH_2- CH_3 + CH_2 = CH- CH_2- CH_3 + H^+$$

n-Hexane   n-Butene

Zeolites are used in modern cracking as catalysts. Large lattices of charged aluminum, silicon, and oxygen atoms make up these complicated aluminosilicates. They share a bond with sodium ions and other positive ions. Zeolites used in catalytic cracking are chosen to provide high percentages of hydrocarbons containing between 5 to 10 carbon atoms, which are very beneficial for gasoline. In addition, branched alkanes and aromatic hydrocarbons like benzene are produced in large quantities. Fluid catalytic cracking (FCC), one of the most significant conversion processes used in petroleum refineries, is one of the three types of catalytic cracking processes, along with moving-bed catalytic cracking and hydrocracking, as 20–30% crude is processed using FCC (Alsobaai, 2013). Its flow sheet diagram is as below.

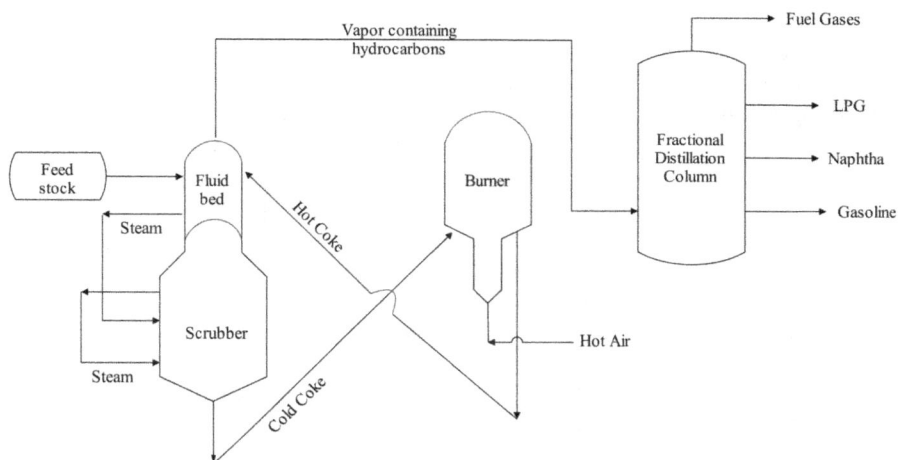

Figure 2.11 Fluid bed catalytic cracking process

2.2.7.8 Catalytic reforming

The chemical process known as catalytic reforming may be used to transform petroleum-derived naphtha from oil refineries-which normally has low octane ratings-into high-octane liquid products known as reformates, which are premium blending stocks for high-octane gasoline. Noble metals like platinum and/or rhenium are present in the regularly employed catalytic reforming catalysts, and they are extremely susceptible to sulfide and nitrogen compound poisoning. In order to remove both the sulfur and therefore the nitrogen compounds, the naphtha feedstock to a catalytic reformer is typically pre-processed. The majority of catalysts need a sulfur and nitrogen level of under 1 ppm. The catalytic reforming process involves a variety of chemical processes, all of which take place in the presence of a catalyst and a high partial pressure of hydrogen. However, four major unit operations are conducted in catalytic reforming: (1) Isomerization (rearrangement), (2) Dehydrogenation, (3) Aromatization and (4) Hydrocracking. These procedures also result in the separation of hydrogen atoms from the hydrocarbon molecules, which generates large quantities of hydrogen gas as a byproduct that can be used in a number of additional procedures in a contemporary petroleum refinery. Additionally, they produce minor amounts of butane, propane, ethane, and methane as byproducts. Advantages of the catalytic reforming process are as follows:

1. This process converts straight run naphtha (C6–C10, Straight chain paraffins and naphthenes) to gasoline components of high antiknock quality.
2. It manufactures Hydrogen and LPG as by-products.
3. The octane of reformates is provided mainly by aromatics having 6 to 10 carbon atoms and light iso-paraffins.
4. Alkanes and cycloalkanes are selectively converted to aromatic hydrocarbons.
5. Transforms low octane naphtha to high octane reformate.

In the reforming process, many reactions are simultaneously conducted; therefore, the catalyst maintains its activity towards all reactions. Using a platinum-containing catalyst, research scientist Vladimir Haensel created a catalytic reforming process in the 1940s while working for Universal Oil Products (UOP). Now-a-days, high purity platinum containing 0.3 to 0.5% acid-based metals is used as the catalyst, because low purity platinum is the non-regenerative type. Also, the acid-based catalyst is responsible for the isomerization and cyclization reactions; while metallic materials favor hydrogenation reactions. Molybdenum oxide and chromium oxide on alumina are other potential catalysts. These are excellent hydrogenation and isomerization catalysts. Moving beds and fluidized beds can use chromium oxides and molybdenum oxides on silica, alumina, or silica-alumina since they are quite abrasion-resistant.

Four key factors in the catalytic reforming process determine how well the unit operates: reactor pressure, reactor temperature, space velocity, and the $H_2/$ oil molar ratio.

(A) Pressure: A drop in reactor pressure boosts the production of hydrogen and reformate, lowers the temperature needed to maintain a constant product quality, and shortens the catalyst cycle by boosting catalyst coking. The reactor pressure decreases during each stage of the numerous reactions as a result of the pressure drop. Reactor pressure is a broad term used to describe the average pressure of the numerous reactors.

(B) Temperature: The reaction temperature is the most crucial factor in catalytic reforming since it has a significant impact on yields and product quality.

(C) Space Velocity: To set a product's octane, space velocity and reactor temperature are frequently used. Either raising the reactor temperature or lowering the space velocity will make the catalytic reforming unit more severe.

(D) $H_2/$oil molar ratio: The hydrogen to oil ratio is expressed in volumetric terms, such as standard cubic feet of hydrogen per barrel of liquid feed ($ft^3/$ bbl).The hydrogen partial pressure increases as the $H_2/$oil ratio goes up, removing coke precursors from the metal sites and lengthening the life of the catalyst. In other words, the $H_2/$oil ratio and hydrogen partial pressure present in the reactor system determine the rate of coke formation on the catalyst and, consequently its stability and life.

The semi-regenerative fixed bed catalytic reformer, the cyclic fixed bed reformer, and the platforming continuous catalytic reformer are three of the main reforming techniques used today. The flow sheet diagram of the semi-regenerative fixed bed catalytic reformer as per Fig. 2.12.

2.2.7.9 Catalytic polymerization

The process of polymerization in the petroleum industry produces hydrocarbons with high molecular weight and higher octane numbers from light unsaturated olefin gases, such as ethylene, propylene, and butylenes. These polymerized compounds could be employed as blending components in gasoline. At lower temperatures and higher pressures, polymerization can occur thermally or

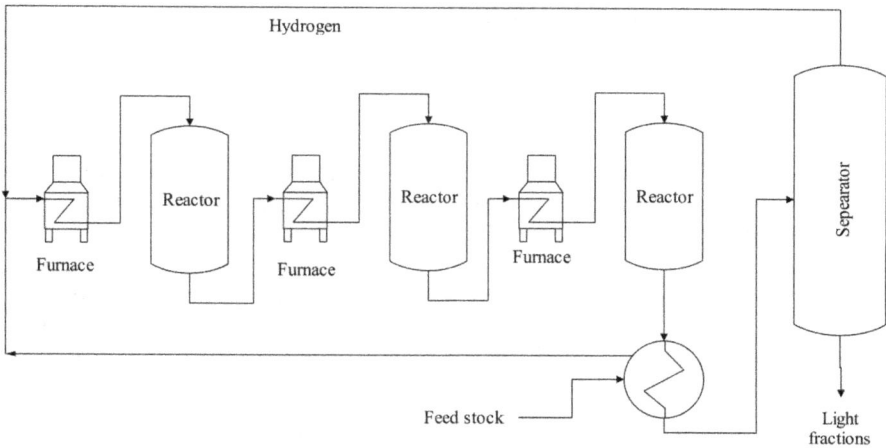

Figure 2.12 Semi-generative fixed bed catalytic reforming process

in the presence of a catalyst, but catalytic polymerization is the preferred method. A phosphate is the main catalyst used in catalytic polymerization; the commercially available catalysts include liquid phosphoric acid, phosphoric acid on kieselguhr, copper pyrophosphate pellets, and phosphoric acid film on quartz. The olefin feedstock is processed in this process to get rid of sulfur and other unwanted substances. After that, the feedstock either comes into contact with liquid orthophosphoric acid or is omitted from a solid orthophosphoric acid catalyst, where an exothermic polymeric reaction takes place. This reaction needs cooling water, thus cold feedstock must be injected into the reactor to

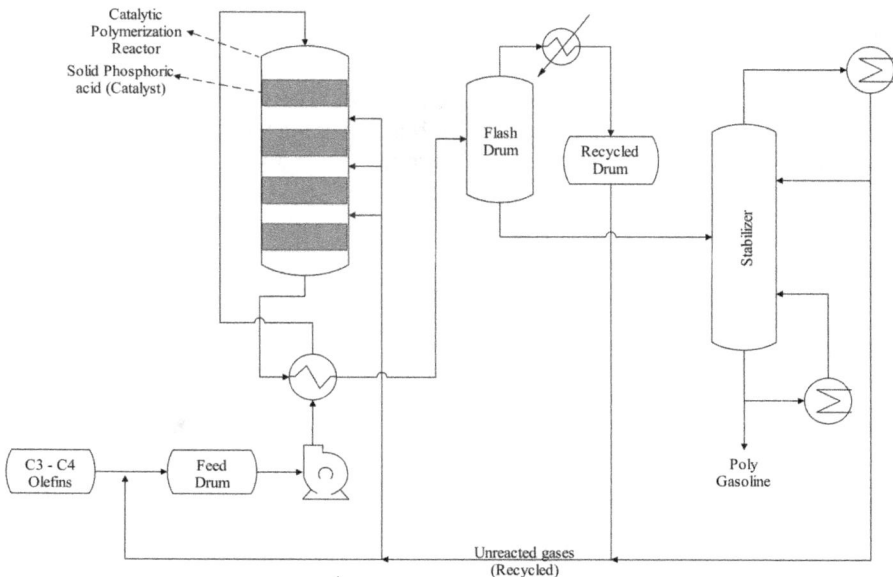

Figure 2.13 Catalytic polymerization process

maintain temperatures between 150 and 230°C and pressures between 15 and 85 atmospheres. A stabilization process is used to extract saturated and unreacted gases from the polymer gasoline product as the reaction products exit the reactor.

2.2.7.10 Catalytic alkylation

In the petroleum industry, alkylation is the combination of low-molecular-weight olefins with isobutene in the presence of a catalyst, such as vitriol or acid. Generally, alkylation is defined as the addition of an alkyl group. Alkylate is the brand name for the product, which is a mixture of high-octane, branched-chain paraffinic hydrocarbons. Alkylate is a premium blending stock because it burns cleanly and has great antiknock qualities. The type of olefins used and the operating environment have a significant impact on the alkylate's octane number. The alkylation process involves the following mechanism.

Step-I: Addition of Proton: The double bond of an olefin is attacked by a proton (H^+) from an acid catalyst, forming a carbonium ion.

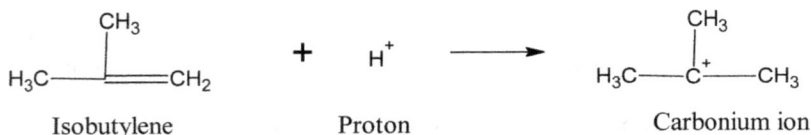

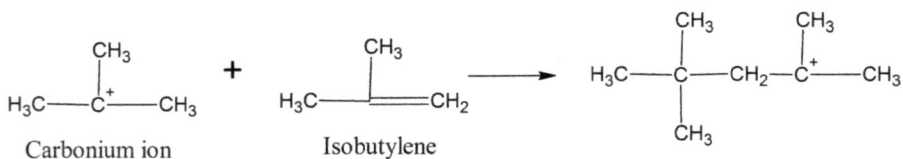

Isobutylene   Proton   Carbonium ion

Step-II: Addition: The double bond of an olefin is attacked by a proton (H^+) from an acid catalyst, forming a carbonium ion.

Carbonium ion   Isobutylene

Step-III Regeneration: The double bond of an the olefin is attacked by a proton (H^+) from an acid catalyst, forming a carbonium ion. A new carbonium ion is formed at that instant to continue the reaction.

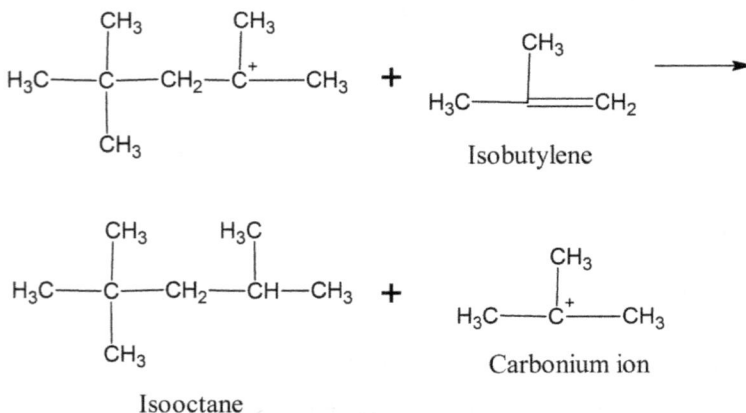

Isobutylene

Isooctane   Carbonium ion

Termination:

$$
\underset{\text{Carbonium ion}}{H_3C-\overset{\overset{\displaystyle CH_3}{|}}{\overset{+}{C}}-CH_3} \longrightarrow \underset{\text{Isobutylene}}{H_3C-\overset{\overset{\displaystyle CH_3}{|}}{C}=CH_2} \quad + \quad \underset{\text{Proton}}{H^+}
$$

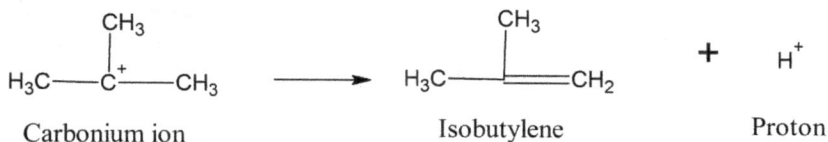

The feedstock (propylene, butylene, amylene, and fresh isobutane) enters the cascade type sulfuric acid (H_2SO_4) alkylation reactor and comes into contact with the concentrated sulfuric acid catalyst (in concentrations of 85 to 95% for good operation and to minimize corrosion). The term "cascade type reactor" refers to a reactor that is separated into zones. Olefins are given to each zone by distributors, while sulfuric acid and isobutanes flow from zone to zone over an up-down type system thus transferring the reaction mixture into the settler. The acid in this case is recycled to the reactor after being separated into hydrocarbon and acid phases in a settler. For pH regulation, the hydrocarbon phase is washed with hot caustic solution in water. The clean liquid goes to the deisobutanizer. Once recovered from the deisobutanizer bottom, the alkylate can either be used right away for mixing motor gasoline or it can be run to create blending stock that is suitable for use in aviation. The recycled isobutene from the top of the deisobutanizer is added to the feed.

2.2.8 Properties of crude oil products

The way that crude oil is processed affects both its chemical and physical characteristics. According to the chemistry perspective, functional groups such as olefins, paraffins, naphthenes, aromatics, and resins should be used to describe the petroleum. In different petroleum processing streams, the predominance of one or more functional group is a sign of the given product's quality and characterization. For instance, only olefins and paraffins are found in the refinery's lighter fractions. On the other hand, gasoline and other similar products should have high octane ratings, which may be a sign of the presence of olefinic and aromatic functional groups in the product stream. It is equally vital to characterize the petroleum physically in terms of its viscosity, density, and boiling point curves. These characteristics also reveal the caliber of both the feed and the product. In order to determine whether a product stream in petroleum processing is diesel, gasoline, or another product, it is important to collect any intermediate or product stream that has various features characterized in detail. Contrary to the chemical process industry, this is the most significant attribute of the petroleum processing industry. The American Society for Testing and Materials (ASTM), the Bureau of Indian Standards (BIS), the German Institute for Standardization (DIN) and the International Organization for Standardization (ISO) have standard techniques for determining the various qualities. This chapter includes the definition and significance of each parameter.

2.2.8.1 API gravity

The relative masses of different types of crude oils are expressed using API (American Petroleum Institute) gravity. The following equation could be used to mathematically determine the API gravity:

$$°API = \frac{141.5}{Specfic\ Gravity\ of\ fraction\ at\ 13.5°C} - 131.5$$

The weight of a particular volume of a substance divided by the weight of a corresponding volume of water at a given temperature is known as specific gravity. In order to determine the mass of crude oil and its byproducts, specific gravity is used. Typically, the volume of crude oil and the liquid products it produces are first measured, and then the related masses are converted to masses using API gravity. According to the equation, in order to determine the API gravity, the first relative density, or the density in relation to water, is first determined using either the hydrometer described in ASTM (American Standard Test Method) D1298 or the oscillating U-tube method described in ASTM D4052. As per ASTM D287, the hydrometer method is also used to directly measure API gravity. In general, oil with an API gravity of 40 to 45° attracts the highest prices. The molecular chains get shorter and less beneficial to refineries above 45 degrees. Depending on its measured API gravity, petroleum is classified as light, medium, or heavy.

1. The API gravity of light crude oil is higher than 31.1°
2. Medium oil's API gravity ranges from 22.3 to 31.1°
3. The API gravity of heavy crude oil is below 22.3°
4. The API gravity of extra heavy crude oil is below 10.0°

The crude oil with an API gravity of less than 10° is referred to as extra heavy oil or bitumen. Bitumen extracted from oil sands in Alberta, Canada, has an API gravity of around 8 degrees. It is frequently diluted with lighter hydrocarbons to produce diluted bitumen, which has an API gravity of only 22.3°, or "upgraded" to an API gravity of 31 to 33 degrees as synthetic crude.

2.2.8.2 Aniline point

The Aniline point of a hydrocarbon or percentage of petroleum is the lowest temperature at which equal amounts of liquid hydrocarbon and aniline are miscible. In order to generate a homogeneous solution for Aniline point analysis, an equal volume of aniline and oil is swirled and heated continuously in a test tube. The heating is turned off, and the tube is allowed to cool. The temperature at which two phases of oil separate is referred to as the aniline point of oil. Aniline is an aromatic chemical consisting of a benzene molecule with a -NH_2 group replacing one hydrogen atom (C_6H_5-NH_2). The aniline point is essential for characterization of petroleum fractions and molecular type analysis. In any homologous series, the aniline point rises as molecular weight increases. It is utilized to assess the aromatic content of crude oil in the following manner:

Aromatic oil with a 75 percent aromatic content and an Aniline point between 32.2 and 48.9°C.

Type naphthenic with a 40% aromatic content and an Aniline melting point between 65.6 and 76.7°C.

Paraffin oil with a 15% aromatic concentration and an Aniline melting point between 93.3 and 126.7°C.

So, it is concluded that Aniline value is increasing with decreasing aromatic content in crude oil.

2.2.8.3 Diesel Index (DI)

Diesel index is described as a measurement of a diesel fuel's ignition quality. Diesel index is defined as:

$$Diesel\ Index\ (DI) = \frac{(API)\ (Aniline\ Point)}{100}$$

With a higher diesel index, the ignition quality of the fuel is improved. Reduced values will produce smoky exhaust. This property can be utilized to forecast the cetane number based on correlations specific to each crude and manufacturing process.

2.2.8.4 Cloud point

The temperature at which the production of tiny particles of an incompatible material generates observable haze or cloudiness in a liquid combination cooled under controlled conditions. The cloud point is the temperature at which, under typical conditions, wax crystals begin to form upon slow cooling. At this temperature, the oil gets hazy and the first wax crystals become visible. The cloud point is determined using the procedures outlined in ASTM D 2500, IP 219, and ISO 3015. The cloud point of petroleum oils at low temperatures increases as their molecular weights rise. Cloud points are assessed for oils that include paraffins in the form of wax; hence, cloud point statistics are not supplied for light fractions such as naphtha or gasoline. Although the temperature differential could range from 0 to 10°C, cloud points often occur about 4–5°C above the pour point.

2.2.8.5 Pour point

Pour point (Tp) is the lowest temperature at which a petroleum fraction will flow or pour when chilled without stirring under conventional cooling conditions. Pour point is the absolute minimum temperature at which an oil may be stored and still flow under gravity. Pour point is one of the criteria of coolness for heavy fractions. When the temperature of a petroleum product is below the pour point, it cannot be stored or transmitted through a pipeline. The cloud point of petroleum fractions is determined using the ASTM D 97 (ISO 3016 or IP 15) and ASTM D 5985 procedures. For commercial formulation of engine lubricants, the pour point can be reduced to between –25 and –40°C. This is accomplished by using pour point depressant chemicals that prevent the formation of wax crystals in the oil. The presence of wax and heavy chemicals raises the petroleum fractions' pouring point. Pour point has limited use in wax and paraffinic heavy oils to assess the degree of solidification and wax content, as well as the minimum temperature necessary to ensure oil fluidity. Oils with greater weight and viscosity have higher pour points.

2.2.8.6 Flash point

The flash point of petroleum fractions is the lowest temperature at which oil vapors ignite when exposed to a flame under specific conditions. Consequently, the flash point of a fuel shows the highest temperature at which it can be stored without posing a significant fire hazard. In relation to the volatility of a fuel, the presence of sunlight, and the existence of volatile components, the upper vapor pressure equates to lower flash points. For crude oils with Reid Vapor Pressure (RVP) greater than 0.2 bar. The flash point is typically below 20°C. The flash point is an essential property of petroleum fractions and products in a high-temperature environment, and it is directly related to the safe storage and management of these petroleum products. Several techniques exist for determining the flash points of petroleum fractions. The flash point of petroleum products is determined by heating a given volume of liquid at a standard rate of temperature rise until sufficient vapor is produced to produce a flammable mixture with air in an enclosed space (i.e., the closed flash point temperature) or with air in an open cup (i.e., the upper open flash point temperature) upon application of a small flame.

For evaluating the closed cup flash point of petroleum products with a flash point between 19°C and 49°C, the Abel apparatus or Cleveland Open Cup method (ASTM D 92) is utilized. The Pensky-Martens apparatus (ASTM D 93) is used to determine the flash point of fuel oils and lubricating oils, as well as bitumen other than cut back bitumen with a flash point greater than 49°C. For determining the flash point of petroleum products other than fuel oils and those with an open cup flash point below 79°C, the Cleveland apparatus or Closed Tag method (ASTM D 56) is utilized.

2.2.8.7 Fire point

The fire point is the lowest temperature at which a crude oil will continue to burn for five seconds when tested with flash point equipment. The method of determination of the fire point is closely related to the flash point, where it is performed in open cup rather a closed cup. Also, vapor-air mixture flashes as well as liquid continues to burn at the fire point (While at the flash point, only the vapor-air mixture continues to burn). For household applications, a very high flash point exceeding 50 degrees Celsius is undesirable. Only volatiles are responsible for the ease of ignition, and a decrease in volatiles raises the flash point, making ignition more difficult.

2.2.8.8 Smoke point

The tendency of aviation turbine fuels and kerosene to burn with a smoky flame is indicated by a fuel's smoke point. A greater proportion of aromatics in a fuel result in a smoky flame and energy loss due to thermal radiation. The smoke point (SP) is the highest flame height at which a typical wick-fed lamp may burn a fuel without producing smoke. It is measured in millimeters, and a high value implies a fuel with a low tendency to produce smoke. Methods for measuring the smoke point are outlined in ASTM D t322 (US) or IP 57 and ISO 3014. For the same fuel, the IP technique of measuring smoke point is 0.5–1 mm higher than the ASTM method for smoke point values in the range of 20–30 mm.

2.2.8.9 Octane Number

Octane number is a significant characteristic of spark engine fuels such as gasoline and jet fuel, as well as fractions used to manufacture these fuels (e.g., naphtha), and it represents the antiknock property of a fuel. The octane number of isooctane (2,2,4-trimethylpentane) is 100, whereas the octane number of n-heptane is 0. The percentage of isooctane in their blends determines the octane number. Isoparaffins and aromatics have higher octane levels than n-paraffin and olefins. Therefore, the octane rating of gasoline is dependent on its molecular structure, particularly its isoparaffin content. It is measured using the standard tests ASTM D 908 and ASTM D 357. By adding additives such as tetra-ethyl lead (TEL), alcohols, and ethers such as ethanol, methyl-tertiary-butyl ether (MTBE), ethyl-tertiary-butyl ether (ETBE), and tertiary-amylmethyl ether (TAME), the octane number is enhanced (TAME). However, increasing the octane number of a fuel would result in a decrease in engine power loss, an increase in fuel efficiency, a decrease in environmental pollutants, and a decrease in engine damage. For these reasons, octane number is one of the most essential gasoline quality characteristics. In addition, it is highly relevant in the refining industry's reforming, isomerization, and alkylation operations. These mechanisms make it possible for effective reactive transformations to produce long side chain paraffins and aromatics with higher octane ratings than the feed constituents, which comprise a greater proportion of constituents with open chain paraffins and non-aromatics (naphthenes).

2.2.8.10 Cetane number

The fuel must possess a characteristic that promotes auto-ignition in diesel engines. The ignition delay period is frequently measured using a fuel characterization component known as the Cetane number (CN). A diesel fuel's behavior is tested by comparing it to two pure hydrocarbons: n-Cetane or n-Hexadecane (n-$C_{16}H_{34}$), which is assigned the number 100, and a-methylnaphthalene, which has a cetane number of 0. A diesel oil has a cetane number of 60 if it resembles a blend of 60 capacity unit cetane and 40 capacity unit-methylnaphthalene. Instead of a-methylnaphthalene, heptamethylnonane (HMN), a branched isomer of n-cetane with a cetane number of 15, is used in practice. In practice, the cetane number is defined as follows:

$$CN = \text{Volume \% (n-cetane)} + 0.15 \text{ (Volume \% HMN)}$$

ASTM D 613 is frequently used to determine the cetane number of diesel oils. The shorter the ignition delay period the upper CN value. Higher cetane number fuels reduce combustion noise and permit improved combustion control leading to increased engine efficiency and power output. Higher cetane number fuels tend to result in easier startups and faster warm-ups in weather.

2.2.8.11 Softening point

The softening point is the temperature at which bitumen reaches a specific degree of softening according to the test specifications. The softening point is evaluated using the ASTM D36 standard Ball and Ring test. First, crude oil is melted and then cast into two-numbered discs in standard rings. A steel ball of known weight and diameter is placed on top of each barrel of crude oil. The entire stand that holds these two discs with balls is submerged in a bath of water or glycol. As heating continues, softening ensues. The temperature at which the sample separates from the discs is the crude oil's melting point. A higher softening point indicates greater bitumen purity.

2.2.8.12 Spontaneous Ignition Temperature (SIT)

It is the temperature at which a substance spontaneously ignites in the absence of a spark or flame. It is also referred to as the "auto ignition point." This temperature is necessary for providing the activation energy required for combustion. Increases in pressure or oxygen concentration reduce the temperature at which a chemical will ignite. Typically, it is applied to an ignitable fuel mixture. Typically, the auto-ignition temperatures of liquid chemicals are measured using a 500 ml flask placed within a temperature-controlled oven in accordance with ASTM E659. The auto-ignition temperature can also be measured for plastics under elevated pressure and 100% oxygen concentration. The resulting value is used as a predictor of serviceability in high-oxygen environments. The procedure follows ASTM G72. The time required for a material to reach its auto-ignition temperature, i.e., t_{ig}, when exposed to a heat flux q^n is given by,

$$t_{ig} = \frac{\pi}{4} \kappa \rho c \left[\frac{T_{ig} - T_o}{q^n} \right]^2$$

Here k = thermal conductivity, ρ = density, and c = specific heat capacity of the material of interest, T_o is the initial temperature of the material (or the temperature of the bulk material).

2.2.8.13 Penetration index (PI)

This is one of bitumen's characteristics. The penetration index is a reliable indicator of the ability of bitumen to withstand repeated variations in pavement temperature. The penetration ratio is a fundamental type of Penetration Index. It is very similar to the penetration index, with the exception that the sample is tested with 100 g on the needle at 25°C and 200 g at 4°C. During the derivation of Penetration Index and Penetration Ratio values, it is assumed that the properties of bitumen vary linearly over the entire temperature range of operation and application. In the case of specific bitumens or modified bitumens, however, this assumption may not be entirely accurate.

2.2.8.14 Viscosity

The viscosity of petroleum can be defined as its resistance to gradual deformation under shear or tensile stress. It corresponds to the slang term "thickness". Viscosity may be a bulk property that is measured for all types of liquid petroleum fractions. In this case, the viscosity of petroleum fractions increases as API gravity decreases, as well as for residues and heavy oils with API gravity less than 10. The following are classifications of crude oil according to viscosity.

1. The viscosity of light crude oil is less than 870 kg/m³.
2. Viscosity of medium oil ranges from 870 to 920 kg/m³.
3. Viscosity of heavy crude oil ranges from 920 to 1000 kg/m³.

Extra heavy oil has a viscosity greater than 1000 kg/m³. It is measured using a capillary U-type viscometer (Cannon-Fenske viscometer) in accordance with ASTM D 445, which corresponds to the ISO 3104 method. Another useful property in petroleum production, refining, and transportation is viscosity. It is utilized in reservoir simulators to estimate the flow rate and production of oil and gas. It is required for the calculation of the power required in mixers or fluid transfer, the amount of pressure drop by a pipe or column, flow measurement devices, and the design and operation of oil/water separators.

2.2.8.15 Viscosity Index (VI)

The viscosity index (VI) is an arbitrary measurement of the variation of viscosity with temperature. It is calculated by following formula.

$$VI = 100 \; \frac{L - U}{L - H}$$

U is the kinematic viscosity of the oil at 40°C, and L and H are values based on the kinematic viscosity of the oil at 100°C. L and H are the viscosity values at 40°C for oils of VI 0 and 100, which have the same viscosity at 100°C as the oil whose VI we are attempting to determine. The values for L and H can be found in ASTM D2270. The lower the viscosity index, the greater the oil's viscosity change with temperature, and vice versa. It is used to describe the viscosity changes of lubricating oil in relation to temperature.

2.2.8.16 Characterization Factor (K_w)

This is one of the earliest parameters, initially defined by Watson et al. of the Universal Oil Products (UOP), and is used extensively in refinery calculations. This factor provides essential information about the fraction, from its physical properties to its tendency to crack. This factor is calculated using the following formula.

$$k_w = \frac{\sqrt[3]{T_B}}{\rho}$$

Here, T_B is the average boiling point in degrees R derived from five temperatures corresponding to 10, 30, 50, 70, and 90% vaporization by volume and ρ represents specific gravity. Typically, the Watson characterization factor ranges from 10.5 to 13 for various crude streams. Commonly, highly paraffinic crude has a K_B factor of 13. In contrast, the K_B factor for highly naphthenic crude is 10.5 Consequently, Watson characterization factors are frequently utilized to determine the quality of petroleum based on the predominance of paraffinic or naphthenic components.

2.2.8.17 Kinematic viscosity (v)

Kinematic viscosity is the ratio of absolute (dynamic) viscosity to absolute density ρ at the same temperature.

$$v = \frac{\mu}{\rho}$$

At temperatures between 15 and 100°C, kinematic viscosity is measured using a capillary U-type viscometer (Cannon-Fenske viscometer) in accordance with ASTM D 445, which is equivalent to ISO 3104 method. It is a useful characterization parameter for heavy fractions for which boiling point data is unavailable due to distillation-induced thermal decomposition. Kinematic viscosity values for pure liquid hydrocarbons are typically measured and reported in cSt at 38°C and 99°C as reference temperatures. However, kinematic viscosities of petroleum fractions are also reported at reference temperatures of 40°C, 50°C, and 60°C. The viscosity of a liquid decreases as its temperature rises. Kinematic viscosity is a useful characterization parameter, particularly for heavy fractions where the boiling point may not be accessible (Chinenyeze and Ekene, 2017; TNAP, 2016; Clayton, 2005; Santos et al., 2014; Schobert, 2013).

2.3 Coal

Coal is one of the world's most valuable natural resources and the second most abundant energy source after crude oil.

2.3.1 Concept of coal

Coal is a naturally occurring, solid, nonrenewable energy source that is combustible. Coal is one of the cheapest and most important energy sources, accounting for 41% of global electricity production and over 60% of steel production. Access to modern energy services contributes not only to economic activity and household incomes, but also to the enhanced quality of life that results from improved education and health services. Future energy demand will require the use of all energy sources, including coal. Globally, coal has many important applications. Coal's most important applications include electricity generation, production, cement manufacturing, and use as a liquid fuel. Global coal consumption has increased more rapidly than any other fuel type. Since

4,000 years ago, coal has been an important source of energy for heating and cooking. In the nineteenth and twentieth centuries, it was utilized for electricity generation and as a chemical feedstock. According to the BP Statistical Review of World Energy 2020, published in 2019, the world's total coal production is 1070 billion tons. The top five coal reserves are located in the United States (249 billion tons, 23 percent), Russia (162 billion tons, 15 percent), Australia (149 billion tons, 14 percent), China (42 billion tons, 13 percent), and India (106 billion tons, 10 percent). The remaining 40% of the world's coal reserves are located elsewhere. The use of coal to generate electricity and heat increased by 3.3%, or 166 Mt, in 2018 compared to 2017. The industrial sector as a whole, including the iron and steel industry, consumes less coal than the previous year (IEA, 2007).

2.3.2 Occurrence and origin of coal

It is acknowledged globally that the origin of coal is plant debris, such as ferns, trees, bark, leaves, roots, and seeds. Peat was initially gathered and deposited in an unconsolidated state in swamps. This peat was stacked layer by layer, and the bottom layer underwent coalification, a metamorphic process involving pressure and heat. There are two main theories supporting the formation of coal.

2.3.2.1 In situ theory

According to this theory, the plants that comprise coal accumulated over many thousands of years in large freshwater swamps or peat bogs. The autochthonous theory is a theory that postulates growth in place of plant material. Consistent with this theory, coal seams can be found where forests once grew. Because the land was sinking gradually, the accumulated plant matter submerged gradually and did not decompose or become destroyed. During your time, the rate of land subsidence accelerated, and coal forests were submerged beneath the water. After a sufficient amount of time, the land adjacent to the coal forest reemerged from the water, and this cycle repeated itself, resulting in the formation of coal strata and seams. The evidence listed below supports this theory

- Relative purity & constancy in thickness and composition of coal seams (i.e., no major transport)
- Presence of erect & rooted fossil tree trunks with roots in the Underclays
- Underclays below coals are generally poor in alkalis and lime.
- Wide lateral extent of coal belts

Subsidence, or movement of the Earth's surface, balances accumulations of substance and associated mineral matter, typically clays and sands, within the area where these materials are accumulating. Therefore, this type of coal has coal and inorganic sedimentary rock layers arranged sequentially.

2.3.2.2 Drift theory

This theory proposes that coal strata accumulated from plants that were rapidly transported and deposited during a flood. This theory is known as the allochthonous theory because it presupposes the transport of plant matter. According to this theory, plant material was transported with flowing water from one location to another, and was subsequently deposited in wetlands, lakes, seas, and estuaries with suitable conditions, such as a supply of sediments. The Indian coal seams are of drift origin (Oviedo et al., 1995; Stephen et al., 2007). The following are some evidence of this theory.

- High levels of ash in coals
- Absence of plant fossils in coal seam roofs.
- The digitization/division of coal seams
- Lateral coal to carbonaceous shale transition
- Currently, peat/brown coals are produced in river deltas (e.g., Ganga, Mississippi)
- Fish remains in coals suggest coastal or marine waters.

2.3.3 Composition of coal

Coal's weight-based composition is roughly 99.0% carbon, hydrogen, oxygen, nitrogen, and sulfur. While the chance of oxygen and hydrogen diminishes with position, the proportional proportion of carbon increases. The components that make up between 1% and 0.01% of coal's weight are considered minor. Common trace elements found in coal include sodium, magnesium, aluminum, silica, phosphorus, potassium, calcium, titanium, manganese, and iron. While the bulk of minor elements are connected to the minerals in coal, phosphorus is also connected to the organic coal matrix. Coal often contains trace amounts of minor elements (Miller, 2004).

2.3.4 Analysis of coal

Coal is a fossil fuel produced over millions of years from decomposed plant matter at high temperatures and pressures. Coal analysis is necessary to determine the quantity and quality of coal used to generate heat (energy).

Coal is analyzed in two ways:

2.3.4.1 Ultimate analysis

Elements in coal, i.e., carbon, hydrogen, oxygen, sulfur, nitrogen are considered in ultimate analysis.

1. **Carbon and Hydrogen:** In the presence of excess air, a sample of powdered, dried coal is burned in the presence of its precise weight. The coal's carbon and hydrogen are converted into carbon dioxide gas and water, respectively. CO_2 and H_2O are subsequently absorbed in KOH solution and dry $CaCl_2$ tubes, respectively. The weights of KOH and $CaCl_2$ are weighted before and after combustion.

$$C \ + \ O_2 \rightarrow CO_2 \qquad\qquad 2KOH + CO_2 \rightarrow K_2CO + H_2O$$

$$12 \qquad\qquad 44$$

$$H_2 \ + \ \frac{1}{2}O_2 \rightarrow H_2O \qquad\qquad CaCl_2 + 7H_2O \rightarrow CaCl_2 \cdot 7H_2O$$

$$2 \qquad\qquad 18$$

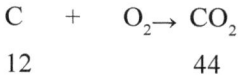

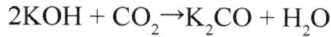

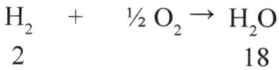

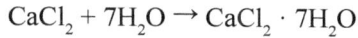

$$\textit{Percentage of Carbon} = \frac{\textit{Increasing in weight of KOH tube X 12 X 100}}{\textit{weight of coal taken X 44}}$$

$$\textit{Percentage of Hydrogen} = \frac{\textit{Increasing in weight of CaCl}_2 \textit{ tube X 2 X 100}}{\textit{weight of coal taken X 18}}$$

Significance of Carbon and Hydrogen

1. The relationship between a fuel's calorific value and its carbon content is direct.
2. Increased carbon content reduces the size of the combustion chamber.
3. The proportion of hydrogen in coal increases its calorific value. From peat to bituminous coal, the proportion of hydrogen ranges between 4.5 and 6.5 percent.

2. **Nitrogen:** For determination of Nitrogen in coal, Kjeldahl method or Kjeldahl digestion is used, which is developed by Johan Kjeldahl in 1883. In this method, approximately 1 g of powdered dry coal sample is heated in a long-necked flask (known as Kjeldahl's flask) containing a solution of concentrated sulfuric acid and potassium sulfate. After the solution has become transparent, it is treated with an excessive amount of KOH to release ammonia. The released ammonia is distilled and absorbed in a standard acid solution of known concentration. The remaining acid is then determined through back titration with a standard KOH solution. Then, based on the amount of acid used, we determine the Nitrogen in the coal sample by the following formula.

$$\textit{Percentage of Nitrogen} = \frac{\textit{Volume of acid used X Normality of acid X 1.4}}{\textit{weight of coal taken}}$$

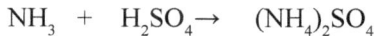

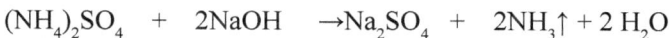

$$NH_3 \ + \ H_2SO_4 \rightarrow \ (NH_4)_2SO_4$$

$$(NH_4)_2SO_4 \ + \ 2NaOH \ \rightarrow Na_2SO_4 \ + \ 2NH_3{\uparrow} + 2H_2O$$

Significance of Nitrogen

During combustion of coal, nitrogen is converted into Nitrogen oxides (NO_x), which is a major environmental pollutant. Also, it has no calorific value, so, the required quantity of nitrogen is least in the coal sample.

4. **Ash:** The noncombustible residue left over after coal has been burned makes up its ash component. It is the mineral substance that is left over after combustion has burned off the carbon, oxygen, sulfur, and water

(including water from clays). The residue left over after the elimination of volatile materials is heated for 30 minutes without a cover at 700 ± 50°C to determine its ash concentration.

$$Percentage\ of\ Ash = \frac{Weight\ of\ residue\ left\ after\ heating\ \ X\ 100}{weight\ of\ coal\ taken}$$

Significance of ash content

1. A high ash content is undesirable. It decreases coal's calorific value.
2. Ashes may restrict airflow and reduce the rate of combustion in the furnace grate.
3. High ash levels result in significant heat losses and the formation of ash lumps.
4. Coal's efficacy is also affected by the composition of ash and fusion range.

5. **Sulfur:** The remaining material in the crucible of the bomb calorimeter is measured to determine the calorific value. Approximately 100 ml of distillated water is used to wash the remaining material of known weight. The washing is treated with a solution containing barium chloride, and barium sulfate is precipitated. The precipitates are filtered, washed with deionized water, dried, and weighed.

$$S \quad + \quad 3O_2 \xrightarrow{\text{Bomb calorimeter}} 2SO_3 \xrightarrow{BaCl_2} BaSO_4$$
$$\phantom{S \quad + \quad 3O_2 \xrightarrow{\text{Bomb calorimeter}} 2SO_3 \xrightarrow{BaCl_2}} 32 \qquad\qquad 233$$

$$Percentage\ of\ Sulphur = \frac{Weight\ of\ BaSO_4\ obtained\ X\ 32\ X\ 100}{Weight\ of\ coal\ sample\ taken\ in\ bomb\ X\ 233}$$

Significance of nitrogen

During combustion of coal, sulfur is changed into oxides of sulfur (SO_x), which are major environmental pollutants. So, the required quantity of Sulfur is least in the coal sample.

6. **Oxygen:** It is determined by the following formula.

Percentage of Oxygen = 100 - Percentage of (C + H + S + N + Ash)

Significance of oxygen

The quantity of oxygen decreases the calorific value. Also, it increases the moisture content and decreases the coking power. So, a high percentage of oxygen is highly undesirable in coal.

2.3.4.2 Proximate analysis

It is termed proximate analysis because the data varies depending on the method used. It provides information regarding the practical application of coal. Coal's moisture content, ash, volatile matter, and fixed carbon make up its approximate analysis.

1. **Moisture content:** Coal moisture content is determined by heating a known quantity of coal in an electric hot air oven to 105–110°C for approximately one hour. After removing the coal from the oven, it is cooled in a desiccator and weighed.

$$Moisture\ content = \frac{Loss\ in\ weight\ after\ drying\ X\ 100}{weight\ of\ coal\ taken}$$

Significance of moisture content

1. The addition of moisture to coal is undesirable.
2. Moisture reduces the calorific value of coal and removes a significant portion of the liberated heat as latent heat of vaporization.
3. Coal that has an excessive amount of surface moisture may be difficult to manipulate.
4. In a furnace, excessive moisture extinguishes the flame.

2. **Volatile matter:** Except for moisture, volatile matter refers to the parts of coal that leak at high temperatures when there is no air present. Usually, this consists of a combination of sulfur, aromatic hydrocarbons, and both short and long-chain hydrocarbons. It is made up of a complicated mixture of liquid and gaseous byproducts of the thermal breakdown of coal. It is determined by heating a known mass of moisture-free coal in a platinum crucible at 950 ± 20°C for seven minutes.

$$Volatile\ matter = \frac{Loss\ in\ weight\ due\ to\ voltile\ matter\ X\ 100}{weight\ of\ coal\ taken}$$

Significance of volatile matter

1. A significant amount of fuel is burned as gas when there is a large proportion of volatile stuff.
2. Long flames, a lot of smoke, and low heating values are the results of the high volatile content.
3. Secondary air must be supplied in order for the outgoing combustible gases to be burned efficiently.
4. In coal gas production, a high volatile matter content is desired because volatile matter indicates the percentage of coal that will be heated and turned into gas and tar.

3. **Ash:** The ash content of coal is measured as per the ultimate analysis.
4. **Fixed carbon:** The sort of carbon that is still present after volatile substances have been eliminated is known as fixed carbon content. Due to the particular carbon's departure with the volatile, it differs from the final carbon content.

The amount of coke that will form from a sample of coal is estimated using fixed carbon. By deducting the mass of volatiles identified by the volatility test from the coal sample's initial mass, fixed carbon is calculated. The amount of fixed carbon rises from lignite to anthracite. The percentage of fixed carbon is given by:

Percentage of fixed carbon = 100 − Percentage of (moisture content + volatile matter + ash)

Significance of fixed carbon

1. The greater the proportion of fixed carbon, the higher the calorific value.
2. The proportion of fixed carbon helps with furnace and firebox designs since only fixed carbon burns in the solid state (Speight, 2005; Khan, 1992; Warwick, 2005; Kabe et al., 2004).

2.3.5 Classification of coal

Several methods are used to classify coal, including visual characteristics, origins, chemical properties, Parr's classification, A.S.T.M. Classification, and Indian classification. We will see them one by one.

2.3.5.1 Visual characteristics

Coal is classified using its visual attributes and flame. Coal is classified into three types: (1) Brown coal/lignite is brown in color and has a woody structure, (2) Bituminous coal is black in color and its flame is smoky yellow, and (3) Anthracite coal is black and lustrous in characteristics and it burns without a flame.

2.3.5.2 Sources genesis (physical properties)

The coals are divided into two major categories based on their physical characteristics: humic and sapropelic.

(A) Humic coal: It is composed primarily of microscopic plant debris and is banded. It contains the four well-known lithotypes found in coal: Vitrain, Clarain, Durain, and Fusain.

(B) Sapropelic coal: It is non-banded, predominantly made up of minute plant debris, spores, pollen, and algae, and most frequently displays conchoidal fracture. Cannel coal, boghead coal, and the transition between two-cannel-boghead coal and boghead-cannel coal are further subcategories of sapropelic coal. Cannel coal is rich in spores, whereas boghead coal is rich in algae. When spores outweigh algae, the coal is known as boghead-cannel coal, and when spores outnumber algae, it is known as cannel-boghead coal. Cannel shale and boghead shale are terms for coal that is highly polluted with silica and clay particles. In comparison to coals, this shale is denser, more reflective, and darker in color. It is referred to as cannel coal ironstones when cannel coal contains siderite.

Two types of coal, namely those containing clay minerals, mica, and quartz, are referred to as carbonaceous shale. They are black, dull, hard, and compact, and they weigh more than coal. In the majority of instances, carboargillite (composed of 20–60% clay minerals, mica, and quartz) may contain carbosilicate and carbopyrite.

2.3.5.3 Chemical parameters

Coal is divided into four types based on its chemical properties: peat, lignite, bituminous, and anthracite. It is determined by carbon, ash, and water content.

(A) Peat: Coal is not identical to peat. It is the beginning of the transition and is distinguished by the buildup of partially decomposed plant or organic debris. It only occurs in naturally occurring places known as peatlands, which are the most effective carbon sinks on the globe because peat and plants trap CO_2 that is naturally emitted from peat and keep a balance. Even though many other plant species can also contribute, sphagnum moss, often known as peat moss, is one of the most prevalent components of peat. Peat forms in wetland areas when floods or standing water prevent oxygen from the atmosphere from reaching the decomposing organisms, slowing down the process. A sufficient amount of volatile stuff, a lot of moisture, less than 40 to 55 percent carbon, and contaminants make up peat.

(B) Lignite: Lignite, sometimes referred to as brown coal, is an intermediate step of the coalification process made out of naturally compressed peat and is a soft, brown, flammable sedimentary rock. Due to its low total heat, it is regarded as coal of the lowest quality. It has a carbon content between 60 and 70%, a high moisture content of up to 75%, and an ash content between 6 and 19%. Due to its high volatile content, lignite spontaneously combusts.

(C) Bituminous: Bituminous coal or black coal is the most widely available and utilized soft coal. Its moisture and volatile content is lower than that of lignite coals; the moisture content starts at 5–45% and the carbon concentration ranges from 69% to 86%. Because they burn with a reasonably lengthy flame and have a high heating value and volatile content, they are easily combustible when ground into a fine powder. The calorific value of these coals is extremely high and they are dense and compact. They are employed in coke and gas production. Geologically, bituminous coals range from the Carboniferous to the Cretaceous periods and are widely distributed across the globe.

(D) Anthracite: The significant level of metamorphism that anthracite has experienced makes it a top ranking coal. The highest fixed-carbon content and most brittle, jet-black material is anthracite (approximately 86 to 98 percent). Being low in volatile matter (2–12%), anthracite has a slow combustion rate. Low in sulfur and volatiles, anthracite burns with a blue flame that is both hot and clean. Due to these properties, anthracite is sometimes utilized in domestic and other industrial applications that require smokeless fuels.

2.3.5.4 Parr's classification

In 1922, Parr revised the classification of coal according to its volatile matter and calorific value.

Table 2.2: Classification of coal according to Parr's classification

Sr. No.	Class	Volatile Matter (%)	Calorific Value (BTU)*
1	Anthracite	0–8	15000–16500
2	Semi Anthracite	8–12	
3	Bituminous-A	12–24	
4	Bituminous-B	25–50	
5	Bituminous-C	30–55	14000–15000
6	Bituminous-D	35–60	12500–14000
7	Lignite	35–60	11000–12500
8	Peat	55–80	9000–11000
Calorific Value (British Thermal Unit—BTU) is defined as heat liberated while 1 unit gram of coal is burnt in excess air			

2.3.5.5 ASTM classification

For high rank coals, the A.S.T.M. (American Society for Testing and Materials) classification employs fixed carbon or volatile matter, and for low rank coals with significant moisture content, it uses the calorific value of moist, mineral-free coal. There are several classes and groups of coals. In this numbered classification, beginning with the higher rank coals, as mentioned in Table 2.3.

Table 2.3: Classification of coal according to ASTM standard

Sr. No.	Class	Group	Limits			Physical Properties
			Fixed carbon	Volatile matter	Calorific Value (BTU)	
1	Anthracite	Metal anthracite	98	2	-	Non-agglomerating
		Anthracite	98–92	2–8	-	
		Semi anthracite	92–86	8–14	-	
2	Bituminous	Low volatile	86–78	14–22	-	Either agglomerating or Non-weathering
		Medium volatile	78–69	22–31	-	
		High volatile A	69	31	-	
		High volatile B	-	-	14000–13000	
		High volatile C	-	-	13000–11000	
3	Sub-bitumious	Sub-bitumious A	-	-	11000–13000	Both weathering and Non-agglomerating
		Sub-bitumious B	-	-	9500–11000	
		Sub-bitumious C	-	-	8300–9300	
4	Lignite	Lignite	-	-	< 8300	Consolidated
		Brown coal	-	-	< 8300	Unconsolidated

2.3.5.6 Indian classification

Coal is also classified according to Indian standard, which is based upon three or four parameters like caking nature of the coals, moisture content, volatile matter and calorific value (on dry mineral matter free basis) as follows (Kernot, 2000; Newhouse, 2009).

Table 2.4 Classification of coal according to Indian standard

Sr. No.	Type	Sub-vision		% volatile at 900 ± 15°C	Calorific Value (Kcal/kg) A	% Moisture B	
1	Anthracite	Anthracite	A1	3–10	8330–8670	2–4	1–3
		Semi-anthracite	A2	10–15	8440–8780	1.5–3.0	1–2
2	Bituminous	Low volatile (caking)	B1	15–20	8670–8890	1.5–2.5	0.5–1.5
		Medium volatile	B2	20–32	8440–8780	1.5–2.5	0.5–2.0
		High volatile (caking)	B3	32+	8280–8610	5–10	1.–3
		High volatile (semicaking)	B4	32+	8060–8440	10–20	3–7
		High volatile (noncaking)	B5	32+	7500–8060	-	-
3	Sub-bituminous coal	No caking slacking on weathering	B6	32+	6940–7500	20–30	10–20
4	Lignites or brown coal	Normal lignite	L1	45–55	6100–6940	30–70	10–25
		Canneloid lignite	L2	55–65	6940–7500	30–70	10–25

2.3.6 Mining and transporting of coal

Coal mining is referred to as the physical extraction or removal of coal from the Earth's Surface. There are mainly two methods available for coal mining.

1. **Surface mining:** This method is utilized when coal is less than 200 feet below the surface. Several steps are considerd in sequence for surface mining: (1) Make land trees and vegetation free, (2) Eliminate and accumulate the top soil, i.e., top layers of the unconsolidated soil, (3) Bore the hard strata over the coal seam, (4) Blast the hard strata with explosives, (5) Segregate the coal seam and clean the top of coal, (6) Convert the small pieces by drilling and blasting and sieving as required, (7) Transport the coal to the plant, and (8) Restore land impacted by mining activity. Because surface mining is less expensive than underground mining, about two-thirds of U.S. coal production comes from surface mines. The surface mining techniques are further categorized as follows: (1) Contour strip mining, (2) Area strip mining, (3) Open-pit mining, and (4) Auger mining.

2. **Underground mining:** Deep mining is the name given to this type of mining. The coal seam and the strata above and below it totally enclose the operating area in the mining operation. The construction of service facilities for such crucial functions such as people and material transportation, ventilation, water treatment and drainage, and power is made possible via a network of

roads pushed into the coal seam. The coal seam is accessed through suitable openings on the surface. Further classifications of underground mining include (1) Room-and-pillar, (2) Long-wall, (3) Short-wall, and (4) Thick-seam.

After mining, coal is transported to the mining house and to the plants by different means like, conveyors, trams, and trucks, used when the distance of the mining house to plants is short to middle. Trains are selected for long-distance transportation. Barges and ships are required for water transportation. In some cases, the transportation cost of coal is much higher than that of mining coal (Fosdyke, 2008).

2.3.7 Pre-treatment of coal

After mining and transportation, it is important for coal undergo pre-treatment, as it has various impurities, such as plant residue, water, clay and dirt; and different sizes. For pre-treatment, there are five distinct levels of cleaning, each of which is an improvement over the previous.

Level 0: This type of coal is referred to as raw coal and is delivered directly to the customer.

Level 1: In this level, coarse coal is crushed to a finer consistency. Raw coal is that, that has been cleansed of impurities such as tramp iron, wood, and hard rocks.

Level 2: Level 2 is sieved coal larger than 12.5 mm and fine coal (smaller than 12.5). After sieving, it is cleaned to remove contaminants.

Level 3: Level 3 is sieved into two products: (1) intermediate coal, which is coal larger than 0.5 mm, and (2) small coal, which is coal smaller than 0.5 mm. After sieving, it is cleaned to remove contaminants.

Level 4: Coal is sieved to less than 0.1 millimeters and all impurities are removed after delivery. These leveling processes of coal are conducted by different processes.

1. **Jigging process:** Jigging is a conventional particle stratification technique based on the alternate expansion and compaction of a particle bed by a vertically oscillating fluid flow. Jigging is the process of separating particles of varying relative density, size, and shape by placing them on a perforated surface (or screen) through which a pulsating fluid is formed. Jigs used in coal cleaning processes are different from the jigs used in the metallurgical industry, which are effectively shaking tables. The coal jig is a tank of water, which is separated lengthways and has one half enclosed with the other open to the atmosphere. The pressure in the enclosed half can be increased or reduced back to atmospheric pressure, causing the water in the tank to move up and down around the central axis. Water has caused a stratification of the feed into high- and low-density fractions. The low-density fraction containing the coal forms the top layer and floats off the top of the jig whilst the higher density fraction falls to the bottom for removal using a screw

conveyor or other techniques. The process is not very efficient and requires relative densities in excess of 1.45. Moreover, it is also size dependent with optimal sizes generally between 10 mm and 200 mm, although particles of around 100 mm are probably separated most efficiently. Fine fractions of coal restrict the capacity of the jig, particularly below 10 cm, although on average most jig machines are likely to be able to process 200 t/hour of raw coal feed.

There are mainly four types of jigs used for coal: (1) Baum jig, (2) Batac jig, (3) Feldspar fine coal jig and (4) ROM Jig.

2. **Heavy medium baths:** A more efficient method of coal sizing is the heavy medium bath. For most of cases, fluid is utilized to distinguish the coal and discarded rock which is blended with coal, and either fine-grained magnetite or other solids such as shale, sand or barite. The advantage of using magnetite is that it can be recovered easily using a magnetic separator and can therefore be re-used in the process.

3. **Cyclones:** Cyclones are extensively utilized in the mining industry in order to separate samples on the basis of either size or density. A cyclone is a conically shaped tank with an angled inlet pointed towards the top, and two central outlets, one at the top and one at the bottom. The less dense and finer particles in a cyclone are drawn into a central helical vortex created by the angle of the inlet pipe and rise through the top of the tank. Those particles that are larger or denser fall to the sides and the bottom of the tank to be extracted through the underflow. Whilst cyclones are often used as sizing apparatus instead of screens, they also find applications as dense medium separators, being able to treat particles with sizes of between 0.5 mm and 30 mm. Improving efficiencies also mean that they can be effective at both larger and smaller sizes. The difference between the sizing cyclones and people used for density separation is the incorporation of an important medium, normally magnetite, within the liquid. This helps the density separation effect of the cyclone. Although it is still possible to use just water, this reduces the general efficiency of the machine.

4. **Froth Flotation:** Froth flotation is a method for separating hydrophobic substances from hydrophilic substances. In the bottom of froth flotation tanks, air and finely ground coal are pumped. As the air rises to the top of the liquid in the tank, it carries the hydrophobic coal with it, while the hydrophilic rock sinks to the bottom and must be removed by other methods. Similar to the metal working industry, the froth flotation process works best on small sized particles, generally below 0.5 mm. A new invention of froth flotation is the column flotation procedure, invented by Microcel, that replaces the tank by a column and is more preferable.

In order to improve the quality of the final product, it is necessary to remove any additional water after the levelling process. As the area of the coal increases, and its size reduces, the cost of dewatering also tends to rise. Coal above about 10 mm in diameter is generally dewatered on a shaking screen, which removes the water through vibrations. Smaller particles

(0.5 ± 10 mm) are often dewatered using centrifuges, where the solid particles coat the interior of the centrifuge and may be extracted employing a screw conveyor. For even finer coal the fabric must be dewatered employing a vacuum filter, although this normally leaves the water content relatively high at 17 ± 25%.

2.3.8 Processing of coal (chemicals from coal)

Worldwide, coal is used for numerous crucial purposes. The main notable applications for coal include the production of cement, the production of power, and the use of coal as a liquid fuel. More coal has been consumed globally than any other fuel. Figure 2.14 represents complete diagram of chemicals and their usages that are produced from coal using different unit processes/operations including oxidation, carbonization, distillation and liquification. (Bartis, 2008; Suajrez-Ruiz and Crelling, 2008; Horn, 2010, Laskowski, 2001, Smoth and Smith, 1985).

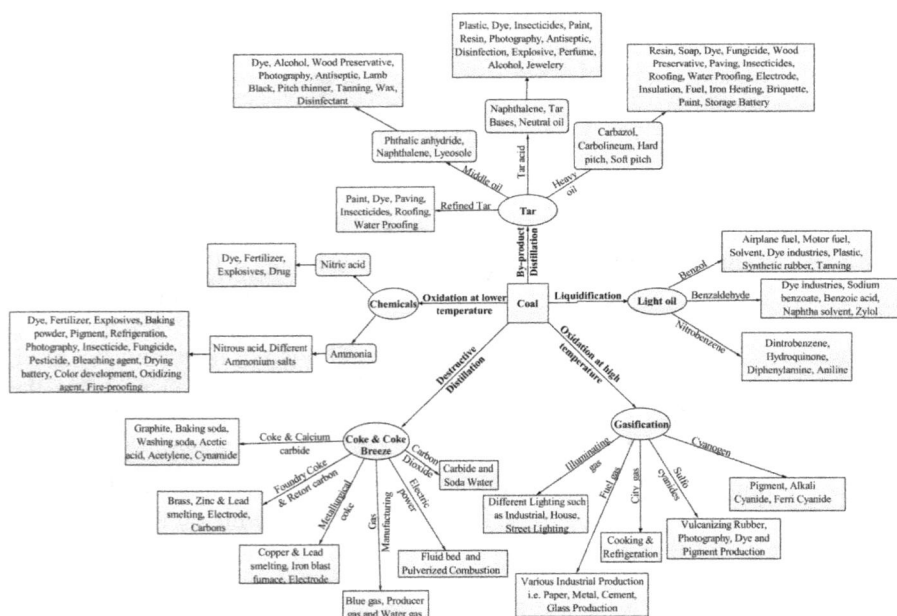

Figure 2.14 Application of coal in various fields

2.4 Metal

Metals are the second most abundant products, after coal, obtained from the Earth's crust. We all know that elements are divided into three kinds: 1) Metals 2) Non-metals and 3) Semi-metals, off which metals are the most useful. Metals possess many useful properties like high malleability, ductility, luster and good electrical conductivity. They are not available in pure state, but found in combined state with other non-metals and impurities in the form of minerals or ores. The pure metal is extracted from its ore. But metals have some characteristics that make them practically less useful like if they are very soft thy can be easily bent. If they are very reactive they are pron to corrosion. If they are highly malleable and ductile they possess low shock and wear resistance.

2.4.1 Physical properties of metals

Metal have some physical properties that make them suitable for particular uses.

1. **Hardness:** Hardness is the resistance of a metal to piercing and abrasion by another metal or substance. Hardness and tenacity are necessary for sustaining significant impacts. In as much as a metal's toughness decreases as its hardness increases, its ease of machining is limited by its hardness. Generally, the hardness of a metal can be controlled through heat treatment.

2. **Tensile strength:** This is a material's capacity to endure a force that might otherwise tear it apart. The greatest tensile stress that a material can sustain without cracking or irreversible deformation, is known as its tensile strength. The purpose of a fabric's transition from flexible to deformed is defined by its tensile strength. It is the lowest force per unit area (tensile stress) necessary to cause material fracture.

3. **Ductility:** The capacity of a metal to be permanently drawn or stretched without rupturing or breaking. Less ductile metals will crack or break prior to bending.

4. **Elasticity:** It is the property that enables a material to retain its original form following the removal of the load producing the deformation.

5. **Plasticity:** It is the opposite of elasticity, i.e., failure to retain its original shape after the load producing the deformation is removed.

6. **Weldability:** It is the capacity of a material to be welded using heat and/or pressure.

7. **Stiffness:** It is the property due to which a material is capable of resisting deflection by an externally applied force. It is of great importance in the design of girders and bridges.

8. **Electrical conductivity:** It is the power of a metal to conduct electricity. A metal which allows current to pass through it with minimum loss, is called a genuine conductor of electricity.

9. **Thermal conductivity:** It is the ability of a material to conduct heat with minimum loss.

10. **Malleable:** If a metal can be bent or twisted into almost any pattern without shattering or breaking, as well as pounded or rolled into incredibly thin sheets with external force, it is malleable. This property is the reverse of

ductility. Basically, it has the ability to deform permanently in all directions without cracking.

Other important physical characteristics are specific gravity, specific heat, fusibility, fluidity, porosity, toughness, machinability, refractoriness, shear-strength, fatigue-resistance, impact-resistance, creep-resistance, resilience, formability and rollability (Wutting and Lium, 2004; Zhigalskii and Jones, 2003; Liddle, 2015).

Metals are not available individually or in pure form in the environment. There are different minerals or ores, obtained from the Earth's crust, which are the main sources of metals. The ores and minerals are natural or rock sediments, which contain various salts (oxides, carbonates, halides or sulfates) of metal. These ores are obtained from the Earth through mining, thereafter undergo various treatments such as purification and smelting for the extraction of pure metals. A detailed description of each process is mentioned below.

2.4.2 Mining of ore

In this process, the raw ore having impurities such as dirt, soil and sand, known as gangue is collected from the Earth. There are different methods of mining available, but mainly three types are preferred: (1) Underground mining, (2) Surface (open pit) mining, and (3) Placer mining. The method selection depends on various factors such as ore grade, strength of rock, location and shape of deposit and mining cost.

1. **Underground mining:** In this method, the rock is drilled upto 300–500 meters below the surface level to form large tabular shaped ore bodies. Thereafter, it is blasted as it is very hard; and the raw ore is collected by a belt conveyor or elevator. It is further transported by truck for ore separation. This method is very expensive, but a very high grade metallic ore is obtained by this process. Ores of copper, aluminum, chromium, gold, platinum and iron are collected by the underground mining process.

2. **Surface mining:** The surface mining technique extracts ore from 12 to 14 meters beneath the Earth's surface. Five types of surface mining exist: (1) open pit mining, (2) strip mining, and (3) mountaintop removal, (4) dredging and (5) highwall mining. Open pit mining is preferred over other techniques since it is less expensive in comparison. In open pit mining, the ore is extracted from the soil by removing it from an open pit or large hole, while a small strip is prepared and the ore is collected using bucket-wheel excavators in strip mining. In mountaintop removal, the ore is extracted by simple drilling and blasting. This technique is used for very hard rock. Rare metals such as gold, silver, and platinum are dredged from the bottoms of lakes, rivers, harbors, and other water bodies. Special types of floating dredges (a barge fitted with conveyor belts and scoops) are used in this technique. Highwall mining is specially used for mining coal.

3. **Placer mining:** Utilizing water to excavate, transport, concentrate, and recover heavy mineral placer deposits is a very ancient technique. Ores of gold, platinum, iron, zinc, chromium, calcium, magnesium, zirconium, titanium, aluminum, manganese, cerium, and lanthanum are collected using this technique.

2.4.3 Purification of ore

The mined ore contains impurities, i.e., gangue which has to be removed. This process is known as ore concentration or ore dressing. There are various ore dressing processes, depending upon the type of ore. Some processes are discussed herein.

1. **Froth flotation process:** This process is specially used for separating hydrophobic materials from hydrophilic particles. In this procedure, powdered ore, water, and pine oil are combined to create foam. By blowing compressed air into the foam, this combination is churned ferociously. The foam is manufactured. Ore particles, which are hydrophobic, adhere to it, while gangue particles, which are hydrophilic, are retained in froth. The foam is slowly extracted and collected from the top of the froth to obtain ore concentrate. This method is used for sulfide ores like zinc blende (ZnS), and copper pyrites (CuFeS$_2$). The advantage of this process is that it can be used for most of ores by alternating foaming material. A disadvantage of the flotation process is that it is quite complex and expensive.

2. **Gravity separation process:** The principle of this method is the variation between the densities of gangue and ore particles, as the ore particles are denser than the water soluble gangue. It is called hydraulic washing, in which the powdered ore is washed with a high flow of water on the sloping floor or platform. The lighter sand and other water soluble impurities are washed down and the heavier ore particles are retained on the platform.

3. **Electromagnetic separation:** In this method, magnetic gangue particles are removed from the non-magnetic ore. The powdered ore is passed through the conveyer belt having electromagnetic rollers. The magnetic gangues stick on the conveyer belt due to attraction, which are collected at a distance from the roller. On the other hand pure non-magnetic ores particles fall in near the roller and collected. This method is applicable for non-magnetic tin ore (Tinstone).

4. **Chemical method:** The ores are treated with a selected dilute alkaline or acidic solution. The ore gets solubilized, while gangues remain solid. Thereafter, gangue is removed by filtration. Aluminum ore (Bauxite-Al$_2$O$_3$) is separated from SiO$_2$ and Fe$_2$O$_3$ by this method, in which it is treated with NaOH solution (Spitz and Trudinger, 2008; TAIMM, 2014, Rajak and Gupta, 2020).

2.4.4 Ore to pure metal

After general purification of the ore, it undergoes a number of unit processes to give the pure precious metal. The processes discussed below remove sulfur, carbon, hydrogen, oxygen and moisture from the ores.

First, metal ores are converted into their oxides using calcination or/and roasting process.

1. **Calcination:** In this process, the pure ore is heated at high-temperature ranging 550–1150°C in the absence of air in a special type of furnace or rotating kiln. During this process, organic matter, volatile impurities and moisture are removed from the ore. Also, carbonate ores give lime stone and carbon dioxide.

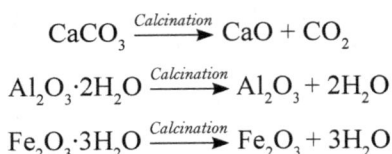

$$CaCO_3 \xrightarrow{Calcination} CaO + CO_2$$

$$Al_2O_3 \cdot 2H_2O \xrightarrow{Calcination} Al_2O_3 + 2H_2O$$

$$Fe_2O_3 \cdot 3H_2O \xrightarrow{Calcination} Fe_2O_3 + 3H_2O$$

2. **Roasting:** This process is useful in removing sulfur present in the ore in the form of a sulfide, oxide or sulfate. Also, it removes easily oxidizable materials and volatile matters from the ore. The ore is heated to about 450–550°C in the presence of air in this process, so, the sulfide is converted to an oxide, and sulfur is released as sulfur dioxide. The roasting process is generally carried out in a reverberatory or blast furnace. Examples are:

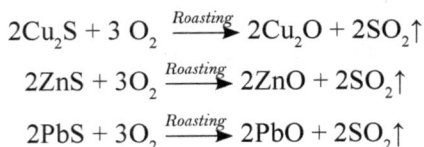

$$2Cu_2S + 3\,O_2 \xrightarrow{Roasting} 2Cu_2O + 2SO_2\uparrow$$

$$2ZnS + 3O_2 \xrightarrow{Roasting} 2ZnO + 2SO_2\uparrow$$

$$2PbS + 3O_2 \xrightarrow{Roasting} 2PbO + 2SO_2\uparrow$$

These metal oxide ores further undergo smelting and sintering processes.

3. **Smelting Process:** After calcination or/and roasting process, metal oxide ores are heated upto 1000–2000°C in the presence of coke, charcoal or coal as a reducing agent. In this process, reduction of metal oxides is performed, so, the oxide and other impurities are removed as slag or matte and pure molten metal is obtained. The reduction is carried out in a reverberatory or blast furnace in a controlled supply of air. The following chemical reactions occur in the smelting process.

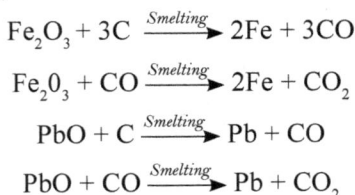

$$Fe_2O_3 + 3C \xrightarrow{Smelting} 2Fe + 3CO$$

$$Fe_2O_3 + CO \xrightarrow{Smelting} 2Fe + CO_2$$

$$PbO + C \xrightarrow{Smelting} Pb + CO$$

$$PbO + CO \xrightarrow{Smelting} Pb + CO_2$$

4. **Refining Process:** In this optional process the pure metal is converted into ultra-pure. There are several refining processes, depending on the chemical and physical properties of the metal, such as liquidation, electrolytic refining, chromatography and zone refining. Electrolytic refining is the most preferable method due to its applicability to most metals. In this method, the impure metal is the anode and the pure metal is the cathode. The metal's solution is taken as the electrolyte and an electric current is passed through it. The pure metal from the anode dissolves in the electrolyte and deposits on the cathode. The impurities are at the bottom of the anode.

5. **Sintering Process:** In this method, the metal is fused into one solid using a combination of pressure and heat without melting the ore. The pure metal is heated below its melting point, where diffusion of pure metal occurs under high pressure and temperature. It creates one solid piece and which is easily transferred (Curtis, 2013; Pan and Porterfield, 1995; Hay, 2000).

References

Abbas, S. 1996. The non-organic theory of the genesis of petroleum, Current Science 71(9): 677–684.

Abdel-Raouf, M. 2012. Crude Oil Emulsions—Composition, Stability, Characterization, Intechopen Publication. DOI: 10.5772/2677

Alsobaai, A.M. 2013. Thermal cracking of petroleum residue oil using three level factorial design, Journal of King Saud University 25(1): 21–28. https://doi.org/10.1016/j.jksues.2011.06.003

Babalola, F.U. and Susu, A. 2019. Pre-Treatment of Heavy Crude Oils for Refining. In book: Processing of Heavy Crude Oils—Challenges and Opportunities. DOI: 10.5772/intechopen.89486

Bartis, J.T. 2008. Producing Liquid Fuels from Coal: Prospects and Policy Issues, Rand Publishing.

Bluemle, J.P. 2000. Gazing into a crystal ball—A brief look at our future energy resources: NDGS Newsletter 27(2).

Carter, K.M. 2004. Early petroleum discoveries in Washington County, Pennsylvania: Pennsylvania Geology, Oil-Industry History 33(3): 12–19.

Chinenyeze, M.A.J. and Ekene, U.R. 2017. Physical and Chemical Properties of Crude Oils and Their Geologic Significances, International Journal of Science and Research 6(6): 1514–1521.

Clayton, C. 2005. Reference Module in Earth Systems and Environmental Sciences, Encyclopedia of Geology, Chemical and Physical Properties 248–260.

Collins, G.C., McKinnon, W.B., Moore, J.M., Nimmo, F., Pappalardo et al. 2009. Tectonics of the Outer Planet Satellites, Planetary Tectonics. DOI: 10.1017/CBO9780511691645.008

Cronquist, C. 2001. Estimation and Classification of Reserves of Crude Oil, Natural Gas, and Condensate, Society of Petroleum Engineers.

Curtis, K. 2013. Gambling on Ore: The Nature of Metal Mining in the United States, University Press of Colorado, USA.

Fosdyke, G.B. 2008. Coal Mining: Research, Technology and Safety, Nova Science Publication.

Hay, R.W. 2000. Reaction Mechanisms of Metal Complexes, Woodhead Publishing, India.

Horn, G.M. 2010. Coal, Oil, and Natural Gas (Energy Today), Infobase Publishing.

IEA - International Energy Agency, 2007. Coal Information, OECD publishing.

Islam, M.R., Chhetri, A.B. and Khan, M.M. 2010. The Greening of Petroleum Operations, 1-20. https://doi.org/10.1002/0471238961.greeisla.a01

Jones, D.S.J. 2015. Introduction to Crude Oil and Petroleum Processing, Handbook of Petroleum Processing. https://doi.org/10.1007/978-3-319-14529-7

Kabe, T., Ishihara, A., Qian, E.W., Sutrisna, I.P. and Kabe, Y et al. 2004. Coal and Coal-Related Compounds: Structures, Reactivity and Catalytic Reactions, Studies in Surface Science and Catalysis, Elsevier Science.

Kernot, C. 2000. Coal Industries, Woodhead Publishing Limited.

Khan, M.R. 1992. Clean Energy from Waste and Coal, ACS Symposium Series 515, American Chemical Society.

Krauskopf, K.B. and Bird, D.K. 1995. Introduction to geochemistry: New York, McGraw-Hill.

Laskowski, J. 2001. Coal Flotation and Fine Coal Utilization, Developments in Mineral Processing, Elsevier Academic Press.

Liddle, S.T. 2015. Molecular Metal-Metal Bonds: Compounds, Synthesis, Properties, 1st Edition, Willey. DOI: 10.1002/9783527673353

Miller, B.G. 2004. Coal Energy Systems, Sustainable World, Academic Press.

Newhouse, T.V. 2009. Coal Mine Safety, Nova Science Publication.

Oviedo, J.M.D., Tascon, J.A., Pajares, J.M.D., Tascon, J.A. and Pajares et al. 1995. Spain International Conference on Coal Science, Elsevier Science & Technology.

Pan, G. and Porterfield, B. 1995. Large-scale mineral potential estimation for blind precious metal ore bodies, Natural Resources Research 4(2).

Rajak, D.K. and Gupta, M. 2020. An Insight Into Metal Based Foams: Processing, Properties and Applications, 1st Edition.

Reddy, H., Pillay, Y. and Mohammadi, A.H. 2019. An Introduction to Petroleum Exploration Methods.

Santos, R.G., Loh, W., Bannwart, A.C. and Trevisan, O.V. 2014. An Overview of Heavy Oil Properties and its Recovery and Transportation Methods, Brazilian Journal of Chemical Engineering 31(3): 571–590. https://doi.org/10.1590/0104-6632.20140313s00001853

Schobert, H. 2013. Chemistry of Fossil Fuels and Biofuels, Chapter 11-Composition, classification, and properties of petroleum, Cambridge University Press. https://doi.org/10.1017/CBO9780511844188.012

Serge-Bertrand, A. 2020. Crude Distillation Unit (CDU), Analytical Chemistry - Advancement, Perspectives and Applications, Intechopen Publication. DOI: 10.5772/intechopen.90394

Smoot, L.D. and Smith, P.J. 1985. Coal Combustion and Gasification (The Plenum Chemical Engineering Series), Springer.

Speight, J.G. 2005. Handbook of Coal Analysis, Chemical Analysis: A Series of Monographs on Analytical Chemistry and Its Applications, Wiley-Interscience.

Spitz, K. and Trudinger, J. 2008. Mining and the Environment: From Ore to Metal, CRC Press.

Stephen, A., Janos B., John D., Denny E., Friedmann, S.J. et al. 2007. The Future of Coal, MIT Publication.

Suajrez-Ruiz, I. and Crelling, J.C. 2008. Applied Coal Petrology: The Role of Petrology in Coal Utilization, Academic Press. https://doi.org/10.1007/s12594-009-0181-y

TAIMM - The Australasian Institute of Mining and Metallurgy 2014. Mineral Resource and Ore Reserve Estimation—The AusIMM Guide to Good Practice, 2nd Edition.

Tissot, B.P. and Welte, D.H. 1984. Petroleum formation and occurrence, 2nd Edition, New York, Springer-Verlag.

TNAP - The National Academies Press, 2016. Spills of Diluted Bitumen from Pipelines: A Comparative Study of Environmental Fate, Effects, and Response. Washington, DC. https://doi.org/10.17226/21834.

Walters, C. 2007. The Origin of Petroleum, Practical Advances in Petroleum Processing, 79–101.

Warwick, P.D. 2005. Coal Systems Analysis (GSA Special Paper 387), Geological Society of America. https://doi.org/10.1130/0-8137-2387-6.1

Wuttig, M. and Liu, X. 2004. Ultrathin Metal Films: Magnetic and Structural Properties, 1st Edition, Springer-Verlag Berlin Heidelberg. https://doi.org/10.1007/b55564

Zhang, M., Qiao, J., Zhao, G and Lan, X. 2019. Regional gravity survey and application in oil and gas exploration in China, China Geology 2(3): 382–390. https://doi.org/10.1016/S2096-5192(19)30188-0

Zhigalskii, G.P. and Jones, B.K. 2003. Physical properties of thin metal films, Taylor & Francis. https://doi.org/10.1201/9780367801113

Zhijun, J., Guoping, B. and Mansoori, G.A. 2004. An introduction to petroleum and natural gas exploration and production research in China 41(1–3): 1–7.

CHAPTER 3

C – 1 Organic Chemicals

3.1 Introduction

Chemicals are mostly divided into two categories: (1) Organic and (2) Inorganic. The compounds, with carbon as a major component are called organic compounds. Organic compounds have covalent bonds. Most of the inorganic compounds do not contain carbon atoms. Inorganic compounds have ionic or hydrogen bonds. Here, we are discussing organic compounds containing only one carbon atom in one molecular formula unit. Organic chemicals containing one carbon atom are called basic chemicals and they are abundant C-1 feedstock. These basic chemicals are obtained from natural resources such as petroleum and coal which are fossil fuels. Very few chemicals available in this C1 series have functional groups like carboxylic acid, alcohol, amine, chloride, aldehyde and nitro groups. Thus, they are more stable than C2–C4 organic chemicals. They are mostly in liquid or gaseous state. Their manufacturing process includes a liquid phase reactor with stirring and heating facilities or a vapor phase tubular reactor. They have lower boiling and melting points. They are mostly used as domestic fuels and in the manufacture of motor solvents and other C2–C5 compounds and in the polymer, petroleum, pharmaceutical, and agricultural industries (Mackay, 2006).

3.2 Methane

In the Earth's atmosphere, a lesser quantity of methane gas is available. It is the simplest hydrocarbon that contains 1 carbon atom and 4 hydrogen atoms. It is a greenhouse gas that has a potential equivalent to 27.9 times higher than that of carbon dioxide.

3.2.1 Properties

Its main properties are as follows.

Table 3.1 Physical & Chemical Properties of Methane

Sr. No.	Properties	Particular
1	IUPAC name	Methane
2	Chemical and other names	Methyl hydride, Marsh gas, Biogas, Carbane, Tetrahydrocarbon
3	Molecular formula	CH_4
4	Molecular weight	16.04 gm/mol
5	Physical description	Colorless odorless gas
6	Solubility	Insoluble in water, Slightly soluble in acetone; soluble in ethanol, ethyl ether, benzene, toluene, methanol
7	Melting point	−182°C
8	Boiling point	− 161°C
9	Flash point	− 188°C
10	Density	0.554

3.2.2 *Manufacturing process*

It is prepared by the Sabatier reaction or Sabatier process. The manufacturing process of methane was invented in 1897 by French chemist Paul Sabatier and Jean-Baptiste Senderens. In this process, hydrogen and carbon dioxide are reacted at a high temperature in the presence of nickel or ruthenium on alumina as catalyst. This reaction is also called methanation of carbon dioxide. Thereafter, Bugante et al., 1989 prepared methane using a gas mixture of carbon monoxide and carbon dioxide instead of carbon dioxide only. Methane can be prepared by the degradation of septic systems, natural materials, sewers and commonly in landfills and marshes.

3.2.2.1 *Raw materials*

Carbon dioxide, Hydrogen and Catalyst

3.2.2.2 *Chemical reactions*

The mixture of hydrogen and carbon dioxide undergoes to get methane methane.

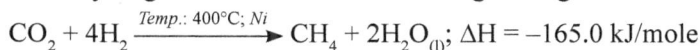

$$CO_2 + 4H_2 \xrightarrow{\text{Temp.: 400°C; } Ni} CH_4 + 2H_2O_{(l)}; \Delta H = -165.0 \text{ kJ/mole}$$

Also, carbon monoxide is also manufactured during this reaction as an unwanted product.

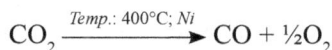

$$CO_2 \xrightarrow{\text{Temp.: 400°C; } Ni} CO + \tfrac{1}{2}O_2$$

Further, this carbon monoxide is reacted with water vapor to produce carbon dioxide and hydrogen. This reaction is called Reserve water-gas shift reaction.

3.2.2.3 *Quantitative requirement*

For the production of 1000 kg methane, we required 3000 kgs of carbon dioxide, 750 kgs of hydrogen and 500 kgs of catalyst.

3.2.2.4 Flow chart

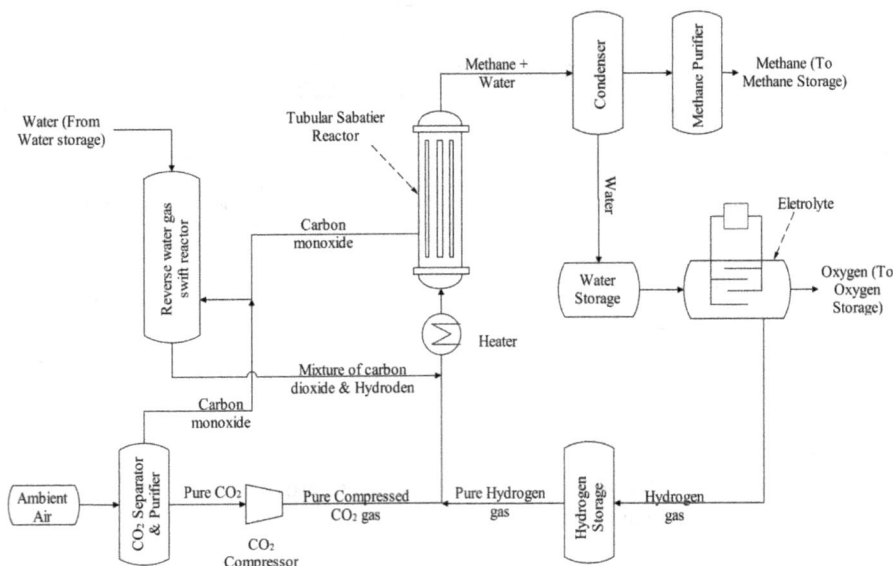

Figure 3.1 Flow sheet diagram of production of Methane gas using Sabatier process

3.2.2.5 Process description

As per Fig. 3.1, air undergoes is separated to hydrogen and carbon dioxide. Carbon dioxide is compressed to get pure compressed carbon dioxide. Recycled hydrogen gas is mixed with carbon dioxide and heated upto 400°C using a heater. A hot mixture of hydrogen and carbon dioxide is sent to the bottom of the tubular Sabatier reactor impregnated with catalyst, where both gases are reacted to get water and methane. A temperature of 400–420°C is maintained using hot oil. Hot vapors of water and methane are collected from the top of the reactor and undergo condensation. Here, water is separated from the bottom and sent to for storage. Methane gas is further purified and stored. Collected water undergoes electrolysis to give hydrogen and oxygen gas. Hydrogen gas is recycled. Now, carbon monoxide coming from the separator and the Sabatier reactor is transferred into the reverse water gas swift reactor. Water is added from the top of the swift reactor. Here, water and carbon monoxide are reacted to get mixture of hydrogen and carbon dioxide and recycled into the Sabatier reactor.

3.2.2.6 Major engineering problems

The reaction is very endothermic. For this reaction, we require a large amount of energy. The energy must either come from the earth in the form of a nuclear reactor, or collected in large solar panel arrays, so, the reaction occurs.

3.2.3 Uses

It is utilized mainly as fuel in gas turbines or steam generators. It is also used to manufacture several organic chemicals like methanol, hydrochloric acid, chloroform, ammonia and carbon black using different unit processes. As it is used in different chemical industries, it is transferred as refrigerated liquid (liquefied gas, or LNG). Methane is utilized as a rocket fuel (Yasunori et al., 2013; Bibby et al., 1998; Nguyen, 1996).

3.3 Methanol

Naturally, it is prepared by the anaerobic metabolism of several bacteria, so, it is present in very small amounts in the environment in vapor form. Thereafter, it is easily oxidized by sunlight to form carbon dioxide and water in a couple of days. If a higher amount of methanol is consumed, than it may cause accidental, suicidal, and epidemic poisoning, leading to death or permanent sequelae. Dehydrogenase (ADH) and aldehyde dehydrogenase is prepared by the metabolization of methanol.

3.3.1 Properties

Its main properties are as follows.

Table 3.2 Physical & Chemical Properties of Methanol

Sr. No.	Properties	Particular
1	IUPAC name	Methanol
2	Chemical and other names	Methyl alcohol, Carbinol, Wood alcohol, Methylol, Wood spirit, Colonial spirit
3	Molecular formula	CH_4O
4	Molecular weight	32.04 gm/mol
5	Physical description	Colorless liquid with a characteristic pungent odor
6	Solubility	Miscible with water, ethanol, ether, benzene, most organic solvents and ketones
7	Melting point	–98°C
8	Boiling point	65°C
9	Flash point	9°C
10	Density	0.792 gm/cm³ at 20°C

3.3.2 Manufacturing process

Synthesis gas ($CO + H_2O$) undergoes reactions at high temperature and pressure, and gets converted to methanol. Thereafter, it is purified through several columns.

3.3.2.1 Raw materials

Synthesis gas (mixture of hydrogen and carbon dioxide with a molar ratio of 2:25)

3.3.2.2 Chemical reactions

Manufacturing of methanol from synthesis gas is exothermic, which is as follows.

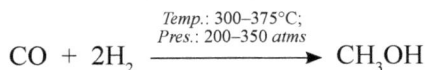

$$CO + 2H_2 \xrightarrow[\text{Pres.: 200–350 atms}]{\text{Temp.: 300–375°C;}} CH_3OH$$

Some side reactions also occurred with synthesis gas.

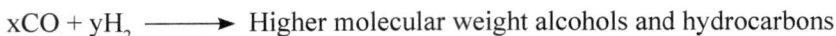

$$CO + 3H_2 \longrightarrow CH_4 + H_2O$$

$$2CO + 2H_2 \longrightarrow CH_4 + CO_2$$

$$xCO + yH_2 \longrightarrow \text{Higher molecular weight alcohols and hydrocarbons}$$

3.3.2.3 Quantitative requirement

For the production of 1000 kgs of methanol, we required 900 kgs of carbon monoxide and 150 kgs of hydrogen gas.

3.3.2.4 Flow chart

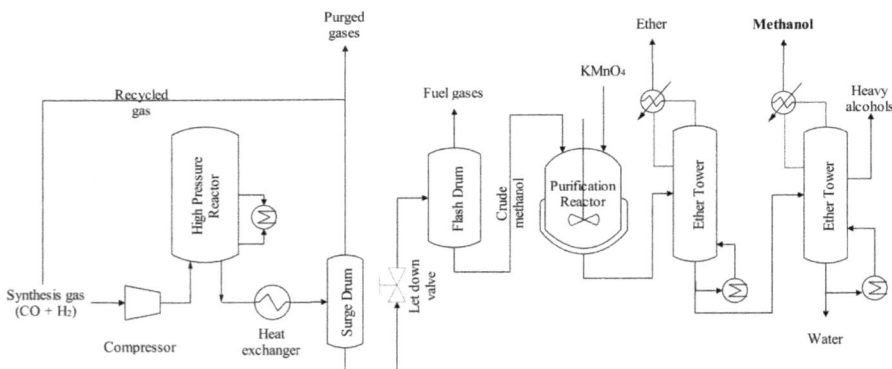

Figure 3.2 Flow sheet diagram of production of Methanol using synthesis gas

3.3.2.4 Process description

Synthesis gas having a mole ratio of 2:25 of hydrogen and carbon monoxide gas is taken for preparing methanol and compressed to 200–350 atmosphere using a compressor, mixed with recycle gas (unreacted gas). Compressed feed is inserted to a high pressure jacketed converter (reactor). The MOC (material of Construction) of the reactor is copper containing zinc, chromium, manganese or aluminum oxides as catalyst. Steam is passed through the jacket for maintaining a temperature of 300–375°C. As the reaction is exothermic, the volume of reaction mass decreases and pressure increases continuously. These conditions are favorable for faster reactions as per Le Chatelier's principle. Thereafter, the vapor mass is cooled using a heat exchanger and undergoes phase separation in the surge drug. The liquid phase containing methanol and vapor phase containing high molecular weight components is separated in drug. Vapor is recycled back after the removal of inert compounds. Liquid containing methanol is again pressured to about 14 atmosphere and subjected into a second phase separator. The liquid phase containing highly pure methanol and vapor phase containing dissolved fuel gas are separated. The dissolved fuel gases are utilized to generate steam in the boiler or furnace. The liquid methanol is sent to the jacketed reactor having agitator. Here, potassium permanganate enters the reactor to purify the methanol. Purified methanol is collected from the bottom of the reactor and sent into the ether column for ether separation. Final purification is conducted in a distillation column to separate out water and heavy alcohols from methanol. The yield is about 98–99% methanol.

3.3.2.5 Major engineering problems

The following engineering problems arise while manufacturing methanol.

1. **Temperature and pressure:** Various temperatures, i.e., 260, 300, 340, 380°C are selected to calculate the equilibrium constant under normal pressure for an exothermic reaction. For temperatures less than 260°C, the process is very economical, but the reaction rate is very low, even though the catalyst is active. However, 50% conversion is achieved under optimum conditions at 240 atmospheres pressure and a temperature of 300°C.

2. **Reactor design:** The reactor design plays an important role in manufacturing methanol. A thick-walled pressure vessel with a copper lining is selected, to prevent the formation of iron carbonyl $Fe(CO)_5$, a volatile compound which is poisonous and may create corrosion problems.

3. **Heat exchanger:** The heat exchanger helps to control the temperature of the reaction mixture in several ways. In the Uhde design, the compressed gas enters an annulus along the shell wall which serves to preheat the gas and cools the heat-treated steel pressure shell. The gases then move in the opposite direction through the shell side of a tubular array, picking up heat. The last pass is through the catalyst on the inside of the tubes. The catalyst is placed in zones with further temperature control accomplished by direct blasts of cold reactant gases which bypass the preheat zone. In the simpler Montechcatini-fauser design, a combination of circulating high pressure water and waste heat boiler acts as the principal heat control.

4. **Catalyst fouling:** Catalyst fouling (i.e., accumulation of unwanted material on the catalyst surface) occurs, if the proper ratio of carbon monoxide and hydrogen gas, i.e., 2:25 is maintained.

5. **Inert gas accumulation:** When a high recycle load is persisted with, the possibility of accumulating inert gas is avoided by maintain a side-stream purge on the recycle gas.

3.3.3 Uses

1. A major portion of methanol produced is utilized as an industrial solvent in inks, resins, adhesives for wooden items, and dyes. It is also utilized as a

solvent in the preparation of cholesterol, streptomycin, vitamins, hormones, and other pharmaceuticals.

2. Mainly methanol is utilized to manufacture a variety of compounds such as methyl tertiary butyl ether (MTBE), formaldehyde, acetic acid and dimethyl ether (DME), a clean-burning fuel with similar properties to propane.

3. It is key part of a renewable fuel, biodiesel instead of diesel fuel.

4. Methanol is used as an antifreeze agent in automotive radiators also a petrol ingredient (Jean-Paul, 1997; Olah et al., 2006; Hanson et al., 1992; Cheng and Kung, 1994).

3.4 Formaldehyde

It is the simplest of the aldehydes (R−CHO).

3.4.1 Properties

Its main properties are as follows:

Table 3.3 Physical & Chemical Properties of Formaldehyde

Sr. No.	Properties	Particular
1	IUPAC name	Formaldehyde
2	Chemical and other names	Formalin, Formol, Methanal, Oxomethane, Methylene oxide, Formic aldehyde, Paraform
3	Molecular formula	CH_2O
4	Molecular weight	30.02 gm/mol
5	Physical description	White solid with a light pungent odor
6	Solubility	Soluble in water, ethanol and chloroform
7	Melting point	−92°C
8	Boiling point	−20°C
9	Flash point	83°C
10	Density	0.815 gm/cm³ at −20°C

3.4.2 Manufacturing process

Industrially, formaldehyde is manufactured by the catalytic oxidation of methanol. Here, silver metal or a mixture of iron and molybdenum or vanadium oxides is the preferred catalyst. This is known as the Formox process.

3.4.2.1 Raw materials

Methanol, Air (Oxygen) and Catalyst

3.4.2.2 Chemical reactions

In the Formox process, formaldehyde is prepared by reacting methanol and oxygen at 250–400°C in the presence of a catalyst which is iron oxide in combination with molybdenum and/or vanadium according to the following chemical equation:

$$2CH_3OH + O_2 \xrightarrow[\text{\textit{Catalyst}}]{\text{\textit{Temp}.: 250–400°C;}} 2CH_2O + 2H_2O; \ \Delta H = -154.8 \ kJ$$

Side reaction: Complete combustion of methanol,

$$CH_3OH + 3/2O_2 \longrightarrow 2H_2O + 2CO_2; \ \Delta H = -677.8 \ kJ$$

3.4.2.3 Quantitative requirements

For the production of 1000 kgs of methanol, we require 1100 kgs of methanol, 550 kgs of air (oxygen) and 750 kgs of catalyst.

3.4.2.4 Flow chart

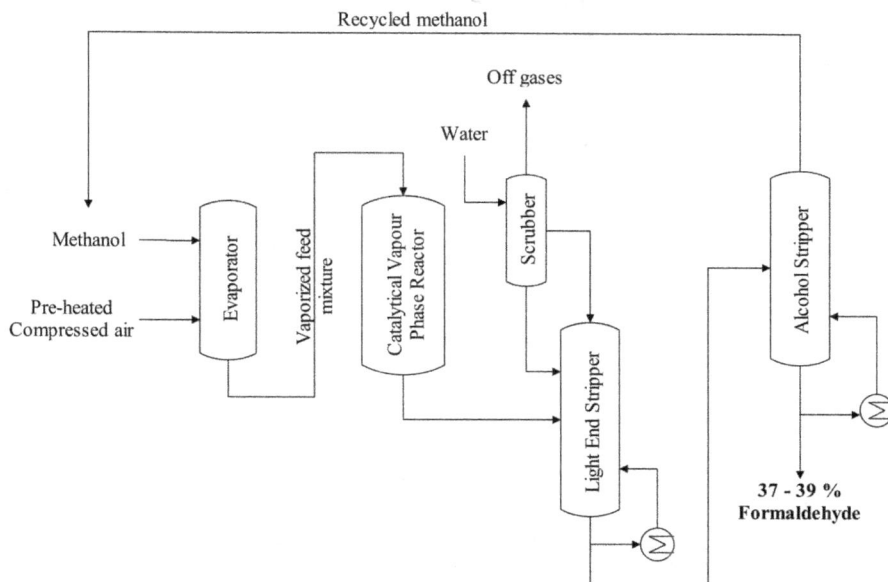

Figure 3.3 Flow sheet diagram of production of Formaldehyde from Methanol

3.4.2.5 Process description

First, preheated compressed air (oxygen) and methanol is inserted into the evaporator at a ratio range of 30–50% at a temperature of 55°C. The vapor mixture at 55°C is fed into the catalytical vapor phase reactor. The feed ratio is about 30–50% for CH_3OH: O_2. The reaction is exothermic, so the product is a vapor mixture with temperature of 450–900°C after the reaction. After the completion of the reaction, the gaseous product is cooled with the heat incorporation concept and finally, fed into the absorption tower. In this tower, water having a high amount of formaldehyde is obtained and recycled. Another stream of formaldehyde and methanol rich water is formed at the bottom and fed into the alcohol stripper. Un-reacted methanol vapor is collected from the top of the stripper, purified and recycled. About 37% formaldehyde is collected from the bottom of stripper. Here, the pure formaldehyde obtained is not stable, as it tends to polymerize. It is stable only with water. Therefore, a 37% solution of formaldehyde with 3–10% stabilizer (preferably methanol) is prepared and sold in the market.

3.4.2.6 Major engineering problems

1. **Pyrolysis of Methanol:** At a higher temperature, methanol is undergoes pyrolysis to give formaldehyde directly ($CH_3OH \rightarrow HCHO + H_2$) in the presence of a mixed catalyst of Mo, V and Fe. It gives a yield of 90–95% with no methanol to be recycled. So, the process is modified with a higher molar ratio of 16:1 for methanol: oxygen. Further, the reaction is endothermic, so, the temperature is maintained in the tubular reactor by oil recirculation.

2. **Choice of space velocity:** Moderate space velocity is maintained, because high combustion of CO to CO_2 occurs at a lower velocity and low methanol conversion is achieved at a higher space velocity.

3.4.3 Uses

Formaldehyde is mainly used to manufacture phenol-formaldehyde resin and melamine. It has been utilized in a variety of industries for various purposes such as: building materials, fiber board, cigarette smoke, fuel burning appliance and kerosene heater. It is also used as a preservative or sterilizer in pharmaceutical products and as a major element in glues and as a supplement in permanent-press (Masamoto et al., 1993; Cheng and Lin, 2012; Zhnag, 2018).

3.5 Chloromethanes

There are four types of chloromethanes: (1) Methyl chloride (monochloromethane), (2) Dichloromethane (methylene chloride), (3) Trichloromethane (chloroform), and (4) Tetrachloromethane. The chlorine atoms are progressively increasing in the methane molecule. The electron-withdrawing capacity of the chlorine atom, electrophilicity and toxicological profile of the carbon atom significantly increases due to the presence of the carbon-chlorine bond. The polarity and dipole moments in chloromethanes decreases with the substitution of a hydrogen atom with an additional chlorine atom.

3.5.1 Properties

The properties of chloromethane, i.e., methyl chloride are as per Table 3.4.

3.5.2 Manufacturing process

Monochloromethane was first prepared by J. Dumas and E. Peligot using sodium chloride and methanol in the presence of sulfuric acid. Thereafter, it was prepared

Table 3.4 Physical & Chemical Properties of Monochloromethane

Sr. No.	Properties	Particular
1	IUPAC name	Chloromethane
2	Chemical and other names	Methyl chloride, Monochloromethane, Artic, Clorometano
3	Molecular formula	CH_3Cl
4	Molecular weight	50.48 gm/mol
5	Physical description	Colorless gas with a faint sweet odor
6	Solubility	Soluble in water, ethanol, ethyl ether, acetone, benzene and chloroform
7	Melting point	–98°C
8	Boiling point	–24°C
9	Flash point	–45°C
10	Density	0.911 gm/cm³ at 20°C

by C. Groves in 1874 using hydrogen chloride and methanol in the presence of zinc chloride. Initially industrial vapor phase chlorination of methane was conducted in 1923 by Hoechst. Lately, Chloromethanes are being prepared by the following two routes, i.e., thermal chlorination and catalytic oxychlorination of methane. Out these two processes, the first process is mostly utilized for manufacturing all chloromethanes.

3.5.2.1 Raw materials

Methane, Chlorine gas and Sodium hydroxide

3.5.2.2 Chemical reactions

Industrially, all chloromethanes are prepared by the chlorination of methane in the presence of sun-light at 380–390°C. The substitution reaction takes place and all the hydrogen atoms are replaced by chlorine atoms gradually.

$$CH_4 + Cl_2 \xrightarrow{hv;\ Temp.:\ 380–390°C} CH_3Cl + HCl;\ \Delta H = 103.5\ kJ/mole$$

$$CH_3Cl + Cl_2 \xrightarrow{hv;\ Temp.:\ 380–390°C} CH_2Cl_2 + HCl;\ \Delta H = 102.5\ kJ/mole$$

$$CH_2Cl_2 + Cl_2 \xrightarrow{hv;\ Temp.:\ 380–390°C} CH_3Cl + HCl;\ \Delta H = 99.2\ kJ/mole$$

$$CH_3Cl + Cl_2 \xrightarrow{hv;\ Temp.:\ 380–390°C} CCl_4 + HCl;\ \Delta H = 94.8\ kJ/mole$$

3.5.2.3 Quantity requirement

If the methanol/chlorine ratio is 1.8/1, then monochlormethane is the major product, but, if methanol is the limited reactant, then tetrachloromethane and trichloromethane are the major products. Thus, the feed ratio plays an important role for the required product output.

3.5.2.4 *Flow chart*

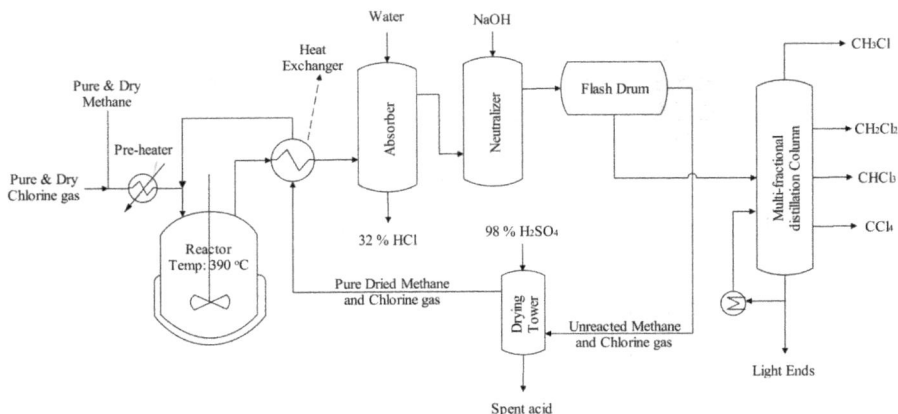

Figure 3.4 Flow sheet diagram of production of Chloromethanes

3.5.2.5 *Process description*

Pure dry methane and chlorine gas are mixed, pre-heated in a heater and transferred to the reactor. The reactor is facilitated with a stirrer and hot oil jacket. The reaction mixture is heated upto 380–390°C with stirring. Here, the substitution reaction of methane occurs to form chloromethanes. These chloromethanes are collected from the top of the reactor and quenched in the heat exchanger. The cooled gaseous mass is passed through the absorber. Water is sprayed from the top of absorber. The HCl is removed from the mass. It is then neutralized with NaOH to removal carbon dioxide and traces of HCl. Thereafter, the gaseous mass is sent to flash drug to recover unreacted methane and chlorine gas. The unreacted mass is further dried with 98% sulfuric acid, pre-heated and recycled. Raw chloromethanes undergo separation to each chloromethane in the multi-fractional distillation column.

3.5.3 *Uses*

1. Usually, chloromethanes are utilized as intermediates in different industrial segments such as pharmaceuticals, refrigerants, coolants, agrochemicals, automobiles, domestic cookware, electronics and water treatment.
2. Methyl chloride is employed in surfactants, pharmaceuticals and dyes. A little fraction is employed as a coldness solvent for the assembly of butyl rubbers.
3. Dichloromethane is employed as a solvent in the preparations for metal degreasing, polycarbonates, and triacetates. It is also used as a solvent in chemicals and pharmaceuticals.
4. Chloroform is employed as a staple for the assembly of HCFC 22 (R 22), which is used as a refrigerant and pioneer for fluoropolymers (PTFE and

derivatives). Further, thereon it is often used as a biofuel for methyl and ethyl orthoformate.

5. Tetrachloromethane is merely used as a staple preparation in pharmaceutical, polymer, agro-based chemical and refrigerant manufacture. It is also used as a solvent in specialty chemical industries (Polyakov et al., 2004; Clever, 2003).

3.6 Methylamines

There are four types of methylamines such as monomethylamine (MMA), dimethylamine (DMA), and trimethylamine (TMA). Initially methylamine was prepared by Charles-Adolphe Wurtz in 1849 by the hydrolysis of methyl isocyanate.

3.6.1 Properties

The properties of one methylamine, i.e., monomethylamine are as follows.

Table 3.5 Physical & Chemical Properties of Methylamine

Sr. No.	Properties	Particular
1	IUPAC name	Methanamine
2	Chemical and other names	Aminomethane, Monomethylamine, methylamine hydride
3	Molecular formula	CH_5N
4	Molecular weight	31.06 gm/mol
5	Physical description	Yellow to colorless gas or a liquid with pungent fishy odor
6	Solubility	Soluble in water, ethanol, benzene, and acetone; miscible with ether
7	Melting point	−93.6°C
8	Boiling point	−6.3°C
9	Flash point	−10.0°C
10	Density	0.693 gm/cm³ at 20°C

3.6.2 Manufacturing process

Previously, methylamines were prepared by the vapor phase ammoniation of methanol with ammonia using the Leonard process variable. Commercially, they are prepared by the reaction of ammonia with methanol.

3.6.2.1 Raw materials

Ammonia, Methanol and Catalyst

3.6.2.2 Chemical reactions

Industrially, they are prepared by the reaction of ammonia with methanol in the presence of an amorphous aluminosilicate catalyst at a temperature of 400–420°C and super-atmospheric pressure.

$$CH_3OH + NH_3 \xrightarrow[\text{Catalyst}]{\textit{Temp.: } 400–420°C} CH_3NH_2 + H_2O$$

$$CH_3NH_2 + CH_3OH \xrightarrow[\text{Catalyst}]{\textit{Temp.: } 400–420°C} (CH_3)_2NH + H_2O$$

$$(CH_3)_2NH + CH_3OH \xrightarrow[\text{Catalyst}]{\textit{Temp.: } 400–420°C} (CH_3)_3N + H_2O$$

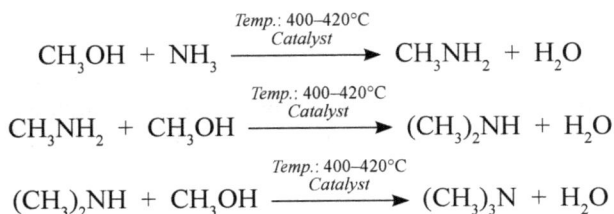

3.6.2.3 Quantitative requirements

The quantity of raw materials is depends upon the required methanolamine required.

3.6.2.4 Flow chart

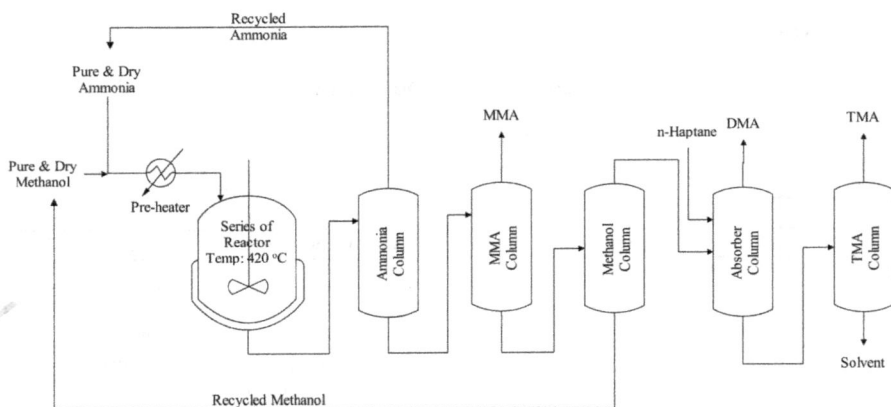

Figure 3.5 Flow sheet diagram of production of Methylamines

3.6.2.5 Process description

Pure dry methanol and ammonia gas are mixed, pre-heated in the heater and transferred into the top of a series of adiabatic, fixed bed reactors. The series reactors are taken due to the fact that the reaction rates for the ammonia consuming reactions are much smaller than those consuming ethanol. The reactors are already impregnated with amorphous aluminosilicate catalyst. The reaction mixture is heated upto 400–420°C with stirring and high pressures (> 20 atmospheres). Formed methylamines and the un-reacted mass are collected from the bottom of the reactor and transferred into the ammoniation column. Un-reacted ammonia is separated out and recycled. Methylammonium mass is transferred into the MMA column to separate out monomethylamine (MMA). The mixture of methanol, TMA and MMA which were obtained from the bottom of MMA column is used as the feed for the methanol column. In this column methanol is obtained at the bottom of the column and; DMA and TMA are collected from the top of column. Two methylamines are transferred into absorber. Also, the solvent, i.e., n-heptane is charged to separate the DMA as an overhead product. Finally, the solvent containing mass is transferred into the TMA column to separate TMA as an overhead product and the solvent at the bottom of column.

3.6.3 Uses

1. Generally, Methylamines are used in interim products in animal nutrients, catalysts, electronics, agro-based chemicals, explosive and fuel additives.
2. Monomethylamine (MMA) is especially used as an intermediate in the production of several herbicides, pesticides, and insecticides, for the assembly of Tovex (water gel explosive), the solvent N-methyl-2-pyrrolidone (NMP), methyldiethanolamine (MDEA) a solvent for hydrocarbon processing, the soil disinfectant metal sodium and a few pharmaceuticals.
3. Dimethylamine (DMA) has the maximum requirement in the global market in terms of methylamine. It is used, among other uses, for the production of the solvents dimethylformamide (DMF) and dimethylacetamide (DMAC), water-treating agents, surfactants, rubber-processing compounds, and a range of agrochemicals.
4. Trimethylamine (TMA) is used in the manufacture of intense sweeteners, choline salts, disinfectants, cationic starches, ion-exchange resins and flotation agents. (Fetting and Dingerdissen, 1992; Kiyoura and Terada, 1995).

3.7 Formic Acid

Formic acid (methanioc acid) is the simplest carboxylic acid. First, it was isolated from some ants and named after the Latin formica, meaning "ant".

3.7.1 Properties

The properties of Formic acid are as follows.

Table 3.6 Physical & Chemical Properties of Formic acid

Sr. No.	Properties	Particular
1	IUPAC name	Formic acid
2	Chemical and other names	Methanoic acid, Formylic acid, Aminic acid, Bilorin, Formisoton
3	Molecular formula	CH_2O_2
4	Molecular weight	46.02 gm/mol
5	Physical description	Colorless liquid with a pungent, penetrating odor
6	Solubility	Miscible with water, ether, acetone, ethyl acetate, methanol, ethanol; partially soluble in benzene, toluene, xylenes
7	Melting point	8.3°C
8	Boiling point	101°C
9	Flash point	–69.2°C
10	Density	1.22 gm/cm³ at 20°C

3.7.2 *Manufacturing process*

It is prepared by several routes, which are mentioned below.

(1) **From oxidation of methanol:** It is prepared by the vapor phase oxidation of methanol.

(2) **From methanol (Kemira Process) via methyl formate:** First, methanol and carbon monoxide are reacted in liquid phase at a high temperature (80–100°C) and pressure (40–50 atmosphere) to give methyl formate.

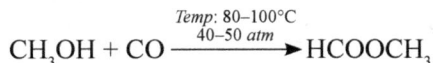

$$CH_3OH + CO \xrightarrow[\text{40–50 atm}]{\textit{Temp: 80–100°C}} HCOOCH_3$$

Thereafter, methyl formate undergoes hydrolysis in an acidic medium to give formic acid.

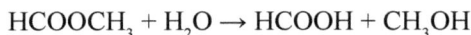

$$HCOOCH_3 + H_2O \rightarrow HCOOH + CH_3OH$$

(3) **From methyl formate:** Methyl formate is treated with ammonia to yield formamide, which is acid hydrolyzed with sulfuric acid to give formic acid.

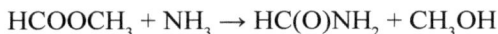

$$HCOOCH_3 + NH_3 \rightarrow HC(O)NH_2 + CH_3OH$$

$$2\ HC(O)NH_2 + 2H_2O + H_2SO_4 \rightarrow 2HCOOH + (NH_4)_2SO_4$$

(4) **From carbon monoxide and sodium hydroxide:**
Carbon monoxide and sodium hydroxide are reacted to give sodium formate, which is acid hydrolyzed with sulfuric acid to give formic acid.

$$CO + NaOH \rightarrow HCOONa \rightarrow HCOOH$$

Industrially, it is prepared by the oxidation of methanol.

3.7.2.1 *Raw materials*

Methanol, Air (Oxygen) and Catalyst

3.7.2.2 *Chemical reactions*

Two step vapor phase oxidation reactions of methanol are performed to give formic acid.

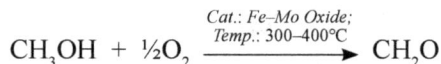

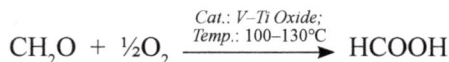

$$CH_3OH + \tfrac{1}{2}O_2 \xrightarrow[\textit{Temp.: 300–400°C}]{\textit{Cat.: Fe–Mo Oxide;}} CH_2O$$

$$CH_2O + \tfrac{1}{2}O_2 \xrightarrow[\textit{Temp.: 100–130°C}]{\textit{Cat.: V–Ti Oxide;}} HCOOH$$

Both reactions are exothermic.

3.7.2.3 *Quantitative requirements*

For the production of 1000 kgs of formic acid, we require 700 kgs of methanol, 1300 kgs of air (oxygen) and 750 kgs of catalyst.

3.7.2.4 Flow chart

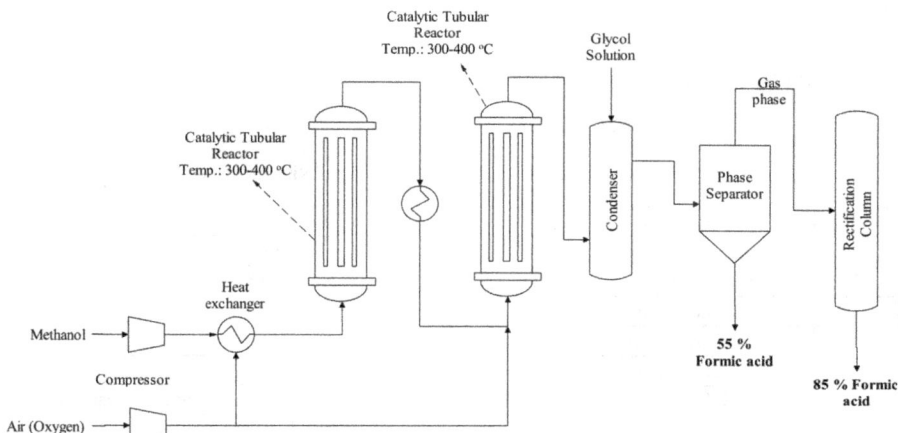

Figure 3.6 Flow sheet diagram of production of Formic acid from Methanol

3.7.2.5 Process description

As per the flow sheet diagram, methanol and air (oxygen) are compressed individually and pre-heated in a heat exchanger to convert the vapor form. These vapors are inserted into a tubular reactor which is already impregnated with iron-molybrade oxide catalyst. The reaction is exothermic, so, cooling water circulation is implemented to maintain a temperature of 300–400°C. After completion of the reaction, the vapor mass is cooled and transferred into another tubular reactor which is already impregnated with vanadium-titanium oxide catalyst. More compressed air is added to the reactor. This reaction is also exothermic, so, cooling water circulation is deployed to maintain a temperature of 100–130°C. The vapor phase reaction mixture is condensed in a condenser using a glycol solution. Then, it is transferred into the phase separator, which separates the liquid phase, containing 55% formic acid in the bottom part. The upper gas phase is further rectified to collect 85% formic acid.

3.7.3 Uses

Most of the formic acid produced is used as a preservative and antibacterial agent in livestock feed. It is also used as a modern fuel source in proton exchange membrane fuel cells at lower temperatures. It is used for tanning and dye fixing as well as a neutralizing agent in the leather and textile processing industries. It is used in the production of metal formate and pentaerythritol. It is also used as a corrosion inhibitor in the hydraulic fracturing process in oil industries (Wang et al., 2008; Fu, 2020; Dubey et al., 2000).

3.8 Nitromethane

It is the simplest organic nitro compound.

3.8.1 Properties

The properties of Nitromethane are given below.

Table 3.7 Physical & Chemical Properties of Nitromethane

Sr. No.	Properties	Particular
1	IUPAC name	Nitromethane
2	Chemical and other names	Nitrocarbol
3	Molecular formula	CH_3NO_2
4	Molecular weight	61.04 gm/mol
5	Physical description	Colorless, oily liquid with a disagreeable odor
6	Solubility	Soluble in water, ethanol, ethyl ether, acetone, carbon tetrachloride, and alkali
7	Melting point	28.7°C
8	Boiling point	101°C
9	Flash point	36°C
10	Density	1.1371 gm/cm³ at 20°C

3.8.2 Manufacturing process

It is prepared by the nitration of methane. It is also a by-product of the nitration of propane in the gas phase at 350–450°C.

3.8.2.1 Raw materials

Methane and Nitric acid

3.8.2.2 Chemical reactions

Nitromethane is prepared by the vapor phase nitration of methane using nitric acid at atmospheric pressure and a temperature of 400–600°C.

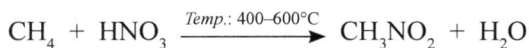

$$CH_4 + HNO_3 \xrightarrow{\textit{Temp.: } 400–600°C} CH_3NO_2 + H_2O$$

Here, a trace amount of formaldehyde is also formed.

3.8.2.3 Quantitative requirements

For the production of 1000 kgs of nitromethane, we require 300 kgs of methane and 1100 kgs of nitric acid.

3.8.2.4 Flow chart

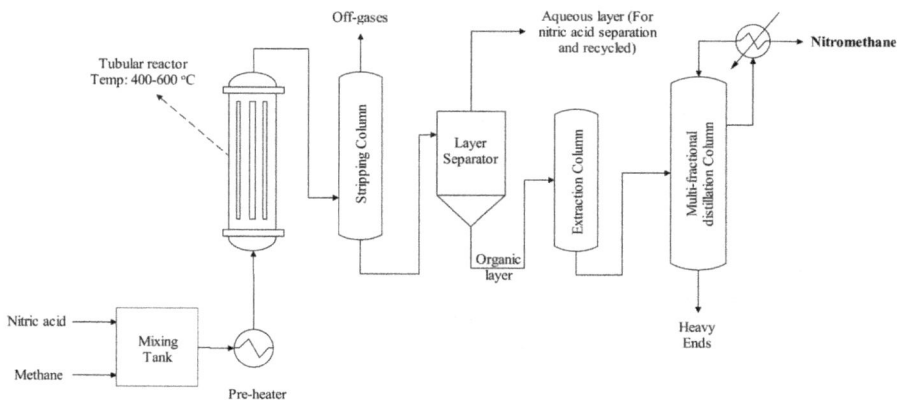

Figure 3.7 Flow sheet diagram of production of Nitromethane from Methane

3.8.2.5 Process description

In the manufacturing process, first nitric acid and methane are mixed in a mixing tank and heated in a pre-heater for vapor phase conversion. This pre-heated vapor mixture is passed through a tubular reactor from its bottom, where a temperature of 400–600°C is maintained using hot oil circulation. After completion of the reaction, the vapor mixture is collected from the top of the reactor and transferred into the stripping column. Here, the vapor mass is cooled and converted to the liquid phase and the off-gases like methane, carbon monoxide, carbon dioxide and oxides of nitrogen are collected from the top of the column. The liquid mass goes into the layer separator, where the lower lighter aqueous layer is separated. On the other hand, the heavier organic layer containing nitromethane and traces of formaldehyde is collected from the bottom. It is then passed into the extraction column and finally, the distillation column to collect nitromethane as the overhead product.

3.8.3 Uses

Nitromethane is primarily used as a fuel additive in various drag racers, motorsports and hobbies, because it requires only about one-ninth as much air as gasoline for complete combustion. It is also used as a solvent in vinyls, epoxies, polyamides and acrylic polymers. It is used as an intermediate in various organic syntheses of biocides, chemicals, fungicides, drugs and agricultural products (Lord-Garcia, 2014; Idar et al., 1999).

References

Bibby, D.M., Chang, C.D., Howe, R.F. and Yurchak, S. 1988. Methane conversion: proceedings of a symposium on the production of fuels and chemicals from natural gas, Studies in Surface Science and Catalysis, Elsevier Science Ltd.

Cheng, C.B. and Lin, F.H. 2012. Formaldehyde: Chemistry, Applications and Role in Polymerization, Nova Science Publication Inc.

Cheng, W. and Kung, H.H. 1994. Methanol production and use, M. Dekker Publication.

Clever, H.L. 2003. The Solubility of the Chloromethanes in Water, MonatsheftefürChemie - Chemical Monthly 134(5). https://doi.org/10.1007/s00706-002-0577-5

Dubey, S., Devra, V., Binyahia, A.R. and Sharma, P.D. 2000. Kinetics and mechanism of oxidation of formic acid by Bismuth(V) in aqueous phosphoric acid medium, International Journal of Chemical Kinetics 32(80): 491–497. https://doi.org/10.1002/1097-4601(2000)32:8<491::AID-KIN7>3.0.CO;2-4

Fetting, F. and Dingerdissen, U. 1992. Production of methylamines over ZK-5 zeolite treated with tetramethoxysilane, Chemical Engineering & Technology 15(3): 202–212. https://doi.org/10.1002/ceat.270150309

Fu, M. 2020. Studies on Green Synthetic Reactions Based on Formic Acid from Biomass, 1st Edition, Springer Singapore. https://doi.org/10.1007/978-981-15-7623-2

Hanson, R.S., Murrell, J.C. and Dalton, H. 1992. Methane and Methanol Utilizers, 1st Edition, Biotechnology Handbooks, Springer US.

Idar, D.I., Asay, B.W. and Ferm, E.N. 1999. Improved Characterization of Nitromethane, Nitromethane Mixtures, and Shaped-Charge Jet Properties, Propellants, Explosives, Pyrotechnics, 24 (1), 1-6. https://doi.org/10.1002/(SICI)1521-4087(199902)24:1<01::AID-PREP1>3.0.CO;2-L

Jean-Paul, L. 1997. Perspectives for Manufacturing Methanol at Fuel Value, Industrial & Engineering Chemistry Research, 36(10): 4282–4290. https://doi.org/10.1021/ie9607762

Kiyoura,T. and Terada, K. 1995. Preparing methylamines, Zeolites 15(4).

Lord-Garcia, J. 2014. Nitromethane, Reference Module in Biomedical Sciences, Encyclopedia of Toxicology, 3rd Edition, 20: 573–574.

Mackay, D. 2006. Handbook of physical-chemical properties and environmental fate for organic chemicals, Volume 1, Introduction and hydrocarbons, 2nd Edition, CRC Press/ Taylor & Francis. https://doi.org/10.1201/9781420044393

Masamoto, J., Iwaisako, T., Chohno, M., Kawamura, M., Ohtake, J. and Matsuzaki, K et al. 1993. Development of a new advanced process for manufacturing polyacetal resins. Part I. Development of a new process for manufacturing highly concentrated aqueous formaldehyde solution by methylal oxidation, Journal of Applied Polymer Science 50(8). https://doi.org/10.1002/app.1993.070500801

Nguyen, Q. 1996. Methane and Its Derivatives, Chemical Industries, CRC Press. https://doi.org/10.1002/cjce.5450750521

Olah, G.A., Goeppert, A. and Prakash, G.K.S. 2006. Beyond Oil and Gas: The Methanol Economy, 1st Edition, Wiley-VCH. DOI: 10.1002/9783527627806

Polyakov, A.M., Starannikova, L.E., Yu, P. and Yampolskii, 2004. Amorphous Teflons AF as organophilic pervaporation materials: Separation of mixtures of chloromethanes, Journal of Membrane Science 238(1–2): 21–32. https://doi.org/10.1016/j.memsci.2004.03.018

Wang, W., Feng, X. and Bao, M. 2008. Transformation of Carbon Dioxide to Formic Acid and Methanol, 1st Edition, Springer Singapore. https://doi.org/10.1007/978-981-10-3250-9

Yasunori, B., Chika, T., Ryoya, W., Fukuda, Nobuyoshi, Y.C. and Yutaka, N et al. 2013. Anaerobic digestion of crude glycerol from biodiesel manufacturing using a large-scale pilot plant: Methane production and application of digested sludge as fertilizer, Bioresource Technology 140: 342–348. DOI: 10.1016/j.biortech.2013.04.020

Zhang, L. 2018. Formaldehyde: Exposure, Toxicity and Health Effects, Royal Society of Chemistry. https://doi.org/10.1039/9781788010269-00001

CHAPTER 4

C – 2 Organic Chemicals

4.1 Introduction

In this chapter, we are discussing organic compounds having two carbons in one molecular formula unit. C2 basic chemicals, i.e., ethane (C_2H_6) and ethene (C_2H_4) are available in petroleum. Another source of C2 chemicals are C1 chemicals and some inorganic chemicals. The number of C2 chemicals is much more than that of C1 chemicals, because they contain alkanes, alkenes and alkynes, having different functional groups such as carboxylic acid, alcohol, amine, ether, ester, chloride, oxide, aldehyde and nitro groups. They are mostly liquid or gas. So, their manufacturing process includes a liquid phase reactor with stirring and heating facilities or a vapor phase tubular reactor. They have moderate boiling and melting points. These chemicals are mostly used as solvents in different industries. They are also used as an intermediate in several industrial segments such as drugs, polymers, petroleum, agricultural and explosives.

4.2 Ethene (Ethylene)

Ethylene is considered as highly explosive, as it is comfortably burnt under prolonged mechanical or chemical shocks.

4.2.1 Properties

Its main properties are as per Table 4.1.

4.2.2 Manufacturing process

Ethylene can be prepared by cracking the fractions acquired from the distillation of natural gas and oil. The processes are:

1. Steam cracking of ethane and propane (from natural gas and from crude oil).
2. Steam cracking of naphtha from petroleum.
3. Catalytic cracking of gas oil from petroleum.

Table 4.1 Physical & Chemical Properties of Ethylene

Sr. No.	Properties	Particular
1	IUPAC name	Ethene
2	Chemical and other names	Ethylene, Acetene, Elayl, Olefiant gas, Athylen
3	Molecular formula	C_2H_4
4	Molecular weight	28.05 gm/mole
5	Physical description	colorless gas with a sweet odor and taste
6	Solubility	Soluble in water, alcohol, ether, benzene
7	Melting point	−169°C
8	Boiling point	−103.7°C
9	Flash point	−100°C
10	Density	0.935 at 0°C

As ethylene can be manufactured using different feedstock (raw materials), the selection of raw materials is determined based on accessibility, expenditure and cracking. It is mainly manufactured by the steam cracking of naphtha or a mixture of ethane and propane, because ethylene (polymer quality) is manufactured by this process. In the cracking of naphtha, two stages of cracking and quenching are required, while one unit of cracking and quenching is adequate for the cracking of the ethane-propane mixture. About 50–55% of ethylene is produced by the cracking of naphtha.

4.2.2.1 Raw materials

Naphtha

4.2.2.2 Chemical reactions

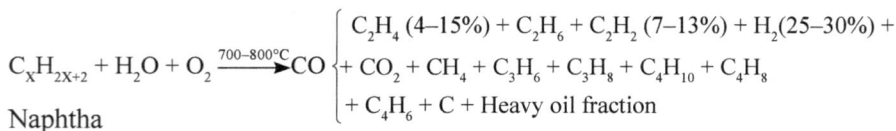

$$C_xH_{2X+2} + H_2O + O_2 \xrightarrow{700-800°C} CO \begin{cases} C_2H_4\,(4\text{–}15\%) + C_2H_6 + C_2H_2\,(7\text{–}13\%) + H_2(25\text{–}30\%) + \\ + CO_2 + CH_4 + C_3H_6 + C_3H_8 + C_4H_{10} + C_4H_8 \\ + C_4H_6 + C + \text{Heavy oil fraction} \end{cases}$$

Naphtha

4.2.2.3 Flow chart

4.2.2.4 Process description

As per Fig. 4.1, ethylene is produced using steam cracking from an ethane-propane mixture. This process is divided into three parts: (1) cracking and quenching; (2) compression and drying; and (3) separation.

(1) **Cracking and quenching:** First, naphtha (hydrocarbon) and superheated steam fed into a pyrolysis furnace. Also, recycled C2–C4 feed is inserted in another pyrolysis furnace. Stream cracking is performed in two different furnaces to get the maximum amount of ethylene and acetylene. A temperature of 700–800°C is maintained using heated coils. Both vapor masses are inserted into individual quenching towers to cool them using cooling water and the produced water stream. Thereafter, solid and heavy

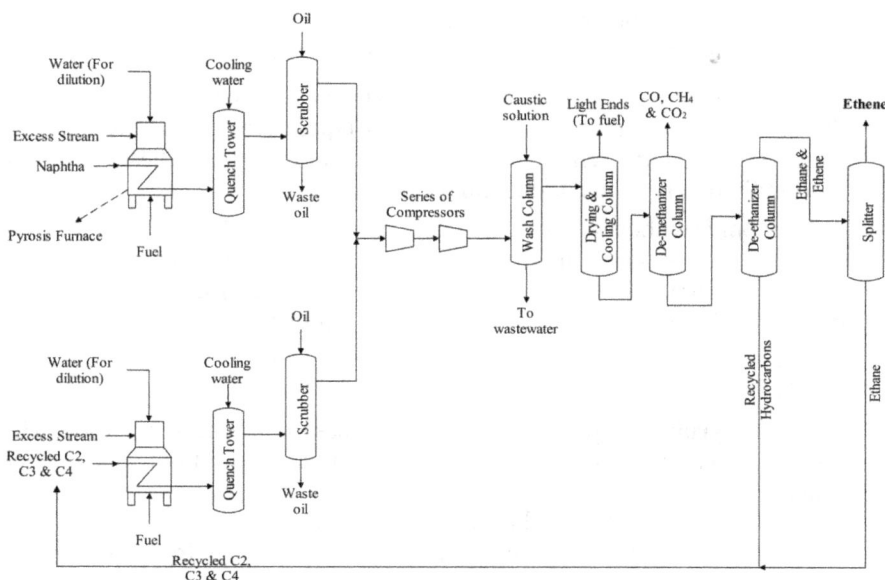

Figure 4.1 Flow sheet diagram of production of Ethylene from Naphtha

hydrocarbons are removed into a separate scrubber from the vapor mass. These solid and heavy hydrocarbons are utilized in LPG and Naphtha furnace. Stream cracking (pyrolysis), quenching and scrubbing process are performed in two different parts to get the maximum amount of ethylene and acetylene.

(2) **Compression and drying:** The cooled gases obtained from the pyrolysis of Naphtha and C2–C4 are mixed and compressed. About 35 atmospheres pressure is maintained using a series of compressors. Here, pressure plays an important role, because separation of light as well as heavy feedstocks depends upon the pressure value. The compressed gas is washed with caustic soda and water to remove carbon dioxide and sulfur. Finally, the gas is passed through the drying and de-methanizer columns respectively to get a mixture of H_2, CO, C1–C3 components only.

(3) **Separation:** The cooled gaseous mixture is inserted into a series of separation (de-ethanizer) columns. Here, the lower part of C3 and C4 hydrocarbons is separated from mixture of ethane and ethylene as over-headed product; and recycled. C2-splitter is separated ethylene from the top of the column. Ethane is collected from the bottom and recycled into furnace.

4.2.2.5 Major engineering problems

There are various engineering problems, which are as follows.

(A) **Feed type:** During the pyrolysis of various feeds in furnaces, different types of by-products with ethylene are produced, such as,

   a. CH_4 or natural gas-gives carbon monoxide and hydrogen gas

b. C2 and C3 hydrocarbons give carbon monoxide, hydrogen and ethane

c. C4 and higher hydrocarbons-give wide ranges of by-products.

(B) Pyrolysis agent: A number of pyrolysis agents ae utilized for cracking for the production of ethylene. Selective feed typesare available according to the pyrolysis agent, such as, heat, catalyst, steamor oxygen. These agents are used individually or in combinations, like

a. Heat only (without catalyst)-Original process, no longer attractive

b. Heat + catalyst-used in Dehydrogenation of butylenes to butadiene

c. Heat + steam (without catalyst)-called Thermal Reforming

d. Heat + steam + catalyst-called Catalytical Reforming

e. Oxygen only (without catalyst)-called Partial Combustion process

f. Oxygen + steam-Modified Partial Combustion process

(C) Type of pyrolysis equipment: Various types of furnaces are used with respect to heating for optimization of temperature, contact time and quench time.

a. Tubular, indirect fire-used for catalytical reactions

b. Coiled type furnace-used for non-catalytical reactions in the absence of air

c. Combustion type burner-used for the combustion process

(D) Heat (steam) recovery: The heat used in theepyrolysis process is recovered by integrating the stack and quench boiler heat transfer surface. Otherwise, heat can be washed out in water or in the oil quenching process. The resulting high pressure steam makes the process completely independent.

4.2.3 Uses

As per Table 4.2, it is used to manufacture various products, which have plenty of industrial applications. Different unit processes like polymerization, oxidation, chlorination, halogenations, involved in production of various chemicals utilize ethylene.

A mixture of 85% ethylene and15% oxygen is used as an anesthetic agent. It is utilized to hasten the ripening process of fruits and as a welding gas (Al-Mefren and Xiao, 2016; George, 2014; Peppel, 1958).

4.3 Ethyne (Acetylene)

It is the first member of alkynes and can be formed by sp hybridization of two carbon atoms. It is used carefully to prevent accidents, as it is unstable in its pure form.

4.3.1 Properties

Its main properties are as per Table 4.3.

Table 4.2 Applications of Ethylene

Sr. No.	Unit Process	Final Products	Uses (Manufacturing of)
1	Polymerization	HDPE, LDPE, LLDPE, etc.	Packaging, carrier bags and precursors, detergents, plasticisers, synthetic lubricants, additives, etc.
2	Oxidation	Ethylene oxide	Bottles, polyester fibers for clothing and furniture, automotive coolants, industrial coolants, heat transfer fluids, detergents and surfactants
		Acetaldehyde	Acetic acid, perfumes, flavors, construction materials, fire retardant paints and explosives
3	Halogenation and hydrohalogenation	Ethylene dichloride, ethyl chloride and ethylene dibromide, etc.	Packaging, carrier bags and precursors, detergents, automotive, industrial coolants, perfumes, flavors, refrigerant, an aerosol spray propellant, an anesthetic, and a blowing agent, etc.
4	Alkylation	Ethylbenzene, precursor to styrene	Rubber, plastic, insulation, fiberglass, pipes, automobile and boat parts, food containers, and carpet backing.
5	Hydroformylation	Propionaldehyde	Propionic acid and n-propyl alcohol.
6	Hydration	Ethanol	Solvent, fuel for motors and a fuel additive in gasoline
7	Dimerization	n-Butenes	Adhesives and sealant chemicals, Fuels and fuel additives, Intermediates, Plasticizers, Process regulators

Table 4.3 Physical & Chemical Properties of Ethyne

Sr. No.	Properties	Particular
1	IUPAC name	Acetylene
2	Chemical and other names	Ethyne, Narcylen, Vinylene, Vinylene, Ethine
3	Molecular formula	C_2H_2
4	Molecular weight	26.04 gm/mole
5	Physical description	Colorless gas with a sweet odor and taste
6	Solubility	Soluble in water and many organic materials
7	Melting point	−80.7°C
8	Boiling point	−84.7°C
9	Flash point	−18.16°C
10	Density	0.91 at 0°C

4.3.2 Manufacturing process

It was invented by Irish chemist Edmund Davy in 1836 during the manufacture of metallic potassium. Thereafter, ethylene was first synthesized by F. Wohler in 1862. Commercially, it is prepared from calcium carbide.

4.3.2.1 Raw materials

Calcium carbide and Water

4.3.2.2 Chemical reactions

Calcium carbide undergoes hydrolysis with water in a gas generator to form acetylene.

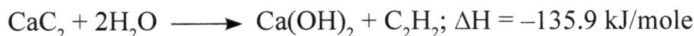

$$CaC_2 + 2H_2O \longrightarrow Ca(OH)_2 + C_2H_2; \Delta H = -135.9 \text{ kJ/mole}$$

During this reaction, traces of H_2S, NH_3 and PH_3 are also produced as impurities.

4.3.2.3 Quantitative requirements

1000 kgs of acetylene requires 2500 kgs of calcium carbide and 1400 kgs of water.

4.3.2.4 Flow chart

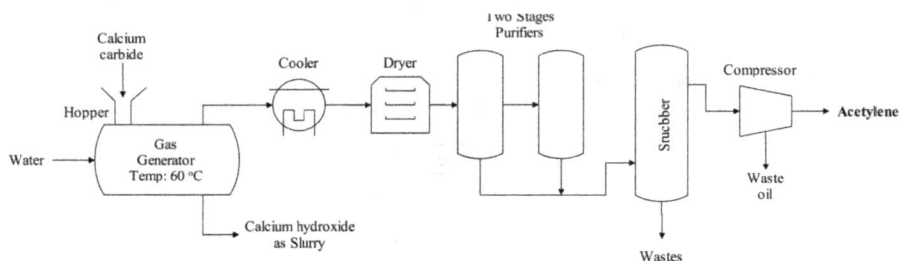

Figure 4.2 Flow sheet diagram of production of Acetylene by hydrolysis of Calcium carbide

4.3.2.5 Process description

As per the simple flow sheet diagram, calcium carbide is fed into a gas generator through a hopper. Also, water is added in this generator. The reaction is exothermic, and the temperature of the generator is 60°C. Hydrolysis of calcium carbide takes place to form calcium hydroxide and acetylene gas. The calcium hydroxide is collected from the bottom of the generator as slurry. Raw acetylene gas is collected from the top of the generator. It is further cooled, dried and purified in two stages. This gas is scrubbed to remove waste. Finally, it is compressed to remove waste oil to get pure acetylene gas.

4.3.3 Uses

1. In the presence of oxygen, it readily burns to given a vast amount of heat, which is used in cutting and welding metals. Welders call this flame an oxy acetylene flame.
2. It is used for the production of a variety of polymers such as poly vinyl chloride (PVC), polypropenonitrile and neoprene.
3. It is also utilized as a source of important organic compounds such as 1,1,2-trichloroethane and 1,1,2,2-tetrachloroethane.
4. It is used in lamps by hunters and miners as a fuel.
5. It is utilized for perfumes and solvents.

6. Acetic acid, several acetylenic alcohols and butanediol are prepared using acetylene.
7. It is employed as an artificial ripening agent and preservative of fruits. (Voronin et al., 2018; Sathya et al., 2019).

4.4 Ethanol

It is a colorless liquid and a major component of alcoholic beverages like beer, wine or whisky.

4.4.1 Properties

Its main properties are as follows.

Table 4.4 Physical & Chemical Properties of Ethanol

Sr. No.	Properties	Particular
1	IUPAC name	Ethanol
2	Chemical and other names	Ethyl alcohol, Absolute Alcohol, Methylcarbinol, Ethyl hydroxide, Ethyl hydrate, Hydroxyethane
3	Molecular formula	C_2H_6O
4	Molecular weight	46.09 gm/mole
5	Physical description	Clear colorless liquid with a bitter astringent taste.
6	Solubility	Miscible with ethyl ether, acetone, chloroform; soluble in benzene
7	Melting point	−114°C
8	Boiling point	78°C
9	Flash point	13°C
10	Density	0.789 gm/cm³ at 20°C

4.4.2 Manufacturing process

About 95% ethanol is produced by the fermentation of biomass, and the remaining 5% is obtained from processes such as,

(A) Catalytic hydration of ethene
(B) Esterification and hydrolysis of ethylene
(C) Oxidation of petroleum

Chemically, it is manufactured by the hydration of ethylene (C_2H_4) using a catalyst. This process was first invented in 1947 in Shell. The reaction is exothermic and reversible.

4.4.2.1 Raw materials
Ethene, Steam and Catalyst

4.4.2.2 Chemical reactions

Ethanol is prepared by passing the mixture of ethene and steam over silicon dioxide coated with phosphoric acid as a catalyst at 300°C and 60–70 atmospheres pressure.

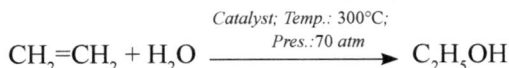

$$CH_2{=}CH_2 + H_2O \xrightarrow[\quad Pres.:70\ atm \quad]{Catalyst;\ Temp.:\ 300°C;} C_2H_5OH$$

Here, acetaldehyde is a by-product of the reaction. Further, acetaldehyde undergoes direct hydrogenation to yield ethanol

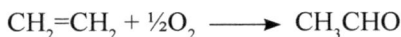

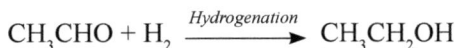

$$CH_2{=}CH_2 + \tfrac{1}{2}O_2 \longrightarrow CH_3CHO$$

$$CH_3CHO + H_2 \xrightarrow{Hydrogenation} CH_3CH_2OH$$

4.4.2.3 Quantitative requirements

1000 kgs of ethanol is requires 710 kgs of ethene and 400 kgs of steam.

4.5.2.4 Flow chart

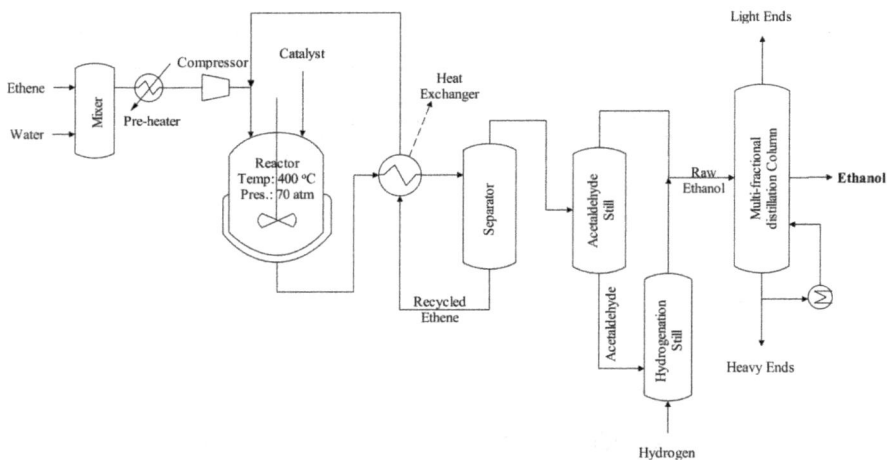

Figure 4.3 Flow sheet diagram of production of Ethanol from Ethene

4.4.2.5 Process description

Fist, ethene is mixed with water in a mixer. The mixture of ethene and water is pre-heated in a pre-heater and compressed in a compressor. This hot compressed mixture is fed into a reactor with stirring and hot oil jacketed facilities. Also, a catalyst is added. The reaction mass is heated upto 400°C with stirring. After completion of the reaction, the reaction mass is collected from the bottom of threactor and cooled in the heat exchanger. The cooled mass is sent into a separator still, in which the unreacted ethane is separated and further recycled into the reactor thought of as a heat exchanger. The ethene free reaction mass is transferred into the acetaldehyde still. Here, acetaldehyde from the raw ethanol is separated out and sent to the hydrogenation still. Hydrogen gas is inserted from the bottom of this still. Acetaldehyde undergoes hydrogenation to yield raw ethanol. Raw ethanol coms from the acetaldehyde still and sent to the hydrogenation still followed by the fractional distillation column. Pure ethanol is separated out from this column.

4.4.2.6 *Major engineering problems*

1. At the first the conversion of ethane to ethanol is only 4.2%, so, un-reacted ethane is continuously recycled through the heat exchanger to give a higher percentage yield of ethanol.
2. Ethanol-water mixture is an azeotropic mixture, which requires special distillation techniques and eventually increases the costs of the plan.
3. Hydrolysis of ethylene is a costly process, as ethanol is more economical than ethylene.
4. As ethene is manufactured from crude oil feedstock, this process is non-renewable.

4.4.3 *Uses*

1. Industrially, it is used to manufacture a variety of chemicals like ethylene dibromide, acetaldehyde, glycols, ethyl acetate, ethyl chlorideand acetic acid.
2. Now-a-days, it is used as a biofuel for engines.
3. It plays an important role in various industries like vinegar, liquid dishwashing detergents, detergents, adhesives, coatings (lacquers), printing inks, liquid soapsand hand sanitizers. It is also used to prepare 99.9% concentrated ethanol.
4. As it has low toxicity and a capacity to dissolve non-polar materials, it is extensively utilized as a solvent in different organic reactions of in the manufacture of dyes, pigments, pharmaceuticals, agricultural products and textiles.
5. Anhydrous ethanol is utilized with iodine as a sterilization agent (Cardona and Sanchez, 2007; Scott, 1998).

4.5 Acetaldehyde

Acetaldehyde, referred to as ethanol or MeCHO, is one among the foremost important aldehydes. It is a naturally occurring compound found in coffee, bread, and ripe fruit.

4.5.1 *Properties*

Its main properties are as per Table 4.5.

Table 4.5 Physical & Chemical Properties of Acetaldehyde

Sr. No.	Properties	Particular
1	IUPAC name	Acetaldehyde
2	Chemical and other names	Ethanal, Ethyl aldehyde, Acetic ethanol
3	Molecular formula	C_2H_4O
4	Molecular weight	44.05 gm/mole
5	Physical description	Colorless liquid with a pungent, fruity odor
6	Solubility	Miscible with water and most common organic solvents like acetone, benzene, ethyl alcohol, etc.
7	Melting point	−123°C
8	Boiling point	20°C
9	Flash point	−38°C
10	Density	0.78 gm/cm³ at 27°C

4.5.2 Manufacturing process

It is manufactured using various routes, as mentioned below.

(A) Oxidation of ethanol (Veba-Chemie Process):
   Acetaldehyde is prepared by the oxidation of Ethanol with oxygen (or air) in the presence of a metal alloy or oxides as a catalyst.

(B) Direct oxidation of ethylene (Wacker process):
   Ethylene undergoes direct oxidation to yield acetaldehyde.

(C) Dehydrogenation of ethanol
   Ethanol undergoes dehydrogenation by passing it through glass tubes at 260°C in the presence of metal oxide as a catalyst to yield acetaldehyde.

$$CH_3CH_2OH \rightarrow CH_3CHO + H_2; \Delta H = +82.45 \text{ kJ/mol}$$

(D) Hydration of acetylene:
   Acetylene reacts with water in the presence of mercury and sulfuric acid as a catalyst to yield acetaldehyde.

$$C_2H_2 + H_2O \xrightarrow{Hg/Sulfuric\ acid} CH_3CHO; \Delta H = -138.2 \text{ kJ/mol}$$

(E) First, methanol and acetylene are reacted in the presence of KOH at a temperature of 150–160°C and a pressure of 16 atmospheres to form methyl vinyl ether. This ether is further hydrolyzed with dilute acid to form acetaldehyde and methanol. This reaction was developed by BASF in 1945.

$$C_2H_2 + CH_3OH \xrightarrow{KOH} CH_3OCH=CH_2 \xrightarrow{Acid\ Hydrolysis} CH_3CHO + CH_3OH$$

(F) Acetic acid is reacted with acetylene in the presence of mercury as a catalyst to form ethylidene diacetate. It is further decomposed at 130–145 °C in the presence of acid to give acetaldehyde and acetic anhydride.

$$2CH_3COOH + C_2H_2 \xrightarrow{Hg} CH_3CH(OCOCH_3)_2 \xrightarrow{Decomposition} (CH_3CO)_2O + CH_3CHO$$

Here, we are discussing important methods such as oxidation of ethanol, oxidation of ethylene and dehydrogenation of ethanol.

4.5.2.1 Oxidation of ethanol (Veba-Chemie Process)

This process was developed and patented by Veba-Chemie AG in 1974, so it is called the Veba-Chemie process.

(A) Raw materials

Ethanol, Air (Oxygen) and Catalyst

(B) Chemical reactions

Acetaldehyde is prepared by vapor phase oxidation using oxygen (or air) at 500–600°C in the presence of metal (copper or silver) oxides or alloys as a catalyst.

$$CH_3CH_2OH + 2O_2 \longrightarrow CH_3CHO + H_2O; \Delta H = -242.0 \text{ kJ/mole}$$

(C) Quantitative requirements

1000 kgs of acetaldehyde requires 1100 kgs of ethanol, 1500 kgs of oxygen (air) and 1000 kgs of catalyst.

(D) Flow chart

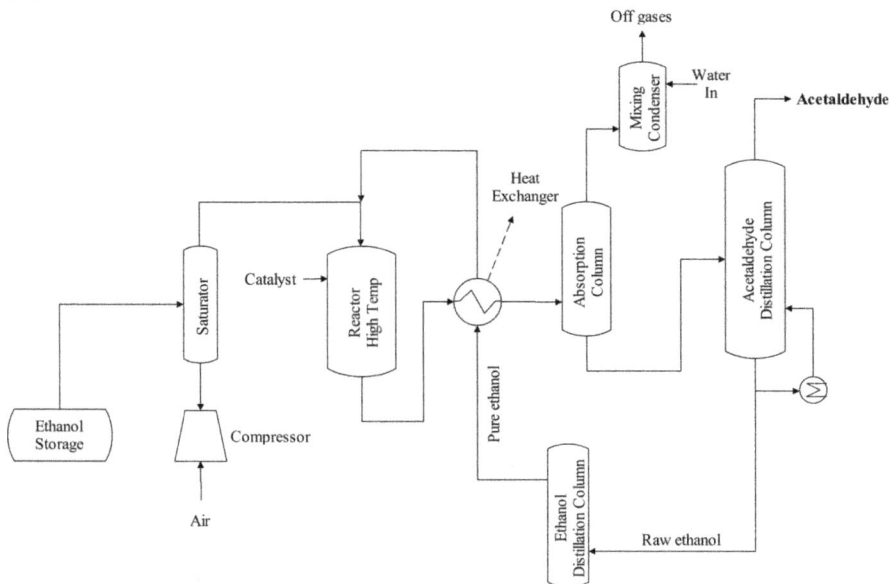

Figure 4.4 Flow sheet diagram of production of Acetaldehyde by oxidation of ethanol

(E) Process description

As per Fig. 4.4, air (oxygen) is compressed in a compressor and fed into a saturator. Also, ethanol (about 85%) is inserted. Now, the mixture of ethanol and air is fed into a horizontal reactor. The reaction of ethanol and air is exothermic and so, acetaldehyde is in vapor form. The reaction mass is cooled in a heat exchanger and fed into an absorber column. Here, gaseous and liquid phases are separated. The gaseous products are scrubbed in the scrubber and vented into the air. The liquid phase containing a mix of acetaldehyde and ethanol is separated into an acetaldehyde column by means of steam. Raw ethanol is purified in the ethanol column using steam, cooled in the heat exchanger and fed into the reactor.

(F) Major engineering problems

By using this process, about 25–30% ethanol is converted into acetaldehyde, so, ethanol is continuously recycled to give a higher percentage yield.

4.5.2.2 Oxidation of ethylene (Wacker process)

Wacker-Chemie and Hoechst developed a process in the years 1957–1959 using a One- and Two-stage method. In the one-stage method, ethylene and oxygen are reacted to form acetaldehyde and; re-generation of catalyst (reduction of $CuCl_2$ and re-oxidation of CuCl) simultaneously. While the two-stage reaction is performed in two stages, i.e., ethylene and oxygen are reacted to form acetaldehyde in the first stage and; re-generation of catalyst in the second stage. The two-stage reaction is more preferable since its cheaper, so, we will discuss two-stage oxidation.

(A) Raw materials

Ethylene, Oxygen (Air) and Catalyst

(B) Chemical reactions

Ethylene undergoes direct oxidation in the presence of palladium (II) chloride and copper (II) chloride as catalyst.

$$C_2H_4 + 2O_2 \xrightarrow{[O], PdCl_2} CH_3CHO; \Delta H = 244 \text{ kJ/mole}$$

Here, palladium (II) chloride and cupric chloride solution is used as catalyst.

(C) Quantitative requirements

1000 kgs of acetaldehyde requires 1800 kgs of ethylene, 2500 kgs of oxygen (air) and 1340 kgs of catalyst.

(D) Flow chart

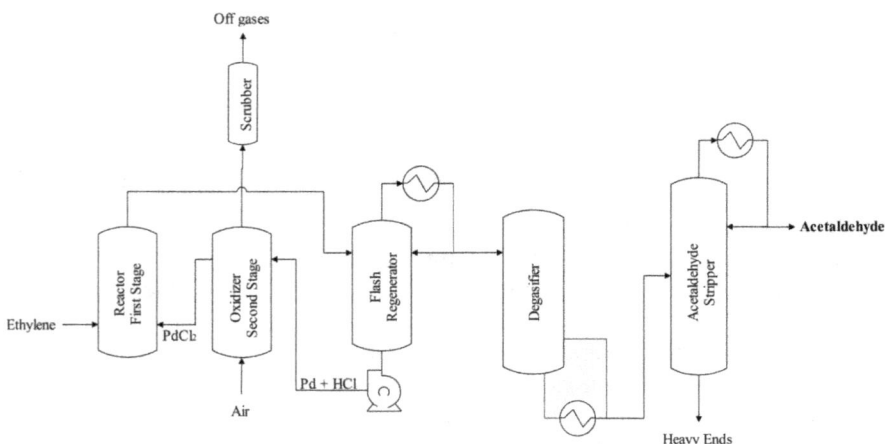

Figure 4.5 Flow sheet diagram of production of Acetaldehyde by oxidation of Ethylene

(E) Process description

In this figure, ethylene is fed into the first reactor at the bottom of reactor. The reactor is already impregnated with palladium chloride and water. Here, some part of the ethylene reacts with air to give acetaldehyde, palladium and hydrochloric acid.

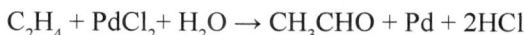

$$C_2H_4 + PdCl_2 + H_2O \rightarrow CH_3CHO + Pd + 2HCl$$

The reaction mass is transferred to the flash regenerator, in which acetaldehyde is separated out, while the reaction mass containing palladium and hydrochloric acid are fed into another reactor called the oxidizer. In this oxidizer, palladium is converted into palladium chloride using copper(II) chloride and hydrochloric acid and sent to the 1st reactor.

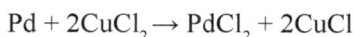

$$Pd + 2CuCl_2 \rightarrow PdCl_2 + 2CuCl$$

$$2CuCl + \frac{1}{2} O_2 + 2HCl \rightarrow 2CuCl_2 + H_2O$$

4.5.2.3 Dehydrogenation of ethanol

Acetaldehyde is also prepared by the dehydrogenation of ethanol.

(A) Raw materials

Ethylene and Catalyst

(B) Chemical reactions

Ethanol is passed through glass tubes at 260°C and dehydrogenated in the presence of metal oxide as the catalyst to yield acetaldehyde.

$$CH_3CH_2OH \xrightarrow[\text{Catalyst}]{Temp.: 260°C} CH_3CHO + H_2; \Delta H = +82.45 \text{ kJ/mole}$$

(C) Flow chart

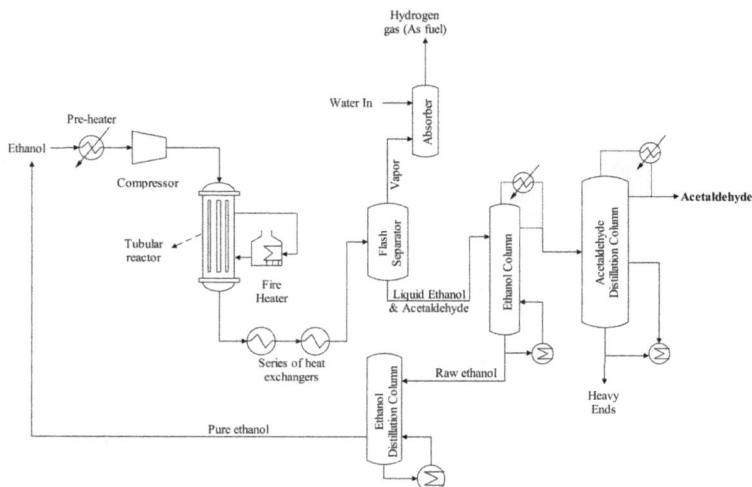

Figure 4.6 Flow sheet diagram of production of Acetaldehyde by dehydrogenation of Ethanol

(D) Process description

Ethanol (purity: 85%) is pre-heated, compressed and fed into a tubular reactor having a catalyst. The ethanol undergoes dehydrogenation to give acetaldehyde and hydrogen gas. Sometimes, ethanol is recycled through a fire heater in the tubular reactor. After completion of the reaction, the reaction mass is cooled using a series of heat exchangers. The cooled mass is transferred into a flash separator, where hydrogen vapor from the top and liquid (acetaldehyde and unreacted ethanol) from the bottom are separated out. The vapor containing hydrogen is purified in an absorber using water and further utilized as a fuel. Now, liquid is collected from the bottom and fed into the ethanol column. Here, unreacted ethanol is separated out by means of steam and recycled. Raw acetaldehyde is purified in the acetaldehyde column and separated as the overhead product.

4.5.3 Uses

1. Acetaldehyde is a key staple for manufacturing different chemical products such as paint binders and plasticizers.
2. Acetaldehyde is employed in the production of construction materials, fire retardant paints and explosives.
3. It is used in the production of various things such as sedatives and tranquilizers in the pharmaceutical and cosmetics industry.
4. It is also used in ethanoic acid, perfumes, dyes and medical industries.
5. It is utilized as an additive in fruit and fish preservatives, flavoring agents and gelatin hardening.
6. Acetaldehyde is utilized as a preservative for fruits and sea-food in the manufacture of vinegar and yeast (Vaca, 1995; Verschueren, 1996; Rao et al., 2014).

4.6 Acetic Acid

The simplest carboxylic acid is known as acetic acid. It has a pungent smell like vinegar and a sour taste. It is moderately dissociated in the presence of solvent. Sometimes it is referred to as glacial acetic acid, because pure, water-free acetic acid looks like an ice like glacier.

4.6.1 Properties

Its main properties are as per Table 4.6.

4.6.2 Manufacturing process

Like ethanol, industrial acetic acid is produced synthetically as well as by bacterial fermentation. Synthetically it is prepared by the following methods:

Table 4.6 Physical & Chemical Properties of Acetic acid

Sr. No.	Properties	Particular
1	IUPAC name	Acetic acid
2	Chemical and other names	Ethanoic acid, Ethylic acid, Glacial acetic acid, Methanecarboxylic acid, Acetasol, Vinegar acid
3	Molecular formula	$C_2H_4O_2$
4	Molecular weight	60.05 gm/mole
5	Physical description	Clear colorless liquid with strong odor of vinegar
6	Solubility	Miscible with ethanol, ethyl ether, acetone, benzene
7	Melting point	17°C
8	Boiling point	118°C
9	Flash point	39°C
10	Density	1.045 gm/cm³ at 27°C

(A) Carbonation of methanol

Acetic acid is prepared by reacting methanol with carbon monoxide in the presence of a catalyst. Sometimes, methanol is manufactured from synthesis gas (mixture of carbon monoxide and hydrogen gas).

(B) Isomerization of methyl formate

Methyl formate is isomerized in the presence of a catalyst to yield acetic acid.

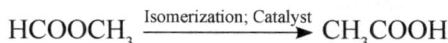

$$HCOOCH_3 \xrightarrow{\text{Isomerization; Catalyst}} CH_3COOH$$

(C) Oxidation of ethylene or ethanol

Ethylene or ethanol undergo gaseous phase oxidation in the presence of a catalyst to acetic acid.

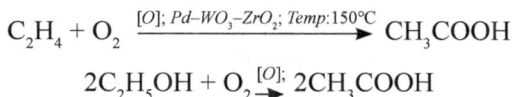

$$C_2H_4 + O_2 \xrightarrow{[O];\ Pd-WO_3-ZrO_2;\ Temp:150°C} CH_3COOH$$

$$2C_2H_5OH + O_2 \xrightarrow{[O];} 2CH_3COOH$$

We are discussing carbonation of methanol

4.6.2.1 Raw materials

Methanol, Carbon monoxide and Catalyst

4.6.2.2 Chemical reactions

Acetic acid is prepared by the reaction of methanol with carbon monoxide in the presence of a catalyst (Rhodium), promoter (methyl iodide) and stabilizer (iodide salt) to give acetic acid.

$$CH_3OH + CO \xrightarrow{Rhodium} CH_3COOH$$

4.6.2.3 Quantitative requirement

1000 kgs of acetic acid is requires 600 kgs of methanol, 550 kgs of oxygen (air) and 410 kgs of catalyst.

4.6.2.4 Flow chart

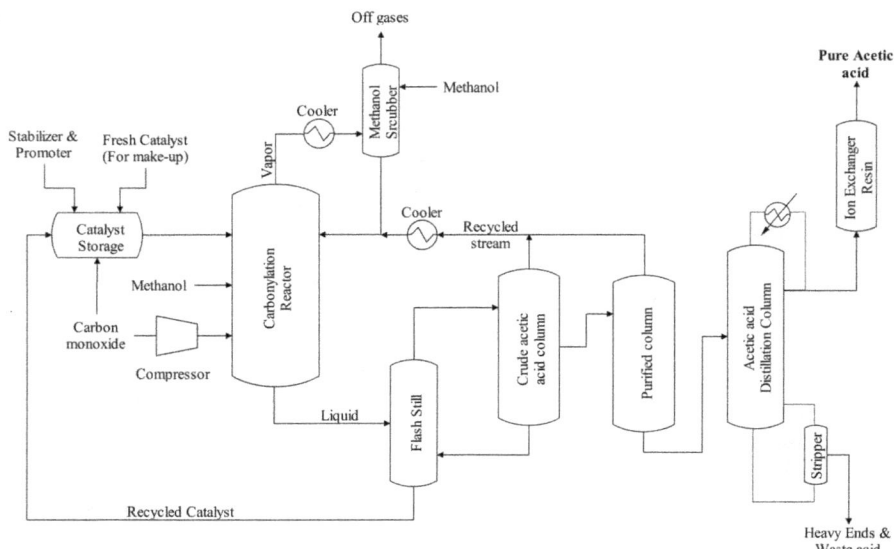

Figure 4.7 Flow sheet diagram of production of acetic acid by carbonation from Methanol

4.6.2.5 Process description

As per Fig. 4.7, initially rhodium catalyst (recycled and fresh), promoter (methyl iodide) and stabilizer (iodide salt) are mixed properly. Using pressurized carbon monoxide, the mixture is dissolved in hot acetic acid. Thereafter, methanol, carbon monoxide, catalyst, promoter and stabilizer are inserted in the carbonylation reaction. Here, acetic acid is prepared by the reaction of methanol with carbon monoxide in the liquid phase. The gases from the reactor are cooled to recover unreacted methanol, methyl iodide and methyl acetate; and sent to the reactor. The liquid reaction mass is collected from the bottom of the reactor and sent to the flash still, where the catalyst solution is collected as an over-head stream and recycled for catalyst preparation. The remaining reaction mass is collected from the top and fed into the crude acetic acid column. In this column, the over-head stream containing different chemicals such as methyl acetate, methyl iodide, water, and a small amount of acetic acid is cooled using a cooler and further, recycled back into the carbonylation reactor. Now, the dissolved catalyst is collected from the bottom and also recycled. The reaction mass is collected from the middle of the column and fed into the drying column. Crude acetic acid is separated from water, methyl iodide and methyl acetate as an over-head stream. This stream is further recycled in the reactor. While acetic acid is collected from the bottom and inserted into the acetic acid distillation column, where it is collected as an overhead product using steam. The bottom part containing heavy ends and waste water is collected from the stripper and discarded. Finally, iodide contaminants are eliminated from the acetic acid by passing it through ion-exchange resin beds.

4.6.3 Uses

There are various uses of acetic acid as follows.

1. Mainly acetic is acid is utilized as a solvent in different industries for manufacturing synthetic fibres, textile inks, dyes, perfumes, soft drinks, bottles, rayon fibre, rubbers and plastics, pesticides and wood glues.
2. It is employed in the development of photographic films.
3. It is also used as a household vinegar to clean indoor climbing holds of chalk.
4. It is also utilized in clinical laboratories.
5. Itis referred to as vinegar and utilized for medication purposes.
6. It is used in ear drop solutions.
7. It is utilized as a food additive (Manyar and Deshmukh, G. 2022; Li et al., 2022).

4.7 Vinyl Chloride

It is a chlorinated hydrocarbon. When it is heated, it decomposes to give toxic fumes of carbon dioxide, carbon monoxide, hydrogen chloride and phosgene.

4.7.1 Properties

Its main properties are as follows.

Table 4.7 Physical & Chemical Properties of Vinyl chloride

Sr. No.	Properties	Particular
1	IUPAC name	Chloroethene
2	Chemical and other names	Vinyl chloride, Chloroethylene, VC, Ethylene monochloride
3	Molecular formula	C_2H_3Cl
4	Molecular weight	62.59 gm/mole
5	Physical description	Colorless gas with characteristic odor
6	Solubility	Soluble in water, ketone, alcohol, nitrobenzene, tetrahydrofuran, methylene chloride
7	Melting point	–154°C
8	Boiling point	–13°C
9	Flash point	–78°C
10	Density	0.912 gm/cm³ at 27°C

4.7.2 Manufacturing process

There are two methods available for preparing vinyl chloride: (1) Hydrochlorination of acetylene and (2) Dehydrochlorination of ethylene dichloride (1,2-dichloroethane), in which the starting material is ethylene. About 85% of vinyl chloride is manufactured using ethylene, chlorine, hydrochloric acid and oxygen (air). Here, we are discussing Dehydrochlorination of ethylene dichloride.

4.7.2.1 Raw materials

Ethylene, Chlorine gas, Iron(III) chloride and Hydrochloric acid

4.7.2.2 Chemical reactions

This process is divided into two parts.

Part – I: Preparation of ethylene dichloride (EDC) from ethylene: It is prepared by direct chlorination of ethylene reacting it with chlorine in the presence of iron(III) chloride as a catalyst. In this process, HCl is an undesirable product, which must be removed.

$$CH_2=CH_2 + Cl_2 \longrightarrow ClCH_2CH_2Cl + HCl$$

Ethylene   Ethylene dichloride

It is also prepared by oxychlorination of ethylene, reacting it with hydrochloric acid and air over a copper(II) chloride catalyst.

$$CH_2=CH_2 + 2HCl + \tfrac{1}{2}O_2 \longrightarrow ClCH_2CH_2Cl + H_2O.$$

Both reactions are exothermic and heat is removed by circulating water or deploying a heat exchanger.

Part – II: Preparation of vinyl chloride from EDC: EDC undergoes decomposition at 480–500°C at 15–30 atmospheres pressure, to produce vinyl chloride and anhydrous HCl.

$$ClCH_2CH_2Cl \longrightarrow CH_2=CHCl + HCl$$

4.7.2.3 Quantity requirement

1000 kgs of vinyl chloride requires 500 kgs of ethylene, 1300 kgs of chlorine gas, 700 kgs of iron(III) chloride and 450 kgs of hydrochloric acid.

4.7.2.4 Flow chart

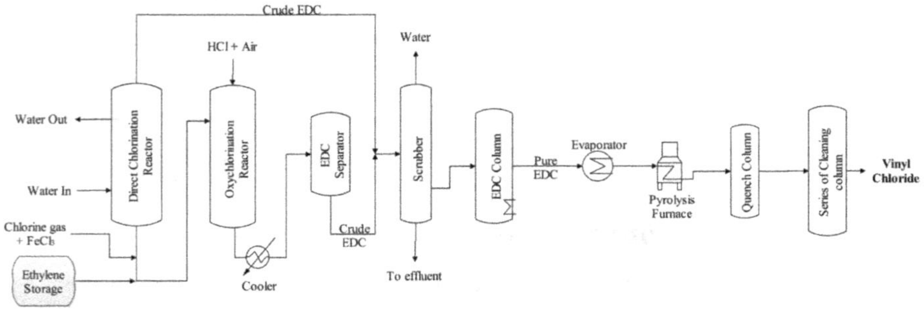

Figure 4.8 Flow sheet diagram of production of vinyl chloride

4.7.2.5 Process description

As previously discussed, first ethylene is inserted at the bottom of the direct chlorination reactor. Also, chlorine gas and catalyst are added. Water is continuously circulating in reactor to remove the heat from reaction mixture. After reaction, the mixture containing crude EDC is collected from the top of thereactor. Further, ethylene is added at the top of the oxychlorination reactor. Hydrochloric acid and air are also added to the reactor. The reaction mixture is collected from the bottom of the reactor and cooled through a heat exchanger. Now, crude EDC from direct chlorination and oxychlorination is than mixed and scrubbed using water. Thereafter, EDC is purified in the column and converted into vapor form in the evaporator by means of steam. It is burnt at 480–500°C at 15–30 atmosphere in a stainless steel tubular cracking furnace to produce vinyl chloride. The conversion is around 50% and the ultimate yield is 95–96%. It is cooled in the quench column and purified by a series of columns.

4.7.2.6 Major engineering problems

1. In the pyrolysis furnace, the reaction occurs at a high pressure drop, causing shut down srequiring periodic cleaning. Increasing conversion beyond 50% with a longer residence time or higher temperature leads to carbon formation and promotes polymerization of the monomer.
2. Excessive corrosion occurs if the system is not kept free from water vapor.
3. An antioxidant is added to stabilize the vinyl chloride monomer so that no interference occurs later in the polymerization process.

4.7.3 Uses

1. Using the polymerization technique, about 90–95% of vinyl chloride is used to manufacture polyvinyl chloride (PVC)
2. Minor quantities of VC are utilized in wall coverings, furniture, automotive parts, automobiles, and housewares (Cohan, 1975).

4.8 Vinyl Acetate

Vinyl acetate is a clear colorless liquid and its vapor is heavier than air.

4.8.1 Properties

Its main properties are as per Table 4.8.

4.8.2 Manufacturing process

Initially, it is manufactured by an oxidation addition reaction of acetic acid and ethylene using palladium as a catalyst. Previously, mercury(I) salt was used as a catalyst in the process. Now-a-days, it is prepared from ethylene.

4.8.2.1 Raw materials

Ethylene, Acetic acid, Palladium chloride and Oxygen (Air)

Table 4.8 Physical & Chemical Properties of Vinyl acetate

Sr. No.	Properties	Particular
1	IUPAC name	Ethenyl acetate
2	Chemical and other names	Vinyl acetate, Ethenyl ethanoate, 1-Acetoxyethylene, Vinyl ethanoate
3	Molecular formula	$C_4H_6O_2$
4	Molecular weight	86.09 gm/mole
5	Physical description	Colorless liquid with a pleasant, fruity odor
6	Solubility	Soluble in water and organic solvents such ethane, acetone, chloroform
7	Melting point	$-136°C$
8	Boiling point	$73°C$
9	Flash point	$-8°C$
10	Density	0.932 gm/cm^3 at 27°C

4.8.2.2 Chemical reactions

It is prepared by vapor phase reaction of ethylene, acetic acid and air (oxygen) in the presence of a palladium catalyst to form vinyl acetate.

$$C_2H_4 + CH_3COOH + \tfrac{1}{2}O_2 \longrightarrow CH_3COOCHCH_2 + H_2O$$

Here, oxidation of ethylene also conducts an unwanted reaction to produce water and carbon dioxide.

$$C_2H_4 + 3O_2 \longrightarrow 2CO_2 + 2H_2O$$

4.8.2.3 Quantity requirements

1000 kgs of vinyl acetate requires 370 kgs of ethylene, 2000 kgs of acetic acid, 350 kgs of palladium chloride and 200 kgs of oxygen.

4.8.2.4 Flow chart

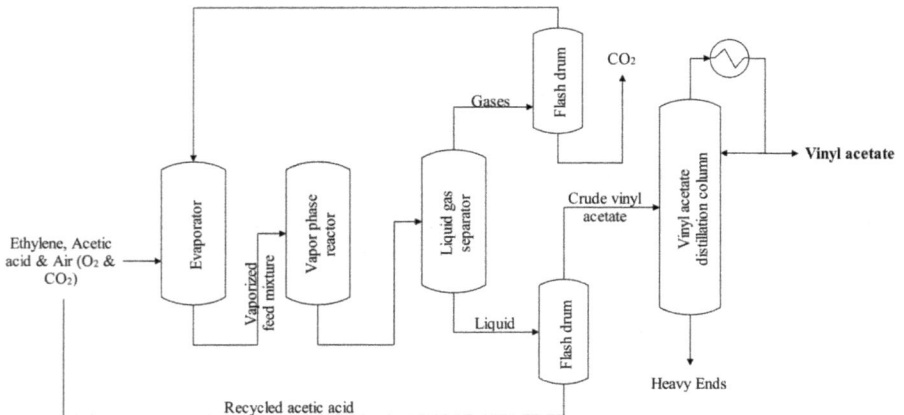

Figure 4.9 Flow sheet diagram of production of vinyl acetate

4.8.2.5 Process description

As per Fig. 4.9, ethylene, acetic acid and air (oxygen and a small amount of carbon dioxide) are vaporized in an evaporator. The vaporized feed mixture is inserted into a reactor. Palladium chloride catalyst is also added. After the reaction, the product mixture is collected from the bottom and sent into a liquid gas separator. Here, liquids are collected from the bottom of the separator and gases are exit the top of column together. The collected gases are further sent to a flash drum, in which they are separated into carbon dioxide and ethylene/oxssygen mixture. The mixture of ethylene/oxygen is recycled and added into the evaporator; and carbon dioxide is drawn off. The liquid is further separated into acetic acid and crude vinyl acetate in a flash drum. Acetic acid is further recycled. Crude vinyl acetate is undergoes purification in a distillation column to get pure vinyl acetate.

4.8.3 Uses

Polymerization of vinyl acetate gives polyvinyl acetate. Also, other polymers such as ethylene-vinyl acetate, vinyl acetate-acrylic acid, PVC acetate, and polyvinylpyrrolidone are also prepared using other monomers. These polymers are water soluble polymers, which are used in manufacturing adhesives, paper coating, emulsifiers, surface coatings, molding materials, lacquers, textile finishing agents, printing auxiliariesand safely glass (Dimian and Bildea, 2008).

4.9 Ethylene Oxide

Ethylene oxide has a three membered ring structure and is the simplest epoxide.

4.9.1 Properties

Its main properties are as follows.

Table 4.9 Physical & Chemical Properties of Ethylene oxide

Sr. No.	Properties	Particular
1	IUPAC name	Oxirane
2	Chemical and other names	Ethylene oxide, Epoxyethane, Dimethylene oxide, Amprolene, Anproline, Oxidoethane
3	Molecular formula	C_2H_4O
4	Molecular weight	44.06 gm/mole
5	Physical description	Clear colorless gas with an characteristic odor
6	Solubility	Soluble in water, benzene, acetone, ethanol, ether
7	Melting point	−112°C
8	Boiling point	11°C
9	Flash point	−18°C
10	Density	0.822 gm/cm³ at 27°C

4.9.2 *Manufacturing process*

First, it was prepared by Charles-Adolphe Wurtz in 1859, by the reaction of 2-chloroethanol and potassium hydroxide.

$$Cl-CH_2CH_2-OH + KOH \rightarrow (CH_2CH_2)O + KCl + H_2O$$

Now-a-days, it is produced by the oxidation of ethylene.

4.9.2.1 *Raw materials*

Ethylene, Air (Oxygen), Silver oxide on alumina, Water

4.9.2.2 *Chemical reactions*

Oxidation of ethylene using air at a temperature of 250–300°C and pressure of 8–20 atmospheres in the presence of silver oxide on alumina catalyst gives ethylene oxide.

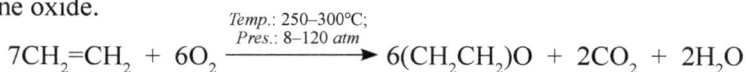

$$7CH_2=CH_2 + 6O_2 \xrightarrow[\text{Pres.: 8–120 atm}]{\text{Temp.: 250–300°C;}} 6(CH_2CH_2)O + 2CO_2 + 2H_2O$$

4.9.2.3 *Quantity requirement*

1000 kgs ethylene oxide requires 750 kgs of ethylene, 400 kgs of oxygen, 550 kgs of silver oxide on alumina and 500 kgs of water.

4.9.2.4 *Flow chart*

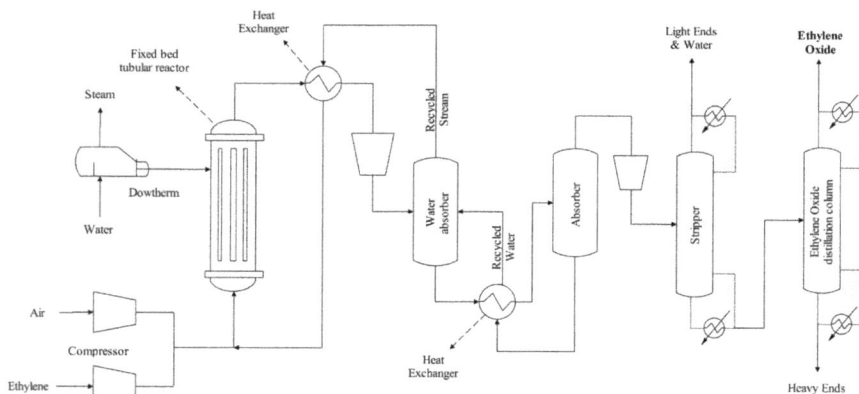

Figure 4.10 Flow sheet diagram of production of Ethylene oxide

4.9.2.5 *Process description*

Air (oxygen) and ethylene are compressed and inserted into a tubular fixed bed reactor from the bottom. This reactor contains the catalyst on the small tubular side. Also, a recycled stream is added. Operating parameters like temperature and pressure of 250–300°C and 8–20 atmospheres respectively are maintained in the reactor. From the side of reactor, dowtherm fluid is also introduced. Dowtherm fluid facilitates heat transfer to maintain the temperature inside the reactor. For generation of steam, hot dowtherm fluid from the reactor is fed to a waste heat recovery boiler. Now, the vapour mixture from reactor is collected from the top. It is further cooled through a heat exchanger, compressed using a compressor and sent to a water adsorbent. Light and heavy ends along with ethylene oxide are absorbed in the absorbent. The rationale for compression is due to the information that absorption is more favored at alow temperature and high pressure. Here, the cooled process is much preferred for a high dissolution of ethylene oxide in water. Water containing a higher amount of ethylene oxide is fed into a stripper to withdraw ethylene oxide and water as vapor. On the other hand the remaining regenerated water is absorbed through a heat integrated exchanger. The vapor mixture of ethylene oxide and water is compressed and fed into the stripper. Here, light ends are collected as the over-head product, while the bottom product containing raw ethylene oxide is further purified in another fractionator. This fractionator separates the ethylene oxide as the over-head product. The heavy ends are collected from the bottom of the fractionator.

4.9.2.6 Major engineering problems

1. **Volume ratio of feedstock:** For ethylene or ethylene oxide, a lower explosion (3%) limit in air is maintained by a specific reactor design. This condition can be maintained by side stream purging which removes CO_2 and H_2O from the absorber with the addition of recycledinert.
2. **Utilization of series reactor:** More conversion of ethylene can be achieved in two or three reactors in series.
3. **Hydration of ethylene oxide to ethylene glycol:** Thermodynamically, hydration of ethylene oxide to ethylene glycol is possible in these conditions, but, commercially hydration is done at 150°C.
4. **Air vs. oxygen:** Oxygen oxidation is more advantageous compared to air oxidation, due to the fact that: (i) The space time yield is increased upto 3–4 times, and (ii) product concentration six times greater than that in an air system.

4.9.3 Uses

Mainly it is employed as an interim for different organic reactions. It is also used to manufacture triethylene glycol, monoethylene glycol, ethylene glycol ethers, diethylene glycol, poly(ethylene) glycols and ethanolamine. It is utilized in the manufacture of water-soluble solvents, flexible and rigid polyurethane foams and brake fluids (Allah et al., 2022; Mubashir et al., 2022).

4.10 Ethylene Glycol

It is a colorless and odorless liquid.

4.10.1 Properties

Its main properties are as per Table 4.10.

Table 4.10 Physical & Chemical Properties of Monoethyleneglycol

Sr. No.	Properties	Particular
1	IUPAC name	Ethane-1,2-diol
2	Chemical and other names	MEG, Ethylene glycol, 1,2-Ethanediol, Monoethylene glycol, 1,2-Dihydroxyethane, Glycol alcohol, 2-hydroxyethanol, 2-hydroxyethanol
3	Molecular formula	$C_2H_6O_2$
4	Molecular weight	62.07 gm/mole
5	Physical description	Clear, viscous, colourless, odorless liquid
6	Solubility	Soluble in water and other organic solvents
7	Melting point	-13°C
8	Boiling point	198°C
9	Flash point	111°C
10	Density	1.115 gm/cm^3 at 27°C

4.10.2 Manufacturing process

It was first prepared in 1856 by French chemist Charles-Adolphe Wurtz by treating ethylene iodide $(C_2H_4I_2)$ with silver acetate, followed by potassium hydroxide hydrolyzation. Industrially, it is prepared by the hydrolysis of ethylene oxide.

4.10.2.1 Raw materials

Ethylene oxide and Water

4.10.2.2 Chemical reactions

Industrially, it is prepared by the hydrolysis of ethylene oxide using acidic water. About 90% yield attained using this process.

$$C_2H_4O + H_2O \longrightarrow HO-CH_2CH_2-OH$$

During this reaction, various oligomers such as diethylene glycol, triethylene glycol, and tetraethylene glycol are prepared as by-products.

4.10.2.3 Quantity requirements

1000 kgs ethylene glycol requires 850 kgs of ethylene oxide and 410 kgs of water.

4.10.2.4 Flow chart

4.10.2.5 Process description

As per figure, recycled as well as make-up water and ethylene oxide are fed into the feed tank. Here, excess water is added to achieve the percentage yield. The mixture is heated in a heat exchanger and sent to a hydrolysis reactor column, where ethylene is reacted with water to yield ethylene glycol. Other glycols such as oligomers diethylene glycol, triethylene glycol, and tetraethylene glycol are also formed as by-products. Steam is continuously inserted into the reactor and water is discharged from the bottom of the reactor. Water is recycled through the heat exchanger. After completion of the reaction, the reaction mass is sent

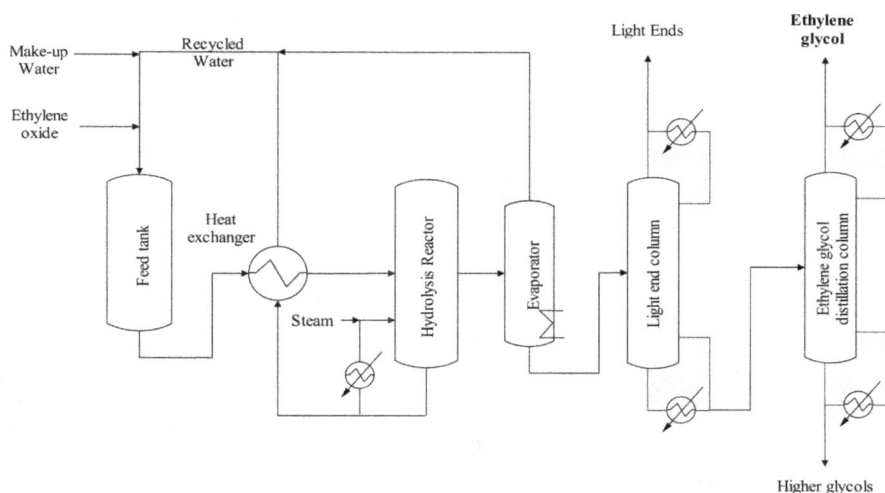

Figure 4.11 Flow sheet diagram of production of Ethylene glycol from Ethylene oxide

to the evaporator. Here, water is separated by means of steam and recycled. The reaction mass is transferred into a light end column, where light ends are separated using steam. Finally, ethylene glycol is separated from heavy glycol and other impurities in the ethylene glycol distillation column.

4.10.3 Uses

A majority of ethylene glycol available is employed as an antifreeze in heating and cooling systems like automobile radiators and boilers. It has also been used as a de-icer for aircrafts and airport runways. It is also used in the preparation of synthetic fibers (Terylene, Dacron), safety explosives, plasticizers and artificial waxes. Also, it is utilized as a solvent in different industries such as pharmaceuticals, paints, drugs and plastics (Dye, 2001).

4.11 Ethanolamines

There are three types of ethanolamines: (1) Monoethanolamine (MEA), (2) Diethanolamine (DEA) and (3) Triethanolamine (TEA)

4.11.1 (Introduction) Properties

The properties of one ethanolamine, i.e., monoethanolamine are as per Table 4.11.

4.11.2 Manufacturing process

It is prepared from ethylene oxide.

4.11.2.1 Raw materials

Ethylene oxide and Aqueous ammonia

Table 4.11 Physical & Chemical Properties of Monoethanolamine

Sr. No.	Properties	Particular
1	IUPAC name	2-Aminoethanol
2	Chemical and other names	Ethanolamine, Monoethanolamine, Colamine, Glycinol, 2-Hydroxyethylamine, 2-Amino-1-ethanol
3	Molecular formula	C_2H_7NO
4	Molecular weight	61.07 gm/mole
5	Physical description	Colorless liquid with unpleasant ammonia-like odor
6	Solubility	Miscible with ethanol, glycerol; soluble in chloroform; slightly soluble in ether, ligroin
7	Melting point	10.5°C
8	Boiling point	171°C
9	Flash point	86°C
10	Density	1.075 gm/cm³ at 25°C

4.11.2.2 Chemical reactions

It is produced by the reaction of ethylene oxide with aqueous ammonia at 150°C temperature and 160 atmospheres pressure. The reaction also produces diethanolamine (DEA) and triethanolamine (TEA).

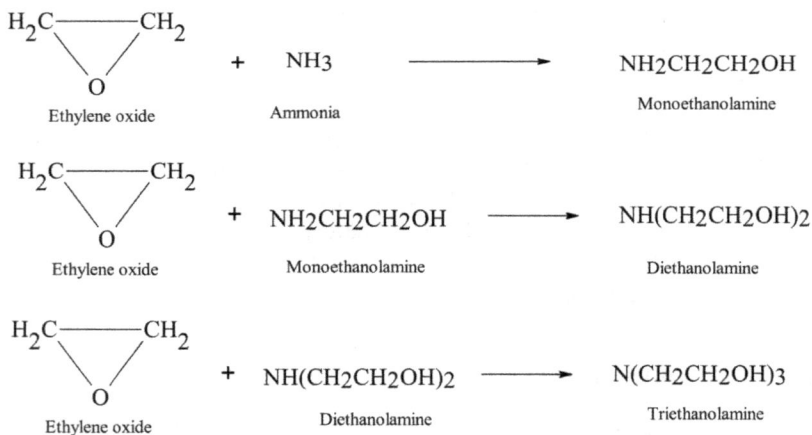

Figure 4.12 Chemical reactions for manufacturing Ethanolamines

4.11.2.3 Quantitative requirements

The ratio of the products can be organized by altering the stoichiometry of the reactants.

4.11.2.4 Flow chart

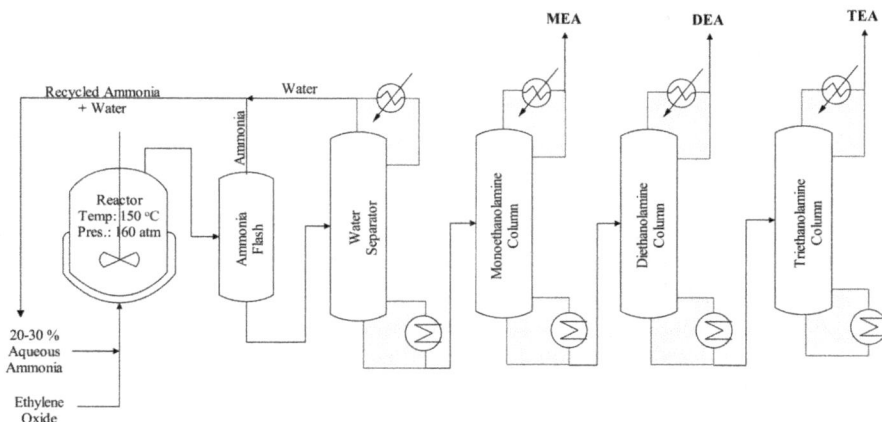

Figure 4.13 Flow sheet diagram of production of Ethanolamines from Ethylene oxide

4.11.2.5 Process description

As per the flow-sheet, ethylene oxide and 20–30% aqueous ammonia with an ammonia recycle stream from the process is mixed and inserted into the reactor facilitated with a stirrer and jacket. The reaction is exothermic, so, cool water is passed into the jacket of the reactor. So, a good liquid phase reaction is maintained in the reactor. The gaseous reaction mixture is fed into a flash column, where ammonia and water are separated from the top and recycled. Liquid having water and ethanolamines is collected from the bottom of the column and fed to the water separation column. The ammonia + water stream is recycled to mix with the fresh ammonia. Reaction mixture is sent for water separation. Again water and traces of ammonia are separated out and recycled. Thereafter, the reaction mixture contains MEA, DEA and TEA, which are separated from each other in three fractional columns.

4.11.2.6 Major engineering problems

(1) **Reactor design:** For the reactor design, the initial step is to determine the rate constant data by changing temperature, pressure and ratio of ammonia: ethylene upto 35–275°C, 1–100 atmospheres and 0.5:3.0 respectively.

(2) **Process alternatives:** When ethylene oxide is added, then a major part of the products are di- or triethanolamine. In such a case, the lower ammonia is recycled to separate the reactants. Further, when ammonia is in very low concentration, than it gives amino-ether in a one-pass reactor. This can be prevented by using an alternative technique, i.e., specific quantity of carbon dioxide is added in the reactor.

(3) **Recovery and purification system:** Colour degradation takes place due to the high boiling points of di- and tri-compounds, which can be minimized by a vacuum fractional system.

(4) **Kinetics of complex series reactions:** Reaction rate and yield are important factors in a series reaction. Reactants' concentrations and specific rate constants are mainly affected by temperature. Higher yield can be achieved

by a higher ammonia to ethylene oxide ratio with higher ammonia recycling. A better settling product ratio is maintained due to proper residence time in the reactor.

4.11.3 Uses

1. Monoethanolamine is used as an ingredient for wood preservative formulations to increase the life time of wood. It is used as feedstock in the production of different chemicals such as pharmaceuticals, detergents, polishes, emulsifiers, chemical intermediates and corrosion inhibitors.
2. Diethanolamine and monoethanolamine are also used in Cocamide MEA, which is utilized in personal care products, such as bath foams, shampoos and soaps.
3. Triethanolamine is used in cement industries to increase cement quality.
4. All the Ethanolamines are used as desulfurized agents from refinery off-gases (Alqaragully, et al., 2015).

4.12 Ethylene Dichloride (EDC)

It is a chlorinated hydrocarbon also known as ethylene dichloride (EDC).

4.12.1 Properties

Its main properties are as follows.

Table 4.12 Physical & Chemical Properties of 1,2-Dichloroethane

Sr. No.	Properties	Particular
1	IUPAC name	1,2-Dichloroethane
2	Chemical and other names	Ethylene dichloride, Ethane, 1,2-dichloro-, Glycol dichloride, Dutch oil
3	Molecular formula	$C_2H_4Cl_2$
4	Molecular weight	98.95 gm/mole
5	Physical description	Colorless liquid with a pleasant, chloroform-like odor.
6	Solubility	Miscible with water, alcohol, chloroform, ether
7	Melting point	−35°C
8	Boiling point	84°C
9	Flash point	13°C
10	Density	1.253 gm/cm^3 at 27°C

4.12.2 Manufacturing process

Ethylene undergoes chlorination using chlorine gas in liquid phase in the presence of a proper catalyst to form Ethylene dichloride. It is also prepared by the reaction of ethylene, anhydrous hydrogen chloride and oxygen (or air) at a temperature of 200–315°C and a pressure of 3–5 atmospheres using cupric chloride as catalyst in a fluid bed reactor.

4.12.2.1 Raw materials

Ethylene, Chlorine and Catalyst

4.12.2.2 Chemical reactions

It is prepared by the reaction of chlorine gas with ethylene in liquid gas in the presence of a catalyst.

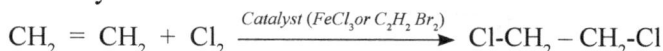

$$CH_2 = CH_2 + Cl_2 \xrightarrow{\text{Catalyst } (FeCl_3 or\ C_2H_2\ Br_2)} Cl\text{-}CH_2 - CH_2\text{-}Cl$$

4.12.2.3 Flow chart

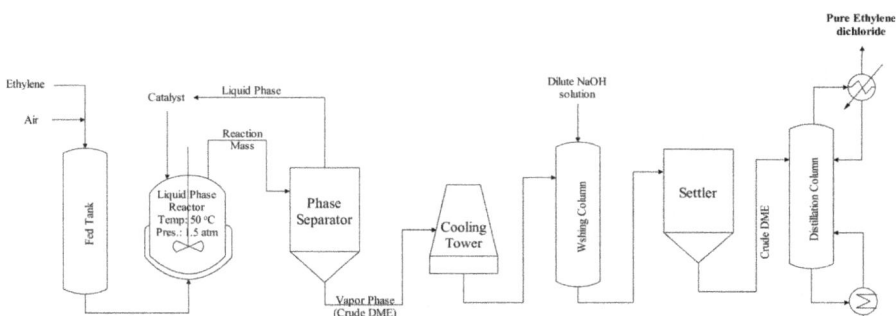

Figure 4.14 Flow sheet diagram of production of Ethylene dichloride from Ethylene

4.12.2.4 Process description

Ethylene and chlorine gas are mixed and fed to the liquid phase reactor. Also, a catalyst is added. Ethylene dichloride is formed. Now, the feed mixture is continuously bubbled through the product, ethylene dichloride. A temperature of 50°C and a pressure of 1.5–2 atmospheres is maintained during the reaction. The reaction is exothermic, thus the temperature is maintained using a cooling jacket or external heat exchanger. Also, the liquid phase is essential to trap different liberated gases and to eliminate higher turbulence during the reaction. These conditions help to produce ethylene dichloride. After completion of the reaction, the liquid phase containing ethylene dichloride with a minor amount of vapor phase is cooled down. Here, the vapor phase is inserted into the refrigeration unit to extract the maximum possible ethylene dichloride. This ethylene dichloride containing un-reacted hydrochloric acid is washed with caustic soda solution and undergoes settling to remove HCl. Raw ethylene dichloride vapor is sent to the distillation column to isolate ethylene dichloride and other heavy end products. Finally, it is purified using NaOH solution.

4.12.2.5 Major engineering problems

The complete gaseous reaction can be achieved by increasing the process temperature upto 85°C. The reaction can be carried out with aluminum chloride or ferric chloride in a packed tubular reactor, but heat control is very difficult is such a modification.

4.12.3 Uses

Essentially, 1,2-Dichloroethane is used for manufacturing vinyl chloride. It is also employed as an interim material in organic reactions. It forms azeotropes with water and other chlorocarbons (Schirmeister et al., 2009).

4.13 Dimethylether (DME)

Dimethyl ether is the simplest ether.

4.13.1 Properties

Its main properties are as follows.

Table 4.13 Physical & Chemical Properties of Dimethyl ether

Sr. No.	Properties	Particular
1	IUPAC name	Methoxymethane
2	Chemical and other names	Dimethyl ether, Methyl ether, Dimethyl oxide, Oxybismethane
3	Molecular formula	C_2H_6O
4	Molecular weight	46.07 gm/mole
5	Physical description	Colorless Gas with Characteristic Odor
6	Solubility	Soluble in water, ether, acetone, chloroform; ethyl alcohol
7	Melting point	$-141.5°C$
8	Boiling point	$-24.8°C$
9	Flash point	$-41.4°C$
10	Density	1.915 gm/cm^3 at 25°C

4.13.2 Manufacturing process

DME is manufactured via the catalytic dehydration of methanol over an acid zeolite catalyst. It is also prepared by the fermentation of lignocellulosic materials.

4.13.2.1 Raw materials

Methanol and Catalyst

4.13.2.2 Chemical reactions

It is prepared by the catalytic dehydration of methanol using acid zeolite catalyst at a temperature of 300–370°C.

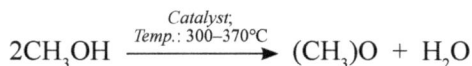

$$2CH_3OH \xrightarrow[\text{Temp.: 300–370°C}]{\text{Catalyst;}} (CH_3)O + H_2O$$

4.13.2.3 Flow chart

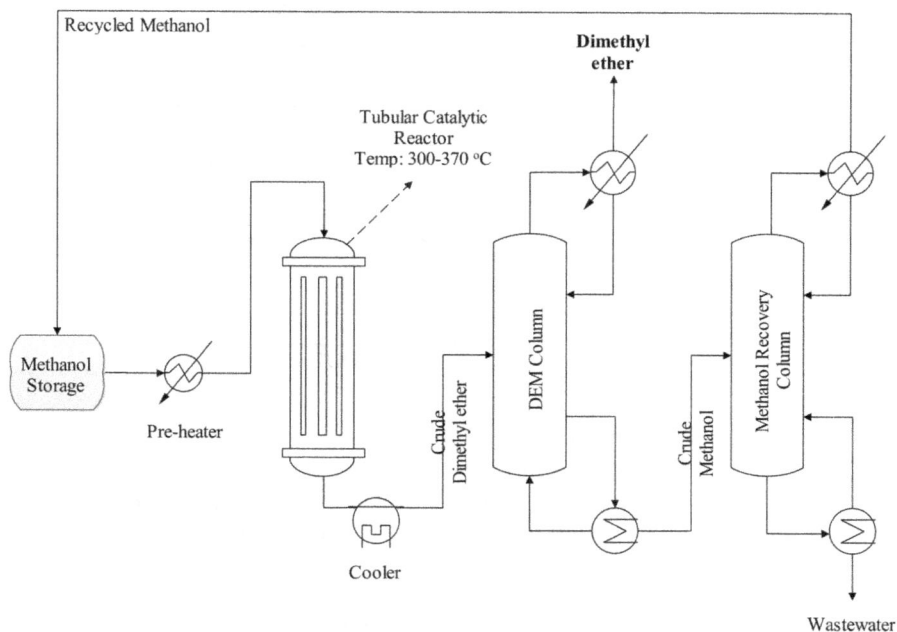

Figure 4.15 Flow sheet diagram of production of Dimethyl ether (DME) from Methanol

4.13.2.4 Process description

Methanol is taken from its storage and is pre-heated to form vapor. This vapor is charged from the top of a tubular catalytic reactor already impregnated with catalyst. This tubular reactor is heated to 400–600°C using a natural gas furnace. After completion of the reaction, crude dimethyl ether is cooled down up to room temperature using a cooler. It is then transferred into the DME column to separate the DME as the overhead product. Remaining water and un-reacted methanol are collected from the bottom. The un-reacted methanol is purified in thee methanol recovery column and recycled.

4.13.3 Uses

DME is widely employed in different chemical industries. Also, it is used as an additive with diesel in compression ignition diesel engines (Makos et al., 2019).

4.14 Methyl Formate

It is the simplest example of an ester. It is the methyl ester of formic acid.

4.14.1 Properties

Its main properties are as follows.

Table 4.14 Physical & Chemical Properties of Methyl formate

Sr. No.	Properties	Particular
1	IUPAC name	Methyl formate
2	Chemical and other names	Methyl methanoate, Methyl ester of formic acid
3	Molecular formula	$C_2H_4O_2$
4	Molecular weight	60.05 gm/mole
5	Physical description	Colorless liquid with a pleasant odor
6	Solubility	Soluble with water, alcohol, chloroform, ether
7	Melting point	−99.5°C
8	Boiling point	31.5°C
9	Flash point	−19°C
10	Density	0.987 gm/cm³ at 27°C

4.14.2 *Manufacturing process*

It is prepared either by condensation of methanol and formic acid or carbonylation of methanol. But industrially, it is manufactured by the carbonylation of methanol with carbon monoxide. This reaction is commercially adopted by BASF.

4.14.2.1 *Raw materials*

Methanol, Carbon monoxide and Sodium methoxide

4.14.2.2 *Chemical reactions*

Methanol undergoes carbonylation with carbon monoxide in the presence of sodium methoxide at a temperature of 80–100°C and a prssur of 3–5 atmospheres.

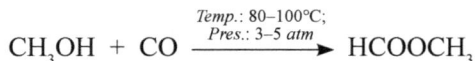

$$CH_3OH + CO \xrightarrow[\text{Pres.: 3-5 atm}]{\text{Temp.: 80-100°C;}} HCOOCH_3$$

Very dry carbon monoxide is essential to avoid side reactions. Also, this reaction is exothermic.

4.14.2.3 *Flow chart*

4.14.2.4 *Process description*

As per Fig. 4.16, carbon monoxide and methanol are individually compressed in the compressor and heated in the pre-heater for conversion to vapor form. The vapor mixture is passed through a vapor phase reactor, which is already impregnated with catalyst. Here, the reaction is exothermic, so, no extra heat is required to maintain the temperature. After completion of the reaction, vapor mass is collected from the top of the reactor, and fed into a light end column to remove carbon monoxide. Then, the reaction mass is undergoes distillation in a column, where raw material methanol is collected fromth bottom, while the final product methyl formate is collected as the over-head product. Both raw materials, i.e., carbon monoxide and methanol are re-purified and recycled.

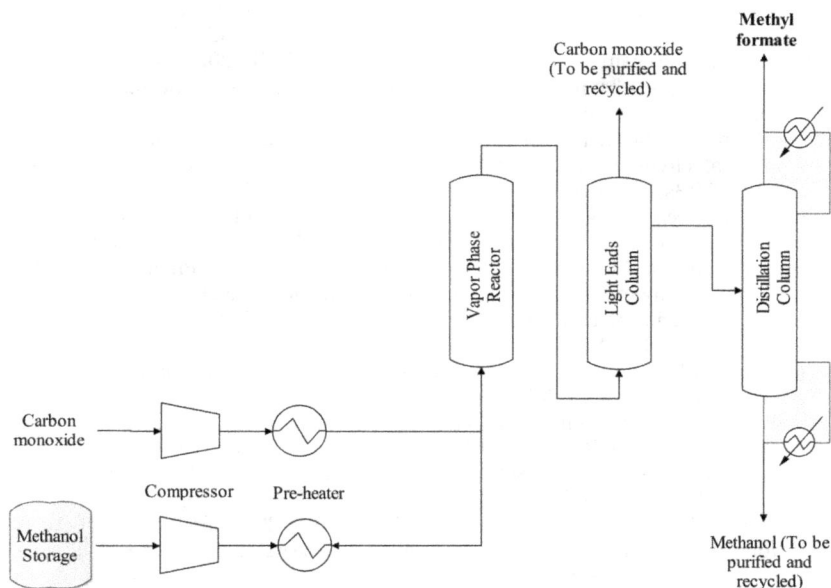

Figure 4.16 Flow sheet diagram of production of Methyl formate by carbonylation from Methanol

4.14.3 Uses

The main use of methyl formate is in manufacturing formic acid and its derivatives. It is used as a replacement refrigerant for CFCs, HCFCs, and HFCs. It is also used as a blowing agent for some polyurethane foam. It is also utilized as an insecticide (Starchevski et al., 2010; Pattanaik et al., 2013).

References

Allah, E.A.A., Kasif, A.E.M.O., Mohamed, Y.A.A. and Mahmoud, A.A.E.H. 2022. Simulation of ethylene oxide production from ethylene cholorhydrin. Proceedings of the Voronezh State University of Engineering Technologies.

Al-Megren, H. and Xiao, T. 2016. Petrochemical Catalyst Materials, Processes, and Emerging Technologies, Engineering Science Reference. DOI: 10.4018/978-1-4666-9975-5

Alqaragully, M.B., AL-Gubury, H.Y., Aljeboree, A.M., Karam, F.F. and Alkaim, A.F et al. 2015. Monoethanolamine: Production Plant. Research Journal of Pharmaceutical, Biological and Chemical Sciences 6(5): 1287–1296.

Cardona, C.A. and Sanchez, O.J. 2007. Fuel ethanol production: Process design trends and integration opportunities. Bioresource Technology 98: 2415–2457. https://doi.org/10.1016/j.biortech.2007.01.002

Cohan, G.F. 1975. Industrial Preparation of PoIy(vinyl Chloride). Environmental Health Perspectives 11: 53–57. DOI: 10.1289/ehp.751153

Dimian, A.C. and Bildea, C.S. 2008. Vinyl Acetate Monomer Process In book: Chemical Process Design: Computer-Aided Case Studies 287–312. DOI: 10.1002/9783527621583

Dye, R.F. 2001. Ethylene Glycols Technology, Korean Journal of Chemical Engineering 18(5): 571–579. https://doi.org/10.1007/BF02706370

George, A. 2014. Control of Ethylene and Propylene Polymerisation Processes. Measurement and Control 47(3): 84–90. https://doi.org/10.1177/0020294014526710

Li, H., Saravanamurugan, S., Pandey, A. and Elumalai, S. 2022. Biomass, Biofuels, Biochemicals: Biochemicals and Materials Production from Sustainable Biomass Resources. Elsevier Science. https://doi.org/10.1016/C2020-0-02067-0

Makos, P., Słupek, E., Sobczak, J., Zabrocki, D., Hupka, J. et al. 2019. Dimethyl ether (DME) as potential environmental friendly fuel, E3S Web of Conferences 116. https://doi.org/10.1051/e3sconf/201911600048

Manyar, H. and Deshmukh, G. 2022. Production pathways of acetic acid and versatile applications in food industry. Intech Open. DOI: 10.5772/intechopen.92289

Mubashir, M., Ahsan, M., Ahmad, I. and Khan, M.N.A. 2022. Process modeling and simulation of ethylene oxide production by implementing pinch and cost analysis, Ain Shams Engineering Journal 13(3): 101585.https://doi.org/10.1016/j.asej.2021.09.012

Pattanaik, B.N. 2013. The Advances in Processes and Catalysts for the Production of Methyl Formate by Methanol Carbonylation—A Review, International Journal of Chemical & Petrochemical Technology 3(2): 55–70.

Peppel, W.J. 1958. Preparation and Properties of the Alkylene Carbonates, Industrial and Engineering Chemical Research 50(5): 767–770. https://doi.org/10.1021/ie50581a030

Rao, K.N., Ratnam, M.V., P. Prasad, R. and Sujatha, V. 2014. Design and Control of Acetaldehyde Production Process, Trends in Chemical Engineering 1(1): 1–11.

Sathya S.S.S.V, Ranganath, R., Shaju, S. and Nayar, A.J. 2019. Production of Acetylene gas for use as an Alternative fuel, International Research Journal of Engineering and Technology 6(4): 3902–3907.

Schirmeister, R., Kahsnitz, J. and Trager, M. 2009. Influence of EDC severity on the Marginal Costs of Binyl Chloride Production, Industrial & Engineering Chemistry Research 48(6): 2801–2809. https://doi.org/10.1021/ie8006903

Scott, K. 1998. Handbook of Industrial Membranes, Elsevier Science.

Starchevski, M.K., Pazderski, Y.A. and Moiseev, I.I. 2010. ChemInform Abstract: Methyl Formate: Production Methods, ChemInform 23(33).

Vaca, C.E., Fang, J.-L. and Schweda, E.K.H. 1995. Studies of the reaction of acetaldehyde with deoxynucleosides. Chemical-biological Interaction 98(1): 51–67. DOI: 10.1016/0009-2797(95)03632-v

Verschueren, K. 1996. Handbook of Environmental Data on Organic Chemicals, 3rd Edition, New York.

Voronin, V.V., Ledovskaya, M.S. Bogachenkov, A.S., Rodygin K.S. and Ananikov, V.P. et al. 2018. Acetylene in Organic Synthesis: Recent Progress and New Uses. Molecules 23(10): 2442. doi: 10.3390/molecules23102442

CHAPTER 5

C – 3 and 4 Organic Chemicals

5.1 Introduction

Two chemicals, i.e., Propylene and butadiene are considered as basic C3 and C4 chemicals respectively. Both chemicals are obtained from petroleum cracking. C3–C4 organic chemicals include the alkanes, alkenes and alkynes having carboxylic acid, alcohol, amine, ether, ester, chloride, oxide, aldehyde and nitro groups with their isomers. So, they contain a variety of chemicals. They are mostly in liquid or gaseous state. So, their manufacturing process includes a liquid phase reactor with stirring and heating facilities or a vapor phase tubular reactor. They have high melting and boiling points. These chemicals are undergo various unit processes such as oxidation, reduction, sulfonation, nitration, halogenation, alkylation, hydrationand dehydration for utilization in different industries (Carey and Sundberg, 2000).

5.2 Propylene

Propylene is a colorless gas, easily ignitable with a petroleum like odor. It is compressed to liquid state and then, shipped

5.2.1 Properties

Its main properties are as per Table 5.1.

5.2.2 Manufacturing process

It is prepared from various sources like (i) the catalytic cracking of propane, (ii) the MTO (Methanol To Olefins) process and (iii) The reaction between ethene and butenes.

Table 5.1 Physical & Chemical Properties of Benzene

Sr. No.	Properties	Particular
1	IUPAC name	Prop-1-ene
2	Chemical and other names	Propylene, Propene, Methylethylene,
3	Molecular formula	C_3H_6
4	Molecular weight	42.08 gm/mole
5	Physical description	Colorless gas with a faint petroleum like odor
6	Solubility	Soluble in water, alcohol, ether, acetic acid
7	Melting point	−185°C
8	Boiling point	−48°C
9	Flash point	−108°C
10	Density	0.514 gm/cm^3 at 27°C

5.2.3 Process description

(i) **Catalytic cracking of propane:** Propylene is prepared by the catalytical cracking of propane.

$$C_3H_8 \quad \rightarrow \quad C_3H_6 + H_2$$

This process is widely used in the United State, as it has a huge natural occurrence of of propane.

(ii) **The MTO (Methanol To Olefins) Process:** The MTO process is an example of the process of converting biomass into methanol via methanol synthesis gas. In this process, methanol is often converted into ethene and propene via dimethyl ether. Methanol vapor is emitted at 330°C and an equilibrium mixture of methanol, dimethyl ether and steam is produced, containing about 25% methanol

$$2\ CH_3OH_{(g)} \quad \rightarrow \quad H_3C - O - CH_{3(g)} \quad + \quad H_2O_{(g)}$$

This is an overall process to make Methanol from Methane. The feed gas methane is passed through a bed of zeolite in a form that encourages high selectivity towards alkenes with 2 to 8 carbon atoms. Using a zeolite treated with acid, the alkene most produced is propene. This mixture of gases is then passed over a bed of zeolite again in another form which only promotes the production of propene. This process can be repeated up to five times. Then following this step, the desired alkenes are purified by cooling to liquid and subjected to fractional distillation which was earlier used in producing gasoline (MTG process).

(iii) **The reaction between ethene and butenes:** Propylene is prepared from ethene and butane using the following chemical reaction:

$$CH_2{=}CH_2 \quad + CH_3CH{=}CHCH_3 \quad \rightarrow \quad 2\ CH_3CH{=}CH_2$$

Using a metal catalyst, bio-alkenes like ethene and butene are prepared by the dehydration of the respective bio-alcohol like bio-ethanol.

Also, ethene undergoes dimerization using hot zeolite along-with a transition metal complex like rhodium, titanium as a catalyst to give but-1-ene as follows.

$$2\ CH_2=CH_2 \quad \rightarrow \quad CH_3CH_2CH=CH_2$$

Then, propene is prepared by passing Schrock catalysts [molybdenum(IV) and tungsten(IV)] and Grubbs' catalysts [organo-ruthenium (II)] with the mixture of ethene and butene into a fixed-bed.

$$CH_2=CH_2 \quad + \quad CH_3CH_2CH=CH_2 \quad \rightarrow \quad 2\ CH_3CH=CH_2$$

The catalyst is activated using heated air in the reactor, as the catalyst gets deactivated after some time.

As per 2010 data 56% and 37% of the total ethene production was manufactured by steam cracking and catalytic cracking of fuel oil respectively. The remaining propene was produced from coal and the cracking of gas oil under vacuum. Recently new methods are being explored for manufacturing 25% of the total propene production.

5.2.4 Uses

Propylene is an important higher-level product, used in the petrochemical industry. After ethylene, it is the second most essential starting material in the manufacture of petrochemical products. It is a raw material for a wide variety of unit processes: polymerization, oxidation, halogenation and hydrohalogenation, alkylation, hydration, oligomerization, hydroformylation and many others. About two-thirds of propylene is used in the production of plastics such as polypropylene. Polypropylene is used to make films and other types of plastic products such as fibers, containers, packaging and caps and closures. Propylene is additionally used to produce various chemicals including propylene oxide (propene oxide), acrylonitrile, cumene and butyraldehyde through the so-called Cumene Process; camphene through catalytic hydrogenation; acrolein-through oxidation with oxygen; propanoic acid-the starting ingredient for fumigation agents; phenol and acetone are manufactured via Propene and Benzene by way of the so-called (Lavrenov et al., 2005; Farshi, 2008).

5.3 Isopropanol

Isopropyl alcohol or 2-propanol or IPA is known as one of the feasible organic alcohols.

5.3.1 Properties

Its main properties are as per Table 5.2.

Table 5.2 Physical & Chemical Properties of Isopropanol

Sr. No.	Properties	Particular
1	IUPAC name	Propan-2-ol
2	Chemical and other names	Isopropanol, Dimethylcarbinol, Rubbing alcohol, Alkolave
3	Molecular formula	C_3H_8O
4	Molecular weight	60.09 gm/mole
5	Physical description	Colorless liquid with the odor of rubbing alcohol
6	Solubility	Miscible with water, alcohol, ether, chloroform, benzene
7	Melting point	–90°C
8	Boiling point	83°C
9	Flash point	12°C
10	Density	0.785 gm/cm³ at 27°C

5.3.2 Manufacturing process

Isopropyl alcohol was first produced in 1920 while studying petroleum by-products. Currently, it is prepared by following three methods.

(1) Indirect hydration: In this process, sulfuric acid is mixed with propene to form an organic and aqueous phase. The mixture of acids and hydrocarbons undergoes hydrolysis by steam resulting in the formation of sulfate esters which are then oxidized to give isopropyl alcohol, a preferred product.

(2) Direct hydration: Direct hydration is the chemical reaction of propene and water at high pressures into a liquid. Some propylene is required for this because it may be expensive or difficult to isolate. The reaction involves either the gas phase or liquid phase reactions, depending on how the product is treated.

(3) Hydrogenation of acetone: The acetone process is important, as it gives an excess acetone production yield. The crude acetone from the acetone process is hydrogenated in the liquid phase over Raney nickel or a mixture of copper and chromium oxide to give isopropyl alcohol.

Out of these three processes, the indirect process is widely used, because this process gives a higher amount of IPA than 1-propanol as it follows Markovnikov's rule.

5.3.2.1 Raw materials

Propene, Sulfuric acid and Water

5.3.2.2 Chemical reactions

$$CH_3CH{=}CH_2 + H_2SO_4 \longrightarrow (CH_3)_2CH(OSO_3H)$$

This reaction is highly exothermic.

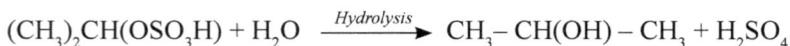

$$(CH_3)_2CH(OSO_3H) + H_2O \xrightarrow{Hydrolysis} CH_3{-}CH(OH){-}CH_3 + H_2SO_4$$

5.3.2.3 Quantity requirements

1000 kgs of isopropanol requires 700 kgs of propene, 110 kgs of sulfuric acid and 300 kgs of water.

5.3.2.4 Flow chart

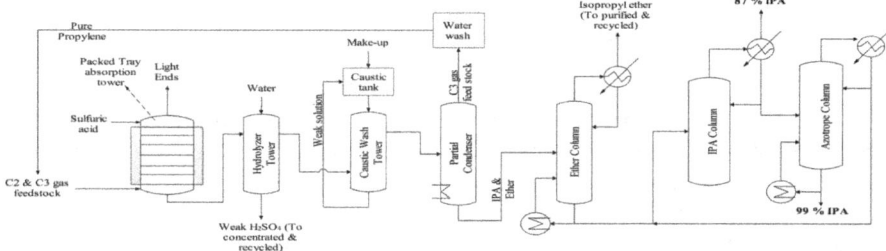

Figure 5.1 Flow sheet diagram of production of Isopropyl alcohol using indirect hydration method

5.3.2.5 Process description

In this process, pure propylene or a mixture of propylene and other C2, C3 components is used as a raw material. This raw material is compressed at about 20–25 atmospheres and fed into the bottom of a packed or sieved tray absorption jacketed tower. About 70% sulfuric acid is also inserted into the top of the tower, i.e., counter current mode, where reactive absorption takes place. The employed sulfonation reaction is extremely exothermic. Thus, continued circulation of refrigerated brine in the jacket is necessary to maintain the temperature of the absorber. After reaction, the gas stream of unreacted light ends such as saturated components is collected from the top of the tower. The rich sulfonated product stream is then sent to a hydrolyzer cum stripper where water is fed from the top. Here, the sulfonated product is hydrolyzed to yield isopropanol. Isopropanol is in vaporized form due to the existing stripper temperatures. Then, elimination of acidic impurities from the isopropanol rich vapor is performed in the caustic wash unit. Unreacted propylene vapor stream from the alcohol ether is separated as the liquid stream by the partial condenser. The vapor stream containing pure propylene is inserted as fresh feed, cooled down and relocated to another unit. Now, the liquid stream containing the alcohol ether mixture is fed into the ether column, where isopropyl ether is separated as the over-head product and recycled. Isopropyl alcohol and water are collected from the bottom and fed into the isopropyl alcohol column. Here, the bottom product, water + heavy ends and the top product, 87% isopropanol-water azeotrope mixture are separated. This azeotrope mixture is fed into the azeotropic distillation column to separate the 99% pure isopropanol from the bottom. Isopropyl ether and water are collected as the over-head product and sent to the reflux column. Here, isopropyl ether is separated from water and further recycled into the IPA column.

5.3.2.6 Major engineering problems

(1) **Sulfonation reaction:** For a C2 chemical as the feed gas, Ethylene is removed before feeding, because it does not get absorbed at lower temperatures and 70% acid concentration. Also, when C4 and other heavy olefins are used, they get absorbed more readily and come through the system.

(2) **Direct hydration reaction:** The process of direct hydration is not commercially feasible, because the reaction requires a high temperature

condition, acid catalyst and vapor phase, which may cause polymerization of polypropylene.

5.3.3 Uses

Isopropyl alcohol is used as a solvent in toiletries, coatings, window cleaning and rubbing alcohol. It is also used for manufacturing acetone, which is an important solvent in various industries. It is also used for the production of electronic parts and metals, and as a solvent and cleaning agent for medical and veterinary products (Xu et al., 2005; Burlage and Hawkins, 1946).

5.4 Acrylonitrile

It is a colorless volatile liquid.

5.4.1 Properties

Its main properties are as follows.

Table 5.3 Physical & Chemical Properties of Acrylonitrile

Sr. No.	Properties	Particular
1	IUPAC name	Prop-2-enenitrile
2	Chemical and other names	Acrylonitrile, Vinyl Cyanide, Propenenitrile, Cyanoethylene, Carbacryl, Acritet, Acrylon
3	Molecular formula	C_3H_3N
4	Molecular weight	53.06 gm/mole
5	Physical description	Colorless to pale-yellow liquid with pungent odor
6	Solubility	Miscible with ethanol, carbon tetrachloride, ethyl acetate, ethylene, toluene, petroleum ether, and xylene
7	Melting point	$-83.5°C$
8	Boiling point	$77°C$
9	Flash point	$-5°C$
10	Density	0.8004 gm/cm^3 at $27°C$

5.4.2 Manufacturing process

The French chemist Charles Moureu synthesized Acrylonitrile initially in 1893. There are several processes for manufacturing acrylonitrile.

(1) From ethylene cyanohydrin: Acrylonitrile is prepared by heating ethylene cyanohydrins in the presence of a catalyst such as an acid sulfate, alumina, active carbon or tin.

$$OH\text{-}CH_2\text{-}CH_2\text{-}C\equiv N \xrightarrow{450–500°C; \ Catalyst} CH_2\text{=}CH\text{-}C\equiv N + H_2O$$

(2) From acetylene: Acetylene is heated at 80–90°C using a cuprous compound as the catalyst to form acrylonitrile.

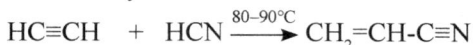

$$HC\equiv CH \ + \ HCN \xrightarrow{80–90°C} CH_2\text{=}CH\text{-}C\equiv N$$

(3) From propionaldehyde: Propanol undergoes ammoniation with ammonia to give acrylonitrile.

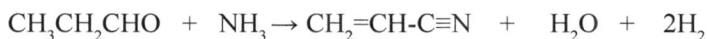

$$CH_3CH_2CHO + NH_3 \rightarrow CH_2=CH\text{-}C\equiv N + H_2O + 2H_2$$

(4) From propylene: Propylene undergone ammoxidation with ammonia and oxygen to give acrylonitrile. This process is called the SOHIO process.

$$2CH_3\text{-}CH=CH_2 + 2NH_3 + 3O_2 \longrightarrow 2CH_2=CH\text{-}C\equiv N + 6H_2O$$

Off these processes, the SOHIO process is widely used because of a better quality product, and an economical and energy efficient process.

5.4.2.1 Raw materials

Propylene, Ammonia, Oxygen and Catalyst

5.4.2.2 Chemical reactions

In the SOHIO process, catalytic ammoxidation of propylene is performed at a high temperature of 400–510°C and pressure of 1–3 atmospheres in the presence of a catalyst to form acrylonitrile and hydrogen cyanide. Here, hydrogen cyanide is recovered and sold.

$$2CH_3\text{-}CH=CH_2 + 2NH_3 + 3O_2 \longrightarrow 2CH_2=CH\text{-}C\equiv N + 6H_2O$$

5.4.2.3 Quantitative requirements

1000 kgs of acrylonitrile requires 1100 kgs of propylene, 500 kgs of sulfuric acid and 30 kgs of water.

5.4.2.4 Flow chart

5.4.2.5 Process description

The SOHIO process is applicable for producing acrylonitrile from propylene, ammonia, and air. In this process, propylene, ammonia, and air are fed into the bottom of a fluidized bed catalytic reactor containing the catalyst at a temperature of 400–510°C and pressure of 1–3 atmospheres. The Cyclone separator is also kept in the fluidized bed reactor in which the catalyst and product gases are separated after fluidization. The contact time for fluidization is in the order of seconds. After ammoxidation, the gaseous reaction mixture is collected from the top of the bed and scrubbed using water in the scrubber to separate acrylonitrile that has been produced in other reactions (ammonia oximation $-NH_3$ conversion) into its lighter ends [carbon monoxide (CO), carbon dioxide (CO_2) and water]. A solution consisting of acrylonitrile, acetonitrile, hydrocyanic acid, ammonium sulfate (from excess ammonia) and heavy ends is inserted intothe acrylonitrile recovery column. A recovery column separates crude acrylonitrile and acetonitrile by distillation. Crude acetonitrile is further purified in the azeotrope column to remove bulk water and it yields crude acetonitrile. Now, crude acrylonitrile is further fed into the light end column to remove hydrogen cyanide and traces of light ends. Thereafter, it is purified in the acrylonitrile column to get pure acrylonitrile. Here, heavy impurities are recycled as feed stock with propylene.

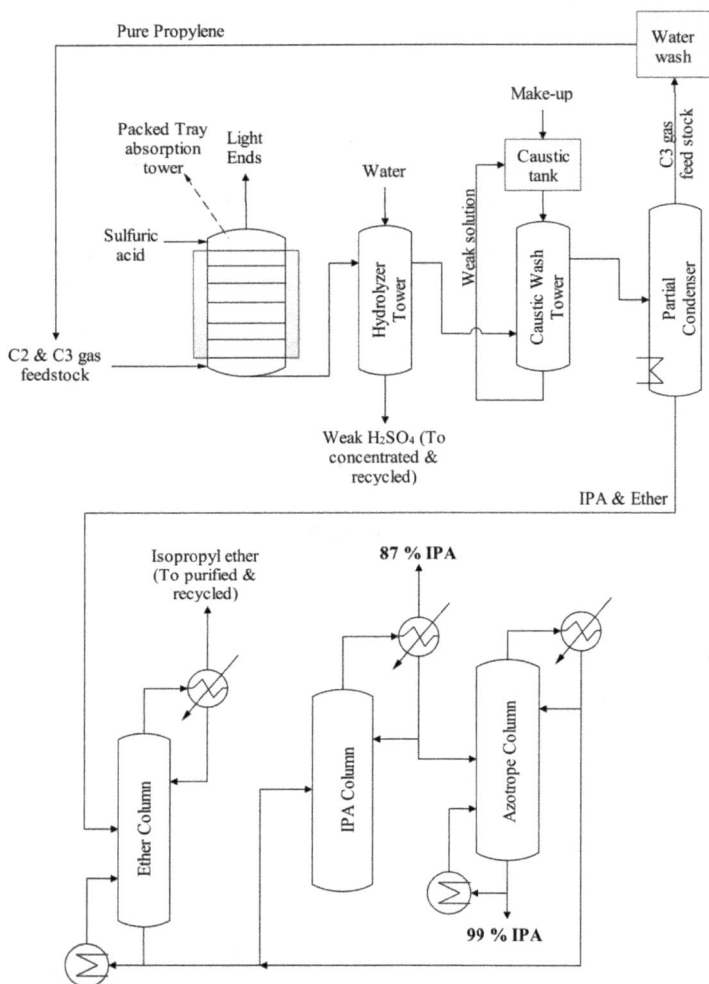

Figure 5.2 Flow sheet diagram of production of Acrylonitrile using SOHIO process

5.4.2.6 Major engineering problems

(1) Catalyst Selection: In the SOHIO process, molybdenum-bismuth (Mo-Bi) catalyst is selected, since it provides a larger adsorption surface to furnish oxygen by surface reaction and transfer. Further, a micro-spherical catalyst size of 0.01–0.03 mm is selected. If desulfurized feedstock is used then, no regeneration is required. And feedstock contains sulfur impurities, so, it gets converted ito sulfite. Therefore, the catalyst must be purified and reutilized.

(2) Feedstock limitation: In this process, propylene concentration must be maintained, so that it is not converted into propane in the presence of a large amount of air. Further, recovery of converted propane from the water scrubber is costly. The feedstock has a propylene concentration of only 30%.

(3) **Purification of product:** Product, ACN, contains impurities of saturated carbonylic cyanohydrins in the final column, which gets dissociated to form volatile contaminants. Oxalic acid is selected as the purification solvent in the final column, because it makes complex and lead towards the heavy ends without dissociation with these impurities. Further, oxalic acid is effective, economical, less corrosive and decreases reboiler fouling.

5.4.3 Uses

a) Mainly acrylonitrile monomers are used for making several rubbers like acrylonitrile butadiene (NBR), acrylonitrile styrene acrylate (ASA), acrylonitrile butadiene styrene (ABS) and styrene-acrylonitrile (SAN).
b) Another use of acrylonitrile is to produce acrylamide, adiponitrile, acrylic acid and several polyamides. It is also useful in Diels-Alder reactions.
c) Small amounts of acrylonitrile are also used as a fumigant (Hansore, 2014; Firouzi et al., 2019).

5.5 Acetone

It is the smallest and simplest ketone.

5.5.1 Properties

Its main properties are as follows.

Table 5.4 Physical & Chemical Properties of Acetone

Sr. No.	Properties	Particular
1	IUPAC name	Propan-2-one
2	Chemical and other names	Acetone, 2-Propanone, Dimethyl ketone, Methyl ketone, Dimethylformaldehyde
3	Molecular formula	C_3H_6O
4	Molecular weight	58.08 gm/mole
5	Physical description	Colorless liquid with a sweetish odor
6	Solubility	Miscible with water, benzene, alcohol, dimethylformamide, ether
7	Melting point	−94°C
8	Boiling point	56.3°C
9	Flash point	−17.0°C
10	Density	0.791 gm/cm³ at 27°C

5.5.2 Manufacturing process

It is prepared by the catalytic dehydrogenation of isopropanol. It is also a co-product of the glycerin-hydrogen peroxide process and phenol manufacture using cumene.

5.5.2.1 Raw materials

Isopropanol

5.5.2.2 Chemical reactions

Isopropanol on dehydrogenation using a metal oxide as a catalyst at a high temperature and pressure gives acetone

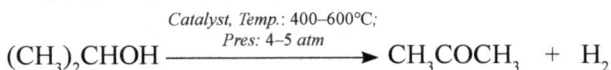

$$(CH_3)_2CHOH \xrightarrow[\text{Pres: 4–5 atm}]{\text{Catalyst, Temp.: 400–600°C;}} CH_3COCH_3 + H_2$$

5.5.2.3 Quantitative requirements

For the production of 1 ton of acetone, we required 1.30 tons of iso propanol.

5.5.2.4 Flow chart

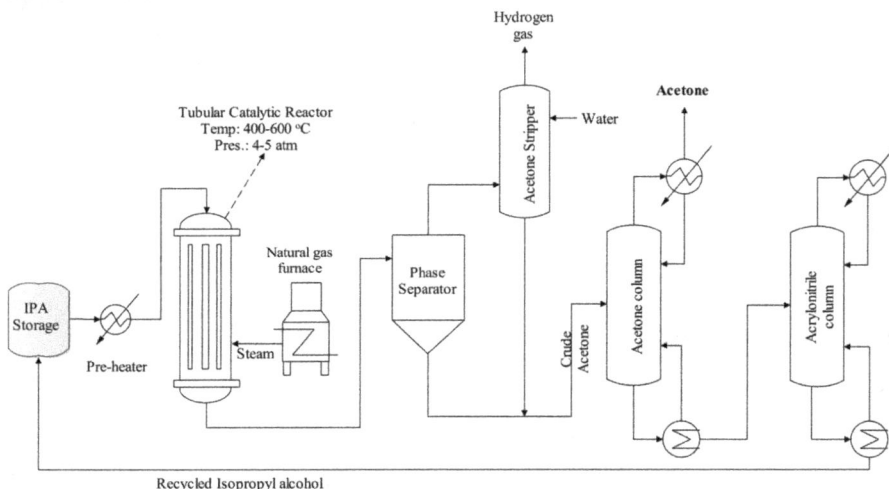

Figure 5.3 Flow sheet diagram of production of Acetone using Isopropyl alcohol

5.5.2.5 Process description

Isopropanol (IPA) is taken from its storage and is pre-heated to form vapor. This vapor is charged from the top of the tubular catalytic reactor, impregnated with catalyst. This tubular reactor is heated to a temperature of 400–600°C using a natural gas furnace. After completion of the reaction, the reaction mixture is collected from the bottom and transferred into the phase separator to separate hydrogen gas from the liquid mixture. The hydrogen gas is collected from the top and further washed with water in the acetone stripper to separate the pure hydrogen. Liquid containing a small amount of acetone is further mixed with the bottom liquor of the phase separator. It undergoes distillation in the acetone column to give acetone as the overhead product. The bottom part containing the unreacted isopropanol is further purified and recycled into isopropanol storage

5.5.2.6 Major engineering problems

(1) The combined process of dehydrogenation and oxidation take place at the same temperature and pressure to give the same percentage yields.

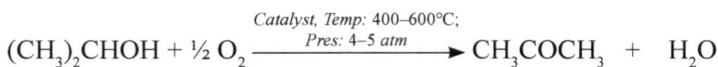

$$(CH_3)_2CHOH + \tfrac{1}{2} O_2 \xrightarrow[\text{Pres: 4–5 atm}]{\text{Catalyst, Temp: 400–600°C;}} CH_3COCH_3 + H_2O$$

(2) The dehydrogenation reaction of isopropanol requires low pressure, but high pressure is maintained during the reaction to reduce the reactor size and water circulation load.

(3) Due to the high pressure and temperature of hydrogen, a chrome steel tubular reactor is more preferable to eliminate embrittlement like blistering, cracking and loss of strength.

5.5.3 Uses

Acetone is employed as a solvent in the plastic and synthetic fiber industries. Due to its volatile nature, it is used as the volatile component of some paints and varnishes. It is also utilized as a nail polish remover. When used as an additive to thin gasoline or diesel fuel, it helps in their movement through the engine. Adding some in to your tank helps prevent it from sticking to the filter and can increase fuel efficiency. It is also widely used in medical applications to clean and sterilize medical tools, instruments and equipment. We use it as a solvent in fillers and active ingredients since its dosages are accurate (Zaoshansky, 2007; Neamah, 2022; Tohaneanu, 2014).

5.6 Propylene Oxide

It is the second smallest oxide having three carbons, six hydrogens and an oxygen atom.

5.6.1 Properties

Table 5.5 Physical & Chemical Properties of Propylene oxide

Sr. No.	Properties	Particular
1	IUPAC name	2-Methyloxirane
2	Chemical and other names	Propylene oxide, 1,2-Epoxypropane, Propene oxide
3	Molecular formula	C_3H_6O
4	Molecular weight	58.08 gm/mole
5	Physical description	Clear colorless volatile liquid with an ethereal odor.
6	Solubility	Miscible with water, acetone, benzene, carbon tetrachloride, methanol, ether
7	Melting point	−112°C
8	Boiling point	34°C
9	Flash point	−37°C
10	Density	0.859 gm/cm³ at 27°C

5.6.2 Manufacturing process

It is prepared by the direct oxidation or chlorohydrin route of propylene. So, here we are discussing both routes.

5.6.2.1 Chlorohydrin route of propylene

(A) Raw materials

Propylene, Chorine gas and Lime water

(B) Chemical reactions

First, propylene undergoes chlorohydrination with hypochlorous acid to get chlorohydrin. It is further reacted with lime water to yield propylene oxide.

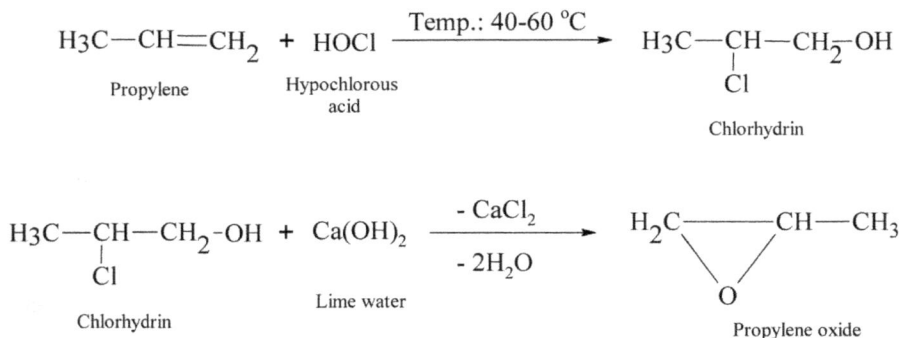

$$H_3C-CH{=}CH_2 \ + \ HOCl \ \xrightarrow{\text{Temp.: 40-60 °C}} \ H_3C-\underset{\underset{Cl}{|}}{CH}-CH_2{-}OH$$

Propylene   Hypochlorous
                acid
                                                              Chlorhydrin

$$H_3C-\underset{\underset{Cl}{|}}{CH}-CH_2{-}OH \ + \ Ca(OH)_2 \ \xrightarrow[- 2H_2O]{- CaCl_2} \ H_2C\underset{O}{\diagdown\diagup}CH-CH_3$$

Chlorhydrin   Lime water
                                                              Propylene oxide

Figure 5.4 Synthesis of Propylene oxide by Chlorohydrine route

The first reaction is exothermic, so, a temperature of 40–60°C is maintained during the reaction.

(C) Quantitative requirements

For the preparation of 1 ton of propylene oxide, we required 0.91 tons of propylene, 2.0 tons of chlorine gas and 1.6 tons of lime.

(D) Flow chart

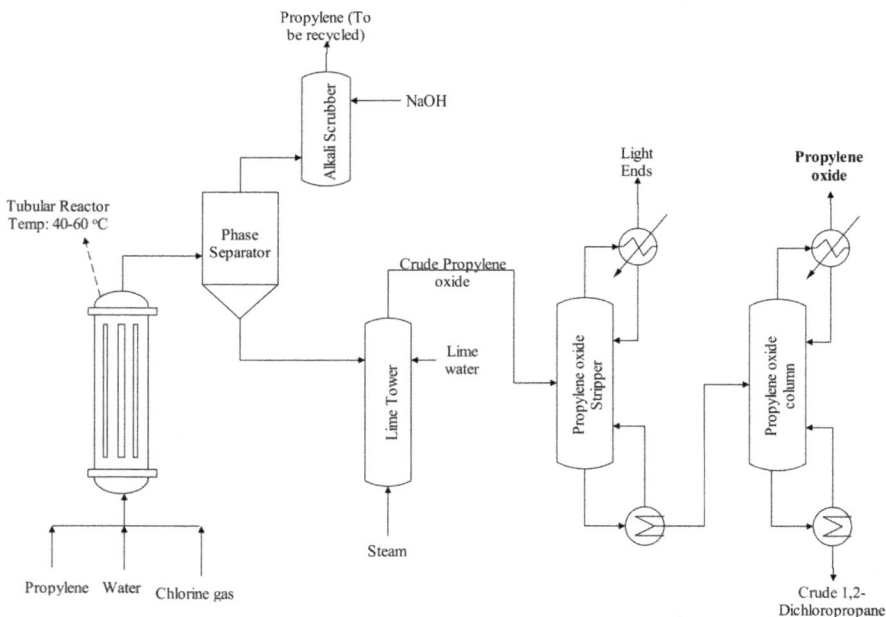

Figure 5.5 Flow sheet diagram of production of Propylene oxide by Chlorohydrine route

(E) Process description

First, propylene gas, chlorine gas and water are fed from the bottom of the tubular reactor. The reaction is exothermic, a temperature of 40–60°C is maintained during the reaction using cooling water circulation. After completion of the reaction, the reaction mass is transferred into the phase separator, where, the gaseous phase is collected as the overhead product. The gaseous phase is scrubbed using alkali to collect the propylene gas and then recycled. Now, the liquid phase containing crude chlorohydrin is collected from the bottom of the phase separator. It is further transferred to the lime tower. Here, steam is used to treat chlorohydrin with lime water. Crude propylene oxide gas (PO) is purified using the PO stripper and column.

5.6.2.2 Direct oxidation route of propylene

This process is called HPPO (hydrogen peroxide-to-propylene-oxide) process, in propylene oxide is prepared by epoxidation of propylene using hydrogen peroxide as an oxidizing agent in the presence of a silver based catalyst. This technology was first developed by German company 'Evonik and ThyssenKrupp'. As, water is the only by-product during the reaction, this process is significantly more environmentally friendly than other competing technologies.

(A) Raw materials

Propylene, Hydrogen peroxide and Catalyst

(B) Chemical reactions

Single step epoxidation of propylene to propylene oxide using hydrogen peroxide as an oxidizing agent in the presence of a silver catalyst and methanol as the solvent is conducted as follows.

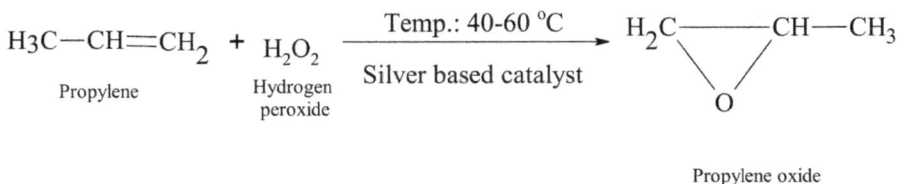

$$H_3C-CH\!=\!CH_2 \;+\; H_2O_2 \xrightarrow[\text{Silver based catalyst}]{\text{Temp.: 40-60 °C}} \underset{O}{H_2C-CH-CH_3}$$

Propylene   Hydrogen peroxide   Propylene oxide

Figure 5.6 Synthesis of Propylene oxide by direct oxidation

(C) Flow chart

(D) Process description

First, propylene gas, hydrogen peroxide and methanol are fed into the bottom of a tubular reactor. The reactor is already impregnated with a silver based catalyst. A temperature of 40–60°C is maintained during the exothermic reaction using circulated cooling water. After completion of the reaction, the reaction mass is transferred into the phase separator, where, the gaseous

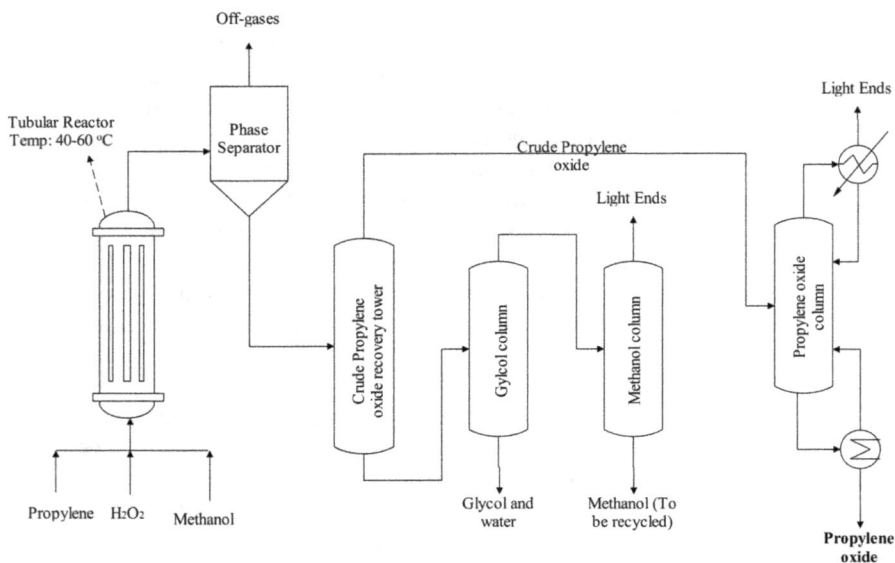

Figure 5.7 Flow sheet diagram of production of Propylene oxide by direct oxidation

phase is collected as off-gases. Now, the liquid phase is transferred into the crude PO column, where, crude PO is collected as the overhead product. It is further purified in the PO column. The bottom mixture from the crude PO column goes to the glycol and methanol columns. The obtained methanol is further recycled.

5.6.3 Uses

PO is used primarily to produce many important chemicals such as polyether polyols, propylene glycols, and propylene glycol ethers. It is also used as a surfactant for gas dispersion, detergency and friction reduction. It is reacted with phosphorous derivatives to create flame retardants. It is also used in detergents, paints, adhesives and cosmetics (Yu et al., 2009).

5.7 Glycerin

Glycerin (or glycerol) is a simple polyol compound. Structurally, three alcoholic groups are jointed with three different carbon atoms of propylene.

5.7.1 *Properties*

Table 5.6 Physical & Chemical Properties of Glycerin

Sr. No.	Properties	Particular
1	IUPAC name	Propane-1,2,3-triol
2	Chemical and other names	Glycerin, Glycerol, 1,2,3-Propanetriol, Glycyl alcohol, Glyceritol
3	Molecular formula	$C_3H_8O_3$
4	Molecular weight	92.09 gm/mole
5	Physical description	Clear, colorless, odorless, syrupy-type liquid
6	Solubility	Miscible with ethanol; slightly soluble in ethyl ether; insoluble in benzene, chloroform, petroleum ether
7	Melting point	18°C
8	Boiling point	290°C
9	Flash point	177°C
10	Density	1.261 gm/cm³ at 27°C

5.7.2 *Manufacturing process*

Two types of glycerin are available in market. The first is natural glycerin, which is produced by the trans-esterification of fatty acids. It will be discussed in the soap manufacturing process. Another is synthetic glycerin, in which propylene undergoes various chemical reactions to yield glycerin.

5.7.2.1 *Raw materials*

Propylene, Water, Air (Oxygen), Isopropanol, Hydrogen peroxide and Catalyst

5.7.2.2 *Chemical reactions*

The vapor phase reaction of propylene with air (oxygen) in the presence of catalyst and water at a high temperature and pressure gives acrolein.

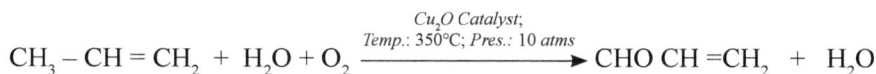

$$CH_3 - CH = CH_2 + H_2O + O_2 \xrightarrow[\text{Temp.: 350°C; Pres.: 10 atms}]{Cu_2O \ Catalyst;} CHO \ CH = CH_2 + H_2O$$

Acrolein undergos a reaction with isopropanol in the presence of MgO-ZnO catalyst at 400°C to give allyl alcohol and acetone.

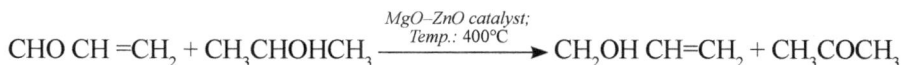

$$CHO \ CH = CH_2 + CH_3CHOHCH_3 \xrightarrow[\text{Temp.: 400°C}]{MgO-ZnO \ catalyst;} CH_2OH \ CH = CH_2 + CH_3COCH_3$$

Further, allyl alcohol is reacted with hydrogen peroxide in the liquid phase in the presence of WO_3 catalyst at 60–70°C to give glycerol.

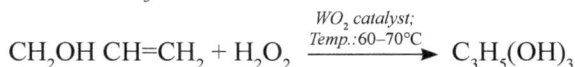

$$CH_2OH \ CH = CH_2 + H_2O_2 \xrightarrow[\text{Temp.:60–70°C}]{WO_2 \ catalyst;} C_3H_5(OH)_3$$

This reaction is exothermic, so, the temperature is maintained by circulating cooling water.

5.7.2.3 Flow chart

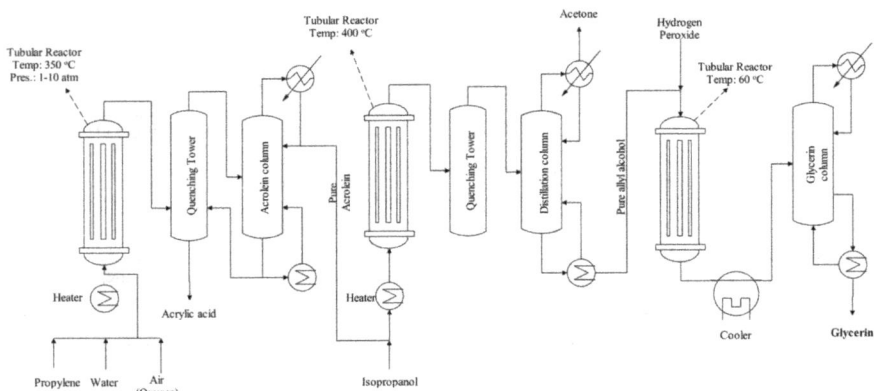

Figure 5.8 Flow sheet diagram of production of Glycerin

5.7.2.4 Process description

As per Fig. 5.8, propylene gas, water and air (oxygen) are pre-heated and fed into the bottom of a tubular reactor. The reactor is already impregnated with copper(I) oxide catalyst. A temperature of 350°C and pressure of 1–10 atmospheres is maintained by heating the tubular coil with hot oil. After completion of the reaction, the mass is quenched and sent to the acrolein column to get pure acrolein. The obtained acrolein is mixed with isopropanol and; fed into a tubular reactor. The reactor is already impregnated with MnO-ZnO based catalyst. The reaction mass is heated by means of hot oil to maintain a temperature of 400°C. After completion of the reaction, the mass is quenched and sent to the distillation column to get pure allyl alcohol and acetone. This alcohol is further mixed with hydrogen peroxide and sent to another tubular reactor. Here, the reactor is already impregnated with the catalyst. The reaction is exothermic, a temperature of 40–60°C is maintained during the reaction using cooling water circulation. After the completion of the reaction, the mass is cooled and sent to the distillation column to separate glycerin from other by-products.

5.7.3 Uses

Glycerin is essentially used in the manufacturing of soaps and lotions. Glycerol is used as a sweetener and preservative in the food industry. It is also used in the pharmaceutical industry for the preparation of cough syrups. It is also utilized as humectant in personal care products, such as toothpastes, hair conditioners, cosmetics, moisturizers and skin moisturizers. Glycerol has good solubility in water which makes it useful as a lubricant and humectant. This molecule has been known to function as a replacement for olive oil in several personal care products such as moisturizers to treat dry, itchy skin and mild skin irritations such as eczema or psoriasis. The skin moisturizing effects are caused by the ability of glycerol to emolliate (soften). This type of moisturizing effect decreases trans epidermal water loss (TEWL) caused by harsh drying interfaces (commercially marketed Cetaphil Moisturizer Body Wash) (Hartshorne, 2012; Tan et al., 2013).

5.8 Nitropropanes

There are two types of nitropropanes: (1) 1-Nitropropane and (2) 2-Nitropropane

5.8.1 Properties

The properties of one nitropropane, i.e., 2-Nitropropane are as follows.

Table 5.7 Physical & Chemical Properties of 2-Nitropropane

Sr. No.	Properties	Particular
1	IUPAC name	2-Nitropropane
2	Chemical and other names	Propane-2-nitronate
3	Molecular formula	$C_3H_6NO_2$
4	Molecular weight	88.09 gm/mole
5	Physical description	Clear, colorless, odorless, syrupy-type liquid
6	Solubility	Miscible with ethanol; slightly soluble in ethyl ether; insoluble in benzene, chloroform, petroleum ether
7	Melting point	18°C
8	Boiling point	290°C
9	Flash point	177°C
10	Density	1.261 gm/cm^3 at 27°C

5.8.2 Manufacturing process

It is prepared by the nitration of propane. Nitropropanes are manufactured as volatile by-products which are separated during Leonard's ring-closure hydantoin preparation. In the production of nitropropanes, 2-nitropropane is the major product, while 1-nitropropane is an impurity.

5.8.2.1 Raw materials

Propane and Nitric acid

5.8.2.2 Chemical reactions

Propane undergoes vapor phase nitration using nitric acid at a temperature of 250–300°C and a pressure of 5–7 atmospheres to give a mixture of nitropropanes.

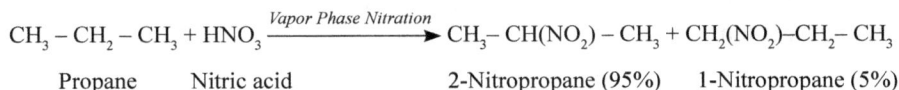

$$CH_3 - CH_2 - CH_3 + HNO_3 \xrightarrow{\textit{Vapor Phase Nitration}} CH_3 - CH(NO_2) - CH_3 + CH_2(NO_2) - CH_2 - CH_3$$

Propane Nitric acid   2-Nitropropane (95%) 1-Nitropropane (5%)

5.8.2.3 Quantitative requirements

For the preparation of 1 ton of nitropropane, we require 0.51 ton of propane and 0.72 ton of nitric acid.

5.8.2.4 Flow chart

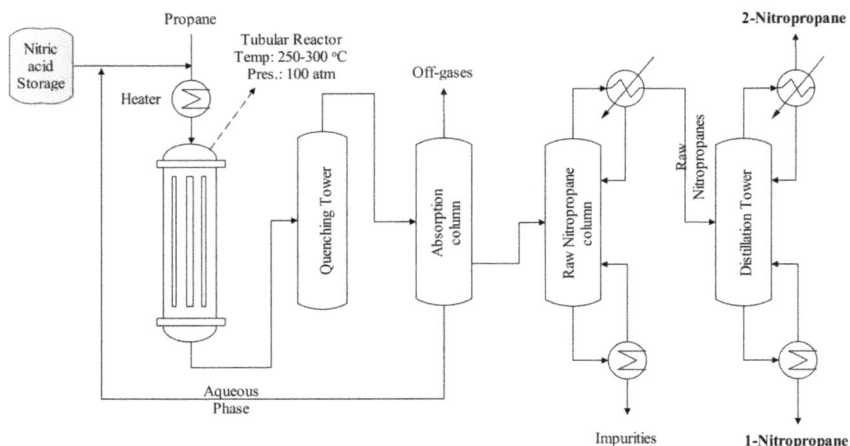

Figure 5.9 Flow sheet diagram of production of Nitropropanes

5.8.2.5 Process description

As per Fig. 5.9, a mixture of propane and nitric acid is pre-heated to the convert vapors. This mixture is fed into a tubular reactor. Here, a temperature of 250–300°C and pressure of 5–7 atmospheres is maintained using hot oil circulation in the tubes. After 30 seconds of reaction time, the reaction mixture is quenched and sent to the absorption tower. The three phases are differentiated in this tower. First, off-gases are removed from top. The aqueous phase containing un-reacted propane and nitric acid is collected from the bottom and recycled. On the other hand the oily phase containing nitropropane is separated from the middle of the column and sent to the nitropropane column. The raw mixture of nitropropanes is collected as the overhead product, while impurities like nitromethane, nitroethane and 2,2-dinitropropane are separated from the mixture. This mixture is sent for fractional distillation, where 2-nitropropane is separated from 1-nitropropane.

5.8.3 Uses

1-Nitropropane is utilized in solvents, chemical synthesis, rocket propellants and gasoline additives. It is also used as an oxidant in the Hass–Bender oxidation process. It is utilized in the manufacturing of pharmaceuticals and other industrial products (Trauns et al., 2008).

5.9 Isopropylamine

It is an isomer of n-propylamine.

5.9.1 Properties

The properties of isopropylamine are as follows.

Table 5.8 Physical & Chemical Properties of Isopropylamine

Sr. No.	Properties	Particular
1	IUPAC name	Propan-2-amine
2	Chemical and other names	Isopropylamine, 2-Propanamine, 2-Aminopropane, 1-Methylethylamine, Isopropyl amine
3	Molecular formula	C_3H_9N
4	Molecular weight	59.11 gm/mole
5	Physical description	Colorless liquid with an ammonia-like odor
6	Solubility	Miscible with water, ethanol, ether, acetone, benzene, chloroform
7	Melting point	–95°C
8	Boiling point	31.8°C
9	Flash point	–18.6°C
10	Density	0.687 gm/cm³ at 27°C

5.9.2 Manufacturing process

The starting materials for manufacturing isopropylamine are isopropanol and ammonia.

5.9.2.1 Raw materials

Isopropanol and Ammonia

5.9.2.2 Chemical reactions

Isopropylamine is manufactured by ammoniating isopropyl alcohol with ammonia in the presence of a catalyst at a temperature of 180–200°C in the presence of a Nickel based catalyst.

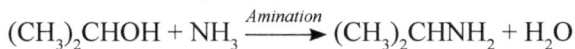

$$(CH_3)_2CHOH + NH_3 \xrightarrow{\text{Amination}} (CH_3)_2CHNH_2 + H_2O$$

5.9.2.3 Quantitative requirements

For the preparation of 1 ton of isopropylamine, we require 1.1 tons of isopropanol and 0.28 ton of ammonia.

5.9.2.4 Flow chart

5.9.2.5 Process description

As per the simple flow chart, isopropanol (IPA) and ammonia are pre-heated using a heater and fed into a tubular catalytic reactor. The reactor is already impregnated with anickel-based catalyst. After completion of the reaction, the mixture is goes to the phase separator, where the gaseous phase is collected from the top and fed to the stripper. Here, ammonia is separated and recycled. The liquid phase is collected from the bottom of the separator and goes to the IPA

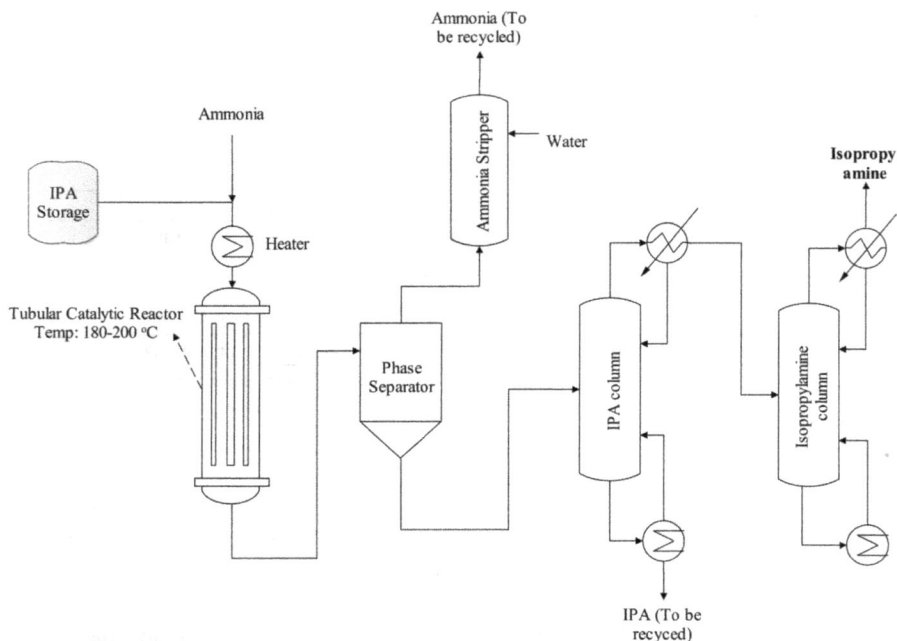

Figure 5.10 Flow sheet diagram of production of Isopropyl amine from Isopropyl alcohol

column. Unreacted IPA is separated from the reaction mass from the bottom and recycled. The overhead product is sent to the isopropyl amine column, where the final product is separated out.

5.9.3 Uses

It is mainly used as a solvent, de-hairing agent and as an intermediate in the preparation of many organic chemicals. It is used in various industries such as pharmaceuticals, dyes, insecticides, soaps and textiles, etc. (Luyben, 2011).

* n-Propylamine is manufactured by reacting 1-propanol and ammonium chloride at a high temperature and pressure in the presence of a a Lewis acid catalyst such as ferric chloride as shown in the flow diagram.

5.10 Trimethylamine

This is the smallest tertiary amine.

5.10.1 Properties

The properties of trimethylamine are as per Table 5.9.

Table 5.9 Physical & Chemical Properties of Trimethylamine

Sr. No.	Properties	Particular
1	IUPAC name	N,N-Dimethylmethanamine
2	Chemical and other names	Trimethylamine, N-Trimethylamine, Dimethylmethaneamine
3	Molecular formula	C_3H_9N
4	Molecular weight	59.11 gm/mole
5	Physical description	Colorless gas with a fishy, amine odor
6	Solubility	Soluble in alcohol, ether, benzene, toluene, xylene, ethylbenzene
7	Melting point	−117.8°C
8	Boiling point	2.8°C
9	Flash point	−7°C
10	Density	0.604 gm/cm³ at 27°C

5.10.2 Manufacturing process

It is prepared by the ammonolysis of methanol with ammonia in the presence of catalyst. It is also prepared by the reaction of ammonium chloride with paraformaldehyde. Here, we are discussing the ammonolysis method.

5.10.2.1 Raw materials

Methanol, Ammonia and Catalyst

5.10.2.2 Chemical reactions

It is prepared by the ammonolysis of methanol with ammonia in the presence of a metal catalyst at a high temperature.

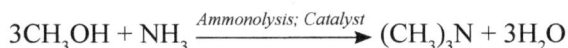

$$3CH_3OH + NH_3 \xrightarrow{Ammonolysis; \ Catalyst} (CH_3)_3N + 3H_2O$$

Also, dimethylamine and methylamine are produced in the side reactions. The chemical reactions are as follows.

Side Reactions:

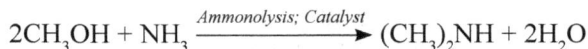

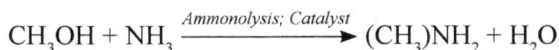

$$2CH_3OH + NH_3 \xrightarrow{Ammonolysis; \ Catalyst} (CH_3)_2NH + 2H_2O$$

$$CH_3OH + NH_3 \xrightarrow{Ammonolysis; \ Catalyst} (CH_3)NH_2 + H_2O$$

5.10.2.3 Quantitative requirements

Here, the molar ratio of methanol and ammonia is fixed at 6:1 during the reaction to prevent side reactions.

5.10.2.4 Flow chart

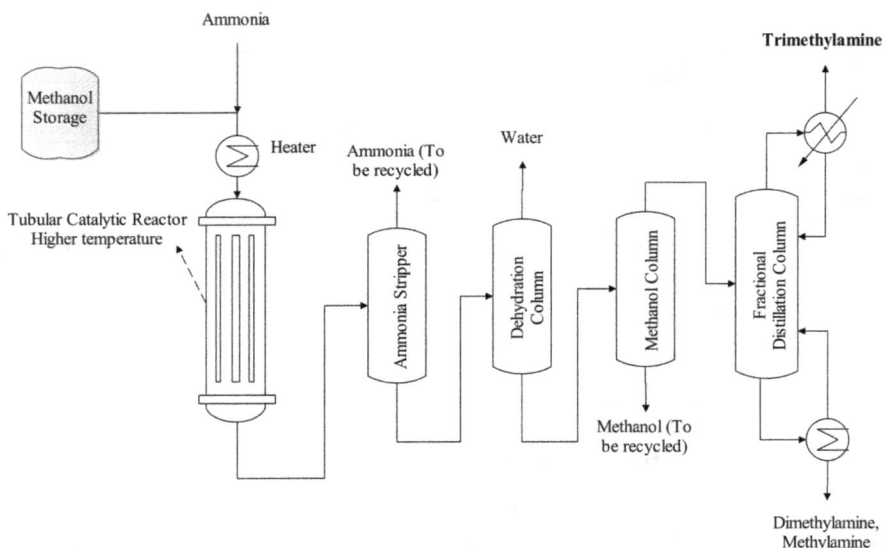

Figure 5.11 Flow sheet diagram of production of Trimethylamine

5.10.2.5 Process description

First, methanol and ammonia are pre-heated and fed into the tubular reactor. The reactor is already impregnated with anickel-based catalyst. A high temperature is maintained in the tubular reactor by passing hot oil. After completion of the reaction, un-reacted ammonia and methanol and; water are separated from the reaction mass using individual reactosr. The remaining mass goes to the fractional distillation column to yield trim ethylamine as the overhead product. By-products, dimethylamine and methylamine are collected from the bottom of the column.

5.10.3 Uses

It is used in a number of chemical reactions as a nucleophile or Lewis base in a variety of industries. It is used in the preparation of chlorides and tetramethyl ammonium hydroxide. It is also utilized as a plant growth regulator or herbicide and dye leveling agent in the dye industry (March and MacMillan, 1979).

5.11 Propionitrile

Propionitrile is a simple aliphatic nitrile having three carbon atoms, five hydrogen atoms and a nitrogen atom.

5.11.1 Properties

The properties of Popionitrile are as per Table 5.10.

Table 5.10 Physical & Chemical Properties of Propionitrile

Sr. No.	Properties	Particular
1	IUPAC name	Propanenitrile
2	Chemical and other names	Ethyl cyanide, Propiononitrile, Cyanoethane, Propylnitrile, Propionic nitrile
3	Molecular formula	C_3H_5N
4	Molecular weight	55.08 gm/mole
5	Physical description	Colorless liquid with a pleasant, sweetish odor
6	Solubility	Miscible with water, ether, alcohol
7	Melting point	–91.8°C
8	Boiling point	97.2°C
9	Flash point	16°C
10	Density	0.802 gm/cm³ at 27°C

5.11.2 Manufacturing process

First, acrylonitrile (ACN) is prepared and thereafter, propionitrile is manufactured by hydrogenation.

5.11.2.1 Raw materials

Propylene, Ammonia, Oxygen, Catalyst and Hydrogen

5.11.2.2 Chemical reactions

There are several processes for manufacturing acrylonitrile.

(1) From ethylene cyanohydrin:

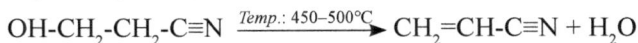

$$OH\text{-}CH_2\text{-}CH_2\text{-}C≡N \xrightarrow{Temp.: 450–500°C} CH_2=CH\text{-}C≡N + H_2O$$

(2) From acetylene:

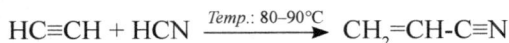

$$HC≡CH + HCN \xrightarrow{Temp.: 80–90°C} CH_2=CH\text{-}C≡N$$

(3) From propionaldehyde:

$$CH_3CH_2CHO + NH_3 \longrightarrow CH_2=CH\text{-}C≡N + H_2O + 2H_2$$

(4) From propylene:

In the SOHIO process, catalytic ammoxidation of propylene is performed at a high temperature of 400–510°C and pressure of 1–3 atmospheres in the presence of a catalyst to form acrylonitrile and hydrogen cyanide. Here, hydrogen cyanide is recovered and sold.

$$2CH_3\text{-}CH=CH_2 + 2NH_3 + 3O_2 \rightarrow 2CH_2=CH\text{-}C≡N + 6H_2O$$

Out of these processes, the SOHIO process is widely used because of a better quality product, economy and energy efficiency.

After manufacturing acrylonitrile, it undergoes hydrogenation in the presence of a zinc-based catalyst at 85–90°C.

$$CH_2=CH\text{-}C≡N \quad + \quad H_2 \rightarrow CH_3\text{-}CH_2\text{-}C≡N$$

5.11.2.3 Flow chart

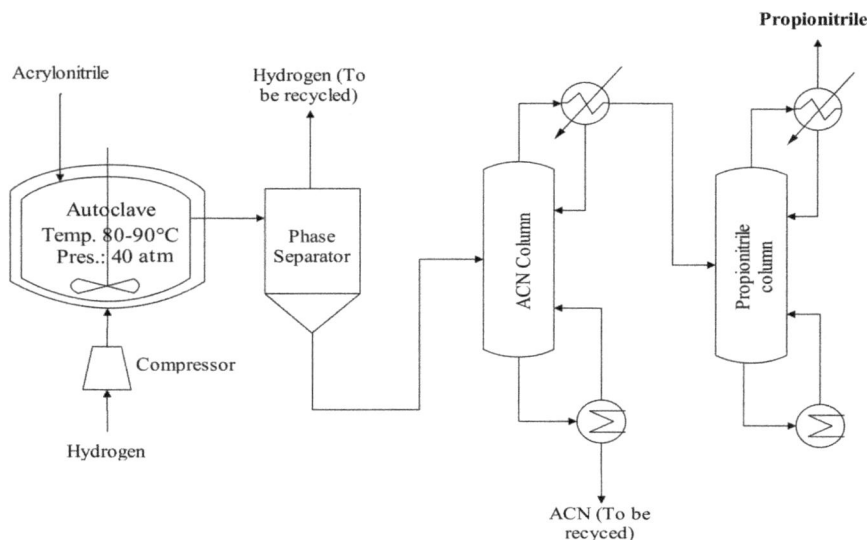

Figure 5.12 Flow sheet diagram of production of Propionitrile from Acrylonitrile

5.11.2.4 Process description

Acrylonitrile is first prepared by the SOHIO process as mentioned previously. Purified acrylonitrile is fed into an autoclave. Also, hydrogen gas is compressed and fed into the bottom of the autoclave. The autoclave is already impregnated with azinc-based catalyst. The reaction mixture is continuously heated upto 85–90°C at a stirring speed of 800 rpm and a pressure of 40 atmospheres. After completion of the reaction, un-reacted hydrogen gas and acetonitrile are separated using a phase separator and ACN column respectively. The remaining mass goes to the propionitrile column to separate out as the final product in the overhead column.

5.11.3 Uses

Propionitrile is a C-3 building block in the preparation of the drug flopropione by the Houben-Hoesch reaction. The compound was used as a testing agent for which it does not provide any indication for this compound in the control experiment. It is also useful as a precursor to propylamines by hydrogenation (Li et al., 2017).

5.12 Butadiene

It is the simplest conjugated diene. Structurally, it may seem to be a combination of two vinyl groups. First, French chemist E. Caventou isolated butadiene from the pyrolysis of amyl alcohol in 1863. Thereafter, Henry Edward Armstrong isolated it from the pyrolysis products of petroleum in 1886. The Russian chemist

Sergei Lebedev polymerized butadiene in 1910 and found that the resulting polymer had rubber-like properties.

5.12.1 Properties

Its main properties are as follows.

Table 5.11 Physical & Chemical Properties of Butadiene

Sr. No.	Properties	Particular
1	IUPAC name	Buta-1,3-diene
2	Chemical and other names	1,3-Butadiene, Butadiene, Divinyl, Vinylethylene, Biethylene, Erythrene
3	Molecular formula	C_4H_6
4	Molecular weight	54.09 gm/mole
5	Physical description	Colorless gas with mild aromatic or gasoline-like odor
6	Solubility	Soluble in water, ethanol, ether, benzene, acetone
7	Melting point	−108.9°C
8	Boiling point	−4.6°C
9	Flash point	−76°C
10	Density	0.621 gm/cm³ at 27°C

5.12.2 Manufacturing process

It is prepared by several methods.

(1) Extraction from C4 hydrocarbons
(2) Dehydrogenation of n-butane
(3) Dehydrogenation-hydration of ethanol
(4) Oxydehydrogenation of butylene
    We are discussing all the processes one-by-one.

5.12.2.1 Extraction from C4 hydrocarbons

Butadiene is the main constituent of C4 hydrocarbons containing propylene, C4 acetylenes, butanes, C3 acetylenes, butenes, 1,2-butadiene, and C5+ hydrocarbons.

(1) Raw materials
    Crude C4 hydrocarbons

(2) Flow chart

(3) Process description
    C4 feedstock is heated into a cracking furnace and sent to the first extraction column, followed by a rectifier column. Then the reaction mixture goes to a second extraction column and purified products are collected over-head from the column. It is then sent to the propylene column where propylene is separated from the mixture. The mixture is then sent to the butadiene column. Here, butadiene is collected by means of steam as an over-head product. Heavy ends are collected from the bottom of the column.

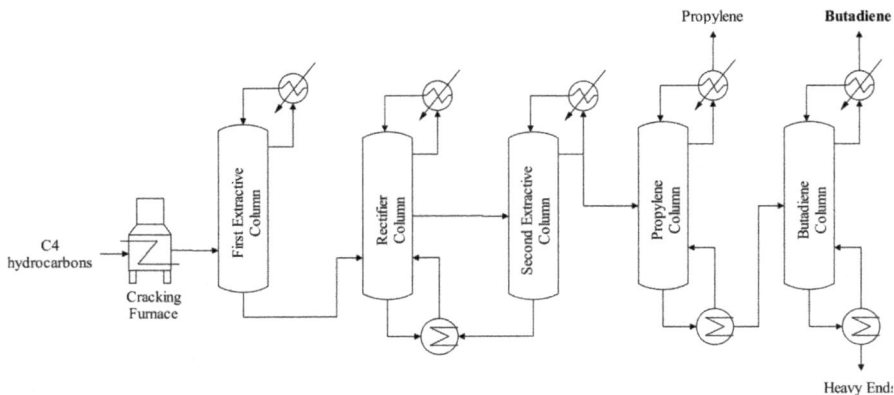

Figure 5.13 Flow sheet diagram of production of Butadiene via extraction from C3 hydrocarbons

5.12.2.2 Dehydrogenation of n-butane

Normal butane (n-butane) undergoes catalytic dehydrogenation to give butadiene. It is a single step process, also called as Houdry Catadiene process, which was developed during World War II.

(1) Raw materials

n-Butane and Catalyst

(2) Chemical reaction

N-butane is undergoes dehydrogenation using chromium/alumina catalysts at a temperature of approximately 600–680°C to yield 1,4-butadiene.

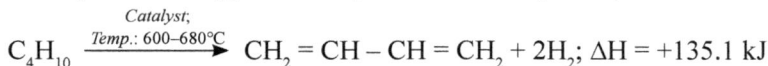

$$C_4H_{10} \xrightarrow[\text{Temp.: 600–680°C}]{\text{Catalyst;}} CH_2 = CH - CH = CH_2 + 2H_2; \Delta H = +135.1 \text{ kJ}$$

This is an exothermic reaction.

(3) Flow chart

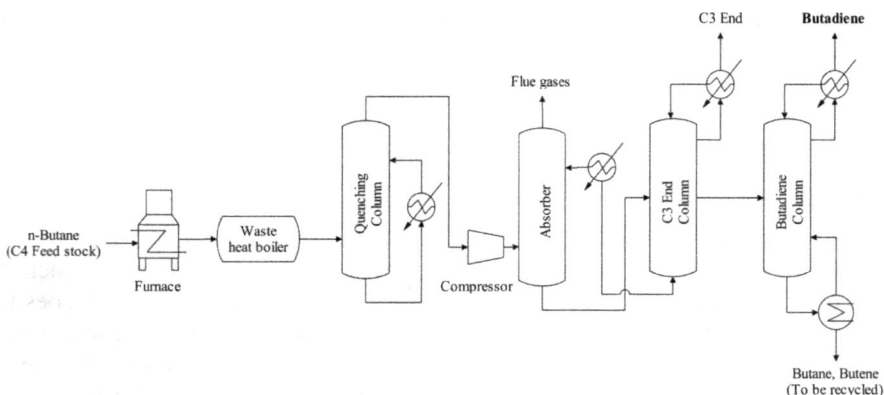

Figure 5.14 Flow sheet diagram of production of Butadiene via Houdry Catadiene Process

(4) Process description

n-Butane is first heated in a furnace to convert it to butane vapor. Butane vapor goes to the waste heat boiler reactor system which is already impregnated with catalyst, where butane is converted to butadiene. Further, this dehydrogenation reaction is an exothermic reaction, so, heat is liberated during the reaction. This heat is further recovered and re-utilized. The vapor is quenched, compressed and sent to the absorber to remove flue gases. Then the reaction mass is fed into the C3 column, where C3 ends are recovered as an over-head product from the column. The bottoms of the column are recycled to the absorber. The raw butadiene stream is collected from the middle of the C3 column and fed into the butadiene column. Here, pure butadiene is separated from the butane, butene, and other C4 ends.

5.12.2.3 Dehydrogenation-hydration of ethanol

(1) Raw materials

Ethanol and Catalyst

(2) Chemical reaction

First, ethanol is dehydrogenated at a temperature of 260–290°C using a tantala-silica catalyst to get acetaldehyde.

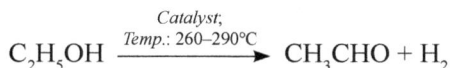

$$C_2H_5OH \xrightarrow[\text{Temp.: 260–290°C}]{\text{Catalyst;}} CH_3CHO + H_2$$

Thereafter, a mixture of ethanol and acetaldehyde is dehydrated at a temperature of 330–350°C in the presence of a silica catalyst.

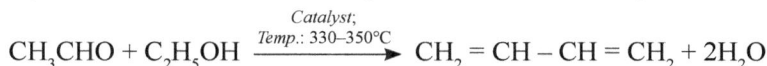

$$CH_3CHO + C_2H_5OH \xrightarrow[\text{Temp.: 330–350°C}]{\text{Catalyst;}} CH_2 = CH – CH = CH_2 + 2H_2O$$

(3) Flow chart

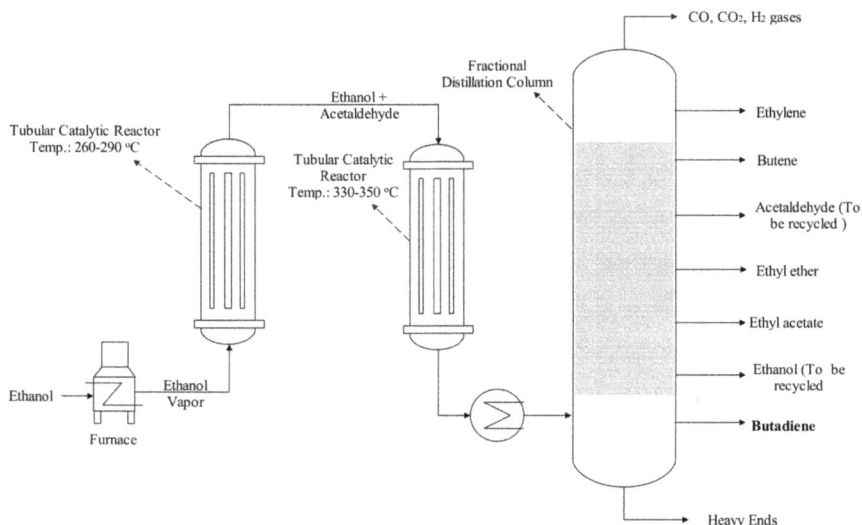

Figure 5.15 Flow sheet diagram of production of Butadiene from Ethanol

(4) Process description

First, ethanol is pre-heated in furnace for vaporization and fed to a tubular catalytic reactor. The reactor is already impregnated with a tantala-silica based catalyst. A high temperature is maintained in the tubular reactor by passing hot oil. After completion of the reaction, the gaseous reaction mixture containing acetaldehyde and un-reacted ethanol is fed to another tubular catalytic reactor. The reactor is already impregnate with a silica-based catalyst. A high temperature is maintained in the tubular reactor by passing hot oil. After completion of the reaction, the reaction mixture contains the main product (butadiene) incorporated with un-reacted ethanol and acetaldehyde and; other byproducts like ethylene, butene, butane, ethyl ether, ethyl acetateand butanol. This reaction mixture goes to the fractional distillation column, where different compounds are separated according to their boiling points. Here, butadiene is separated, whereas un-reacted ethanol and acetaldehyde are recycled.

5.12.3 Uses

Mainly butadiene is used to produce various types of rubbers, resins and latex such as polybutadiene, BUNA-N, BUNA-S, ABS and MMBS. It is also used as a chemical intermediate in the synthesis of adipointrile and chloroprene. It is utilized as a solvent in 1,4-hexadiene sulfonate and 1,5,9-cyclordecatriene (Hou, 2017; White, 2007).

5.13 n-Butanol

2-Butanol is a primary alcohol with a 4-carbon structure having three isomers: isobutanol and 2-butanol. It is one among the group of "fusel alcohols", which has more than two carbon atoms and significant solubility in water.

5.13.1 Properties

Table 5.12 Physical & Chemical Properties of n-Butanol

Sr. No.	Properties	Particular
1	IUPAC name	Butan-1-ol
2	Chemical and other names	1-Butanol, Butanol, n-Butanol, Butyl alcohol, n-Butyl alcohol, 1-Hydroxybutane, Propylcarbinol
3	Molecular formula	$C_3H_8O_3$
4	Molecular weight	74.12 gm/mole
5	Physical description	Colorless liquid with a strong, characteristic odor
6	Solubility	Miscible with water, ethyl ether, benzene, chloroform, petroleum ether, benzene
7	Melting point	−89.8°C
8	Boiling point	177.5°C
9	Flash point	29°C
10	Density	0.8025 gm/cm³ at 27°C

5.13.2 Manufacturing process

n-Butanol is prepared by the fermentation of sugar and other carbohydrate materials as per the next chapter on "Fermentation". But its demand is not fulfilled by fermentation. Synthetically, it is manufactured using propylene in the presence of a homogeneous catalyst.

5.13.2.1 Raw materials

Propylene, Synthesis gas and Hydrogen

5.13.2.2 Chemical reactions

Propylene undergoes a reaction with synthesis gas at a temperature of 110–160°C and a pressure of 250–300 atmospheres in the presence of cobalt carbonyl catalyst to get n-butyraldehyde with a small amount of 2°- butyraldehyde.

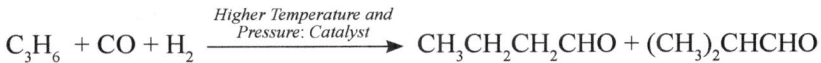

$$C_3H_6 + CO + H_2 \xrightarrow[\text{Pressure: Catalyst}]{\text{Higher Temperature and}} CH_3CH_2CH_2CHO + (CH_3)_2CHCHO$$

n-Butyraldehyde is further hydrogenated at a temperature of 150–160°C and a pressure of 100 atmosphere in the presence of a nickel catalyst to get n-butanol.

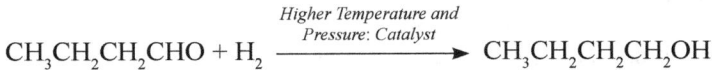

$$CH_3CH_2CH_2CHO + H_2 \xrightarrow[\text{Pressure: Catalyst}]{\text{Higher Temperature and}} CH_3CH_2CH_2CH_2OH$$

5.13.2.3 Quantitative requirements

For the preparation of 1 ton of n-butanol, we require 540 kgs of propylene, 410 kgs of synthesis gas and 30 kgs of hydrogen.

5.13.2.4 Flow chart

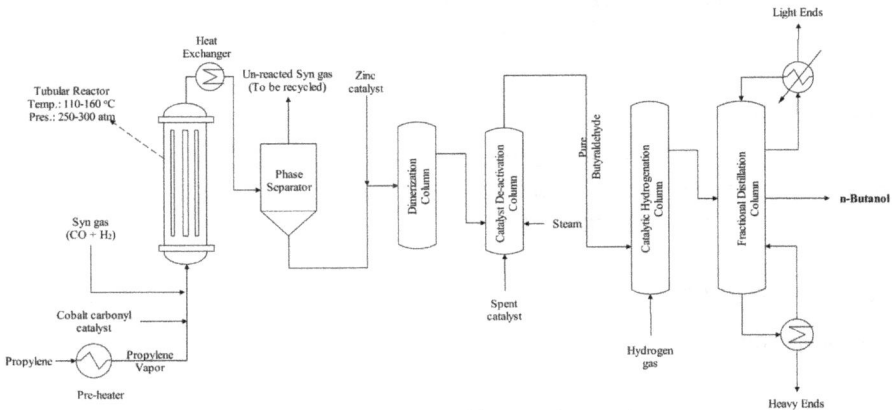

Figure 5.16 Flow sheet diagram of production of n-Butanol from Propylene

5.13.2.5 Process description

First, propylene is pre-heated and fed to the bottom of a tubular column. Also, synthesis gas and cobalt carbonyl catalyst are fed into the bottom of the column. At a temperature of 110–160°C and pressure of 250–300 atmospheres in the presence of cobalt catalyst, propylene and synthesis gas react to give n-butyraldehyde and a small amount of 2°-butyraldehyde. After completion of the reaction, the reaction mass is cooled and transferred into a phase separator, where synthesis gas is separated out from the top and recycled. From the bottom of the separator, the liquid phase is collected and sent for dimerization. If propylene contains trace amounts of ethanol, then it is converted into n-butyraldehyde in the presence of a zinc catalyst. n-Butyraldehyde mass is transferred into the catalyst de-activation column, where, zinc and cobalt catalysts are de-activated by means of steam. Pure n-butyraldehyde is collected from the top of the column and transferred into the catalytic hydrogenation column, which is already impregnated with a nickel catalyst. Hydrogen gas is inserted from the bottom of the column to form n-butanol. The reaction mass is sent to the fractional distillation column, where pure n-butanol is separated from the light and heavy ends.

5.13.2.6 Major engineering problems

(1) Cobalt carbonyl is easily stripped in the tubular column, so, cobalt carbonyl naphthenate is used sometimes.
(2) If Cobalt and zinc catalyst are not deactivated in the catalyst de-activation columnthe hydrogenation process is affected.

5.13.3 Uses

Butyl alcohol employed as a solvent and plasticizer for manufacturing butyl acetate, glycol ester and amine-resin. It is used in several industries such as lubricants, paints, coatings, adhesives, inks, cosmetics, fire retardants, rubbers, polymers, automobiles and textiles (Pereira, 2015; Jiang, 2015).

5.14 Butanone

It is also referred to as methyl ethyl ketone (MEK). It has been banned in some countries due to its potential health and environmental effects and releases, and resulting ambient air concentrations.

5.14.1 Properties

The properties of butanone are per Table 5.13.

5.14.2 Manufacturing process

It is manufactured by the oxidation or dehydrogenation of 2-butanol. Here, we are discussing the dehydrogenation of 2-butanol.

Table 5.13 Physical & Chemical Properties of Butanone

Sr. No.	Properties	Particular
1	IUPAC name	Butan-2-one
2	Chemical and other names	2-Butanone, Methyl Ethyl Ketone (MEK), Butanone, Ethyl methyl ketone, Methylethyl ketone
3	Molecular formula	C_4H_8O
4	Molecular weight	72.11 gm/mole
5	Physical description	Colorless liquid with characteristic odour
6	Solubility	Miscible with ethanol, ether, acetone, benzene, chloroform
7	Melting point	−86.6°C
8	Boiling point	79.5°C
9	Flash point	−9.1°C
10	Density	0.806 gm/cm³ at 27°C

5.14.2.1 Raw materials

2-Butanol and Catalyst

5.14.2.2 Chemical reactions

2-Butanol undergoes dehydrogenation using a copper catalyst at a temperature of 240–260°C.

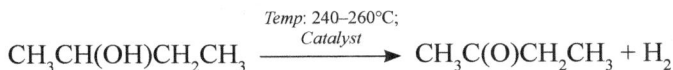

$$CH_3CH(OH)CH_2CH_3 \xrightarrow[\text{Catalyst}]{\text{Temp: 240–260°C;}} CH_3C(O)CH_2CH_3 + H_2$$

5.14.2.3 Quantitative requirements

For the preparation 1 ton of butanone, we require 0.98 ton of 2-butanol.

5.14.2.4 Flow chart

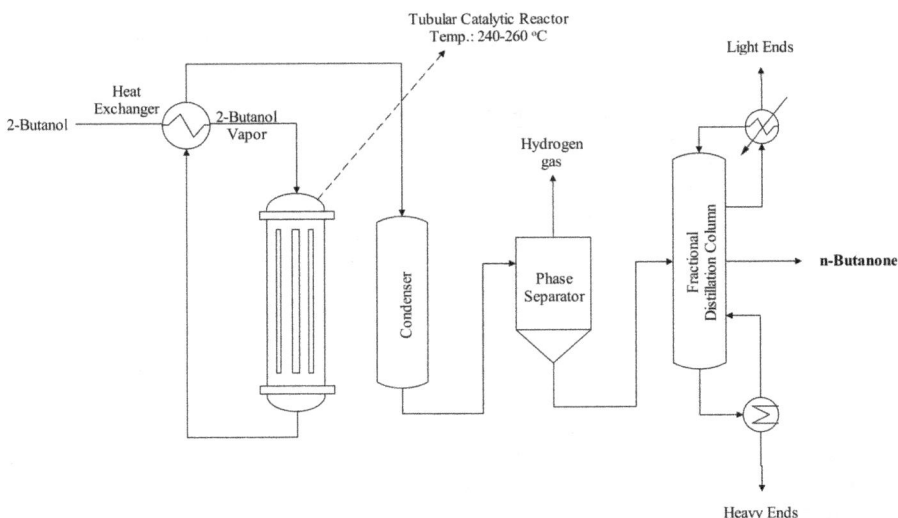

Figure 5.17 Flow sheet diagram of production of n-Butanone from 2-Butanol

5.14.2.5 Process description

As per the figure, 2-butanol is pre-heated and converted into vapor form. It is then charged into the top of the catalytic reactor, which is already impregnated with a copper catalyst. After the reaction, the reaction mass is collected from the bottom and cooled down in the heat exchanger. It is then sent to the phase separator to differentiate hydrogen gas from the reaction mass. Finally, it is fed to the fractional distillation column to separate pure butanone.

5.14.3 Uses

Butanone is utilized as a solvent in the manufacture of resins and gums, as well as cellulose acetate and cellulose nitrate. It is also utilized in glues and as a cleaning agent. Methyl Ethyl Ketone is employed as a plastic welding agent because it has the power to dissolve polystyrene and various other plastics. Butanone is employed in the production of petroleum (Martin and Buisson, 2015).

5.15 Ethyl acetate

Ethyl acetate is one of the simplest carboxylate esters after methyl formate.

5.15.1 Properties

Table 5.14 Physical & Chemical Properties of Ethyl acetate

Sr. No.	Properties	Particular
1	IUPAC name	Ethyl acetate
2	Chemical and other names	Ethyl ethanoate, Acetic acid ethyl ester, Acetoxyethane, Acetic ether, Ethyl acetic ester
3	Molecular formula	$C_4H_8O_2$
4	Molecular weight	88.11 gm/mole
5	Physical description	Colorless liquid with characteristic odor
6	Solubility	Miscible with water, ethanol, ethyl ether, acetone, benzene
7	Melting point	−84.3°C
8	Boiling point	77.1°C
9	Flash point	−4°C
10	Density	0.902 gm/cm³ at 27°C

5.15.2 Manufacturing process

It is prepared following methods.

(1) **Esterification:** One of simplest method is the Fischer esterification reaction of ethanol and acetic acid to synthesize ethyl acetate.

(2) **Tishchenko reaction:** It is prepared by the Tishchenko reaction, in which two moles acetaldehyde combine in the presence of an alkoxide catalyst to give ethyl acetate.

$$2CH_3CHO \rightarrow CH_3COOCH_2CH_3$$

(3) Alkylation: It is also prepared by the alkylation of acetic acid by ethylene using silicotungstic acid.

$$C_2H_4 + CH_3COOH \rightarrow CH_3COOC_2H_5$$

(4) Dehydrogenation of ethanol: It is also prepared by the dehydrogenation of ethanol.

$$2CH_3CH_2OH \rightarrow CH_3COOCH_2CH_3 + 2H_2$$

Here, we are discussing Fischer esterification.

5.15.2.1 Raw materials

Ethanol, Acetic acid and Catalyst

5.15.2.2 Chemical reactions

Ethanol and acetic acid undergo Fischer esterification at a high temperature of 240–260°C in presence of an acid.

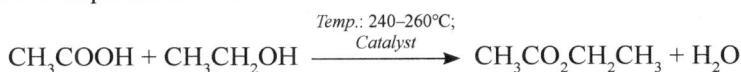

$$CH_3COOH + CH_3CH_2OH \xrightarrow[\text{Catalyst}]{\text{Temp.: 240–260°C;}} CH_3CO_2CH_2CH_3 + H_2O$$

5.15.2.3 Quantitative requirement

For the preparation of 1 ton of butanone, we required 0.55 ton of ethanol, 0.68 ton of acetic acid and 0.5 ton of catalyst.

5.15.2.4 Flow chart

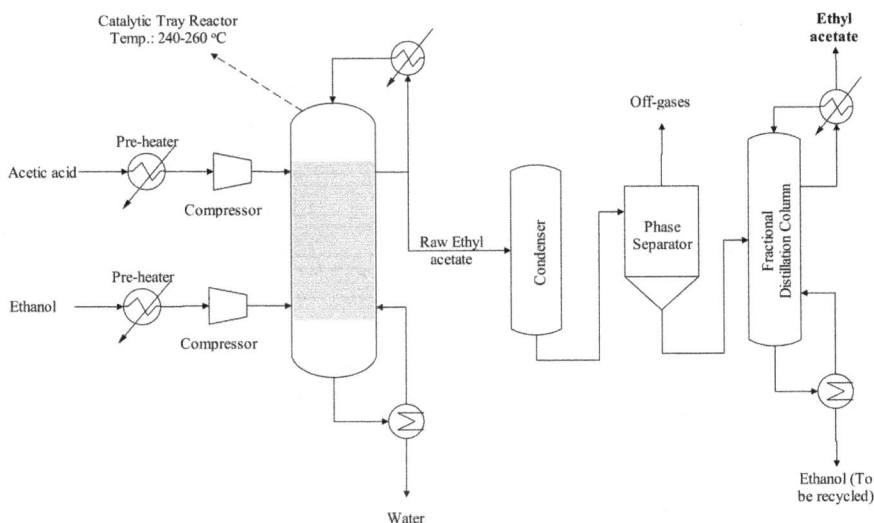

Figure 5.18 Flow sheet diagram of production of Ethyl acetate

5.15.2.5 Process description

As per the simple process flow chart, acetic acid and ethanol are pre-heated in a pre-heater and compressed in a compressor individually. Now, these two compressed gases are fed into a catalytic tray reactor having 30–40 trays. A temperature of 240–260°C is maintained in the reactor. During this reaction, water is liberated, which is collected from the reactor. The remaining mass containing raw ethyl acetate is collected from the top of reactor. It is further cooled in the heat exchanger and condenser. The cooled mass is further transferred into the phase separator, where off-gases are removed from the top. Finally, the mass goes to the fractional distillation column, where pure ethyl acetate is collected as the overhead product, and un-reacted ethanol is collected from the bottom and recycled.

5.15.3 Uses

Ethyl acetate is used as a solvent in printing inks paints, varnishes and perfumes. It is also used for cleaning electric circuit boards and as a nail polish remover. It is utilized as an extraction solvent in the production of pharmaceuticals and food, and as a carrier solvent for herbicides. As it has a fruity smell, it is also used in artificial fruit essences and flavors in confectioneries, ice creams, cakes and pastries (Rao et al., 2015).

5.16 Maleic Anhydride

It is the acid anhydride of maleic acid having four carbon atoms, two hydrogen atoms and three oxygen atoms.

5.16.1 Properties

Table 5.15 Physical & Chemical Properties of Maleic anhydride

Sr. No.	Properties	Particular
1	IUPAC name	Furan-2,5-dione
2	Chemical and other names	Maleic anhydride, 2,5-Furandione, Maleic acid anhydride
3	Molecular formula	$C_4H_2O_3$
4	Molecular weight	98.06 gm/mole
5	Physical description	Colorless or white crystals with pungent odor
6	Solubility	Soluble in acetone, ethyl acetate, chloroform, benzene, toluene, o-xylene, carbon tetrachloride
7	Melting point	52.8°C
8	Boiling point	202.1°C
9	Flash point	110°C
10	Density	1.48 gm/cm³ at 27°C

5.16.2 Manufacturing process

Industrially, it is manufactured by the vapor phase oxidation of benzene or butane.

5.16.2.1 Raw materials

Benzene, Air (Oxygen), Catalyst and Dibutyl phthalate (DBP)

5.16.2.2 Chemical reactions

Benzene undergoes vapor phase oxidation at a temperature of 440–460°C in the presence of a metal oxide catalyst.

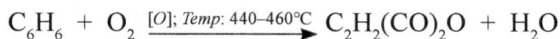

$$C_6H_6 + O_2 \xrightarrow{[O]; \ Temp: \ 440-460°C} C_2H_2(CO)_2O + H_2O$$

5.16.2.3 Quantitative requirements

For the preparation 1 ton of maleic anhydride, we require 0.80 ton of benzene, 0.17 ton of air and 0.42 ton of catalyst.

5.16.2.4 Flow chart

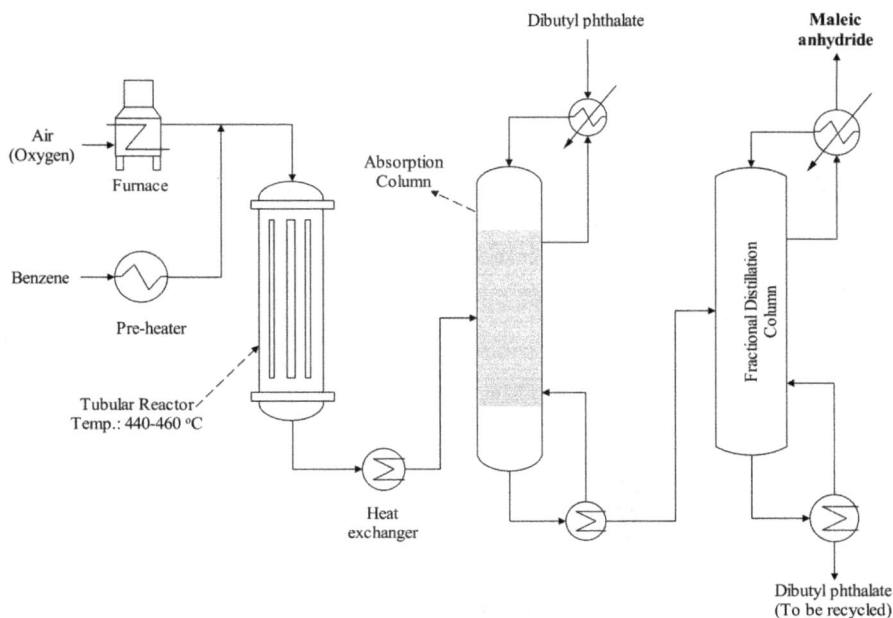

Figure 5.19 Flow sheet diagram of production of Maleic anhydride from Benzene

5.16.2.5 Process description

Benzene is pre-heated and fed into a tubular reactor, which is already impregnated with a catalyst. Also, air (oxygen) is heated upto 450°C and fed into a column. Both reactants are inserted from the top of reactor. After completion of the reaction, the reaction mass is collected from the bottom, cooled down using the heat exchanger and fed into the adsorption column. Dibutyl phthalate (DBP) is added from the top of the column. Here, maleic acid is absorbed in DBP. The Remaining off-gases are removed from the top of the column. Now, DBP in maleic acid is transferred into the distillation column. Here, maleic acid is collected as the overhead product, while DBP is collected from the bottom of the distillation tower and recycled.

5.16.3 Uses

Maleic anhydride is used in the preparation of unsaturated polyester resins (UPR), which are Fiber-reinforced plastics (FRPs). These FRPs are used in pleasure boats, bathroom fixtures, automobiles, tanks and pipes. They are also used in manufacturing coatings, pharmaceuticals, agricultural products, surfactants, and as additivesto plastics (Musa, 2016; Tridevi and Calbertson, 1982).

5.17 Succinic Acid

Its name derives from Latin *succinum*, meaning amber. Because previously it is manufactured by distillation from amber and; thus been known as spirit of amber. Succinic acid is an alpha, omega-dicarboxylic acid.

5.17.1 Properties

Table 5.16 Physical & Chemical Properties of Succinic acid

Sr. No.	Properties	Particular
1	IUPAC name	butanedioic acid
2	Chemical and other names	1,2-Ethanedicarboxylic acid, 1,4 Butanedioic acid, Amber acid, Dihydrofumaric acid
3	Molecular formula	$C_4H_6O_4$
4	Molecular weight	118.09 gm/mole
5	Physical description	Colourless or white, odourless crystals
6	Solubility	Soluble in ethanol, ethyl ether, acetone, methanol; insoluble in toluene, benzene.
7	Melting point	188°C
8	Boiling point	235°C
9	Flash point	160°C
10	Density	1.572 gm/cm³ at 27°C

5.17.2 Manufacturing process

It is prepared by different methods, which are as follows.

(1) **Fermentation of glucose:** This process is discussed in the Fermentation chapter.
(2) **Hydrogenation of maleic acid:** Maleic acid undergoes hydrogenation in the presence of an acid/alkali catalyst.
(3) **Oxidation of 1,4-butanediol:**

$$OHCH_2CH_2CH_2CH_2OH \xrightarrow{[O];Metalcatalyst} HOOCCH_2CH_2COOH$$

(4) **Carbonization of ethylene glycol:**

$$OHCH_2CH_2OH \xrightarrow{Cabronization;KMnO_4} HOOCCH_2CH_2COOH$$

Here, we are discussing hydrogenation of maleic acid.

5.17.2.1 Raw materials

Maleic Acid, Hydrogen gas, Water, Catalyst and Hydrochloric acid

5.17.2.2 Chemical reactions

Maleic acid undergoes liquid phase hydrogenation with hydrogen gas at a temperature of 40–45°C in the presence of a metal preferably nickel and sodium hydroxide catalysta.

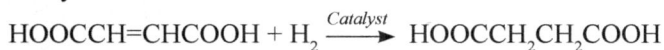

$$\text{HOOCCH=CHCOOH} + \text{H}_2 \xrightarrow{\textit{Catalyst}} \text{HOOCCH}_2\text{CH}_2\text{COOH}$$

In this process, sodium succinate is a by-product.

5.17.2.3 Quantitative requirements

For manufacturing 1000 kgs of succinic acid, we require 980 kgs of maleic acid, 100 kgs of hydrogen gas, 400 kgs of sodium hydroxide and 600 kgs of hydrochloric acid.

5.17.2.4 Flow chart

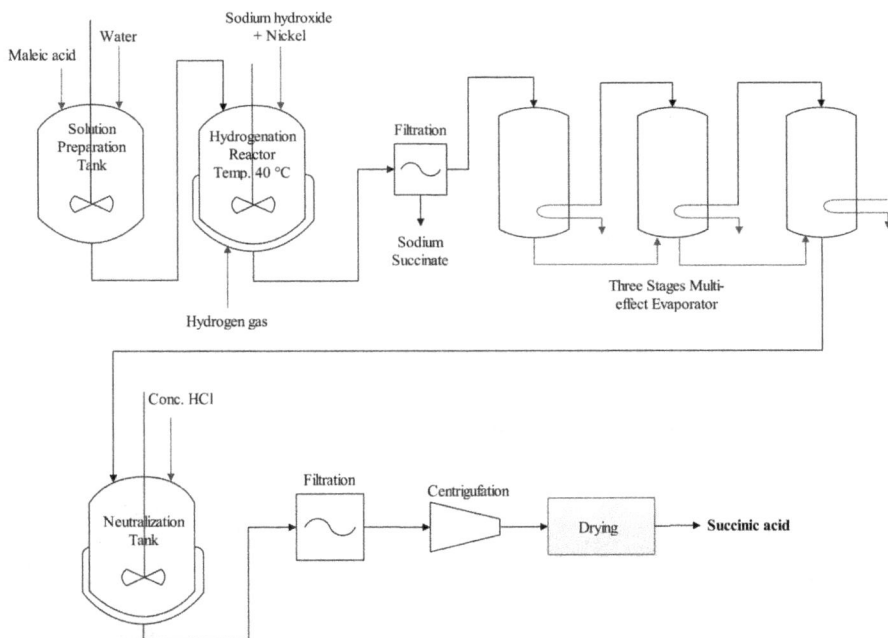

Figure 5.20 Flow sheet diagram of production of Succinic acid from Maleic acid

5.17.2.5 Process description

First, maleic acid is dissolved in water in a solution preparation tank. Maleic acid solution is charged into a hydrogenation reactor. Also, nickel and sodium hydroxide are charged intothereactor. Hydrogen gas is passed into the bottom of the reactor. Here, a temperature of 40–45°C is maintained during the reaction.

After completion of the reaction, the by-product, sodium succinate precipitate is filtered. The remaining filtrate goes to a multi-effect evaporator to concentrate the liquid mass. The concentrated mass is neutralized with concentrated hydrochloric acid, filtered, centrifuged and dried to get pure dry succinic acid.

5.17.3 Uses

It is primary used as a flavoring agent in foods and beverages. It is also utilized as an intermediate for dyes, perfumes, lacquers, photographic chemicals, alkyd resins, plasticizers, metal treatment chemicals, and coatings. It is also utilized as solvent and lubricant (Saxena et al., 2015).

References

Burlage, H.M. and Hawkins, D.B. 1946. Pharmaceutical applications of isopropyl alcohol: As a solvent in pharmaceutical manufacturing, Journal of the American Pharmaceutical Association (Scientific Edition) 35(12): 379–84. https://doi.org/10.1002/jps.3030351208

Carey, F. and Sundberg, R. 2000. Advanced Organic Chemistry. Structure and Mechanisms, Kluwer Publication.

Farshi, A. 2008. Propylene Production Methods and FCC Process Rules. *In*: Propylene Demands, 12th Chemical Engineering Conference, Iran.

Firouzi, E., Hajifatheali, H., Ahmadi, E. and Sefidan, M.M. 2019. An Overview of Acrylonitrile Production Methods: Comparison of Carbon Fiber Precursors and Marketing. Mini-Reviews in Organic Chemistry 16(5). DOI: 10.2174/1570193X16666190703130542

Hansora, D.P. 2014. Industrial manufacturing process of Acrylonitrile. LAP Lambert Academic Publishing.

Hartshorne, H. 2012. A Monograph on Glycerin and Its Uses. Ulan Press. https://doi.org/10.1039/B907530E

Hou, R. 2017. Catalytic and Process Study of the Selective Hydrogenation of Acetylene and 1,3-Butadiene, Springer, Singapore. https://doi.org/10.1007/978-981-10-0773-6

Jiang, Y., Liu, J., Jiang, W., Yang, Y. and Yang, S. et al. 2015. Current status and prospects of industrial bio-production of n-butanol in China. Biotechnology Advances 33(7): 1493–1501. https://doi.org/10.1016/j.biotechadv.2014.10.007

Lavrenov, A.V., Saifulina, L.F., Buluchevskii, E.A. and Bogdanets, E.N. et al. 2015. Propylene production technology: Today and tomorrow. Catalysis in Industry 7: 175–187. https://doi.org/10.1134/S2070050415030083

Li, X., Yang, S., Qian, C. and Chen, X. 2017. Synthesis of Propionitrile by Acrylonitrile Hydrogenation over the Ni Catalyst in the Gas-solid Phase. Chemical Reaction Engineering and Technology 33(1): 15–20.

Luyben, W.L. 2011. Design and Control of the MonoIsopropyl Amine Process, In book: Principles and Case Studies of Simultaneous Design 263–290. DOI: 10.1002/9781118001653

March, E. and MacMillan, C. 1979. Trimethylamine Production in the Caeca and Small Intestine as a Cause of Fishy Taints in Eggs. Poultry Science 58(1): 93–98. https://doi.org/10.3382/ps.0580093

Martin, R. and Buisson, J. 2015. Aromatic Hydroxyketones: Preparation and Physical Properties: Aromatic Hydroxyketones from Butanone (C4) to Dotriacontanone (C32), Springer International Publishing. https://doi.org/10.1007/978-3-319-14185-5

Musa, O.M. 2016. Handbook of Maleic Anhydride Based Materials: Syntheses, Properties and Applications, Springer International Publishing. https://doi.org/10.1007/978-3-319-29454-4

Neamah, A.I. 2022. Acetone Production, The Hilltop Review, Article 12: 13(1).

Pereira, L.G., Dias, M.O.S., Mariano, A.P., Filho, R.M. and Bonomi, A. et al. 2015. Economic and environmental assessment of n-butanol production in an integrated first and second generation sugarcane biorefinery: Fermentative versus catalytic routes. Applied Energy 160: 120–131. https://doi.org/10.1016/j.apenergy.2015.09.063

Rao, K.N., Ratnam, M.V., Reddy, G.K., Prasad, P.R. and Sujatha, V. et al. 2015. Design and Control of Ethyl Acetate Production Process. Emerging Trends in Chemical Engineering, 2(1): 9–20.

Saxena, R.K., Saran, S., Isar, J. and Kaushik, R. 2017. Production and Applications of Succinic Acid, In book: Current Developments in Biotechnology and Bioengineering 601–630. https://doi.org/10.1016/B978-0-444-63662-1.00027-0

Tan, H.W. Raman, A.A.A. and Aroua, M.K. 2013. Glycerol production and its applications as a raw material: A review, Renewable and Sustainable Energy Reviews 27: 118–127. https://doi.org/10.1016/j.rser.2013.06.035

Tohaneanu, M., Valentin, P., Petrica, I. and Bumbac, G. 2014. Simulation and Process Integration of Clean Acetone Plant. Chemical Engineering Transactions 39: 469–474. https://doi.org/10.3303/CET1439079

Traus, E., El. James, D. and Richard. 2008. Nitropropane production process. Patent No. JP2011518162A.

Trivedi, B.C. and Culbertson, B.M. 1982. Maleic Anhydride, Springer US.

White, W. C. 2007. Butadiene production process overview, Chemico-biological Interactions, 166(1-3): 10–14. https://doi.org/10.1016/j.cbi.2007.01.009

Xu, Y., Chuang, K.T. and Sanger, A.R. 2002. Design of a Process for Production of Isopropyl Alcohol by Hydration of Propylene in a Catalytic Distillation Column, Chemical Engineering Research and Design 80(6): 686–694. https://doi.org/10.1205/026387602760312908

Yu, Z., Xu, L., Wei, Y. and Wang, Y. 2009. A new route for the synthesis of propylene oxide from bio-glycerol derived propylene glycol. Chemical Communications 14(26): 3934–6. https://doi.org/10.1039/B907530E

Zakoshansky, V.M. 2007. The Cumene Process for Phenol-Acetone Production. Petroleum Chemistry 47(4): 307–307. https://doi.org/10.1134/S096554410704007X

CHAPTER 6

Aromatic Chemicals

6.1 Introduction

Aromatic compounds are chemical compounds that contain conjugated planar ring systems amid delocalized π-electron clouds *in situ* with individual alternating double and single bonds. The unique stability of these compounds is known as aromaticity. They are additionally referred to as aromatics or arenes. Aromatic compounds are commonly divided into two categories: benzenoids (one containing a benzene ring) and non-benzenoids (those that no longer contain a benzene ring) for example, Furan. Aromatic compounds follow the Huckel rule, which refers to the fact that aromatic compounds have (1) whole delocalization of the π electrons in the ring, (2) Presence of $(4n + 2) \pi$ electrons in the ring where n is an integer (n = 0, 1, 2,...) and in the end (3) a planar structure. Aromatic compounds are immiscible with water and often unreactive (stable). Aromatic compounds are characterized by a sooty yellow flame due to their high ratio of carbon to hydrogen. The basic aromatic compounds are benzene, toluene and xylenes, which are obtained from petroleum. Aromatic compounds are mostly liquid or solid, so, their manufacturing processes are performed in the liquid phase reactor with stirring and heating facilities. They have higher melting and boiling points compared to C1–C5 chemicals. They cover a wide range of chemicals. Actually, industries cannot exist without these aromatic compounds, as they utilize these chemicals (Hepworth et al., 2013).

6.2 BTX Separation

Benzene, toluene and xylenes (BTX) are prepared by various methods, but, the BTX manufacturing process dependson the recapture of aromatics derived from various processes, such as catalytic reforming, coal carbonization, steam cracking of hydrocarbons, dearomatisation of naphtha, Hydro dealkylation and disproportionation. Out of these processes, catalytic reforming of naphtha is considered as one of the best renovation processes in the crude oil industry. This is because, the requisite octane number and the appropriate market price is justified with this process using platinum (Pt), platinum-iridium (Pt-Ir) or platinum-iridium-tin (Pt-Ir-Sn) alloy as a catalyst.

6.2.1 *Properties*

Their main properties are as follows.

Table 6.1 Physical & Chemical Properties of Benzene

Sr. No.	Properties	Particular
1	IUPAC name	Benzene
2	Chemical and other names	Cyclohexatriene, Benzole, Pyrobenzole, Phenyl hydride, Pyrobenzol, Phene
3	Molecular formula	C_6H_6
4	Molecular weight	78.11 gm/mole
5	Physical description	Clear colorless liquid with a petroleum-like odor
6	Solubility	Miscible with water, alcohol, chloroform, ether, acetone, carbon tetrachloride and glacial acetic acid
7	Melting point	5.5°C
8	Boiling point	80°C
9	Flash point	−11°C
10	Density	0.876 gm/cm^3 at 27°C

Table 6.2 Physical & Chemical Properties of Toluene

Sr. No.	Properties	Particular
1	IUPAC name	Toluene
2	Chemical and other names	Methylbenzene, Methylbenzol, Phenylmethane, methacide
3	Molecular formula	C_7H_8
4	Molecular weight	92.14 gm/mole
5	Physical description	Colorless liquid with a sweet, pungent odor
6	Solubility	Miscible with water, alcohol, chloroform, ether, acetone, glacial acetic acid, carbon disulfide
7	Melting point	−94°C
8	Boiling point	111°C
9	Flash point	4°C
10	Density	0.862 gm/cm^3 at 27°C

Table 6.3 Physical & Chemical Properties of m-Xylene

Sr. No.	Properties	Particular
1	IUPAC name	1,3-Xylene
2	Chemical and other names	m-Xylene, 3-Xylene, m-Xylol, 1,3-Dimethylbenzene, m-Methyltoluene
3	Molecular formula	C_8H_{10}
4	Molecular weight	106.17 gm/mole
5	Physical description	Colorless liquid with an sweet odor
6	Solubility	Miscible with water, acetone, alcohol, ether, benzene, chloroform
7	Melting point	−48°C
8	Boiling point	139°C
9	Flash point	25°C
10	Density	0.869 gm/cm^3 at 27°C

6.2.2 Raw materials

Naphtha and Catalyst

6.2.3 Chemical reactions

Various chemical reactions occur in catalytic reforming, but the four mainones are as follows.

(1) Dehydrogenation: Methylcyclohexane (a naphthene) $\rightarrow$ Toluene+ $3H_2$
(2) Isomerization: Normal Octane $\rightarrow$ 2,5-Dimethylhexane (an isoparaffin)
(3) Dehydrocyclisation: Normal Heptanes $\rightarrow$ Toluene + $4H_2$
(4) Hydrocracking: Normal Heptane$\rightarrow$Isopentane +   Ethane

6.2.4 Flow-chart

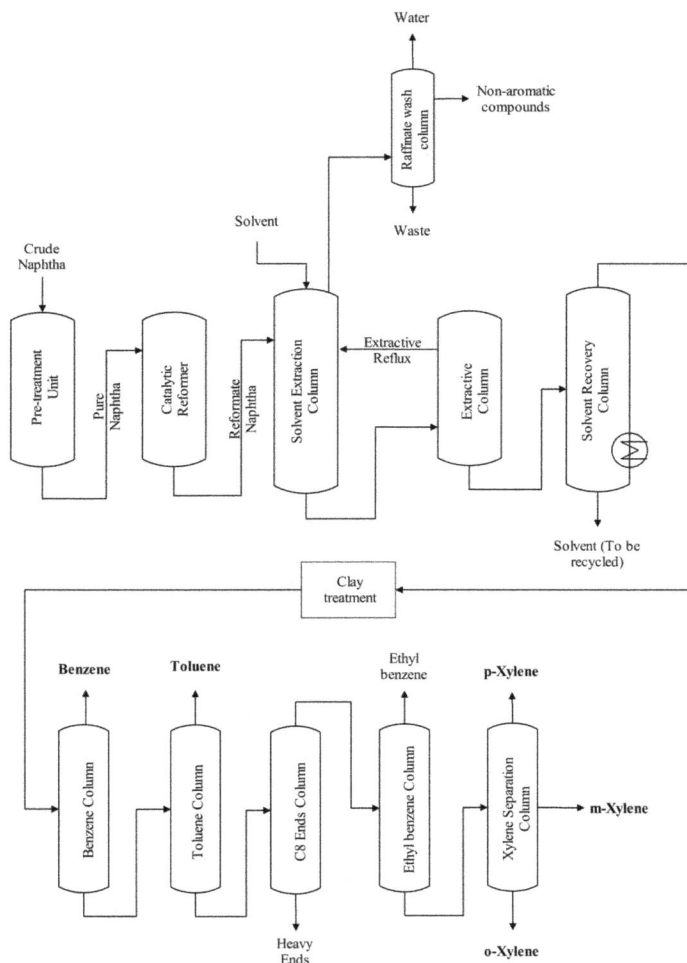

Figure 6.1 Flow sheet diagram of separation of BTX by Naphtha reforming process

6.2.5 Process description

Naphtha containing non-aromatic hydrocarbonswith 11 or 12 carbons undergoes catalytic reforming to produce various products like BTX (C6 to C8 compounds), paraffins and heavier aromatic compounds (C9 to C11 compounds).

It is divided into the following processes.

Naphtha pretreatment unit: In the first part of reforming, pretreatment of naphtha is conducted for the elimination of different impurities that breakdown the catalyst's activities. These impurities include sulfur, nitrogen, halogens, oxygen, water, olefins, di-olefins, arsenic and other metals, which are already present in naphtha.

Catalytic reforming: After removal of impurities, pure naphtha undergoes catalytical reforming at a higher temperature and pressure. There are various reactions simultaneously conducted in the reforming process. This process is classified according to the catalyst regeneration as follows: (1) Semi Regenerative catalytic reforming, (2) Cyclic catalytic reforming and (3) Continuous catalytic reforming (CCR), in which a CCR unit is more useful because lower pressure operations give products with higher octane ratings.

Solvent extraction: After reforming, naphtha is inserted into the solvent extraction column along with the solvent. Non-aromatic compounds are separated from the top of extraction column as raffinate. This raffinate is further washed with water to remove impurities. Aromatic compounds are extractive column and solvent recovery column. The solvent is recovered and further used. The reaction mixture is purified using clay treatment and fed into a benzene and toluene column. Here, benzene and toluene are separated respectively. Now, the heavy ends are removed from the bottom of the C8 column. The remaining methyl benzene and xylenes are collected from the top of the column and fed into the methyl benzene column. Here, methyl benzene is separated from the xylenes. Different xylenes, i.e., ortho, meta and para are separated in the xylene column.

6.2.6 Uses

Benzene

Benzene undergoes various unit processes like nitration, sulfonation, halogenation, hydration, and hydrogenation. It is used mainly as an intermediate to other chemicals, in particular ethylbenzene, cumene, cyclohexane, nitrobenzene, and alkylbenzene. More than half of the total benzene produced is processed into ethylbenzene, a precursor to styrene, which is employed to form polymers and plastics like polystyrene and EPS. Some 20% of the benzene production is employed to manufacture cumene, which is required for phenol and acetophenone production, inputs for the production of resins and adhesives. Cyclohexane consumes about 10% of the world's benzene production; it is primarily utilized in the manufacture of nylon fibers, which are processed into textiles and engineering plastics.

Minimal quantities of benzene are utilized in the production of rubbers, lubricants, dyes, detergents, drugs, explosives, and pesticides.

Toluene

Toluene is mainly used as a pioneer to benzene through hydrodealkylation. It's a feedstock for toluene diisocyanate (used in the manufacture of polyurethane foam), trinitrotoluene (the explosive, TNT), and a range of artificial drugs. Toluene might also be a frequent solvent, e.g., for paints, paint thinners, silicone sealants, many chemical reactants, rubber, printing inks, adhesives (glues), lacquers, leather-based tanners, and disinfectants. Toluene is regularly used as an octane booster in gas fuels for combustion engines. It is employed as an intoxicative inhalant.

p-Xylene

p-Xylene is a crucial chemical feedstock. It is used for manufacturing different polymers. It is a component, especially, in the production of terephthalic acid for polyesters like polyethylene terephthalate. It can also be polymerized to produce parylene.

o-Xylene

Most of the o-Xylene produced is used in the manufacture of anhydride, which is utilized in manufacturing phthalate plasticizers, unsaturated polyester resins and alkyd resins. Small portions are employed for their solvent function in structuring bactericides, soya bean herbicides and lube oil additives. A substitute outlet for o-Xylene is in the manufacturing of polyethylene naphthenate (PEN) polymer, which may additionally be an excessive overall performing polyester utilized in movie films and inflexible packaging (Sweeney and Bryan, 2000; Mohammed and Baki, 2008; McVey et al., 2020; Kolmetz et al., 2003; Niziolek et al., 2016).

6.3 Phenol

It is one of the important chemicals, as it is used in manufacturing various compounds. It is also useful in different chemical industries.

6.3.1 Properties

Its main properties are as per Table 6.4.

Table 6.4 Physical & Chemical Properties of Phenol

Sr. No.	Properties	Particular
1	IUPAC name	Phenol
2	Chemical and other names	Hydroxybenzene, Carbol, Carbolic acid, Benzenol, Oxybenzene, Phenyl hydrate
3	Molecular formula	C_6H_6O
4	Molecular weight	94.11 gm/mole
5	Physical description	Colorless to light-pink, crystalline solid with a sweet, acrid odor
6	Solubility	Soluble in water, alcohol, chloroform, ether, glycerol, carbon disulfide, petrolurm
7	Melting point	41°C
8	Boiling point	182°C
9	Flash point	85°C
10	Density	1.06 gm/cm³ at 27°C

6.3.2 Manufacturing process

Initially phenol was prepared from coal tar, but now-a-days, it is manufactured industrially from petroleum by various processes such as cumene peroxidation-hydrolysis; from benzene and benzenesulfonate; from the direct oxidation of benzene; and the oxidation of toluene of (quaintly: about 7 billion kg/year).

6.3.2.1 Cumene Peroxidation-hydrolysis

Cumene Peroxidation-hydrolysis was first discovered in 1944 by Heinrich Hock, so, this process is also known as Hock process. In this process, benzene and propylene are raw materials.

(A) Raw materials: Cumene, Air, Sulfuric acid and Sodium hydroxide

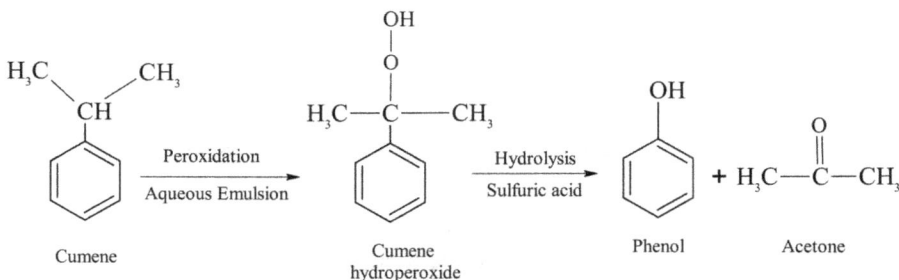

Figure 6.2 Synthesis of Phenol by Cumene process

(B) Chemical reactions: Cumene undergone peroxidation to yield cumene hydroperoxide, which on hydrolysis with sulfuric acid gives phenol.

(C) Quantitative requirements

For manufacturing 1000 kgs of phenol, we require 1450 kgs of cumene, 1630 kgs of air, 100 kgs of sulfuric acid and 250 kgs of sodium hydroxide.

(D) Flow chart

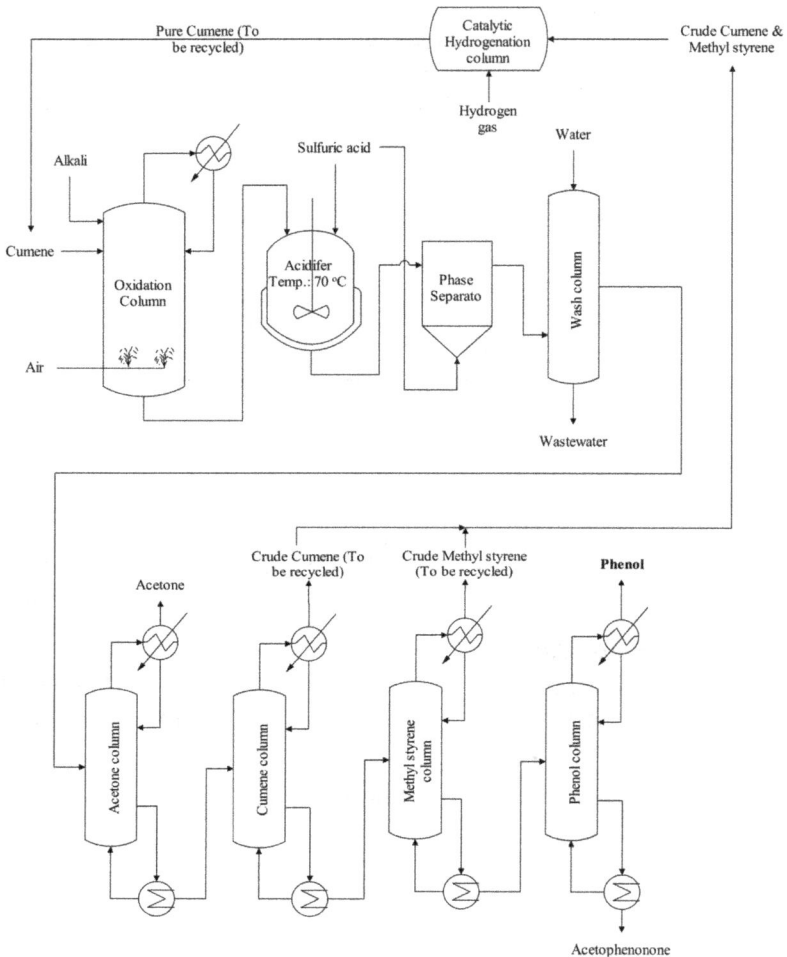

Figure 6.3 Flow sheet diagram of production of Phenol by Cumene process

(E) Process description

As per Fig. 6.3, fresh cumene and purified recycled cumene along with dilute soda ash solution are fed to the oxidation vessel in order to maintain the pH between 6.0 and 8.0. Also, compressed air is added from the bottom of vessel. Here, a mixture of cumene and soda ash is reacted with air at 110–115°C till 20–30% of the cumene is transformed to cumene peroxide. Yields of cumene peroxide may also be expanded by way of running at decreased temperatures (100–110°C) and allowing the cumene to react in a sequence of three to five steps in the oxidation vessel. The yield may be increased upto 80%. Thereafter, the crude combination is accrued from the bottom of the oxidizer and inserted into a reactor in which phenol and acetone are formed from cumene peroxide. The reactor is facilitated with a jacket and stirrer. The reaction typically occurs with prerequisites of a low temperature (70 to 80°C) and pressure in the presence of a small quantity of sulfuric acid.

The reaction mixture is collected from the bottom of the reactor and sent to the separator. The lower layer of cumene peroxide is added to the reactor. The upper layer containing phenol is separated and washed with water. Distillation is carried out followed by washing, to separate acetone as the overhead product and undergoes further purification. Unreacted cumene and methyl styrene are collected from the bottom of the distillation column. Further, unreacted cumene and methyl styrene are separated using vacuum distillation. Here, the byproduct, methyl styrene, is collected as the overhead, undergoes catalytical hydrogenation to yield cumene, which is recycled. The bottoms containing acetophenone and cumene are further separated using a vacuum still. About 90–92% phenol is obtained as the overhead product.

6.3.2.2 From benzene

It is also prepared using benzene.

(A) Raw materials: Benzene, Hydrochloric Acid, Air, Water and Catalyst

(B) Chemical reactions: Benzene undergoes hydrochlorination in the presence of catalyst ($FeCl_3$ + $CuCl_2$) at 240°C to give chlorobenzene. Phenol is prepared by the hydrolysis of chlorobenzene in the presence of silicon oxide at 350°C.

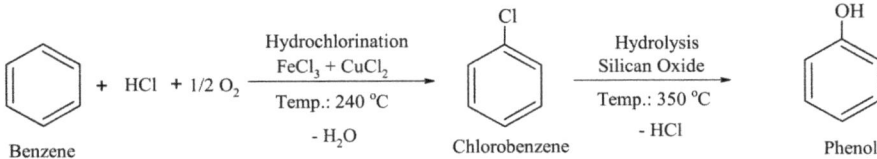

Figure 6.4 Synthesis of Phenol from Benzene

(C) Quantitative requirements: For manufacturing 1000 kgs of phenol, we require 1200 kgs of benzene, 200 kgs of hydrochloric acid, 2400 kgs of air and 5000 kgs of water.

(D) Flow chart

(E) Process description: Pure benzene is pre-heated and fed into a reactor, which is already packed with ferric chloride and cupric chloride catalyst. Also, a mixture of hydrochloric acid and oxygen (air) is fed at 220°C. About 20% of benzyl chloride is obtained from benzene, and thus, conversion is very low. The other part of the reaction mixture contains benzyl chloride and poly benzyl chlorides. The gaseous reaction mixture is fed into a fractional distillation still, where un-reacted benzene and polychlorobenzeneare separated. The crude benzyl chloride is washed with phenol and water to pure benzyl chloride. It is further pre-heated upto 350°C and fed into a tubular reactor, which is packed with a catalyst. Chlorobenzene is converted to phenol at this temperature. It is further washed with benzene and water. Finally, the fractional distillation process is conducted to separate phenol.

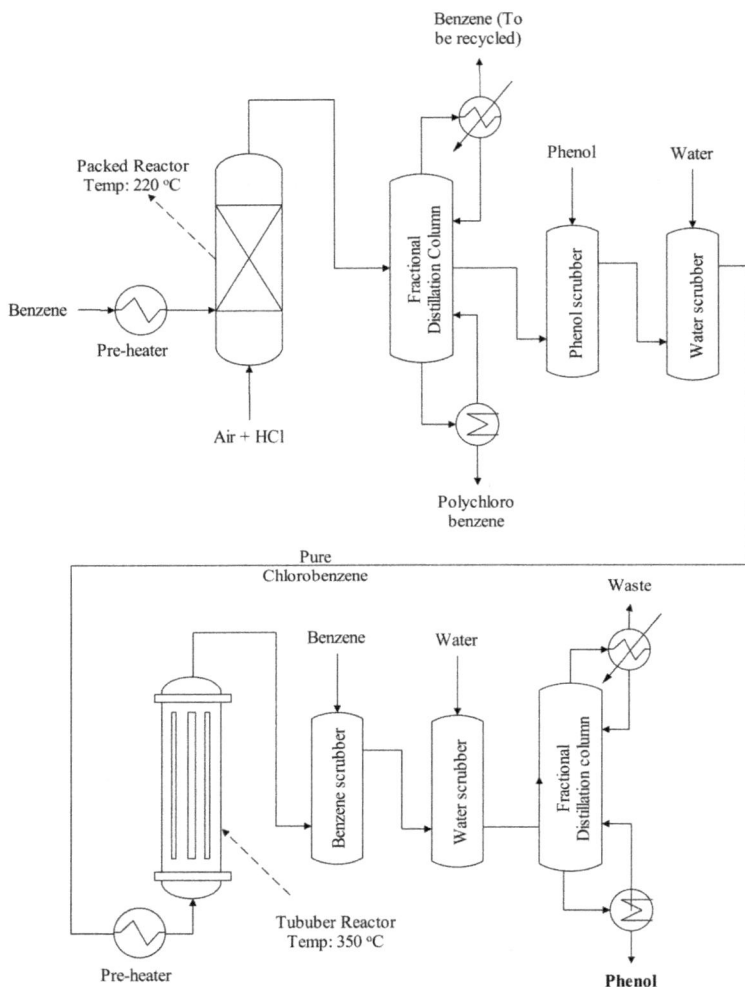

Figure 6.5 Flow sheet diagram of production of Phenol from Benzene

(F) Major engineering problems

1. Polymerization of chlorobenzene is minimized by maintaining reactor design conditions with low residence time and economical balance.

2. Processing of wet hydrochloric acid at a high temperature leads to corrosion in the hydrochloric acid reactor. The ceramics of the reactor have to be protected or a highly corrosion resistant reactor structure is required.

6.3.3 Uses

Due to the many uses of phenol, it is one of the most important constituents of natural compounds. Phenol finds utility in many industries and has been commercially synthesized on massive scales.

a) Two thirds of the total phenol production is used in preparing reagents. These reagents are used in the plastic manufacturing industries. The condensation of phenol with acetone produces Bisphenol A, which is extensively utilized in polymer industries to synthesize various epoxide resins and polycarbonates.
b) Phenolic resins are prepared by the co-polymerization of phenol and formaldehyde. The resulting resin is the phenol-formaldehyde resin, commercially marketed as Bakelite. Bakelite is extensively utilized in electrical switches and automobiles thanks to its property of withstanding extreme conditions of warmth and resistance to electricity and other chemicals. Novolac resin is employed as a binding agent and protective coating.
c) It is also utilized in the extraction of bio-molecules and nucleic acids.
d) Epoxy resins are manufactured from phenol and used to produce paints coatings and moldings. It is also utilized in polycarbonate plastics which may be seen in CDs and domestic electrical appliances. Phenol is additionally utilized in the cosmetic industry in the manufacturing of sunscreens, skin lightening creams and hair coloring solutions. It's an ingredient that won't kill microorganisms. It also prevents or inhibits the expansion and reproduction of microorganisms.
e) Phenol sprays are mostly utilized in the medical industry. Phenol sprays work alright within painful and aggravated areas. They need anesthetic and analgesic combinations which help in giving instant relief from allergies too. Phenol gives an immediate relief from pharyngitis. It is also utilized in the manufacturing of pesticides and insecticidesin the pharma industry (Zakoshansky, 2007; Schmidt, 2005; Baynazarov et al., 2018; Ganji and Tabarsa, 2011).

6.4 Nitrobenzene

Nitrobenzene is a primary aromatic nitrate.

6.4.1 Properties

Its main properties are as follows.

Table 6.5 Physical & Chemical Properties of Nitrobenzene

Sr. No.	Properties	Particular
1	IUPAC name	Nitrobenzene
2	Chemical and other names	Nitrobenzol, Mirbane oil
3	Molecular formula	$C_6H_5NO_2$
4	Molecular weight	123.11 gm/mole
5	Physical description	Pale yellow to dark brown liquid with a pungent odor like paste shoe polish
6	Solubility	Miscible with water, ethanol, ether, acetone, benzene
7	Melting point	6°C
8	Boiling point	211°C
9	Flash point	88°C
10	Density	1.203 gm/cm³ at 27°C

6.4.2 Manufacturing process

6.4.2.1 Raw materials

Benzene, Nitric acid and Sulfuric acid

6.4.2.2 Chemical reactions

It is prepared by the nitration of benzene using nitric acid and sulfuric acid. This reaction is exothermic.

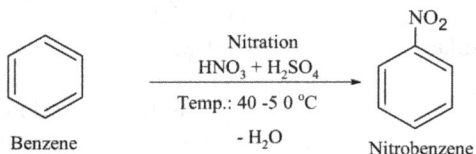

Figure 6.6 Synthesis of Nitrobenzene from Benzene

6.4.2.3 Quantitative requirements

For manufacturing 1000 kgs of nitrobenzene, we require 600 kgs of benzene, 700 kgs of nitric acid and 850 kgs of sulfuric acid.

6.4.2.4 Flow chart

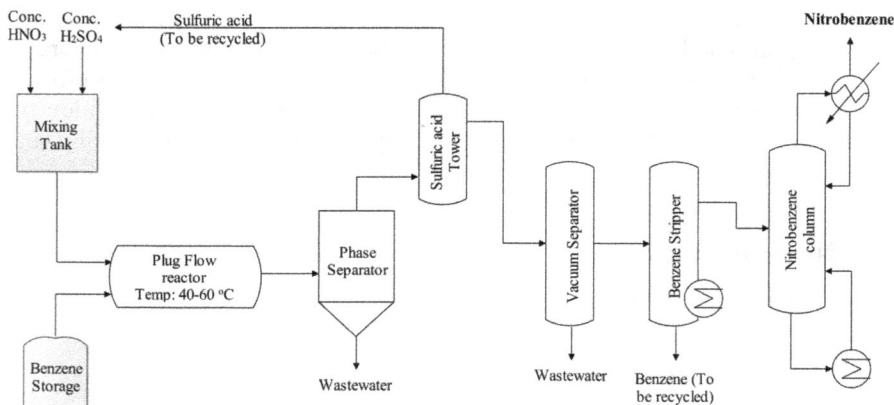

Figure 6.7 Flow sheet diagram of production of Nitrobenzene from Benzene

6.4.2.5 Process description

First, nitric acid and sulfuric acid are mixed and fed into a plug-flow reactor. Also, benzene is fed into this reactor. This reaction is exothermic, so, the temperature is maintained upto 40–50°C. After completion of the reaction, reaction mass undergoes phase and vacuum separation to remove wastewater and un-reacted sulfuric acid from the reaction mass. Then, un-reacted benzene is collected from the benzene stripper and recycled. Finally, it is purified in the nitrobenzene column using steam.

6.4.2.6 Major engineering problems

All equipment including reactor, separators, stripper, etc., are corrected by means of strong acids.

6.4.3 Uses

Most of the nitrobenzene is used to produce aniline by hydrogenation. It is also used to prepare synthetic lubricants, which are used in machinery and motors. It is employed in the manufacture of dyes, drugs, pesticides and synthetic rubber (Meng et al., 2011; Agustriyanto et al., 2017).

6.5 Aniline

It is a primary aromatic amine.

6.5.1 Properties

Its main properties are as follows.

Table 6.6 Physical & Chemical Properties of Aniline

Sr. No.	Properties	Particular
1	IUPAC name	Aniline
2	Chemical and other names	Benzenamine, Aminobenzene, Phenylamine, Cyanol, Benzidam
3	Molecular formula	C_6H_7N
4	Molecular weight	98.16 gm/mole
5	Physical description	Yellowish to brownish oily liquid with a musty fishy odor
6	Solubility	Miscible in water, ethanol, ethyl ether, benzene and acetone
7	Melting point	–6°C
8	Boiling point	184°C
9	Flash point	76°C
10	Density	1.022 gm/cm³ at 27°C

6.5.2 Manufacturing process

Industrially, it is prepared by either the catalytic hydrogenation of nitrobenzene or amination of phenol with ammonia. Here, we are discussing both methods.

6.5.2.1 Hydrogenation of nitrobenzene

(A) Raw materials

Nitrobenzene and Catalyst

(B) Chemical reactions: Nitrobenzene is undergoes vapor phase hydrogenation in the presence of zinc and hydrochloric acid at a temperature of 320–340°C and a pressure of 2–5 atmosphere to give aniline as per Fig. 6.8.

(C) Quantitative requirements: For manufacturing 1000 kgs aniline, we require 1400 kgs of nitrobenzene and 700 kgs of catalyst.

(D) Flow chart

(E) Process description: As per the flow chart, nitrobenzene is heated upto 300–320°C in a heat exchanger and converted to the vapor state. This vapor is

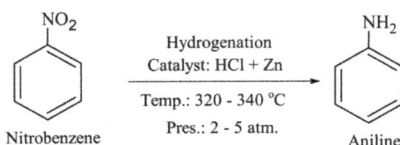

Figure 6.8 Synthesis of Aniline by hydrogenation of Nitrobenzene

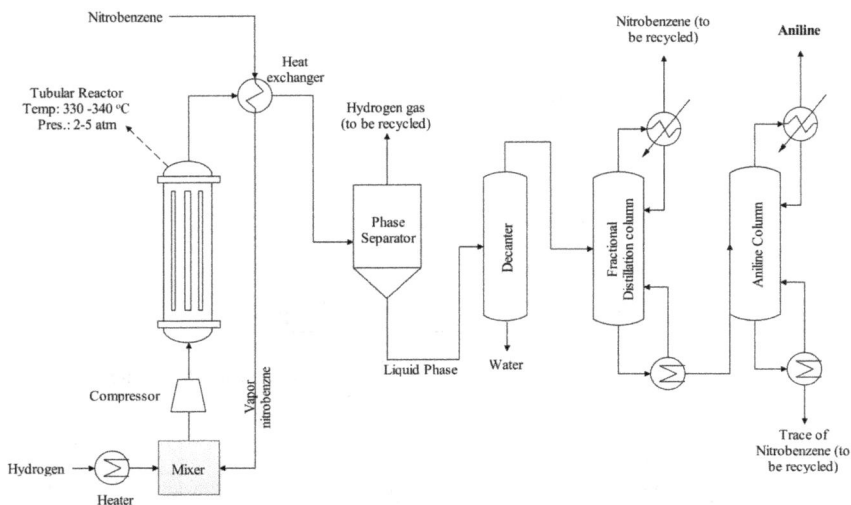

Figure 6.9 Flow sheet diagram of production of Aniline by hydrogenation of Nitrobenzene

mixed with pre-heated hydrogen, compressed and fed into a tubular reactor. The rector is already impregnated with the catalyst (Zn + HCl). A temperature of 320–340°C and a pressure of 2–5 atmospheres is maintained during the reaction. After completion of the reaction, the mass is cooled down to the room temperature and sent to the phase separator. Here, un-reacted hydrogen gas is separated from the top and recycled. Water is removed in a decanter from the liquid phase. Aniline is separated from the unreacted nitrobenzene in two fractionating columns.

6.5.2.1 Amination of phenol

(A) Raw materials

Phenol, Ammonia and Catalyst

(B) Chemical reactions

The direct amination of phenol in the presence of a silica-alumina catalyst using ammonia at a temperature of 280–320°C and a pressure of 6–10 atmospheres gives aniline.

Diphenylamine is a by product of the amination of phenol.

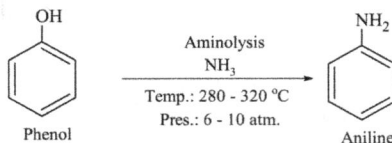

Figure 6.10 Synthesis of Aniline from amination of Phenol

(C) Quantitative requirements: For manufacturing 1000 kgs of aniline, we require 1100 kgs of phenol, 210 kgs of ammonia and 650 kgs of catalyst.

(D) Flow chart

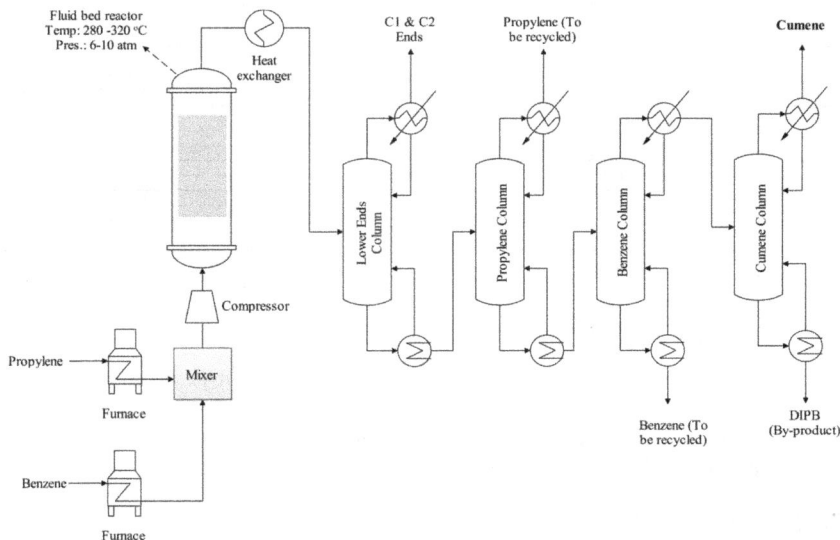

Figure 6.11 Flow sheet diagram of production of Aniline from amination of Phenol

(E) Process description: First, phenol and ammonia are pre-heated using individual furnaces. These vapors are mixed, compressed and fed into a fluid bed reactor. The reactor is already impregnated with catalyst. A temperature of 280–320°C and a pressure of 6–10 atmospheres is maintained during the reaction. After completion of the reaction, the mass is cooled down to room temperature and sent to a phase separator. Here, un-reacted ammonia gas is separated from the top and recycled. Water and the byproduct diphenylamine is removed in a decanter from the liquid phase. Aniline separated from the unreacted nitrobenzene in two fractionating columns.

6.5.3 Uses

Most of the aniline produced is used to manufacture methylene diphenyl isocyanate (MDI), which is used to prepare polyurethane foams as insulators for refrigerators, freezers, and buildings. It is also employed in various industries such as polymers, rubbers, pigments, agro-based chemicals, varnishes and explosives, and dyes (Dutta et al., 2011).

6.6 Cumene

This aromatic hydrocarbon consists of aliphatic substitution.

6.6.1 Properties

Its main properties are as per Table 6.7.

Table 6.7 Physical & Chemical Properties of Cumene

Table 6.7 Physical & Chemical Properties of Cumene

Sr. No.	Properties	Particular
1	IUPAC name	Cumene
2	Chemical and other names	2-phenylpropane, isopropylbenzene, cumol,
3	Molecular formula	C_9H_{12}
4	Molecular weight	120.19 gm/mole
5	Physical description	Colorless liquid with a sharp, penetrating, aromatic odor
6	Solubility	Miscible with water, ethanol, ethyl ether, acetone, benzene, petroleum ether, carbon tetrachloride
7	Melting point	–97°C
8	Boiling point	152°C
9	Flash point	39°C
10	Density	0.866 gm/cm³ at 27°C

6.6.2 Manufacturing process

Friedel-Crafts alkylation of benzene is conducted with propylene to get Cumene. Previously, sulfuric acid was used as a catalyst in this alkylation reaction, but due to drawbacks of neutralization and recycling; and corrosion problem, solid phosphoric acid + alumina is used as the catalyst now-a-days.

6.6.2.1 Raw materials

Benzene, Propylene and Catalyst

6.6.2.2 Chemical reactions

Benzene undergoes vapor phase Friedel-Crafts alkylation with propylene in the presence of a catalyst [solid phosphoric acid (SPA) supported on alumina] at a temperature of 200–220°C and a pressure of 12–20 atmospheres. Also, diisopropyl benzene (DIPB) is a by product of this reaction as per Fig. 6.12.

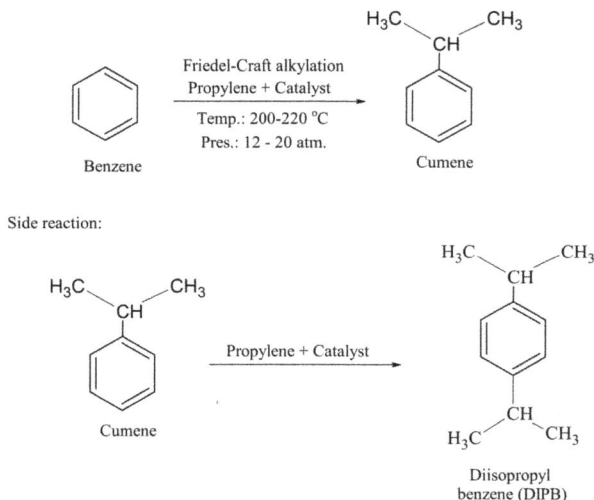

Figure 6.12 Synthesis of Cumene by Friedel Craft alkylation of Benzene

6.6.2.3 Quantitative requirement

For manufacturing 1000 kgs of cumene, we require 800 kgs of benzene, 480 kgs of propylene and 740 kgs of catalyst.

6.6.2.4 Flow chart

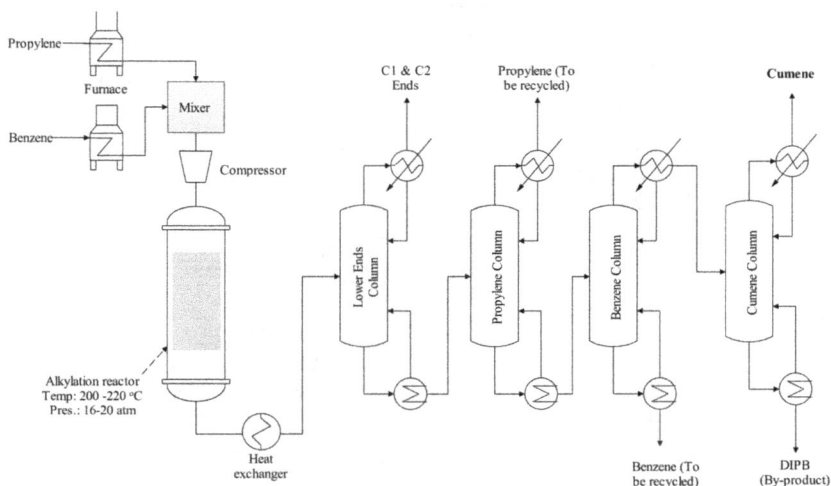

Figure 6.13 Flow sheet diagram of production of Cumene by Friedel Craft alkylation of Benzene

6.6.2.5 Process description

Benzene and propylene are heated in individual furnaces, mixed, compressed and fed intothe top of the alkylation reactor. A temperature of 220–220°C and a pressure of 12–20 atmospheres is maintained in the reactor. After completion of the reaction, the mass is cooled down to room temperature. The cooled mass goes to the lower end of the propylene and benzene columns to remove impurities from the cumene mass. Finally, it is separated from the DIPB to yield pure cumene.

6.6.3 Uses

Mainly, cumene is used for manufacturing phenol and acetone, but has other important industrial applications too. It is additionally used as a solvent in lacquers, paints, and enamels. It is also used in the metal and paper industries (Luyben, 2009; Vinila, et al., 2021).

6.7 Styrene

Styrene, also referred to as vinyl benzene, is a colorless oily liquid. It smells sweet at higher concentrations. Styrene is a precursor to polystyrene and a number of other copolymers.

6.7.1 Properties

Its main properties are as per Table 6.8.

Table 6.8 Physical & Chemical Properties of Styrene

Sr. No.	Properties	Particular
1	IUPAC name	Styrene
2	Chemical and other names	Vinylbenzene, Styrol, Phenylethylene, Cinnamene, Phenethylene
3	Molecular formula	C_8H_8
4	Molecular weight	104.42 gm/mole
5	Physical description	Clear colorless to dark liquid with an aromatic odor
6	Solubility	Immiscible with water; miscible in carbon disulfide, alcohol, ether, methanol, acetone
7	Melting point	−31°C
8	Boiling point	145°C
9	Flash point	35°C
10	Density	0.901 gm/cm³ at 27°C

6.7.2 Manufacturing process

The manufacturing of styrene by the dehydrogenation of ethylbenzene was first accomplished in the 1930s. The manufacturing of styrene had greathistorical significance throughout the 1940s, when it was once referred to as the feedstock for artificial rubber. Since it is manufactured on such a large scale, ethylbenzene is organized on a prodigious scale (by alkylation of benzene with ethylene).

6.7.2.1 Raw materials

Benzene, Ethylene, Anhydrous Aluminum chloride, Sodium hydroxide and Iron oxide

6.7.2.2 Chemical reactions

Benzene undergoes alkylation with ethylene in the presence of anhydrous aluminum chloridecatalyst at 95–100°C to give ethylbenzene, which is dehydrogenated in the presence of FeO catalyst at 800–850°C to give styrene.

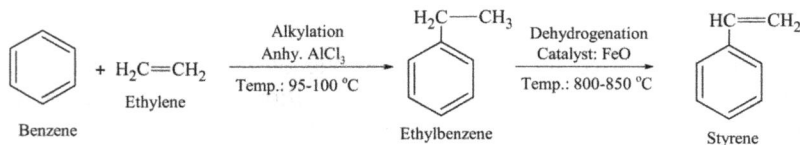

Figure 6.14 Synthesis of Styrene from Benzene

The first reaction is exothermic and the second reaction is endothermic.

6.7.2.3 Quantitative requirements

For manufacturing 1000 kgs of styrene, we require 900 kgs of benzene, 400 kgs of ethylene, 1400 kgs of anhydrous aluminum chloride, 3810 kgs of sodium hydroxide and 1070 kgs of iron oxide.

6.7.2.4 Flow chart

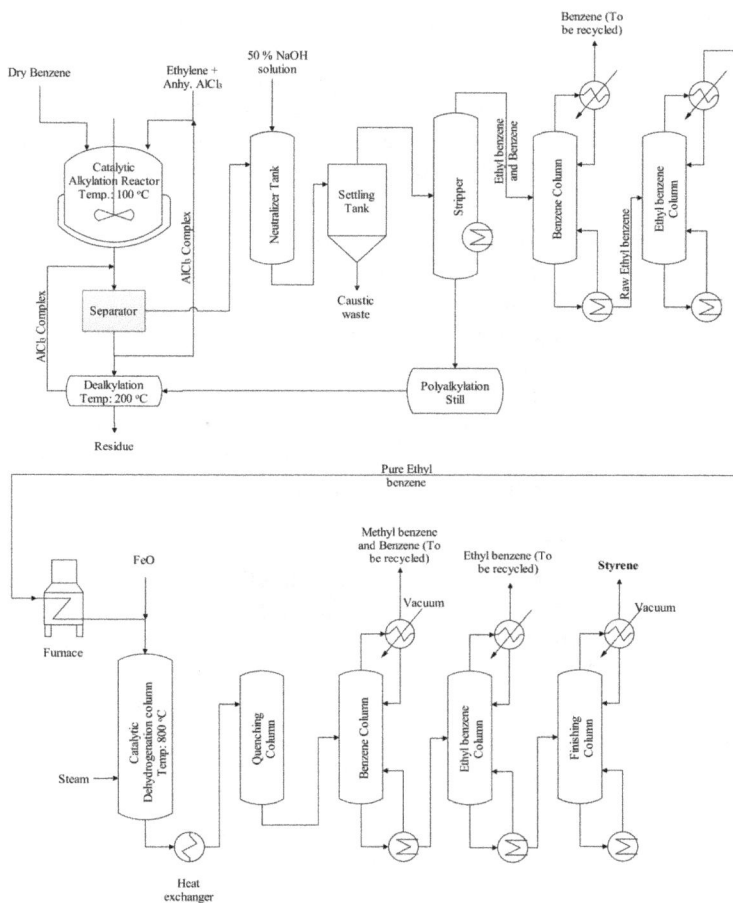

Figure 6.15 Flow sheet diagram of production of Styrene from Benzene

6.7.2.5 Process description

A typical flow-chart of styrene,consisting of two steps is shown in Fig. 6.15.

(1) Preparation of ethylbenzene:

As per Fig. 6.15, first, dry benzene, anhydrous aluminum chloride and ethylene are charged into a jacketed alkylation reactor having a stirrer. As the reaction is exothermic, cooling water is circulated in the jacket to maintain the reactor temperature. After completion of the reaction, the product mixture containing incondensable gases and the liquid product with the $AlCl_3$ complex undergoes separation in a separator. A small part of the $AlCl_3$ complex is recycled into the alkylation reactor to keep a pre-requesite quantity of catalyst. The remaining $AlCl_3$ complex is inserted into a dealkylator unit, where, the catalyst is regenerated at 200°C and recycled. Now, the product consisting of ethylbenzene and acidic impurities is cooled down and neutralized with 50% NaOH. Here, the acidic impurities are removed from ethylbenzene. Then, polyalkylbenzenes are separated from ethylbenzene + benzene in a stripper, polyalkyl still and a heat integrated exchanger. The purified mixture of ethylbenzene + benzene is inserted into a benzene column to separate wet benzene from ethylbenzene. Benzene is further purified and recycled in the alkylation reactor. Now, ethylbenzene is purified by washing it with a caustic solution. It is then converted into vapor using a pre-heater.

(1) **Preparation of styrene:** Pure dried vapor of ethylbenzene alongwith excess superheated steam is fed into a catalytic dehydrogenator. After completion of the reaction, the vapor mass is cooled down in the quench tower. Thereafter, styrene is separated from benzene and toluene. The benzene-toluene distillation column is used to recover benzene as the overhead product. The benzene is recycled to the azeotropic distillation unit. The bottom part of the distillation column consisting of styrene and ethylbenzene is fed into the ethylbenzene column to separate the ethylbenzene and styrene streams. Finally, styrene is further purified in the finishing column.

6.7.2.6 Major engineering problems

1. Anhydrous raw materials are required to eliminate unwanted chemical reactions in the alkylation reactor and catalyst poisoning.
2. Reducing loss of aromatics contained in polyalkylbenzenes by high temperature dealkylation.
3. Minimizing losses of $AlCl_3$ catalyst by extracting residues from the high temperature dealkylator.
4. Control of the dehydrogenation reaction by use of a large mole ratio of superheated steam to ethylbenzene.
5. Prevention of the unwanted polymerization of styrene during purification is accomplished by the inhibitor and refrigerator.

6.7.3　Uses

Styrene undergoes co-polymerization using different monomers to prepare styrene-divinylbenzene (S-DVB), acrylonitrile-butadiene-styrene (ABS), styrene-butadiene rubber (SBR), styrene-butadiene latex, SIS (styrene-isoprene-styrene), S-EB-S (styrene-ethylene/butylene-styrene) and styrene-acrylonitrile resin (SAN). These materials are used in the manufacture of food containers, rubbers, plastics, pipes, insulations, automobiles, boat parts, fiberglass, and carpet backings (Dimain et al., 2019; Miller et al., 1994; Luyben, 2010).

6.8　Phthalic anhydride

It is an anhydride of Phthalic acid.

6.8.1　Properties

Its main properties are as per Table 6.9.

Table 6.9 Physical & Chemical Properties of Phthalic anhydride

Sr. No.	Properties	Particular
1	IUPAC name	2-Benzofuran-1,3-dione
2	Chemical and other names	1,2-Benzenedicarboxylic anhydride, Phthalic acid anhydride, Phthalic anhydride, Isobenzofuran-1,3-dione
3	Molecular formula	$C_8H_4O_3$
4	Molecular weight	148.11 gm/mole
5	Physical description	Clear, colorless to white lustrous needles solid with a characteristic, acrid odor.
6	Solubility	Soluble in water, ethanol, acetone, benzene
7	Melting point	130.2°C
8	Boiling point	295.5°C
9	Flash point	152°C
10	Density	1.53 gm/cm³ at 27°C

6.8.2 Manufacturing process

It is prepared by the oxidation of Naphthalene or O-Xylene.

6.8.2.1 Raw materials

Naphthalene or O-Xylene, Oxygen and Catalyst

6.8.2.2 Chemical reactions

Naphthalene or o-xylene undergoes air oxidation in the presence of vanadium pentoixde as a catalyst at 360°C to give phthalic anhydride.

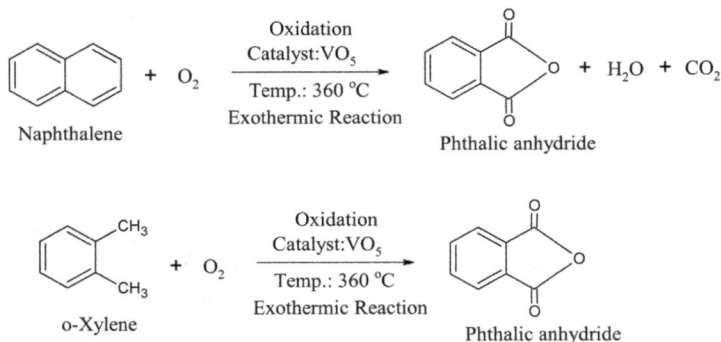

Figure 6.16 Synthesis of Phthalic anhydride from Naphthalene or o-Xylene

6.8.2.3 Quantitative requirements

For manufacturing 1000 kgs of phthalic anhydride we require 1200 kgs or naphthalene or 1000 kgs of o-xylene, 2500 kgs of air and 870 kgs of vanadium pentoxide as a catalyst.

6.8.2.4 Flow chart

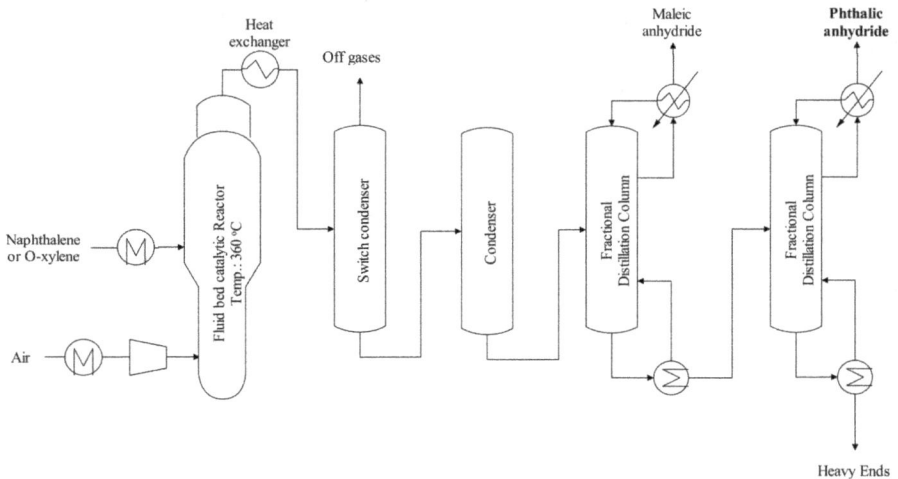

Figure 6.17 Flow sheet diagram of production of Phthalic anhydride from Naphthalene or o-Xylene

6.8.2.5 Process description

A simple flow chart is shown in Fig. 6.17 for the production of phthalic anhydride, in which phthalic acid or o-xylene is heated and inserted into a fluid bed oxidation reactor. Also, air is compressed, heated and fed into the bottom of the reactor. The exothermic heat of reaction is removed using molten salts, which generates high pressure steam. The obtained mixture of gases is cooled down to room temperature and fed to switch condensers. Here, desublimation of phthalic anhydride is conducted using heat exchange with cold oil. The off-gases obtained from the switch condensers are treated in an incinerator. Instead of incineration, sometime the off-gases are treated with water in the scrubber to form a maleic acid aqueous solution, which is then used to prepare fumaric acid or maleic anhydride as a by-product. The remaining reaction mixture is collected from the bottom and fed into the top of the distillation column. Here, the by-product maleic acid is separated from the reaction mixture and collected from the top of the column. Thereafter, the reaction mixture is fed into another distillation column. Heavy end products are separated from the bottom column. Pure phthalic anhydride is collected from the top of the column.

6.8.2.6 Major engineering problems

1. Explosion hazards are minimized by adding excess air to maintain the lower explosive limit.
2. Fluid bed reactor is more compatible than the fixed bed reactor.
3. A convenient catalyst must be developed for high specificity of oxidation.
4. If a fixed bed reactor is developed for the production of phthalic anhydride, then a tubular fixed bed with proper tube sizes that maintain the heat through the heat exchanger is required. This avoids very high temperatures at the center of the reactor. Further, mercury and diphenyl are used as the catalyst and coolant respectively.

6.8.3 Uses

Phthalic anhydride is used for the production of phthalate plasticizers, which are employed in the construction industry. It is used in the resin, pigment and dye industries (El-Gharbawy, 2021; Giarola et al., 2015).

6.9 Benz aldehyde and Benzoic Acid

Benzaldehyde is the simplest aromatic aldehyde, having a formyl substituent. Initially, it was extracted in 1803 by the French pharmacist Martres from the fruit of *Prunus dulcis* in. Then, Friedrich Wohler and Justus von Liebig first prepared it using Toluene.

6.9.1 Properties

Its main properties are as follows.

Table 6.10 Physical & Chemical Properties of Benzaldehyde

Sr. No.	Properties	Particular
1	IUPAC name	Benzaldehyde
2	Chemical and other names	Benzoic aldehyde, Phenylmethanal, Benzenecarbonal, Benzene carbaldehyde
3	Molecular formula	C_6H_7O
4	Molecular weight	106.12 gm/mole
5	Physical description	Colourless to yellow liquid with a sweet, strong almond odour
6	Solubility	Miscible in water, ethanol, ethyl ether, benzene
7	Melting point	−31°C
8	Boiling point	179°C
9	Flash point	27°C
10	Density	1.051 gm/cm³ at 27°C

Table 6.11 Physical & Chemical Properties of Benzoic acid

Sr. No.	Properties	Particular
1	IUPAC name	Benzoic acid
2	Chemical and other names	Dracylic acid, Benzenecarboxylic acid, Carboxybenzene, Benzeneformic acid
3	Molecular formula	$C_6H_7O_2$
4	Molecular weight	122.12 gm/mole
5	Physical description	White crystal scales or needles with a faint urine, almond odour
6	Solubility	Miscible in water, acetone, benzene, carbone tetrachloride, chloroform, ethanol, ethyl ether, hexane
7	Melting point	123°C
8	Boiling point	250°C
9	Flash point	122°C
10	Density	1.268 gm/cm³ at 27°C

6.9.2 Manufacturing process

Both organic chemicals are prepared by various methods; benzaldehyde is manufactured by the partial oxidation of benzyl alcohol and toluene, alkaline hydrolysis of benzal chloride and the carbonylation of benzene. Benzoic acid is prepared by the acid hydrolysis of benzonitrile and benzamide, carboxylation of phenyl magnesium bromide and; oxidation of toluene, benzyl alcohol and benzyl chloride. But commonly, they are manufactured by the oxidation of toluene.

6.9.2.1 Raw materials

Toluene, Acetic acid and Air (Oxygen)

6.9.2.2 Chemical reactions

Toluene undergoes liquid phase oxidation in the presence of Fe and Mn salts at a high temperature to give benzaldehyde and benzoic acid.

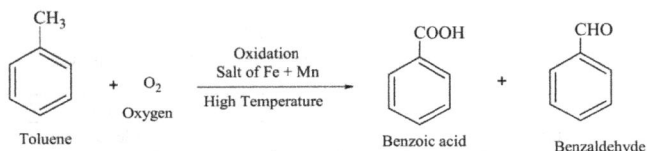

Figure 6.18 Synthesis of Benzoic acid and Benzaldehyde

6.9.2.3 Flow chart

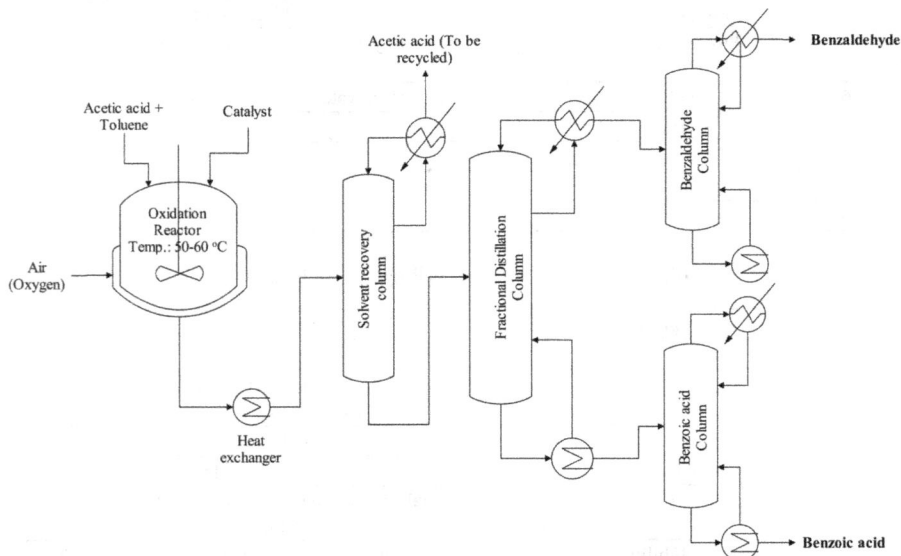

Figure 6.19 Flow sheet diagram of production of Benzoic acid and Benzaldehyde

6.9.2.4 Process description

First, acetic acid and toluene are taken into an oxidation reactor facilitated with a stirrer and hot water jacket. The catalyst is also added. Now, air (oxygen) is passed into the bottom of the reactor with stirring. A temperature of 50–60°C is maintained during the reaction. After completion of the reaction, the mass is collected from the reactor and transferred into a solvent recovery tower, where acetone is recovered and recycled. Thereafter, benzaldehyde and benzoic acid are separated from the fractional distillation tower. They are further purified in individual columns.

6.9.3 Uses

Benzaldehyde is considered as the most useful chemical in the aldehyde group chemicals. As it has almond-like smell, it is used as a flavonoid in food beverages and cosmetic personal care products. It is also used as a bee repellent. It is widely utilized in the production of numerous aromatic chemicals, which are used in the manufacture of pharmaceuticals, plastics and dyes.

Benzoic acid is essentially used as a preservative in acidic food items and beverages. It is also utilized in the production of several aromatic as well as aliphatic compounds, which are used in perfumes, dyes, soaps, detergents, topical medications and insect repellents. Its sodium salt (sodium benzoate) is commonly employed in buffer solutions and the food industry (Renita et al., 2020; Zhao et al., 2018).

6.10 Chlorobenzenes

It is the simplest aromatic halogenated product. There are several types of chlorobenzenes available, but monochlorobenzene and 1,4-dichlorobenzene are utilized more than other chlorobenzenes. So, we are discussing monochlorobenzene and 1,4-dichlorobenzene.

6.10.1 Properties

The main properties of monochlorobenzene are as follows.

Table 6.12 Physical & Chemical Properties of Monochlorobenzene

C	Properties	Particular
1	IUPAC name	Chlorobenzene
2	Chemical and other names	Monochlorobenzene, Phenyl chloride, Benzene chloride, Chlorobenzol
3	Molecular formula	C_6H_5Cl
4	Molecular weight	112.55 gm/mole
5	Physical description	Clear colorless to yellowish liquid with a sweet almond-like odor
6	Solubility	Miscible in water, ethanol, ethyl ether, benzene, carbon tetrachloride
7	Melting point	–45°C
8	Boiling point	132°C
9	Flash point	27°C
10	Density	1.106 gm/cm³ at 27°C

6.10.2 Manufacturing process

It is prepared by the reaction of chlorine gas and benzene.

6.10.2.1 Raw materials

Chlorine gas, Benzene and Catalyst

6.10.2.2 Chemical reactions

Benzene undergoes chlorination in the presence of a catalyst at a temperature of 30–40°C to yield monochlorobenzene and 1,4-dichlorobenzene.

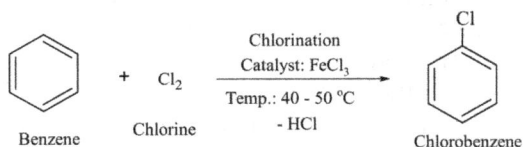

Figure 6.20 Synthesis of Chlorobenzene

This is an exothermic reaction. So, the temperature is controlled upto 30–40°C to minimize the production of byproducts, i.e., other chlorobenzenes. Sometimes, a small quantity of fuller's earth or monochlorobenzene itself is added to reduce side reactions.

6.10.2.3 Quantitative requirements

For manufacturing 1000 kgs of monochlorobenzene, we required 8100 kgs of benzene, 350 kgs of chlorine gas and 410 kgs of catalyst.

6.10.2.4 Flow chart

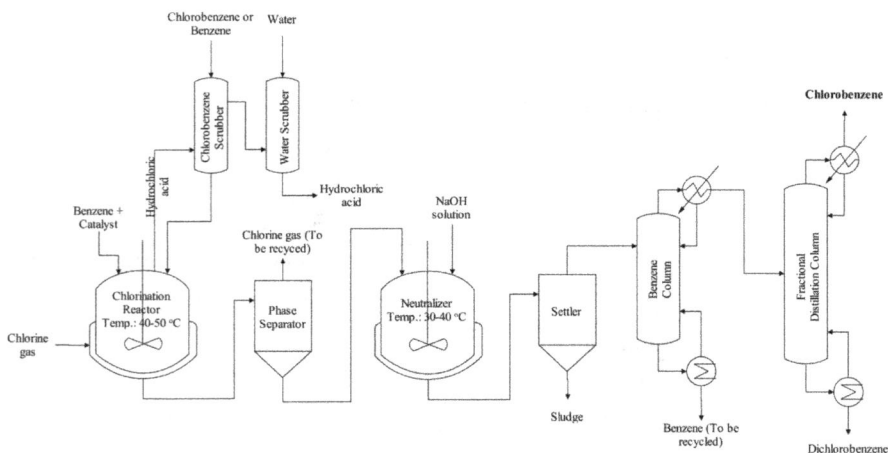

Figure 6.21 Flow sheet diagram of production of Chlorobenzene

6.10.2.5 Process description

First, benzene and catalyst are taken into an oxidation reactor facilitated with a stirrer and cold water jacket. Now, chlorine gas is passed into the bottom of the reactor with stirring. The reaction is exothermic, so, a temperature of 40–50°C is maintained during the reaction. During the reaction, hydrochloric acid is scrubbed into two stages scrubbers, i.e., monochlorobenzene and water scrubber. After completion of the reaction, the reaction mass is transferred into the phase separator to separate chlorine gas and then recycled. The bottom layer of the separator is neutralized with casuistic soda solution and settled down to separate sludge from the reaction mass. Thereafter un-reacted benzene is separated out from the mass. Finally, monochlorobenzene and 1,4-dichlorobenzene are separated from the fractional distillation column.

6.10.3 Uses

Mainly, it is employed in different pesticides like DDT, 1,4-Dichlorobenzene and 1,4,5-Trichlorobenzene. It is also used in different industries like adhesives, paints, paint removers, polishes, dyes, drugs, agro-based chemicals and rubber. (Lui et al., 2021).

[Note: Bromobenzene is prepared using the same process by the bromination of benzene in the presence of a catalyst at a temperature of 30–40°C]

6.11 Acetophenone

Acetophenone is one among the only types of ketones. It is available naturally in different foods, such as bananas, beef, apples, apricots, cheese, and cauliflowers.

6.11.1 Properties

Its main properties are as follows.

Table 6.13 Physical & Chemical Properties of Acetophenone

Sr. No.	Properties	Particular
1	IUPAC name	1-Phenylethanone
2	Chemical and other names	Acetophenone, Methyl phenyl ketone, Phenyl methyl ketone, Acetylbenzene
3	Molecular formula	C_8H_8O
4	Molecular weight	120.15 gm/mole
5	Physical description	Colourless liquid or white crystals with sweet pungent odor
6	Solubility	Slightly soluble in water, Soluble in ethanol, ether, acetone, benzene,
7	Melting point	20°C
8	Boiling point	202°C
9	Flash point	78°C
10	Density	1.028 gm/cm³ at 27°C

6.11.2 Manufacturing process

The starting material for the preparation of acetophenone is benzene.

6.11.2.1 Raw materials

Benzene, Ethane, Anhydrous Aluminum chloride Catalyst, Air (Oxygen), Manganese acetate and Caustic soda

6.11.2.2 Chemical reactions

Benzene undergoes Friedel-crafts alkylation with ethylene in the presence of anhydrous aluminum chloride catalyst at 95–100°C to give ethylbenzene, which undergoes oxidation in the presence of manganese acetate catalystata temperature of 135–145°C and a pressure of 3–5 atmospheres to yield the final product, acetophenone.

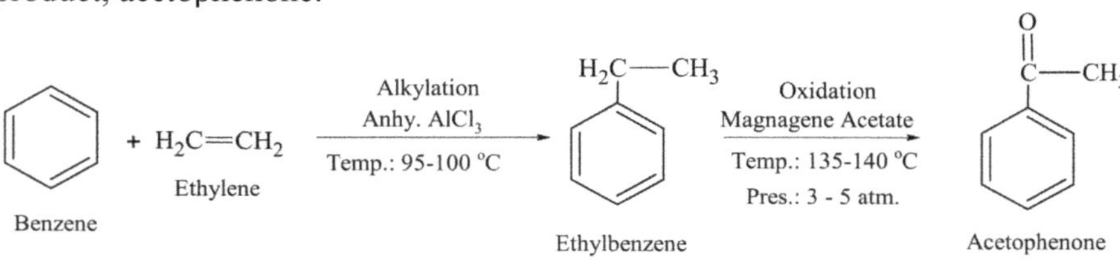

Figure 6.22 Synthesis of Acetophenone from Benzene

Here, the first reaction is exothermic.

6.11.2.3 Quantitative requirements

For manufacturing 1000 kgs of acetophenone, we require 680 kgs of benzene, 270 kgs of ethane, 780 kgs of anhydrous aluminum chloride, 180 kgs of air (oxygen), 880 kgs of manganese acetate and 410 kgs of caustic soda.

6.11.2.4 Flow chart

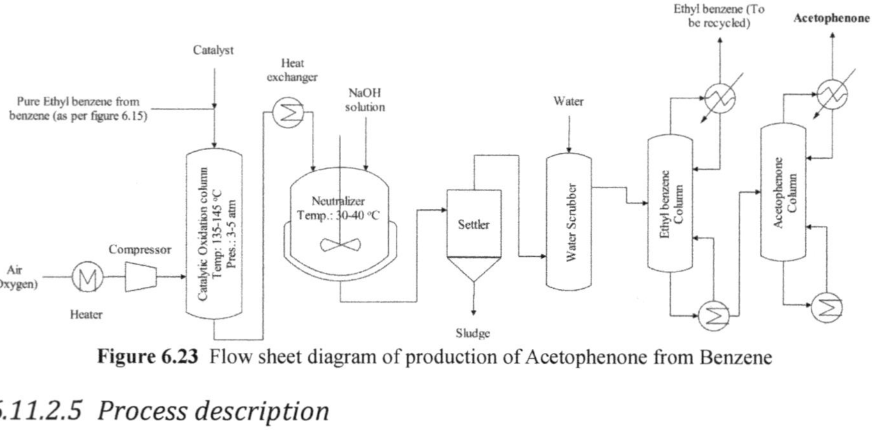

Figure 6.23 Flow sheet diagram of production of Acetophenone from Benzene

6.11.2.5 Process description

As per Fig. 6.23, ethyl benzene is produced and inserted into an oxidation reactor. The catalyst is fed into the reactor. Now, air (oxygen) is pre-heated, compressed and fed into the bottom of the reactor. This reaction is exothermic; a temperature of 135–145°C and a pressure of 3–5 atmospheres is maintained using cold water circulation. After completion of the reaction, the mass is cooled down to the room temperature and transferred into the neutralization reactor. Here, the mass is neutralized by means of a caustic solution. Sludge is removed using a settler and scrubbed in a water scrubber. Now, un-reacted ethyl benzene is separated and recycled. Finally, acetophenone is purified in another column.

6.11.3 Uses

As it has a floral type odor, it is widely used as an essential ingredient for creating fragrances resembling cherry, almond, jasmine, honey-suckle and strawberrys. It is also used for manufacturing styrene and resins. Previously, it was used in medicines as an anti convulsant and hypnotic. Acetophenone also can be used as a catalyst in olefin polymerization. Acetophenone is additionally used as a fragrance ingredient in detergents, soaps, lotions and creams (Sivcev et al., 2012).

6.12 Benzophenone

Benzophenone is the simplest ether containing two benzene rings.

6.12.1 Properties

Its main properties are as follows.

Table 6.14 Physical & Chemical Properties of Benzophenone

Sr. No.	Properties	Particular
1	IUPAC name	Diphenylmethanone
2	Chemical and other names	Benzoylbenzene, Phenyl ketone, Benzophenone
3	Molecular formula	$C_{13}H_{10}O$
4	Molecular weight	182.22 gm/mole
5	Physical description	White solid with flowery odor
6	Solubility	Insoluble in water, ethanol. ethyl ether, acetone, benzene, chloroform
7	Melting point	19°C
8	Boiling point	342°C
9	Flash point	-
10	Density	1.085 gm/cm³ at 27°C

6.12.2 Manufacturing process

Benzophenone is prepared by two different methods: (1) Friedel–Crafts alkylation of benzene with benzoyl chloride followed by oxidation, and (2) Benzene is reacted with tetrachloride to form intermediate diphenyldichloromethane. This intermediate undergone hydrolysis; followed by oxidation of DPM in the presence of chromic acid and nitric acid to get benzophenone. Here, we are discussing the first method.

6.12.2.1 Raw materials

Benzene, Benzoyl chloride, Anhydrous Aluminum chloride Catalyst, Air (Oxygen), Copper naphthalene catalyst and Caustic soda

6.12.2.2 Chemical reactions

Benzene and benzyl chloride are undergo Friedel-crafts alkylation in the presence of anhydrous aluminum chloride at a temperature of 50–70°C to give diphenylmethane. It undergoes further oxidation with air (oxygen) in the presence of copper naphthalene catalyst at a temperature of 175–185°C.

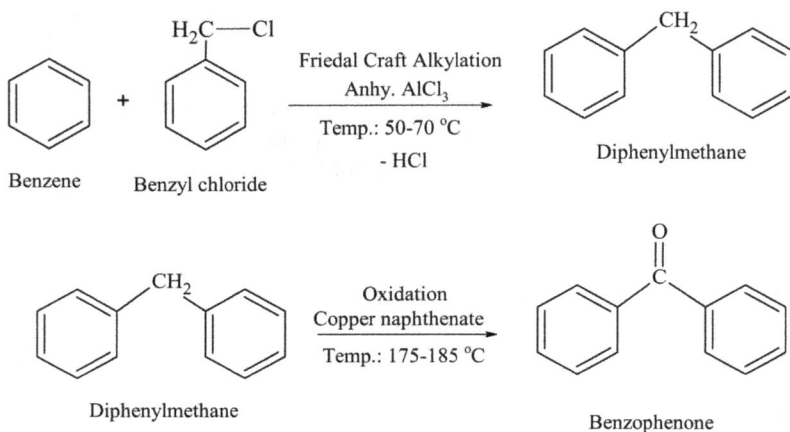

Figure 6.24 Synthesis of Benzophenone

6.12.2.3 Quantitative requirements

For manufacturing 1000 kgs of benzophenone, we require 510 kgs of benzene, 841 kgs of benzoyl chloride, 780 kgs of anhydrous aluminum chloride, 110 kgs of air (oxygen), 740 kgs of copper naphthalene catalyst and 570 kgs of caustic soda.

6.12.2.4 Flow chart

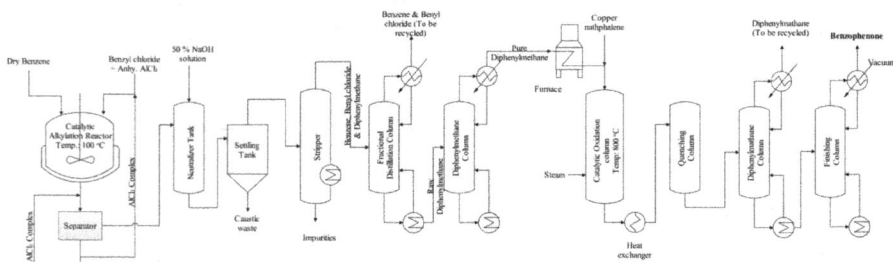

Figure 6.25 Flow sheet diagram of production of Benzophenone

6.12.2.5 Process description

A typical flow-chart for styrene manufacture, is shown in Fig. 6.25, consisting of two steps.

(1) **Preparation of ethylbenzene:** Dry benzene, anhydrous aluminum uncondensable and benzyl chloride are inserted into a jacketed alkylation reactor containing a stirrer. As the reaction is exothermic, water is sed in the jacket to maintain the temperature. After the reaction, the product mixture contains uncondensable gases and the liquid containing the $AlCl_3$ complex which is is separated in a separatorto obtain the pre-requisite quantity of catalyst; the aluminum chloride complex stream is partially recycled to the alkylator. The remaining part of this complex is sent to the dealkylator unit. Here, the feed is heated to 200°C to regenerate the catalyst and recycled. The mass is neutralized with 50% caustic solution and made to settle down for the removal of impurities from the bottom. The supernatant layer then enters a stripper to separate the impurities. Thereafter, un-reacted benzene and benzyl chloride are separated. Finally, diphenylmethane is purified in another column.

(2) **Preparation of benzophenone:** The obtained diphenylmethane is pre-heated and fed to the catalytic oxidation reactor. This reactor is already impregnated with the catalyst. Also, steam enters the reactor to maintain a temperature of 175–185°C. The reaction mass is cooled in the heat exchanger and quenched. Un-reacted diphenylmethane is separated from the mass. Finally, benzophenone is purified in another column.

6.12.3 Uses

Benzophenone is employed as an ultraviolet curative mediator, flavor element, fragrance booster and perfume fixative. It is also used in several industries such as plastics, agricultural chemicals, pharmaceuticals, soaps, dyes, polymers, coatings and adhesive formulations. (Goto et al., 2011).

6.13 Diphenyl Oxide

It is the simplest diaryl ether, with a variety of niche applications.

6.13.1 Properties

Its main properties are as per Table 6.15.

Table 6.15 Physical & Chemical Properties of Diphenyl oxide

Sr. No.	Properties	Particular
1	IUPAC name	Phenoxybenzene
2	Chemical and other names	Diphenyl oxide, Phenoxybenzene, Phenyl ether, Oxydibenzene
3	Molecular formula	$C_{12}H_{10}O$
4	Molecular weight	170.21 gm/mole
5	Physical description	Colorless, crystalline solid or liquid with a geranium-like odor.
6	Solubility	Soluble in water, ethanol, ether, benzene, acetic acid; slightly soluble in chloroform
7	Melting point	38°C
8	Boiling point	258°C
9	Flash point	115°C
10	Density	1.08 gm/cm³ at 27°C

6.13.2 Manufacturing process

It is prepared by two methods: (1) Dehydration of phenol in the presence of a catalyst at a high temperature and (2) By reacting phenol with bromobenzene in the presence of an alkaline solution and a copper catalyst. Here, we are discussing the dehydration of phenol. It is also obtained as a by-product in the manufacture of phenol during the high-pressure hydrolysis of chlorobenzene.

6.13.2.1 Raw materials

Phenol and the catalyst comprising tungsten oxide with alumina, zirconia and titania

6.13.2.2 Chemical reactions

Two moles of phenol undergoee liquid phase dehydration in the presence of a catalyst at a temperature of 350–450°C to yield diphenyl oxide.

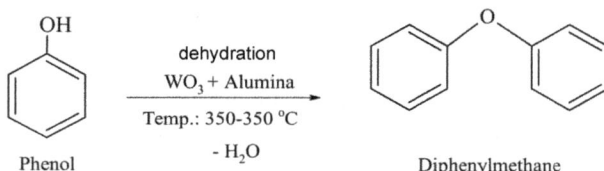

Phenol

dehydration
WO_3 + Alumina
Temp.: 350-350 °C
- H_2O

Diphenylmethane

Figure 6.26 Synthesis of Diphenyl oxide by dehydration of Phenol

6.13.2.3 Quantitative requirements

For manufacturing 1000 kgs of biphenyl oxide, we require 1180 kgs of phenol and 570 kgs of catalyst.

6.13.2.4 Flow chart

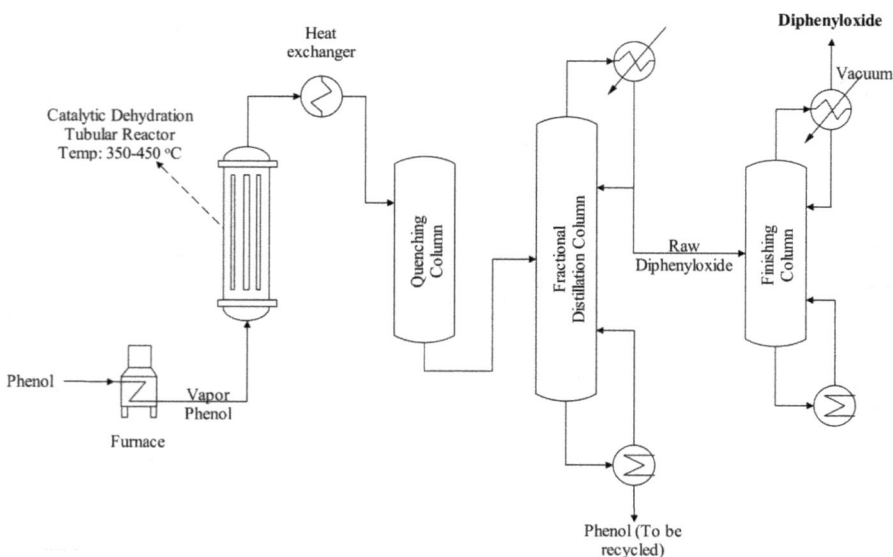

Figure 6.27 Flow sheet diagram of production of Diphenyl oxide by oxidation of Phenol

6.13.2.5 Process description

First, phenol is pre-heated to convert it into vapor form, thereafter phenol vapor is fed into the bottom of adehydration tubular reactor, which is already impregnated with catalyst. Now, steam enters the reactor to maintain a temperature of 350–450°C. After completion of the reaction, the vapor mass is cooled and quenched. Thereafter, the mass goes to the fractional distillation column for the separation of the un-reacted phenol and diphenyl oxide. Further, diphenyl oxide is purified in the finishing column.

6.13.3 Uses

As the odor of diphenyl oxide is like that of geranium, it is essentially used in the perfumery compounding, soaps, detergents and incense sticks. It is used as a high boiling solvent in various industries like drugs, textiles, dyes, mining, petroleum, agriculture and pigments. It is also used as a heat transfer agent and a dye carrier. Diphenyl oxide along with Diphenyl forms a heat transfer fluid, which is used at temperatures as high as 400°C (Yangirov et al., 2021).

6.14 Dimethyl Terephthalate

It is the diester of terephthalic acid and methanol.

6.14.1 Properties

Its main properties are as per Table 6.16.

Table 6.16 Physical & Chemical Properties of Dimethyl Terephthalate

Sr. No.	Properties	Particular
1	IUPAC name	Dimethyl benzene-1,4-dicarboxylate
2	Chemical and other names	Dimethyl p-phthalate, Dimethyl terephthalate, Methyl 4-carbomethoxybenzoate
3	Molecular formula	$C_{10}H_{10}O_4$
4	Molecular weight	194.5 gm/mole
5	Physical description	Colourless odorless crystal
6	Solubility	Soluble in chloroform; slightly soluble in ethanol, methanol
7	Melting point	140°C
8	Boiling point	288°C
9	Flash point	150°C
10	Density	1.24 gm/cm³ at 27°C

6.14.2 Manufacturing process

It is prepared by the direct esterification of terephthalic acid with methanol. It is also prepared by p-xylene. p-Xylene or methyl p-xylene is undergoes oxidation in the presence of a catalyst to give terephthalic acid, which esterifies with methanol to give dimethyl terephthalate.

6.14.2.1 Raw materials

p-Xylene or Methyl p-Xylene, Air (Oxygen), Cobalt/Manganese/Bromine Catalyst, Acetic acid, Methanol and Si/Al catalyst

6.14.2.2 Chemical reactions

p-Xylene or methyl p-xylene is undergoes liquid phase oxidation in the presence of cobalt/manganese/bromine catalyst at a temperature of 220–240°C and a pressure of 100–105 atmospheres to give terephthalic acid. This acid is further esterified with methanol using Si/Al catalyst at 280–300°C and a high pressure to yield dimethyl terephthalate.

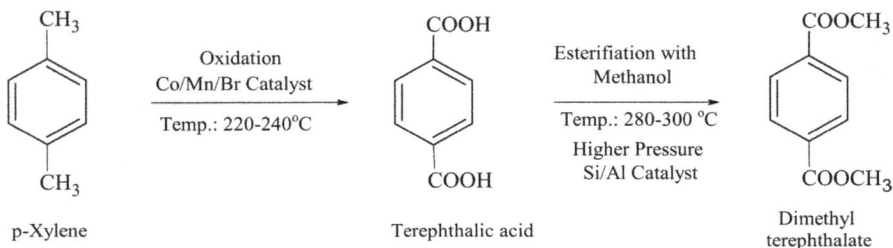

Figure 6.28 Synthesis of Dimethyl terephthalate from p-Xylene

6.14.2.3 Quantitative requirements

For manufacturing 1000 kgs of dimethyl terephthalate, we require 640 kgs of p-xylene, 870 kgs of cobalt/manganese/bromine catalyst, 350 kgs of acetic acid, 185 kgs of methanol and 450 kgs of Si/Al catalyst.

6.14.2.4 Flow chart

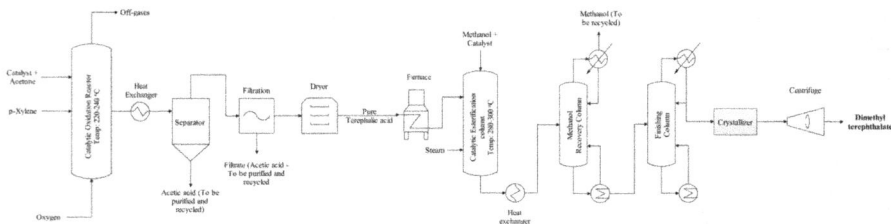

Figure 6.29 Flow sheet diagram of production of Dimethyl terephthalate from p-Xylene

6.14.2.5 Process description

As per Fig. 6.29, p-xylene, acetone as a solvent, and a catalyst are fed into the oxidation reactor. Also, oxygen is added into the bottom of the column. During the reaction, a temperature of 220–240°C and a pressure of 100–105 atmospheres are maintained and the off-gases are withdrawn from the top. After completion of the reaction, the reaction mass is transferred into the separator where acetone is recovered from the bottom and recycled. The upper part is filtered and then, the precipitate of terephthalic acid is dried. This pure dried acid is heated in a furnace and fed to an esterification column. Methanol and the catalyst are also charged into the column. After esterification, methanol is separated and recycled. It is further purified in the finishing column, crystallized and centrifuged to give pure dry dimethyl terephthalate.

6.14.3 Uses

The preliminary DMP is used to produce intermediate polyethylene terephthalate (PET) polymer. It is also used as a plasticizer in paints, inks and adhesives. It is utilized in automotive components like reinforcing beams, bumpers, windscreen wiper blades, electrical systems, and hubcaps. This compound is employed as an intermediate to herbicides (Zeki amd Yilmaz, 2017).

6.15 Toluidines

There are three types of toluidines: (1) o-Toluidine, (2) m-Toluidine and (3) p-Toluidine, in which ortho and para-isomers have industrial applications.

6.15.1 (Introduction) Properties

Most properties of toluidines are the same, so, the main properties of o-Toluidine are as per Table 6.17.

Table 6.17 Physical & Chemical Properties of o-Toluidine

Sr. No.	Properties	Particular
1	IUPAC name	2-Methylaniline
2	Chemical and other names	o-Toluidine, 2-Toluidine, o-Tolylamine, 2-Methylbenzenamine
3	Molecular formula	C_7H_9N
4	Molecular weight	107.15 gm/mole
5	Physical description	Colorless to pale-yellow liquid with aromatic odor
6	Solubility	Miscible in ethanol, diethyl ether, carbon tetrachloride
7	Melting point	$-16.3°C$
8	Boiling point	200°C
9	Flash point	84°C
10	Density	1.01 gm/cm^3 at 27°C

6.15.2 Manufacturing process

The starting material for manufacturing toluidines is toluene.

6.15.2.1 Raw materials

Toluene, Concentrated Nitric acid, Concentrated Sulfuric acid, Catalyst and Hydrogen gas

6.15.2.2 Chemical reactions

The nitration of toluene using conc. nitric acid and conc. sulfuric acid is conducted at a temperature of 30–60°C and atmospheric pressure using zeolite asthe catalyst to give 63–65% o-nitrotoluene, 33–35% p-nitrotoluene and 4%

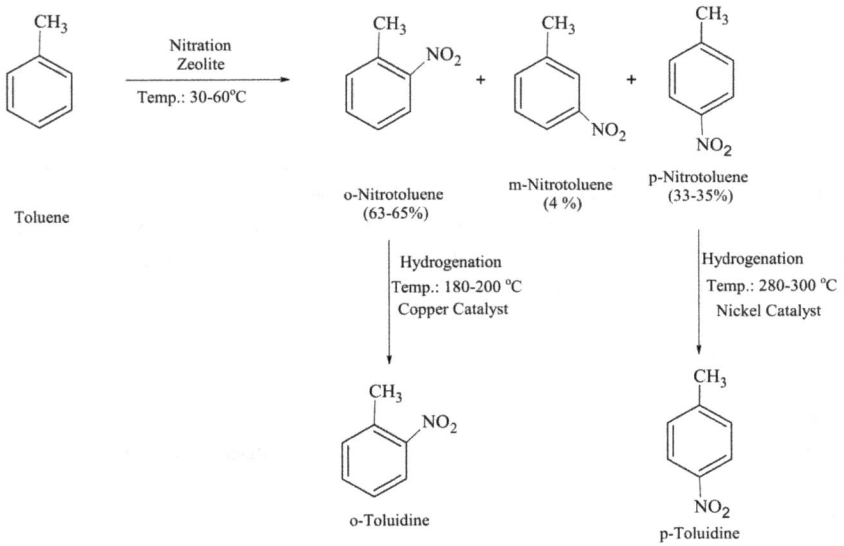

Figure 6.30 Synthesis of Toluidines from Toluene

m-nitrotoluene. Here, increasing the temperature increases p-nitrotoluene yield. Now, o-nitrotoluene undergoes hydrogenation at a temperature of 180–200°C in the presence of copper to o-toluidine. Further, p-nitrotoluene undergoes hydrogenation at a temperature of 280–300°C in the presence of Raney nickel to p-toluidine.

6.15.2.3 Flow chart

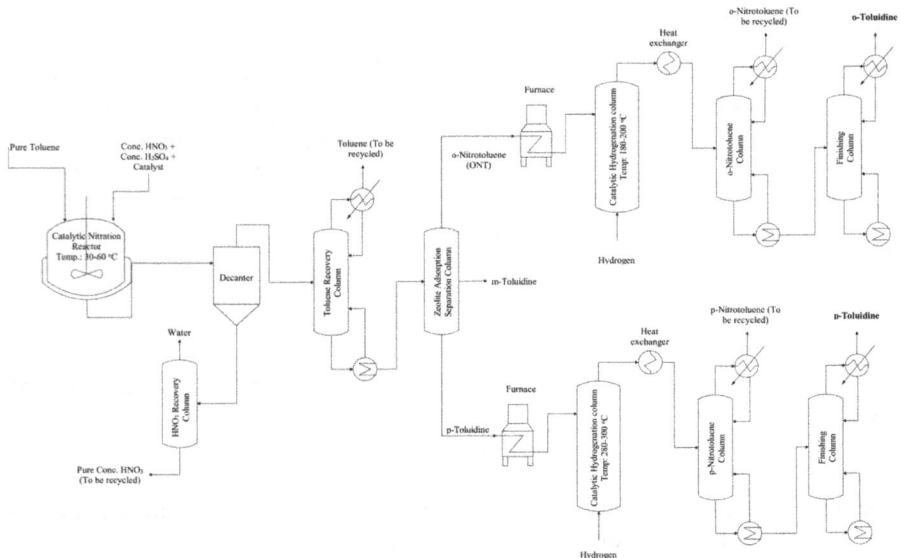

Figure 6.31 Flow sheet diagram of production of Toluidines from Toluene

6.15.2.4 Process description

First, pure toluene and catalyst are added to the jacketed reactor having a stirring facility. Thereafter concentrated nitric acid and concentrated sulfuric acid are slowly added to the reactor. A temperature of 30–60°C is maintained in the reactor. After completion of the reaction, the mass is transferred into a decanter, where nitric acid and water are separated from the bottom. Nitric acid is further separated and recycled. Unreacted toluene is separated from the upper layer of the decanter by means of steam and recycled. The remaining mixture containing o, m and p-nitrotoluene are separated in the zeolite adsorption column. O-Nitrotoluene, collected from the top of the adsorption column, is further heated and fed into a catalytic hydrogenation column. Hydrogen is also fed into the bottom of the column. A temperature of 180–200°C is maintained during the reaction. After completion of the reaction, the mass is cooled, separated from the un-reacted o-nitrotoluene, and finally purified to get o-Toluidine. Similarly, p-Nitrotoluene, collected from the bottom of the adsorption column, is further heated and fed into the catalytic hydrogenation column. Hydrogen is also fed from the bottom of the column. A temperature of 280–300°C is maintained during the reaction. After completion of the reaction, the mass is cooled, separated from un-reacted p-nitrotoluene, and finally purified to get p-Toluidine.

6.15.3 Uses

It is utilized as an intermediate for different industrial segments such as dyes, agro-based chemicals, pigments, rubbers, pharmaceuticals and pesticides. O-Toluidine is additionally utilized in the clinical laboratory as an ingredient in a reagent for glucose analysis, and for tissue staining. Further, p-Toluidine is also used as an intermediate in the preparation of various dyes, organic chemicals and aromatic azo compounds. It is also used as an activator in cyanoacrylate glues. It is used as a bidentate Schiff base ligand through condensation with salicylaldehyde (Hanley et al., 2012).

6.16 Nitrochlorobenzenes

Nitrochlorobenzene has three isomers: (1) o-Nitrochlorobenzene, (2) m-Nitrochlorobenzene and (3) p-Nitrochlorobenzene.

6.16.1 Properties

Most of the properties of nitrochlorobenzenes are the same, so, the main properties of o-Nitrochlorobenzene are as per Table 6.18.

Table 6.18 Physical & Chemical Properties of o-Nitrochlorobenzene

Sr. No.	Properties	Particular
1	IUPAC name	1-Chloro-2-nitrobenzene
2	Chemical and other names	o-Nitrochlorobenzene
3	Molecular formula	$C_6H_4ClNO_2$
4	Molecular weight	157.55 gm/mole
5	Physical description	Yellow-to-green crystals with characteristic odour
6	Solubility	Soluble in alcohol, benzene, ether
7	Melting point	32.3°C
8	Boiling point	246°C
9	Flash point	127°C
10	Density	1.368 gm/cm³ at 27°C

6.16.2 Manufacturing process

Chlorobenzene i undergoes nitration using a mixed acid to obtain three isomers of nitrochloro benzenes.

6.16.2.1 Raw materials

Chlorobenzene, Concentrated Nitric acid, Concentrated Sulfuric acid and Water

6.16.2.2 Chemical reactions

Nitrochlorobenzene is prepared by the nitration of chlorobenzene using an acid mixture consisting of concentrated nitric acid and concentrated sulfuric acid at a temperature of 60–90°C. The nitration products consist of 34–36% o-nitrochlorobenzene, 63–65% p-nitrochlorobenzene and 1% m-nitrochlorobenzene. This reaction is exothermic.

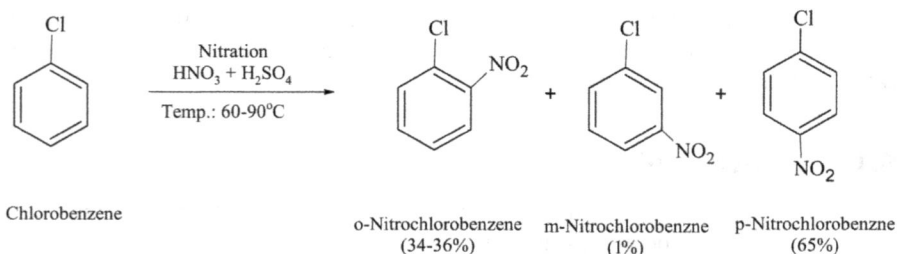

Chlorobenzene   o-Nitrochlorobenzene   m-Nitrochlorobenzne   p-Nitrochlorobenzne
                              (34-36%)   (1%)   (65%)

Figure 6.32 Synthesis of Nitrochlorobenzenes from Chlorobenzene

6.16.2.3 Quantitative requirement

For manufacturing 1000 kgs of nitrochlorobenzene, we required 780 kgs of chlorobenzene, 490 kgs of concentrated nitric acid, 410 kgs of concentrated sulfuric acid and 500 kgs of water.

6.16.2.4 Flow chart

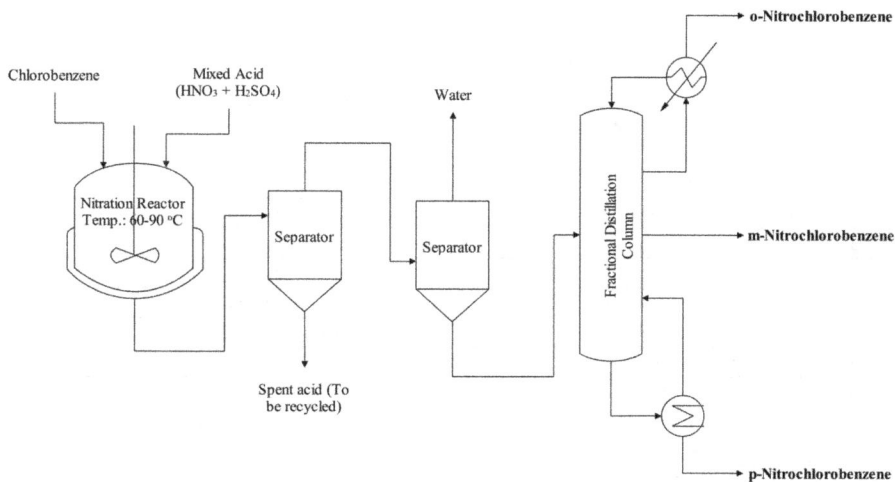

Figure 6.33 Flow sheet diagram of production of Nitrochlorobenzenes from Chlorobenzene

6.16.2.5 Process description

First, chlorobenzene is fed into a jacketed nitration reactor with a stirring facility. Now, a mixture of nitric acid and sulfuric acid is added to the reactor with stirring. This reaction is exothermic, so, a temperature of 60–90°C is maintained by circulating cooling water. After completion of the reaction, mass is transferred into a separator. Here, spent acid is separated and recycled. Then, it is washed with water in another reactor and again transferred into a separator to separate water from the reaction mass. It has undergone fractional distillation to separate individual nitrochlorobenzene isomers. Finally, they are dried individually to get pure solid dried products.

6.16.3 Uses

o-Nitrochlorobenzene is employed in manufacturing derivatives such as o-nitrophenol, 2-chloroaniline, o-amino-phenol, o-phenetidines and azo dyes. It is also utilized as an inputin pesticides (for manufacturing carbofuran) and rubber chemicals.

m-Nitrochlorobenzene is an important intermediate in several industries such as dyes, pharmaceuticals, pigments, drugs and pesticides.

p-Nitrochlorobenzene is used to prepare several fine products such as p-nitrophenol, p-nitroaniline, p-aminophenol, phenacetin, acetominophen and parathion. It is also used in the agriculture, rubber and oil industries (Veretennikov et al., 2001).

6.17 2,4,5-Trichloronitrobenzene

Trichloronitrobenzene has several isomers, in which 2,4,5-Trichloronitrobenzene is the most important chemical.

6.17.1 Properties

Its main properties are as follows.

Table 6.19 Physical & Chemical Properties of 2,4,5-Trichloronitrobenzene

Sr. No.	Properties	Particular
1	IUPAC name	1,3,5-Trichloro-2-nitrobenzene
2	Chemical and other names	2,4,6-Trichloronitrobenzene, 1,3,5-Trichloro-2-nitro benzene
3	Molecular formula	$C_6H_2Cl_3NO_2$
4	Molecular weight	226.4 gm/mole
5	Physical description	White crystal
6	Solubility	Soluble in water, alcohol
7	Melting point	72°C
8	Boiling point	-
9	Flash point	-
10	Density	-

6.17.2 Manufacturing process

1,2,4-Trichlorobenzene is undergone nitration using concentrated nitric acid and concentrated sulfuric acid to get 2,4,5-trichloronitrobenzene.

6.17.2.1 Raw materials

1,2,4-Trichlorobenzene, Concentrated Nitric acid, Concentrated Sulfuric acid and Water

6.17.2.2 Chemical reactions

2,4,5-Trichloronitrobenzene is prepared by the nitration of 1,2,4-trichlorobenzene using concentrated nitric acid and concentrated sulfuric acid at a temperature of 50–70°C.

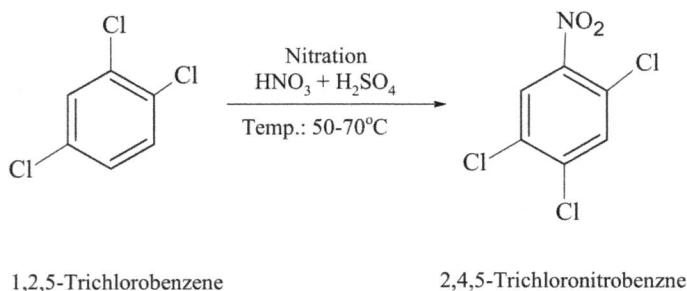

1,2,5-Trichlorobenzene   2,4,5-Trichloronitrobenzne

Figure 6.34 Synthesis of 2,4,5-Trichloronitrobenzene

6.17.2.3 Quantitative requirements

For manufacturing 1000 kgs of 2,4,5-Trichloronitrobenzene, we require 980 kgs of 1,2,4-trichlorobenzene, 390 kgs of concentrated nitric acid, 310 kgs of concentrated sulfuric acid and 650 kgs of water.

6.17.2.4 Flow chart

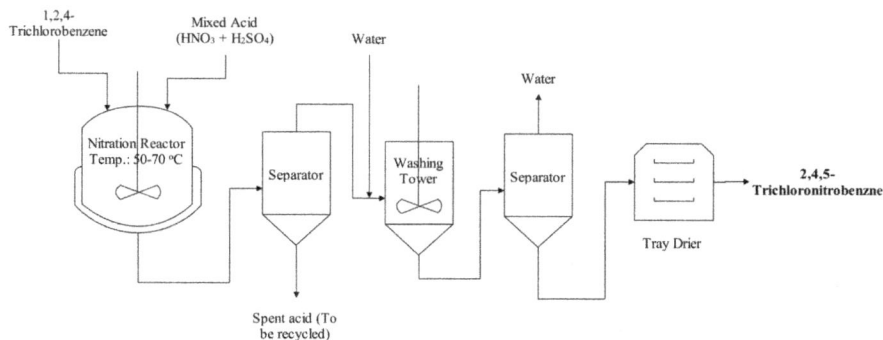

Figure 6.35 Flow sheet diagram of production of 2,4,5-Trichloronitrobenzene

6.17.2.5 Process description

As per the simple flow chart, 1,2,4-Trichlorobenzene is fed into the jacketed reactor having a stirring facility. Now, a mixture of nitric acid and sulfuric acid is added to the reactor with stirring. This reaction is exothermic, so, a temperature of 50–70°C is maintained by circulating cooling. After completion of the reaction, the mass is transferred into a separator to separate the spent acid from the mass. The reaction mass is transferred into a washing reactor, where it is washed with water to remove impurities. Finally, it is tray dried to get pure solid dried 2,4,5-Trichloronitrobenzene.

6.17.3 Uses

It is used as an intermediate in dyestuffs and pigments.

6.18 Acetanilide

It is a flammable colorless solid.

6.18.1 Properties

Its main properties are as follows.

Table 6.20 Physical & Chemical Properties of Acetanilide

Sr. No.	Properties	Particular
1	IUPAC name	N-phenylacetamide
2	Chemical and other names	Acetanilide, Antifebrin, Acetanil, Acetamidobenzene
3	Molecular formula	C_8H_9ON
4	Molecular weight	135.16 gm/mole
5	Physical description	white to gray odorless powder
6	Solubility	Soluble in water, acetone, alcohol, chloroform, 5 ml glycerol, dioxane, benzene
7	Melting point	114.3°C
8	Boiling point	304°C
9	Flash point	170°C
10	Density	1.219 gm/cm³ at 27°C

6.18.2　Manufacturing process

It is prepared by the reaction of aniline and acetic acid.

6.18.2.1　Raw materials

Aniline and Acetic acid

6.18.2.2　Chemical reactions

Aniline is undergoes acetylation using glacial acetic acid in the presence of a small quantity of benzene or acetic anhydride at a temperature of 180–200°C and a pressure of 0.5–0.6 atmosphere to give acetanilide.

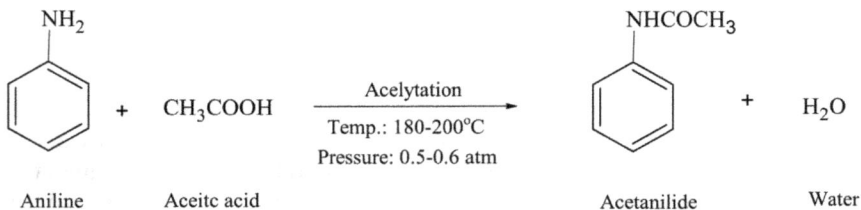

$$NH_2 + CH_3COOH \xrightarrow[\text{Temp.: 180-200°C} \atop \text{Pressure: 0.5-0.6 atm}]{\text{Acelytation}} NHCOCH_3 + H_2O$$

| Aniline | Aceitc acid | | Acetanilide | Water |

Figure 6.36　Synthesis of manufacturing Acetanilide

6.18.2.3　Quantitative requirements

For manufacturing 1000 kgs of acetanilide, we require 850 kgs of aniline, 610 kgs of glacial acetic acid, 50 kgs of benzene or acetic anhydride and 600 kgs of water.

6.18.2.4　Flow chart

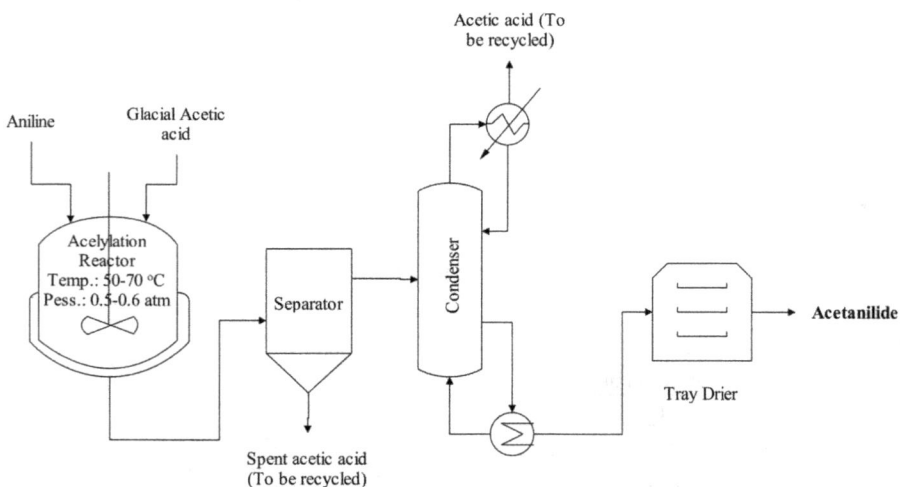

Figure 6.37　Flow sheet diagram of production of Acetanilide

6.18.2.5　Process description

First, aniline is taken in a closed acetylation reactor having a jacket with a stirring facility. Now, glacial acetic acid is heated upto 180–200°C and added with stirring. A temperature of 180–200°C and a pressure of 0.5–0.6 atmosphere is maintained in the reactor. After the completion of the reaction, the reaction mass is cooled and transferred into a separator. Here, glacial acetic acid is separated from the mass and recycled. Thereafter, it is transferred to the condenser to separate out traces of acetic acid and recycled. Finally, it is tray dried to get pure dried acetanilide.

6.18.3 Uses

Acetanilide was used as a substitute for aspirin for many years for the treatment of common complaints such as headaches, menstrual cramps, and rheumatism. Now-a-days, it is used as an anti-pyretic (fever reducing agent). It is also used as an intermediate in the drugs, dyes, paints, chemical intermiediates and electroplating industries. It is also a pioneer input for the manufacture of penicillin and other pharmaceuticals (Singh et al., 2018).

References

Agustriyanto, R., Sapei, L., Rosaline, G. and Setiawan, R. 2017. The Effect of Temperature on the Production of Nitrobenzene. IOP Conference Series Materials Science and Engineering 172(1): 012045.

Baynazarov, I.Z., Lavrenteva, Y.S., Akhmetov, I.V. and Gubaydullin, I.M. 2018. Mathematical model of process of production of phenol and acetone from cumene hydroperoxide. Journal of Physics Conference Series 1096(1): 012197. DOI:10.1088/1742-6596/1096/1/012197

Dimian, A.C., Bildea, C.S. and Kiss, A.A. 2019. Styrene Manufacturing, In book: Applications in Design and Simulation of Sustainable Chemical Processes 443–481.

Dutta, S., De, S. and Saha, B. 2011. Recent Advancements of Replacing Existing Aniline Production Process with Environmentally Friendly One-Pot Process: An Overview, Critical Reviews in Environmental Science and Technology 43(1). https://doi.org/10.10 80/10643389.2011.604252

El-Gharbawy, A. 2021. A Review on Phthalic Anhydride Industry and Uses, Oil and Energy Trends 1(1): 1–2. DOI: 10.53902/TPE.2021.01.000505

Ganji, M. and Tabarsa, T. 2011. A Novel Phenol-Based Composite Production: Features and Characterization, Key Engineering Materials. 471–472, 715–720. https://doi.org/10.4028/www.scientific.net/KEM.471-472.715

Giarola, S., Romain, C., Williams, C.K. and Hallett, J. 2015. Production of phthalic anhydride from biorenewables: Process design, Computer Aided Chemical Engineering 37: 2561–2566. https://doi.org/10.1016/B978-0-444-63576-1.50121-7

Goto, M., Konishi, T., Kawaguchi, S. and Yamada, M. 2011. Process Research on the Asymmetric Hydrogenation of a Benzophenone for Developing the Manufacturing Process of the Squalene Synthase Inhibitor TAK-475, Organic Process Research and Development 15(5): 1178–1184. https://doi.org/10.1021/op2001673

Hanley, K.W., Viet, S. Hein, M.J. and Carreon, T. 2012. Exposure to o-Toluidine, Aniline, and Nitrobenzene in a Rubber Chemical Manufacturing Plant: A Retrospective Exposure Assessment Update. Journal of Occupational and Environmental Hygiene 9(8): 478–90. DOI: 10.1080/15459624.2012.693836

Hepworth, J.D., Waring, D.R. and Waring, M.J. 2013. Aromatic Chemistry, Wiley-RSC.

Kolmetz, K., Chua, M., Desai, R., Gray, J. and Sloley, A. et al. 2003. Staged modifications improve BTX extractive distillation unit capacity. Oil and Gas Journal 101(39): 60–65.

Liu, X., Yang, L., Wang, M. and Minghui, Z. 2021. Insights into the Formation and Profile of Chlorinated Polycyclic Aromatic Hydrocarbons during Chlorobenzene and

Chloroethylene Manufacturing Processes, Environmental Science and Technology 55(23): 15929–15939. https://doi.org/10.1021/acs.est.1c05688

Luyben, W.L. 2009. Design and Control of the Cumene Process. Industrial & Engineering Chemistry Research 49(2): 719–734. https://doi.org/10.1021/ie9011535

Luyben, W.L. 2010. Design and Control of the Styrene Process, Industrial & Engineering Chemistry Research 50(3): 1231–1246. https://doi.org/10.1021/ie100023s

McVey, M., Elkasabi, Y. and Ciolkosz, D. 2020. Separation of BTX chemicals from biomass pyrolysis oils via continuous flash distillation, Biomass Conversion and Biorefinery 10(1): 15–23. https://doi.org/10.1007/s13399-019-00409-1

Meng, T., Wang, D. and Zhang, L. 2011. New preparation process of nitrobenzene and reaction mechanism, Speciality Petrochemicals 28(1): 54–56.

Miller, R.R., Newhook, R. and Poole, A. 1994. Styrene Production, Use, and Human Exposure, Critical Reviews in Toxicology 24: S1–10.

Mohammed, A. and Baki, M.K. 2008. Separation Benzene and Toluene from BTX using Zeolite 13X, Iraqi Journal of Chemical and Petroleum Engineering 9(3).

Niziolek, A.M., Onel, O. and Floudas, C.A. 2016. Production of Benzene, Toluene, and the Xylenes from Natural Gas via Methanol, Computer Aided Chemical Engineering In book: 26th European Symposium on Computer Aided Process Engineering 2349–2354. https://doi.org/10.1002/aic.15144

Renita, A., Salla, S. and Lakshmi, S. 2020. Synthesis of Acid Free Benzaldehyde by Highly Selective Oxidation of Benzyl Alcohol Over Recyclable Supported Palladium Catalyst. Combinatorial Chemistry & High Throughput Screening 25(2): 284–291. DOI: 10.2174/1386207323666201230091613

Schmidt, R.J. 2005. Industrial catalytic processes-phenol production, Applied Catalysis A: General 280(1): 89–103. https://doi.org/10.1016/j.apcata.2004.08.030

Singh, R.K., Kumar, A. and Mishra, A.K. 2018. Chemistry and Pharmacology of Acetanilide Derivatives: A Mini Review. Letters in Organic Chemistry 15(1): 6–15. DOI: 10.2174/1570178615666180808120658

Sivcev, V.P., Volcho, K., Salakhutdinov, N.F. and Anikeev, V. 2012. Transformations of acetophenone and its derivatives in supercritical fluid isopropanol/CO_2 in a continuous flow reactor in the presence of alumina. Journal of Supercritical Fluids 70: 35–39. https://doi.org/10.1016/j.supflu.2012.05.012

Sweeney, W.A. and Bryan, P.F. 2000. BTX Processing. In book: Kirk-Othmer Encyclopedia of Chemical Technology.

Veretennikov, E.A., Lebedev, B.A. and Tselinskii, I.V. 2001. Nitration of Chlorobenzene with Nitric Acid in a Continuous Installation. Russian Journal of Applied Chemistry 74(11): 1872–1876. https://doi.org/10.1023/A:1014840627266

Vinila M.L., Kallingal, A. and Sreekumar, S. 2021. Modelling and performance analysis for cumene production process in a four-layer packed bed reactor. International Journal of Chemical Reactor Engineering 20(8). https://doi.org/10.1515/ijcre-2021-0177

Yangirov, T.A., Abdullin, B.M., Fatykhov, A.A. and Zakharova, E.M. 2021. Diphenyl Oxide Copolyarylenephthalides with Different Ratio of Phthalide and Diphthalide Groups, Polymer Science Series B 63(1): 13–21. https://doi.org/10.1134/S1560090421010085

Zakoshansky, V.M. 2007. The Cumene Process for Phenol-Acetone Production, Petroleum Chemistry 47(4): 307–307. DOI: 10.1134/s0965544107040135

Zeki, O. and Yılmaz, T. 2018. Removal of Acetic Acid from Dimethyl Terephthalate Manufacturing Wastewater with Ion Exchange. CLEAN - Soil Air Water 46(8). https://doi.org/10.1002/clen.201700436

Zhao, Y., Yu, C., Wu, S. and Zhang, W. 2018. Synthesis of Benzaldehyde and Benzoic Acid by Selective Oxidation of Benzyl Alcohol with Iron(III) Tosylate and Hydrogen Peroxide: A Solvent-Controlled Reaction. Catalysis Letters 148(10): 3082–3092. https://doi.org/10.1007/s10562-018-2515-0

CHAPTER 7

Dye & Pigment Industries

7.1 Introduction

7.1.1 Light and color

In scientific terms, no special consideration is given to wave lengths in the visible range since those outside it are considered colorless. Some animals like deer can see other wave lengths, but human cannot. We consider them colorless. Based upon this consideration, we can define the color as: a psychological sensation which is the byproduct of a certain wave length reaching the retina of the eye. This supports the idea that color depends on and changes depending on the type of light that illuminates the matching substance. Simply put, an object absorbs a portion of the radiation in the visible range that is less over the object, making the object look colored.

Normal day-light or ordinary light is a mixture of electromagnetic radiations of varying wave lengths. Off these we can respond to certain wave lengths. White light is categorized into three types: Ultraviolet (UV): 1000–4000°A, White (visible): 4000–7500°A and Infrared: 7500–100000°A. As the visible range of the human eyes is 4000–8000°A, we respond to wave lengths in it. These wavelengths are responsible for giving a definite color to a particular substance. Wave lengths of the visible range are composed of seven different colors, violet, indigo, blue, green, yellow, orange and red [VIBGYOR]. Wave length radiations below 400°A lie in the ultraviolet region and above 800°A lie in the infrared region. Both these ranges are in the invisible range.

Appearance of color is dependent on the light incident on the substance as follows:

(1) The substance appears bright when white light is completely reflected.
(2) The substance appears black, when white light is completely absorbed.
(3) In the case of a single narrow band, all wave lengths of white light are absorbed, and the color of the substance appears as the color of the reflected band. For example: when a single narrow band such as green (5100 °A) is reflected and all other wavelengths are absorbed by the substance, the color

Table 7.1 Absorbed colour and visualized colour compose to wave length vicinity

Wave length (°A)	Absorbed colour	Visualized colour
4000–4350	Violet	Yellow-green
4350–4800	Blue	Yellow
4800–4900	Green-blue	Orange
4900–5000	Blue-Green	Red
5000–5600	Green	Purple
5600–5800	Yellow-green	Red
5800–5950	Yellow	Blue
5950–6050	Orange	Green–blue
6050–7500	Red	Blue–Green

of the substance appears green. Table 7.1 shows the absorbed colour and visualized colour compose to wave length vicinity.

(4) The substance will appear to have the complimentary color of the absorbed band if only one white light band is absorbed. For example: if light of 5900 °A (region of yellow color) is absorbed, blue color is produced. And if a composite of the remaining wavelengths is reflected blue color is produced.

Thus, blue and orange are said to be complementary colors, because the absorption of one from white light gives the other.

7.1.2 Bathochromic and Hypsochromic effect

Bathochromic effect: Any group that produces the deepening of color in the sequence of Yellow→Orange→Red→Purple→Violet→Blue→Green, is known as bathochromic. And, deepening of the color is known as bathochromic effect. In the bathochromic effect, the structure of the dye molecule changes which shifts the absorption from the lower to higher wavelengths and deepens the color. In such a phenomenon, the bathochromic group is introduced in the dye, responsible for increasing the resonance, which in turn decreases the energy gap of the ground state-exited transitions with visible color appearance.

Hypsochromic effects: Any group that produces the deepening of color in the sequence of Green→Blue→Violet→Purple→Red→Orange→Yellow, is known as a Hypsochromic group. And, deepening of the color is known as the hypsochromic effect. In the hypsochromic effect, the structure of the dye molecule changes which shifts the absorption from higher to lower wavelengths and lightens the color. Here, the introduced hypsochromic group in a dye is responsible for decreasing resonance of the π orbitals. A change in the structure of a dye due to an increase in the absorption intensity is said to be a hypochromic effect and a decreasein the intensity of absorption is termed as hypochromic effect. Bathochromic, hypsochromic, hydrochromic and hyperchromic effects are represented by the absorption spectra by plotting the absorption intensity vs. wavelength (Arnkil et al., 2012; Butyrskaya et al., 2004).

4000	4500	5000	5500	6000	7000	
Ultra-Violet	Violet	Blue	Green	Yellow	Orange	Red	Infra-Red

(Bathochromic) High ◄——— Increasing Energy ———► Low (Hypsochromic)

Figure 7.1 Bathochromic and Hypsochromic shift

7.1.3 Color and chemical constitution

The compound's color is associated with its chemical constitution, which is understood as per the following example:

(i) Benzene is colorless, but its isomers are colored.
(ii) Compounds become colorless with the reduction process of colored organic compounds, which oxidation of color organic compound turns into colorless compound.

Different theories show the relationship between color and chemical constitution.

(1) Witt's theory
(2) Aronstong's theory
(3) Baeyer's theory
(4) Nietzki's theory
(5) Watson's theory

Also, recent modern theories of color and constitution are as follows.

(1) Quantization of light energy
(2) Absorption of radiation by a molecule
(3) Dipole moment, which consists of:
    (a) Valence Bond (VB) theory (Resonance theory)
    (b) Molecular Orbital theory (MOT)

Now, diazomethane contains an unsaturated group such as the azo group, so the yellow color appears. But, on the reduction of the azo group methylenehtdrazine is produced, which is colorless. This is because of the absence of the unsaturated group.

Quinoid Theory (Chromophore-Auxochrome Theory) An early theory of dyes that Otto Witt first proposed in 1876 gave a fundamental grasp on the interaction between color and molecular structure. This theory is known as 'chromophore-quxochrome Theory'. It was noticed by O. Witt in 1876 that:

The color in an organic compound is associated with the presence of certain multiple unsaturated bonded groups called chromophores. Also, compounds containing chromophoric groups are called chromogens. As per the Greek meaning, Chroma means color and phores means bearing. These groups are

- NO	- NO$_2$	- N = N -	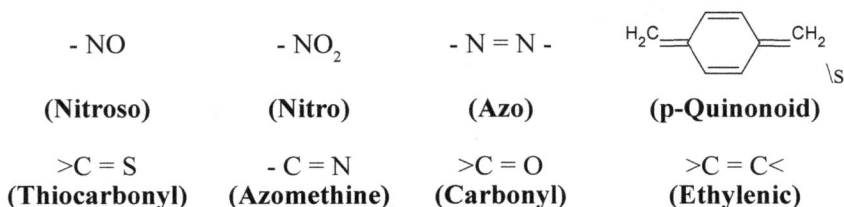
(Nitroso)	**(Nitro)**	**(Azo)**	**(p-Quinonoid)**
>C = S	- C = N	>C = O	>C = C<
(Thiocarbonyl)	**(Azomethine)**	**(Carbonyl)**	**(Ethylenic)**

There are two types of chromophores represented as follows:

(A) Independent chromophore: When a single chromophore is required to impact the color, it is called an independent chromophore, e.g., nitro, nitroso, p-quinonoid.

(B) Dependent chromophore: When more than one chromophore is required to impact the color of dye, it is called a dependent chromophore, e.g., carbonyl, ethylenic.

Now, the atomic configuration of chromophores affects the energy in a delocalized system. They are composed of atoms joined in alternate single bond and double bond sequences. Double bond are of two types. If the atoms with the double bond are not adjacent, then it is known as an isolated double bond, and they not interact with each other. If the atoms with the double bond are adjacent, then it is referred as a conjugated double bond, and they interact with each other. The chromophore configuration consists of multiple units having conjugated double bonds, which are more effective. This is due to the interaction between double bonds, which causes partial delocalization of the electrons. While isolated double bonds give color.

For example,

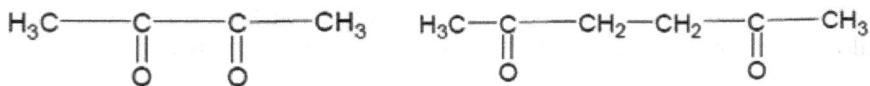

is colorless.

Acompound having a chromophoric group, is known as chromgen. For example, in nitrobenzene, the nitro group is the chromophore and benzene is the chromogen.

Auxochrome: Certain groups do not produce color themselves, but are able to intensify it when a chromophore is present. These groups are known as auxochromes. The Greek meaning of auxin means to increase and chroma means color. So, the basic meaning of auxochrome is increasing color. A chromogen may be colored, but it does not represent a dye without an auxochrome. These auxochromes may be acidic (hydroxyl, sulphonic, carboxyl) or basic (amino, alkylamino, dialkylamino). Some other auxochromes are chloro, methyl, methoxy, cyano, acetyland acetamido.

Auxochromes mainly result in two functions, such as:

(A) Increasing the intensity of a color. For example,
e.g., azobenzene has red color, but p-hydroxyazo benzene is brilliant red, in which the auxochrome is the hydroxyl group.

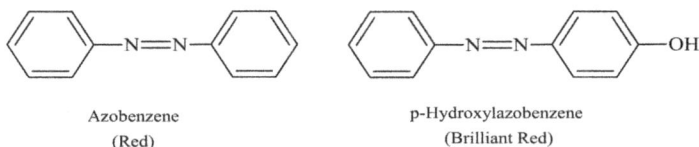

Azobenzene   p-Hydroxylazobenzene
(Red)   (Brilliant Red)

Figure 7.2 Structure of Azobenzene and p-Hydroxylazobenzene

Also, nitrobenzene is pale yellow in color, while p-nitrobenzene having anauxochromic hydroxyl group is deep yellow.

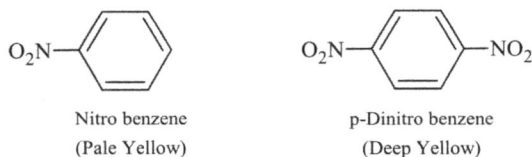

Nitro benzene   p-Dinitro benzene
(Pale Yellow)   (Deep Yellow)

Figure 7.3 Structure of Nitrobenzene and p-Dinitrobenzene

(2) An auxochrome is able to bind fibers, thus a colored substance carrying it functions as a dye. Also, it associates or dissociates by salt formation. Auxochromes are mainly of two types.

(A) **Bathochromic auxochromes:** Their main function is to darken the color. The presence of auxochromes is responsible for the absorption maxima from violet to red, which lead to depth in color. So, it is known as a red shift. This shift occurs when hydrogen atoms are replaced by R (alkyl or aryl group) in an amino ($-NH_2$) group in the compound.

(B) **Hypsochromic auxochromes:** The presence of this type of auxochromeis responsible for decreasing the color. This happens due to the absorption maxima from red to violet, which results in a blue shift. When the hydrogen atom of the amino or hydroxyl group is replaced by an acetyl group, then a blue shift occurs (Masuda, 2016; Lye, 2002; Daniel, 2012).

7.1.4 Primaries

Aromatic compounds are raw materials for the manufacture of dyes, like carbozole benzene, toluene, pyrene, pyridine, phenol, anthracene and naphthalene, referred to as primaries. They have been obtained almost exclusively from the distillation of coal tar in previous years. Now-a-days, they are available from crude petroleum and natural gas, as they have increased quantities of primaries, mainly benzene and toluene. Synthetic dyes are still frequently associated with the term "coal tar dye".

The following are the main sources of primaries:

7.1.4.1 Coal tar

Fractional distillation results in coal gas and coal tar from coal. Coal gas is mainly utilized as a fuel and coal tar is subjected to fractional distillation to produce several products as tabulated in following table.

Table 7.2 Fractional distillation of coal tar

Sr. No.	Name of fraction	Temperature	Main constituents
1	Light oil or crude naphtha	Upto 170°C	Benzene, Toluene, Xylene
2	Middle oil or carbolic acid	170–230°C	Naphthalene, Carbolic acid, Phenol
3	Heavy oil or creosote	230–270°C	Cresols
4	Anthracene oil or green oil	270–360°C	Anthracene, Phenanthrene and Carbozol

After distillation the black residue left is known as pitch consisting of 92–94% carbon and small amounts of pyrene and chrysene. Isolation of this compound from the pitch is done by steam distillation followed by fractional distillation and crystallization from the solvent such as naphtha.

(A) **Light oil:** This is a very important product obtained from coal tar, which undergoes further fractional distillation to yield three fractions as follows.

  (a) Low boiling fraction: It contains fatty hydrocarbons, carbon disulfide and acetonitrile

  (b) Middle boiling fraction: This fraction is first treated with concentrated sulfuric acid and extra pyridine is the basic substance collected. After sequential water washing and caustic treatment of here maining part phenol is collected. The remaining part is treated with excess concentrated sulfuric acid to remove acidic impurities. Again, water washing and fractional distillation are performed, which results in the following fractions.

   (i) 90% Benzol: About 70% Benzene and 30% toluene is obtained at 80–110°C.

   (ii) 50% Benzol: 46% Xylene along with different proportions of Benzene and toluene is obtained at 110–140°C.

   (iii) Solvent Naphtha or Benzine: Xylene and mesi-tyleneare obtained inthe temperature range of 140–170°C.

  (B) **Middle oil:** Naphthalene and phenol are the main components of middle oil. This oil is crystallized and separated by centrifugation. The basic impurities are removed by melting the crude crystals and sulfuric acid treatment. These crystals are further washed with aqueous sodium hydroxide to remove phenol and excess sulfuric acid. The naphthalene is distilled out to get a pure form.

  (C) **Heavy oil:** Cresol is the main component, which is used to prepare various types of dyes.

  (D) **Anthracene oil:** Various compounds such as anthracene, phenanthrene and carbozol are used to manufacture different dyes.

7.1.4.2 Petroleum

The importance of benzene, toluene and xylene as primary petroleum sources has increased recently. These vital primaries are obtained by the cracking and hydrogenation processes.

7.1.4.3 Other sources

Methane, water gas and olefins generate aromatic hydrocarbons by a catalytic process, which are used as primaries for dyes.

7.1.4.4 Inorganic raw materials

Different inorganic materials such as nitric acid, sulfuric acid, caustic soda, oleum, fulfuric acid, bromine, sodium sulfide, sodium hydrosulfite, sodium dichromate, hydrochloride acid, acetic acid and sodium boron tetrahydrate are utilized. These materials are for performing various unit processes like sulfonation, nitration, oxidation, reduction, halogenation, hydrolysis and hydrogenation.

7.1.5 Intermediates

In dye synthesis, primaries are not directly utilized. At first, primaries are converted into a number of derivatives, which there after, turn into dyes. These derivatives are known as intermediates. Different reactions like sulfonation, nitration, oxidation, reduction, halogenation, hydrolysis and hydrogenation are conducted to produce intermediates. These reactions are responsible for substituted hydrocarbon formations which are capable of undergoing further chemical reactions. Various intermediates are required for the manufacture of dyes, which are divided into the following categories.

(a) Aliphatic compounds
(b) Aromatic compounds
(c) Heterocylic compounds

7.1.6 Dye and pigment

A dye is a colored substance with a strong attraction for the surface on which it has to be applied. Organic dyes are used to add color to a variety of substrates, such as paper, leather, fur, medicines, cosmetics, waxes, plastics, greases, and textiles. A pigment is a substance that has undergone wavelength-selective absorption to alter the color of transmitted or reflected light. Dyes are often organic or inorganic substances that are water-soluble or water dispersible and capable of being absorbed into the substrate while obliterating the substance's crystal structure. Organic and inorganic pigments are nearly always applied in aggregated or crystalline, insoluble form, necessitating the use of a binder to create a coating on the surface of a substrate. The dye molecule is often chemically attached to the surface and ingrained into the substance. A pigment, on the other hand, has no interaction with the substrate and does not damage the crystal structure of the material. It works well with practically all surfaces that need to be dyed. The pigments link with a limited range of suitable substrates.

The term "pigment" refers to an insoluble organic or inorganic component that is frequently used in surface coatings. In order to add color, they are also used in inks, plastics, rubbers, ceramics, paper, and linoleum industries. Contrary to popular belief, the pigment industry is distinct from the paint industry. Pigments are typically manufactured or mined from the Earth's surface for the purpose of making paints for commercial use. Dyes are mainly used for coloring textiles and fabrics. Also, dyes have an affinity for cellulosic hydroxyl groups, so they can be used to color paper; these types of dyes are called paper dyes. A number of dyes are routinely used in analytical chemistry as pH and redox indicators. Colored organic compounds are used in photography, i.e., cyanine or isocyanine. A few anthraquinone pigments are used for producing colored smoke. These compounds are highly volatile and produce thermally stable, colored vapors (Gurses et al., 2016).

7.2 Classification of Dyes

Dyes are classified in different ways such as:
- Method of application
- Chemical constitution
- Type of materials to be dyed
- Intermediate from which they are prepared.

But it is mainly classified according to its application and chemical constitution.

7.2.1 Classification of dyes based on application

Classification of dyes according to methods of application to the fiber are considered as the most reliable as they are used for Color Indices, published by the Society of Dyes and Colorists, England and the American Association of Textile Chemists and Colorists. Each dye is given an individual unique number and listed along with its name and properties. This provides reliability to the identification of a dye. Main achievement of color Indices is that they arrange dyes according to their structure with the most important feature being their chromophoric group. For example, all nitroso dyes have their unique numbers between 10000–10299. Nitro dyes have their unique numbers between 10300–10999. So, there are 31 groups having unique numbers up to 18000.

7.2.1.1 Acid dyes

They are also called anionic dyes. These are water soluble dyes of sodium salts with acidic groups like, -SO_3H or phenolic groups like -OH. They have an affinity for protein fibers like wool, silk and leather so, these types of dyes are used to dye animal as well as synthetic fibers. They are used to some extent for paper, leather, ink-jet printing, food and cosmetics. While dying, these types of dyes are soluble in aqueous solvents to give colored anions.

$$R - SO_3Na + H_2O \rightarrow R - SO_3 + Na^+$$

$$R - SO_3Na + H_2O \rightarrow R - SO_3 + Na^+$$

Under acidic conditions, proteins and polyamide fibers create cationic sites in water; as the acidity of the solution increases, more cationic sites are created under these extremely acidic circumstances. The acid dye anions can thus connect with these cationic sites via hydrogen bonding, Vander Waals forces, or ionic bonding. Since these links can be broken, dyeing occurs quickly. The dyeing temperature is generally 35–60°C and sometimes, dyeing leveling agents are used.

$$R - SO_3 + {}^+NH_3 - R' \rightarrow R - SO_3 - NH_3 - R'$$

Acid dyes are classified into the following chemical groups based on their structural makeup: azo, anthraquinone, triphenylmethane, pyrazolone, azine, nitro, and quinoline. Now-a-days, the azo dye group is considered to be the third largest essential group followed by anthraquinone and tetramethylene group. Further, acid dye are divided into three groups according to alterations in affinity which are primarily a function of the molecular size:

(a) **Leveling dyes** are relatively small chemicals that bind to protein fibers in a way that is similar to a salt bond.

(b) **Milling dyes** are large-volume dye molecules for which the adsorption forces between the dye molecule's hydrophobic sections and those of the protein fiber predominate while salt formation with the fiber only plays a minor role.

(c) **Super Milling dyes** with intermediate molecular sizes not only create a salt-like connection with the wool fiber, but they are also held to it by intermolecular forces. These dyes have qualities that fall between those of levelling milling dyes.

These dyes provide extremely vivid hues and a wide variety of fastness qualities, from very low to high. Acid dyes are used to color basic-group fibers including wool, silk, and polyamides. These dyes can be removed from fibers by washing. The rate of removal depends upon the rate of water diffusion washing conditions. Also, the diffusion rate depends on temperature, shape and size of the dye molecules and the number and kinds of linkages formed with fibers. Some examples of these dyes are as follows.

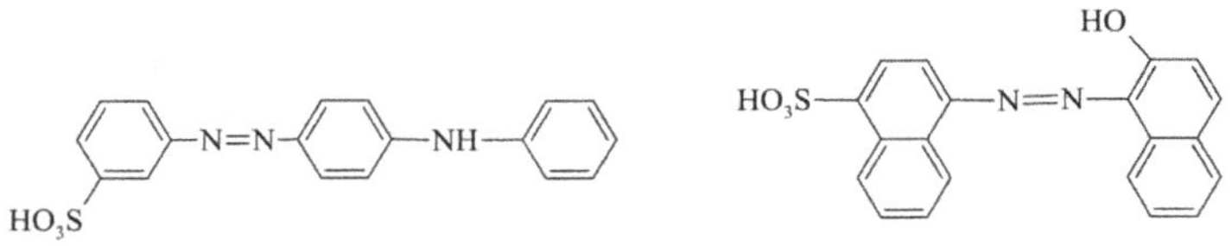

Acid Red - 36   Acid Red - 88

Figure 7.4 Examples of Acid dye.

7.2.1.2 Basic dye

Perkins manufactured the first basic dye, mauve, in 1856. These dyes are water soluble consisting of mainly amino and/or substituted amino derivatives. Amino groups are protonated under acidic conditions and salt linkages are made between the acidic groups of the fiber and amino groups of the dye. They often have low light fastness qualities yet are given strong and dazzling hues. Paper, polyacrylonitrile, modified nylons, and modified polyesters can all be dyed using them. They are fluorescent, which accounts for their extraordinary brightness. These colors work well on paper, wool, silk, leather, and acrylic fibers. It is also used for making inks, typewriter ribbons and dyeing leather. However, they have poor affinity for cellulosic fibers. Hence, cellulosic fibers are first treated with an intermediate compound called mordant such as tannic acid and then it is dyed. The mordanting process usually requires a buffering system of pH 5–6. Today urea is used for this purpose. A basic dye is a water-soluble dye and dissociates into two types of ions: (1) anions and (2) colored cations. So, these types of dyes are also called "cations". The cations form salts when they come into contact with an acidic group (sulfonic or carboxylic). Since these connections are so strong, washing fastness is frequently excellent and light fastness varies greatly depending on the dyestuff. Basic dyes are typically applied in batches, and package, skein, and stock dyeing are more common than piece dyeing on becks or jet machines. After completion of the dyeing process, the remaining basic dye is removed by treating the material with dilute acetic and hydrochloric acids.

Diazahemicyanine, triarylmethane, cyanine, hemicyanine, thiazine, oxazine, and acridine are the main chemical classes. Some common colors exhibit biological activity and are applied as antiseptics in medicine. Crystal Violet and Auramine O are examples of basic dyes.

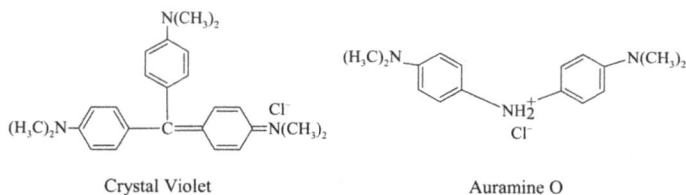

Crystal Violet   Auramine O

Figure 7.5 Examples of Besic dye.

7.2.1.3 Azoic dye

This dye is manufactured inside textile fibers by azo coupling. This dye is firmly occluded and is fast to washing. A variety of hues can be achieved by the proper choice of diazo and coupling components. Finally, dyes are treated with soap and rinsed to produce an insoluble azo dye. Cotton being in primary use, provides high standard colors and fastness to light and wet processes. They produce vivid, dramatic colors, especially in yellow, orange, and red tones. Two aromatic rings, have an azo (-N = N-) group between them that serves as a chromophore. The majority of synthetic dye are azo dyes, which offer a huge range of hues. They are further divided into monoazo, diazo, and other azo group count dyes. They are directly applied to fibers usually cotton, rayon and polyester.

Azo dye are prepared by the following steps.

Step 1: A primary amine is converted into a diazonium compound using sodium nitrite in excess hydrochloric acid (Diazotization).

Step 2: The obtained diazonium compound is unstable, and thus, readily reacts with other aromatic compounds like phenols, naphthols, or other basic aromatic amine solutions.

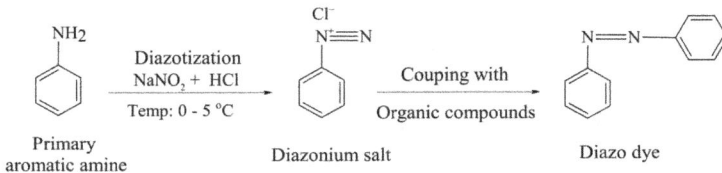

Figure 7.6 Preparation of Diazo dye.

Since, diazonium compounds are typically unstable, the aforementioned reactions take place at low temperatures (0–5°C). The obtained dye is an aromatic molecule with an azo group acting as a chromophore and an auxochrome in the form of a hydroxyl or amino group.

Mechanism of azoic dye manufacture: First, a reaction between sodium nitrite and hydrochloric acid occurrs to obtain hydrogen nitrite, which is subjected to protonation to yield a nitrite ion. This nitrite ion attacks a primary aromatic amine to give a diazonium salt.

Due to the resonance effect, aromatic amine diazonium salts are more stable than aliphatic amine diazonium salts. The diazonium ion undergoes resonance as shown below.

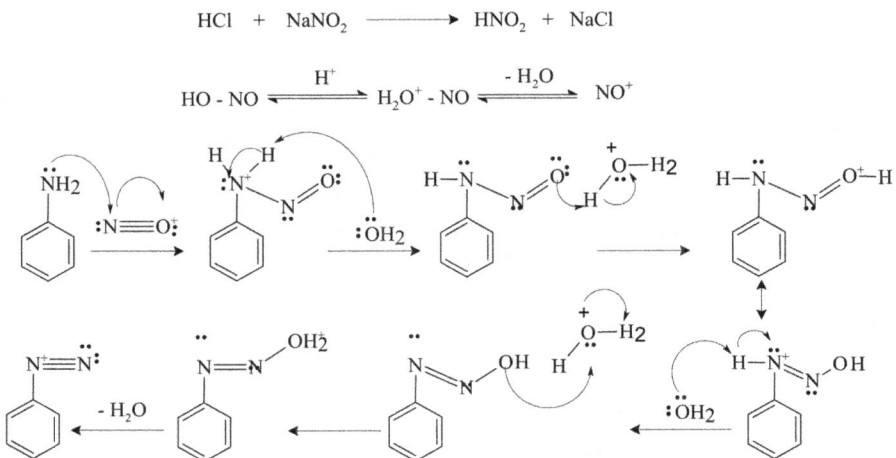

Figure 7.7 Mechansim of Diazo dye.

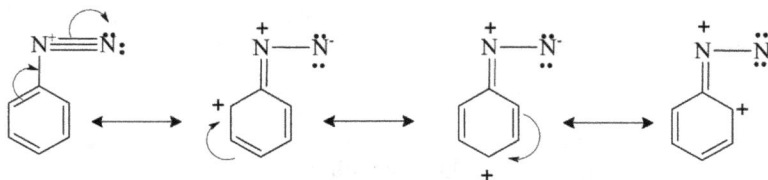

Figure 7.8 Stability of Diazoium salt.

The following are examples of azoic dyes.

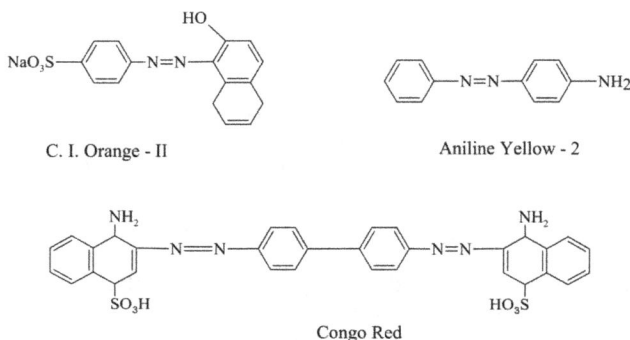

C. I. Orange - II

Aniline Yellow - 2

Congo Red

Figure 7.9 Examples of Azoic dye.

Effect of substituents on the diazotization process: Primary aromatic amines undergo diazotization to yield diazonium salts and couplings with various substitute organic molecules. The coupling reactions depend on the nature and position of the substituents. Due to the presence of an electron-withstanding NO_2 group, p-nitro aniline and 2,4-dinitroaniline are significantly less basic than aniline. requiring unique techniques for their diazotization.

Additionally, challenges with diazotization may arise from their limited solubility in aqueous acids, the presence of easily replaceable groups like $-SO_3H$ and $-NO_2$ or easily oxidizable groups like $-OH$ and $-CHO$.

7.2.1.4 Direct dyes

This dye usually consists of bear sulfonic acid groups that are soluble in water. Because these groups are not used for fiber attachment, these dyes are not regarded as acid dyes. Direct dyes are large, flat and linear swollen molecules of amorphous cellulose and orient themselves along the crystalline regions. Common salt or Glauber's salt is often used to promote the dyeing process. Hence, they are called salt or substantive dyes. When there are too many sodium ions present, the dyeing process favors equilibrium being established with the least amount of dye remaining in the dye solution. This dyeing process is reversible, which has an affinity for the cellulosic fiber and gets attached to it by adsorption. Unless treated with fibers and dye fixing agents, this dye has poor light and washing fastness. But they are extensively used due to their low cost and simple dyeing process. This dye is applicable for wool and silk. Direct-azo dyes are employed in cotton-wool or cotton-silk combinations. They are also used for dyeing the regenerated cellulose and paper leather. A special type of direct dye having free amino groups to diazotized coupled in the fiber. Hence, it improvised fastness toward washing.

A special type of direct dye having free amino groups to diazotized coupled in the fiber. Hence, it improvised fastness toward washing. Direct Black A (Zambesi Black D) primarily used to color plain grounds then latter to be printed in a pattern with vat dyes.

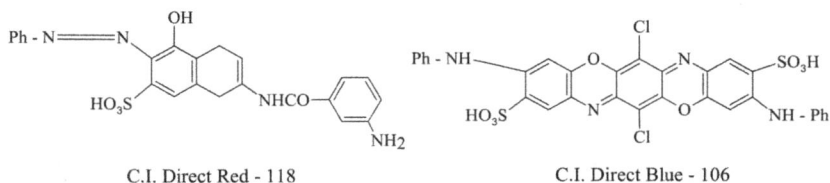

C.I. Direct Red - 118   C.I. Direct Blue - 106

Figure 7.10 Examples of Direct dye

They are further classified into three classes according their dyeing behavior.

(A) Class A dyes have good migration or levelling qualities and are self-leveling.

(B) Class B dyes (also known as salt controllable dyes) are not self-leveling, but they may be made to achieve levels by salt addition.

(C) Class C dyes are not self-leveling and are extremely sensitive to salts; the process of controlling the exhaustion of these dyes cannot be satisfactorily regulated by the addition of salts alone.

7.2.1.5 Reactive dyes

This dye belongs to a new class of colors that bond with fibers that include amino or hydroxyl groups to form covalent bands. The reactive dye molecule consists of dyes (with chromophore) attached to a fiber reaction system.

The reactive dye molecule consists of a dye (with chromophore) attached to a fiber reactive system. These are mainly used for cotton and cellulosic fibers. The reactive system is a heterocyclic compound containing groups, which chemically react with hydroxyl groups of cellulose. Hence, the dye molecule gets attached to the fiber by means of covalent bonds. These dyes form covalent bonds with the cellulosic fiber. This produces dye fibers with extremely high washing fastness properties.

Reactive dyes are obtained from azo, anthraquinone and phthalocyanine dyes. The most common fiber reactive system is cyanuric chloride (i.e., triazine trichloride) in which the chlorine atoms are reactive groups and can be replaced. This type of dye is soluble in water due to the presence of the sulfonic acid or other anionic groups. Thus, the cotton fibers are impregnated with this solution and then heated under mild alkaline conditions. The remaining chlorine atoms of the reactive dye react with the hydroxyl groups of cellulose forming covalent bonds. Thus, the dye and fiber are attached through an ether linkage. Finally, the fibers are washed with soap to remove the partially hydrolyzed reactive dye molecules.

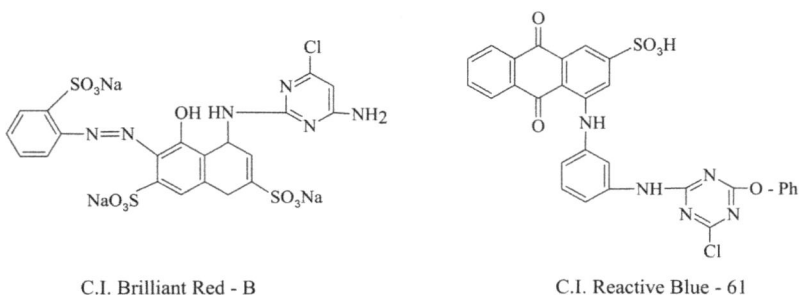

C.I. Brilliant Red - B   C.I. Reactive Blue - 61

Figure 7.11 Examples of Reactive dye

7.2.1.6 Vat dyes

The usage of vat dyestuffs is traced to ancient times: a blue coloring matter known as 'Indigo' is one among the oldest present vat dyes and has been known to Indian people for around 5000 years. Indigo was first recovered from *glucoside indican*, naturally found within the indigo plants in India. The term vat dyes relates to dyes of any chemical class that are applied by the vat process. This type of dye is insoluble in water and cannot be utilized for dyeing directly. However, their reduced forms are soluble. Initially the cloth is immersed in the vat dye, in which the reduction of dye molecules takes place. After reduction the vat dye is adsorbed on the fiber. And thereafter, it is oxidized with air or chemicals or sunlight to form original insoluble forms. Dyeing carried out in this way is very fast to washing, fast to light and bleaching and has excellent fastness properties. This dye is mostly used in the cotton fabric. The dyeing process is costly and time consuming, but these dyes are extremely important for certain textiles because of their superior fastness properties. Vat dye is applied to cellulosic fibers, wool, silk, nylon and cellulosic acetate fibers.

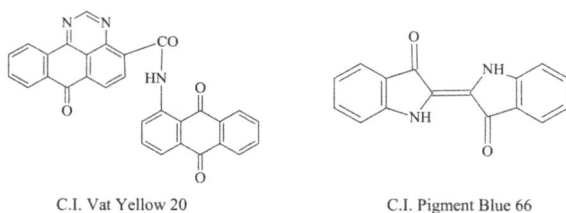

C.I. Vat Yellow 20   C.I. Pigment Blue 66
Figure 7.12 Examples of Vat dye

They are divided according to their structure, i.e., indigoid, thioindigoid and anthraquinone.

7.2.1.7 Disperse dyes

To provide limited water solubility at dyeing temperatures, disperse dye molecules are typically tiny and contain a hydroxyl or amino group. These dyes are usually grounded upto a particle size of 1–4 μ in aqueous solution containing a dispersing agent so that they stabilize in dye suspensions and act as resisting and retarding agents. These dispersing agents enhance the solubility of the dye and fiber. Several kinds of dispersing agents like alkyl sulfates and alkylaryl sulfonates are used. High pressures and temperatures or organic compounds are used for absorption into the fiber. It is preferable to employ organic compounds that have been used with disperse dyes, such as lignin sulfonate, fatty alcohol or amine + ethylene oxide condensation products and naphthalene sulfonic acid + formaldehyde condensation products.

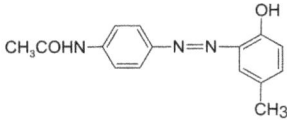

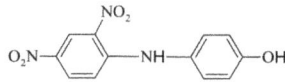

C.I. Disperse Yellow - 3   C.I. Disperse Yellow - 9
Figure 7.13 Examples of Disperse dye

Disperse dyes used in the industry are based on a variety of chromophore systems. A quarter of all products are anthraquinone dyes and around a third of all products are azo dyes. The remaining products are quinophthalone, methine, naphthalimide, naphthoquinone, and nitro dyes. These dyes are principally used on cellulose acetate nylon, polyester, polyacrylonitrile fiber and dacron fibers. These dyes are usually applied by a dry heat process to polyester fiber under pressure and with an organic swelling agent. These dyes are also utilized for dyeing the woolen sheepskins.

7.2.1.8 Mordant dye

The fabric affinity of these dyes is weak. In order to bond the dye, this dye necessitates pretreating the fiber with a mordant substance. The dye and mordant mix to create an insoluble colored complex after the mordant attaches to the fiber (dye-mordant-fabric complex). The name of this complex is lake. Typically, aluminum, iron, and chromium oxides can be used as mordants. Wool, silk, and cotton are all dyed with this substance. Mordants are metal salts that are electrically cationic. Mordants are used for dyes which have a low affinity for the fibers, i.e., they are used to increase the dye-fiber affinity. The mordant dyes have a good fastness property.

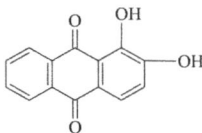

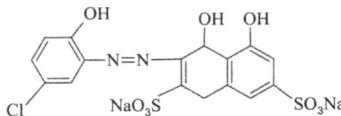

Allizarin   C.I. Mordant Blue 13
Figure 7.14 Examples of Mordant dye

Only azo dyes have a good spectral colour range, making them more conspicuous. Mordant dyes are chemically classed as azo, anthraquinone, oxazine, xanthene, triphenyl methane, nitroso, and thiazine kinds. Chelate groups or groups that can hold the metal in a stable combination must be present in a dye to have mordant dyeing capabilities. Metallic salts or lake are applied directly on the fiber by the use of aluminum, chromium and iron salts, which cause precipitation *in situ*. Different mordants, i.e., salts are used for different purposes. Chromium salts are used for dyeing wool and cotton with a mordant. Cotton is dyed and printed with alizarin using aluminium salts. O-nitroso phenols are used to print cotton using iron salts.

7.2.1.9 Sulfur dyes

These dyes are insoluble in water. They become soluble and show affinity for cellulose fibre when reduced with sodium sulfide. Due to adsorption, this dye acts as direct dye, but when it is reoxidized by exposure to air, it becomes the original insoluble dye inside the fiber. Thus, they become very resistant towards washing. In these dyes, sulfur acts as an integrated part of the polysulfide chains and the chromophore. The actual makeup of the majority of sulfur dyes is unknown. These colours are reasonably priced and have good washing stability. Their brightness and quickness to bleach are frequently poorer, though.

7.2.1.10 Solvent dyes

These dyes do not contain any sulfo or other groups that are water soluble, but they are soluble in organic solvents, and their nature varies depending on the application. These dyes are employed in the production of stains, varnishes, inks, lacquers, copy paper, ribbons for typewriters, candles, polishes, soaps, and cosmetics, among other products.

7.2.2 Classification of dyes based on the structure

7.2.2.1 Nitro or nitroso dyes

They are nitro or nitroso groups as the chromophores with the hydroxyl groups as the auxochromes. They are further classified into two categories: (1) Acid dyes and (2) Disperse dyes. Acid dyes have a solubilizing group such as sulfonic acid.

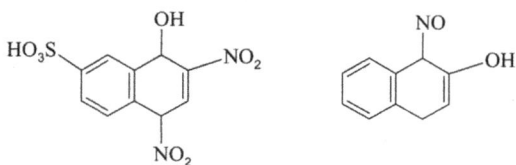

Naphthol Yellow - S   Naphthol Green - G
Figure 7.15 Examples of Nitro and Nitroso dye

These colors were once used to dye natural animal fabrics like wool and silk. They are easily synthesized from affordable intermediates, which makes them cost-effective, but because the colors aren't particularly quick, they aren't of great industrial significance.

7.2.2.2 Azo dyes

They have an azo (-N = N-) group between two aromatic rings that serves as a chromophore. The majority of synthetic dyes are azo dyes, which provide a wide range of hues. They are further divided into groups based on the quantity of azo groups, into monoazo, diazo dyes and more. They are directly applied to fibers usually cotton, rayon and polyester.

7.2.2.3 Anthraquinone dye

They are derivatives of anthraquinone, with a p-quinone system as the chromophore. It contains a group such as -OH, $-NH_2$, -NHR, $-NR_2$, -Ar and -NHCOR, as an auxochrome. They are used as modern and vat dyes. Anthraquinone acid dyes are available in different shades of violet and converted into shades of blue to green by adding azo dyes to them. Because of their brilliance, excellent light fastness, and chromophore stability in both acidic and basic environments, anthraquinone-based dyes are important. Anthraquinone dyes are used to color paper, leather, wool, silk, nylon, and cotton. An example of an anthraquinone dye is alizarin.

7.2.2.4 Triarylmethane dyes

In this type of dye, a central carbon is added to three aromatic rings, one of them in the quinoid as a chromophore. The common auxochromes are $-NH_2$, $-NR_2$ and -OH. However, triarylmethane dyes are not light- or washing-fast unless they are used on acrylic fibers. This type of dye is mainly used in paper dyeing and typewriter ribbons, where fastness to light is not so important. They are colorless compounds called leuco bases. These compounds on oxidation give colored bases, which on treatment with acid give a salt. Crystal Violet is an example of a triarylmethane dye.

7.2.2.5 Indigo dyes

This is one of the oldest dyes having an enedione ($\overset{O}{\underset{C}{\|}}-C=C-\overset{O}{\underset{C}{\|}}$) group as the chromophore. It is used extensively due to its special hue and excellent fastness properties. They are important vat dyes. They are used to dye polyester, wool, and cotton. These dyes cannot be used for dyeing other fabrics because they are insoluble in water. However, when reduced to a leuco form, they become soluble in the presence of an alkali and develop an affinity for cellulosic fibers. A leuco compound solution can then be used for dyeing or printing. The original insoluble dye is then built into the fiber's structure after re-oxidation, which is typically accomplished by exposure to air. For the dye particles to get firmly set and the shade to fully emerge, aggregation or crystallization occurs with a final hot soap or detergent treatment. Indigo is exemplified by C.I. Pigment Blue 66.

7.2.2.6 Xanthene dyes

They are derivatives of xanthenes (dibenzo-1,4-pyran) having quinonoid as the chromophore and -OH, $-NHR_2$, -COOH substituent groups. They are produced by condensing phenols with phthalic anhydride in the presence of zinc chloride, sulfuric acid, or anhydrous oxalic acid. Rhodamine B, eosin, and fluorescein are a few examples of this dye. They give brilliant shades with fluorescence, but they lack light fastness. They are given bright colors with metal salts.

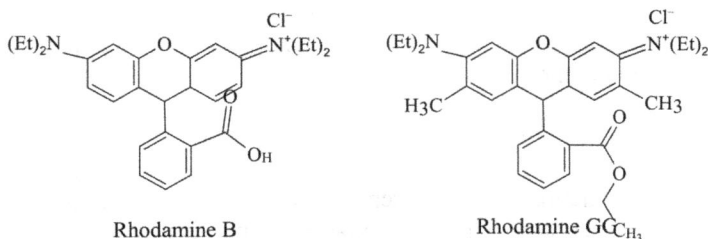

Rhodamine B   Rhodamine GGCH_3

Figure 7.16 Examples of Xanthene dye

7.2.2.6 Heterocyclic dyes

They are derivatives of heterocyclic rings with -NH_2 and -NR_2 groups as auxochromes. They contain hetero atoms like nitrogen, sulfur and oxygen in the ring. They are all basic dyes and their salts are colored. They are used for dyeing silk, wool leather and also cotton with a mordant. They are further classified on the basis of the type of heterocyclic rings present like azine, thizine, oxazine, acridine, thiazoleand quinoline. The ring itself acts as the chromophore.

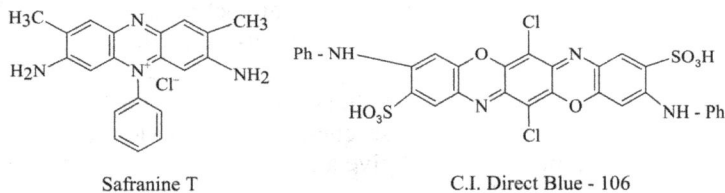

Safranine T   C.I. Direct Blue - 106

Figure 7.17 Examples of Heterocyclic dye

7.2.2.7 Cyanine dyes

They contain two quinoline rings attached through a methine (=CH-) or polymethine (=CH-CH=CH-) group. These attributes have led to their use in many biological applications, including DNA sequencing, immune assays, fluorescence microscopy, and single-molecule detection. They are also used as photographic sensitizers. They are further classified on the basis of their positions through which they are linked like isoeyunine, pseudocyunlne, carboeyanine and kryptocyaninc dyes.

7.2.2.8 Some miscellaneous dyes

(A) Azine Dye: Azine dyes are the oldest known dyes and mauveine among them is considered as the oldest synthetic dye. The azine structure ring structure is found in Safranine-T.

(B) Oxazine Dyes: Oxazine dyes consist of different types of dyes like (1) oxazine basic dye, (2) oxazine mordant dye and (3) oxazine direct dye. Also, they containdiazine pigments. Generally, these dyes have a blue shade. These dyes are obtained by the condensation of p-nitroso dialkyl aniline with suitable phenols in alcoholic solution in the presence of zinc chloride. One of the oxazine dyes, i.e., Capra Blue (Basic dye) is prepared by the condensation of m-diethyl aminophenol and p-nitroso dimethyl aniline.

These dyes possess brilliant shades, very good fastness to light and moderate fastness to washing.

(C) Thiazine Dyes: In this type of dye, thiazine is the chromophore and methylene blue is an important constituent. The starting material of methylene blue is aniline. First, aniline undergoes nitration with the help of concentrated nitric acid and concentrated sulfuric acid to get p-animoanilne, followed by reduction to yield p-aminodimethyl aniline. Thereafter, it is converted into thiosulfate by treatment with an aqueous solution of thiosulfate, sodium dichromate and sulfuric acid. Then, thiosulfate is treated with dimethylaniline under oxidation conditions to obtain Indamine thiosulfonic acid. Now, this acid is further oxidized with hydrochloric acid and zinc duct to the yield final product, i.e., Methylene blue. Methylene blue dye acts as a basic leather dye (Benkhaya et al., 2020; Beal, 2008; Kiernan, 2001).

7.3 Chrome Blue Black

7.3.1 Introduction

The Chrome blue black dyes are the mordant group of dyes, in which mordant is required for dyeing process. The mordant for chrome dyes is potassium dichromate. The examples of chrome blue black are

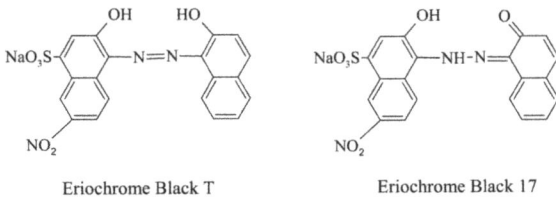

Eriochrome Black T   Eriochrome Black 17
Figure 7.18 Examples of Chrome Blue Black dye

7.3.2 Manufacturing process

The starting raw material for chrome blue black manufacture is 1-amino-2-naphthol-4-sulfonic acid (1,2,4-acid) or its derivatives.

7.3.2.1 Raw materials

1-Amino-2-naphthol-4-sulfonic acid (1,2,4-acid) or its derivatives, β-Naphthol, Caustic soda, Sodium nitrite, Hydrochloric acid, Salt, Water and Copper sulfate

7.3.2.2 Chemical reactions

In the manufacturing process, 1,2,4-acid or its derivatives undergo the diazotization process in the presence of sodium nitrite and hydrochloric acid at 0–5°C. Thereafter it coupls with β-naphthol or another substance. The general chemical reaction is:

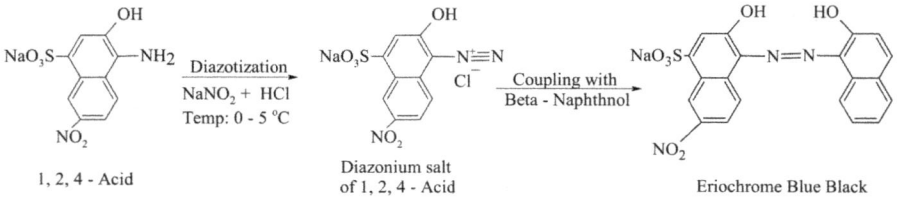

Figure 7.19 Synthesis of Chrome Blue Black dye

7.3.2.3 Quantitative requirements

3500 kgs (yield: 80–85%) of chrome blue black are obtained using 1200 kgs of 1,2,4-acid, 730 kgs of β-naphthol, 600 kgs of caustic soda, 370 kgs of sodium nitrite, 500 kgs of hydrochloric acid, 2000 kgs of salt, 8500 kgs of water and 15 kgs of copper sulfate.

7.3.2.4 Flow sheet diagram

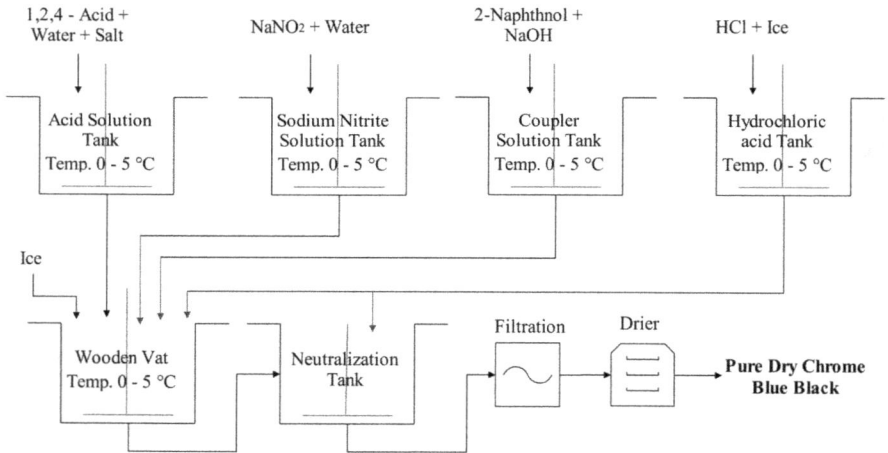

Figure 7.20 Flow sheet diagram of production of Chrome Blue Black dye

7.3.2.5 Process description

In the manufacturing process, the diazotization and coupling reactionsare conducted in one vat (vessel) and thereafter filtration and drying is carried out.

(A) Diazotization and coupling: As discussed, aromatic amines ($-NH_2$) are converted diazo salts ($-N=N^+\ Cl^-$). Additions of chemicals in the vat is done in the following order. Firstly, 1-amino-2-naphthol-4-sulfonic acid (1,2,4-acid) is mixed with water and salt in a vessel. Secondly, sodium nitrite solution in water is prepared in another vessel. Thereafter, β-naphthol solution is prepared by stirring and heating at 60°C with caustic soda using steam and cooled overnight. A mixture of hydrochloric acid and water is carried in another vessel. All these solutions are cooled down to 5°C and added slowly into the vat in their respective order. Addition of solutions is done in such a way that the temperature of the reactor is maintained at 0–5°C for 2 hours. Also, ice is added to maintain the temperature. In this vat, first the diazotization of 1,2,4-acid takes place and thereafter, it couples with β-naphthol, thus a sodium salt of chrome blue black is formed.

(B) Purification and separation: A sodium salt of chore blue black is neutralized with hydrochloric acid to precipitate chore blue black. Precipitates of chrome blue black are collected by filtration using a plate frame filter press, where solids and liquids are separated. The paste is then dried with a Tunnel drier at 75°C, powdered in a mill, mixed with a salt and packed.

7.3.3 Uses

It is mostly used for dyeing fabrics made of wool, silk, and nylon after they have been chrome-dyed, but it may also be used to dye fabrics made of wool and a range of blended fibers, producing hues ranging from black to navy blue. This type of dye has its advantages that dyeing process is conducted in a single step and requires less time, and also, easily matches the shades. But its disadvantage is that the range of shades is limited (Balaban et al., 2005).

7.4 Koch Acid

7.4.1 Properties

It is an important dye intermediate. It has the following properties.

Table 7.3 Physical & Chemical Properties of Koch acid

Sr. No.	Properties	Particular
1	IUPAC name	1-Naphthylamine 3,6,8-trisulphonic acid
2	Chemical and other names	Koch acid, T-acid
3	Molecular formula	$C_{10}H_9NO_9S_3$
4	Molecular weight	383.3 gm/mole
5	Physical description	Light brown coloured crystalline powder
6	Solubility	Soluble in diluted alkaline solution
7	Melting point	Not Available
8	Boiling point	> 94°C
9	Flash point	Not Available
10	Density	1.974 gm/cm³ at 27°C

7.4.2 Manufacturing process

It is prepared from naphthalene.

7.4.2.1 Raw materials

Naphthalene, Sulfuric acid, 62% Oleum, Nitric acid, Iron, Acetic acid and Hydrochloric acid

7.4.2.2 Chemical reactions

Naphthalene is sulfonated with 62% oleum and sulfuric acid to give naphthalene 3,6,8-trisulfonic acid at 180°C, which is then nitrated with a mixture of nitric acid and sulfuric acid to give 1-nitro naphthalene 3,6,8-trisulfonic acid. This compound undergoes Bechamp reduction with iron and acetic acid to give 1-naphthyl amine 3,6,8-trisulphonic acid (T-acid or Koch acid). Iron is removed as sludge from the reaction mixture by adding soda ash solution and filtering it. The filtrate containing Koch acid is isolated by adding hydrochloric acid and filtered.

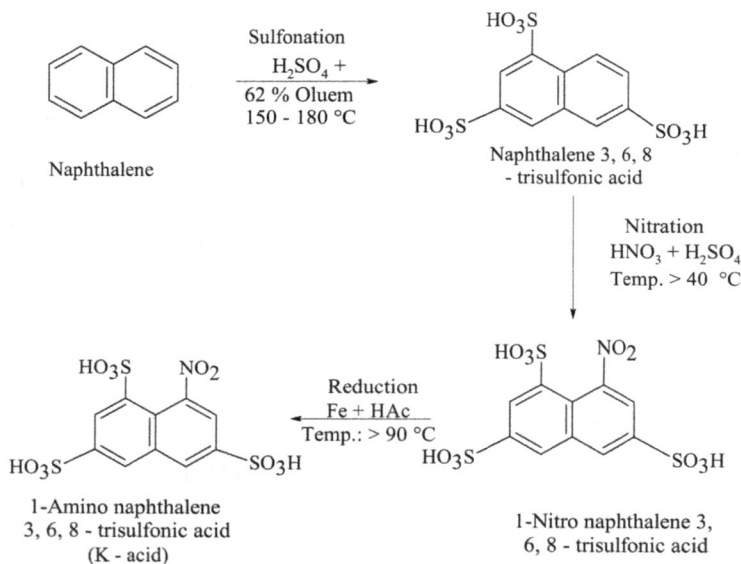

Figure 7.21 Synthesis of K-acid

7.4.2.3 Quantitative requirements

1000 kgs of Koch-acid is obtained using 1000 kgs of naphthalene, 3000 kgs of oleum, 2300 kgs of sulfuric acid, 900 kgs of nitric acid, 1000 kgs of iron and 2000 kgs of hydrochloric acid.

7.4.2.4 Flow sheet diagram

7.4.2.5 Process description

Thefollowing major steps are involved in the Koch acid manufacturing process.

(A) Sulfonation: 62% oleum and sulfuric acid are taken into a cast iron jacketed sulfonation reactor with an anchor type agitator. The mix is cooled to 20°C by circulating the chilled water in the jacket. Add naphthalene (powder or flask) into the reactor over a time of period of 2–2½ hours in such way that the temperature does not exceed 35°C. After addition of naphthalene, heat the reaction mixture upto 85°C using steam in the jacket. Maintain this

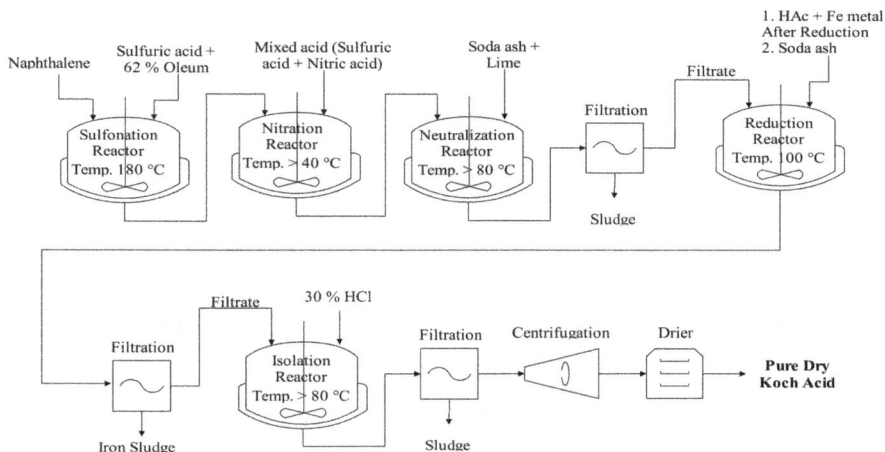

Figure 7.22 Flow sheet diagram of production of K-acid

temperature for one hour. Thereafter, heat the solution to 150–180°C and maintain for 30 hours to prepare naphthalene 3,6,8-trisulfonic acid. Cool the reaction mixture upto 40°C using cool water. Collect the sulfonation mass ready for nitration from the bottom of reactor.

(B) Nitration: This sulfated naphthalene is transferred into a cast iron jacketed sulfonation reactor with an anchor type agitator. The nitrating mixture of nitric acid and sulfuric acid is added to the reactor with continuous stirring and cooled water circulation in the jacket. Addition of acids in such way that the temperature does not exceed 40°C over a time period for 12–13 hours. Maintain the temperature of 40°C for 3–4 hours to prepare 1-nitro naphthalene 3, 6, 8-trisulfonic acid. Then, the reaction mixture is transferred into a neutralization reactor.

(C) Neutralization and Filtration: This reaction mixture, having a pH of about 1.3–1.5, is transferred into a jacketed neutralization tank with an agitator. A solution of soda ash and lime is added slowly with continuous stirring for 3 hours maintaining the temperature of 80°C. The batch is continuously stirred for 1–2 hours and the pH is adjusted upto 6.5–7.5. Extra soda ash and lime solution is added, if required. Thereafter, the sludge is removed by filtration and the filtration mass is used for further processing.

(D) Reduction: First, glacial acetic acid and iron are added to a jacketed reactor with an agitator. Heat the solution upto 90°C with continuous stirring. Add the reaction filtration mass with stirring. Maintain a temperature of 90–100°C for 3–4 hours to complete the reduction process. After completion of the reaction, soda ash solution is added to the reduction reactor with constant stirring over a time period of 3–4 hours and a temperature of 90°C to precipitate iron from the reaction mass. It is filtered to remove iron sludge and mother liquor containing Koch acid which is used for further processing.

(E) Precipitation and filtration: After filtration, this filtration mass, having a pH of 8–10, is transferred into a precipitation tank having a stirring facility. 30% hydrochloric acid is added to the tank in a controlled matter such that the temperature does not rise upto 80°C over a time period of 4 hours. Then the solution is stirred for 1 hour to precipitate Koch acid. Extra acid is added to achieve a pH of 3.3–3.5 and complete the precipitation of Koch acid. It is filtered, centrifuged, dried and packed.

7.4.3 Uses

It is mainly used as a dye intermediate and in the manufacture of H-acid.

7.5 H-Acid

7.5.1 Introduction

It is an important dye intermediate. It has the following properties.

Table 7.4 Physical & Chemical Properties of H - acid

Sr. No.	Properties	Particular
1	IUPAC name	1-Amino-8-naphthalene-3, 6-disulfonic Acid
2	Chemical and other names	H-acid
3	Molecular formula	$C_{14}H_{21}NO_{11}$
4	Molecular weight	379.3 gm/mole
5	Physical description	Light yellowish brown to grey coloured crystalline powder
6	Solubility	Soluble in water, ether and alcohol
7	Melting point	−11°C
8	Boiling point	122°C
9	Flash point	Non-flammable
10	Density	1.974 gm/cm³ at 27°C

7.5.2 Manufacturing process

It is prepared from naphthalene using various unit processes like sulfonation, nitration and reduction.

7.5.2.1 Raw materials

Naphthalene, Sulfuric acid, 62% Oleum, Nitric acid, Iron, Sodium hydroxide, Acetic acid, Hydrochloric acid, Caustic soda, Lime-water and Soda ash

7.5.2.2 Quantitative requirements

1000 kgs H-acid is obtained using 1000 kgs of naphthalene, 3000 kgs of oleum, 2300 kgs of sulfuric acid, 900 kgs of nitric acid, 1100 kgs of soda ash, 4800 kgs of calcium carbonate, 1000 kgs of iron, 2000 kgs of hydrochloric acid, 2500 kgs of sodium hydroxide, 2300 kgs of caustic lye solution and 10,600 kgs of sulfuric acid (40%).

7.5.2.3 *Chemical reactions*

Naphthalene is sulfonated with 62% oleum and sulfuric acid to give naphthalene 3,6,8-trisulfonic acid at 180°C, which is then nitrated with a mixture of nitric acid and sulfuric acid to form 1-Nitro naphthalene 3,6,8-trisulfonic acid. This compound undergoes Bechamp reduction with iron and acetic acid to give 1-naphthyl amine 3,6,8-trisulphonic acid (T-acid or Koch acid). Iron is removed as sludge from the reaction mixture by adding soda ash solution and filtering it. The filtrate containing Koch acid is then fused with 60% caustic soda solution in an autoclave at 170–220°C under a pressure of 6–8 atmospheres to hydroxylate the sulfonic acid group in position 8, thus converting it to H-acid. Finally, H-acid is isolated with sulfuric acid from the reactor mixture and filtered.

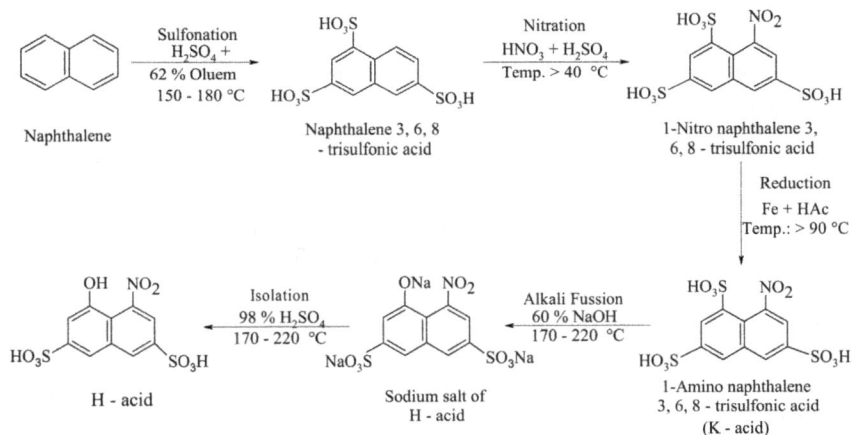

Figure 7.23 Synthesis of H-acid

7.5.2.4 *Flow sheet diagram*

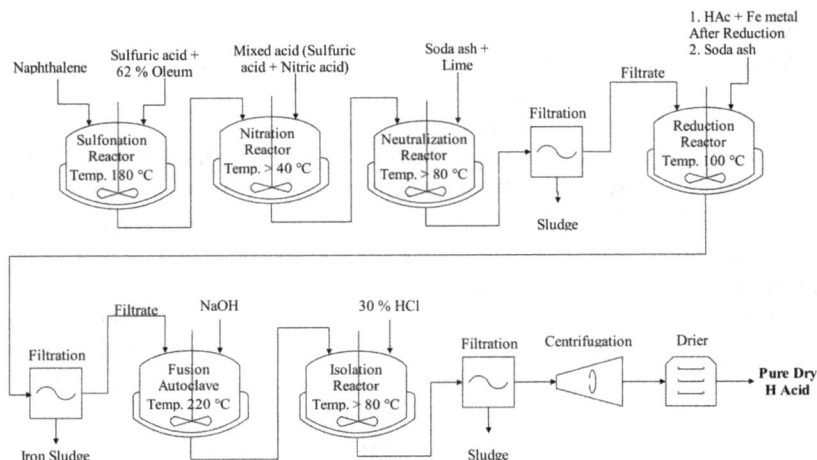

Figure 7.24 Flow sheet diagram of production of H-acid

7.5.2.5 Process description

The following major steps are involved as per the flow sheet diagram in Diagram 7.24.

(A) **Sulfonation:** 62% oleum and sulfuric acid are taken into a cast iron jacketed sulfonation reactor with an anchor type agitator. The acid is cooled to 20°C by circulating chilled water in the jacket. Add naphthalene (powder or flask) into the reactor over a time of period of 2–2½ hours in such way that the temperature does not exceed 35°C. After addition of naphthalene, the reaction mixture is heated upto 85°C using steam in the jacket. Maintain this temperature for one hour. Thereafter, heat the solution to 150–180°C and maintain for 30 hours to prepare naphthalene 3,6,8-trisulfonic acid. Cool the reaction mixture to 40°C using cool water. Collect the sulfonation mass ready for nitration from the bottom of reactor.

(B) **Nitration:** This sulfated naphthalene is transferred into a cast iron jacketed sulphonation reactor with an anchor type agitator. The nitrating mixture of nitric acid and sulfuric acid is added to the reactor with continuous stirring and cooled watercirculation inthe jacket. Acids are added in such a way that the temperature does not exceed 40°C over a time period of 12–13 hours. Maintain the temperature of 40°C for 3–4 hours to prepare 1-nitro naphthalene 3,6,8-trisulfonic acid. Then, reaction mixture is transferred into neutralization reactor.

(C) **Neutralization and Filtration:** This reaction mixture, having a pH of about 1.3–1.5, is transferred into a jacketed neutralization tank with an agitator. A solution of soda ash and lime is added slowly with continuous stirring for 3 hours maintaining the temperature at 80°C. The batch is continuously stirred for 1–2 hours and the pH is adjusted to 6.5–7.5. Extra soda ash and lime solution are added, if required. Thereafter, the sludge is removed by a filtration process and the filtration mass is used for further processing.

(D) **Reduction:** First, glacial acetic acid and iron are added to a jacketed reactor with an agitator. The solution is heated upto 90°C with continuous stirring. Add the reaction filtration mass with stirring. Maintain a temperature of 90–100°C for 3–4 hours to complete the reduction process. After completion of the reaction, soda ash solution is added to the reduction reactor with constant stirring over a time period of 3–4 hours at a temperature of 90°C to precipitate the iron from the reaction mass. It is filtered to remove the iron sludge and mother liquor containing Koch acid which is used for further processing.

(E) **Fusion:** The caustic soda solution is taken into a mild steel jacketed autoclave having an agitator and a pressure gauge. The filtration mass containing Koch acid is added with stirring. The reaction mixture is heated upto 170–220°C. This temperature is maintained for 2 hours at a pressure of 6–8 atmospheres. If the pressure rises, it is controlled upto 6–8 atmospheres using a pressure gauge. 2 hours after the fusion reactions, the reaction mass is cooled and transferred into the neutralization tank.

(F) Neutralization and purification: The reaction mass is taken into a jacketed isolation tank equipped with a stirrer. The 98% sulfuric acid is added into the tank in a controlled manner such that the temperature does not rise to 80°C over a time period of 4 hours. Stir the solution for 1 hour to precipitate H-acid. Extra acid is added to achieve a pH of 1.3–1.5 and complete precipitation of H-acid. Cool the reaction mixture to room temperature. It is filtered, centrifuged, dried and packed.

7.5.3 Uses

H-acid plays a key role in coupling during the manufacture of acid dyes. The light fastness of the dyes made from H-acids is greatly improved by the acylation of the amino group. These dyes have good levelling capabilities and are lightweight and wet fast. With N-toluenesulphonyl-H-acid and N-benzoyl-H-acid, valuable dyes can also be produced. C.I. Acid Red 1, or I. 18050, are examples of rather straightforward structures. H-acid is a crucial dual coupling component employed in diazo acid dyes as well. Throughout this series, there is very little diversity in shades; primarily black, drab brown, and blue colors are obtained. They are the chromophores that are fully conjugated. One of the crucial acid dyes is C.I. Acid Black 1, also known as C.I. 20470. It exhibits great affinity, decent levelling power, and dyes wool in colors of blue-black with excellent light fastness but only fair wet fastness. It still serves as a foundational dye for the conception of black acid dyes (Bokari et al., 2016).

7.6 Tobias Acid

It is also used as a dye intermediate.

7.6.1 Introduction

Its properties are tabulated as under.

Table 7.5 Physical & Chemical Properties of Tobias acid

Sr. No.	Properties	Particular
1	IUPAC name	2-Amino-1-naphthalenesulfonic acid
2	Chemical and other names	Tobias acid, 2-Naphthylamine-1-sulfonic acid
3	Molecular formula	$C_{10}H_9NO_3S$
4	Molecular weight	223.2 gm/mole
5	Physical description	White solid powder
6	Solubility	Slightly soluble in cold water, More soluble in hot water, very slightly soluble in alcohol and ether
7	Melting point	Not Available
8	Boiling point	Not Available
9	Flash point	Not Available
10	Density	1.502 gm/cm^3 at 27°C

7.6.2 Manufacturing process

It is prepared from β-Naphthol.

7.6.2.1 Raw materials

β-Naphthol, 98% Sulfuric acid, Liquid Ammonia, Ammonium sulfate, 40% Sulfuric acid, Common salt, Activated charcoal and Hyflosuperse

7.6.2.2 Chemical reactions

β-Naphthol undergoes sulfonation with 98% sulfuric acid at a low temperature of about −8 to −10°C to give Oxy-tobias acid. This acid is ammoniated with ammonia and ammonium sulfate at 150–160°C to yield 2-amino naphthalene. It is further acidified using 40% sulfuric acid at 40–45°C to give Tobias acid as the final product.

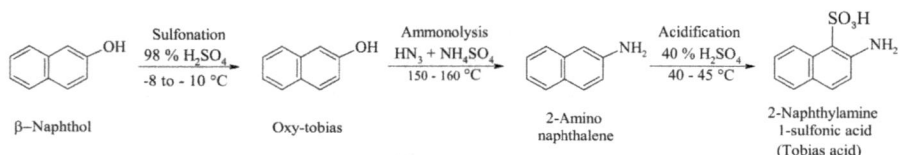

Figure 7.25 Synthesis of Tobias acid

7.6.2.3 Quantitative requirements

1000 kgs of Tobias acid is obtained from 580 kgs of β-naphthol, 1000 kgs of sulfuric acid (98%), 230 kgs of liquid ammonia, 350 kgs of ammonium sulfate, 300 kgs of sulfuric acid (40%), 175 kgs of common salt, 10 kgs of activated charcoal and 10 kgs of hyflosupersel.

7.6.2.4 Flow sheet diagram

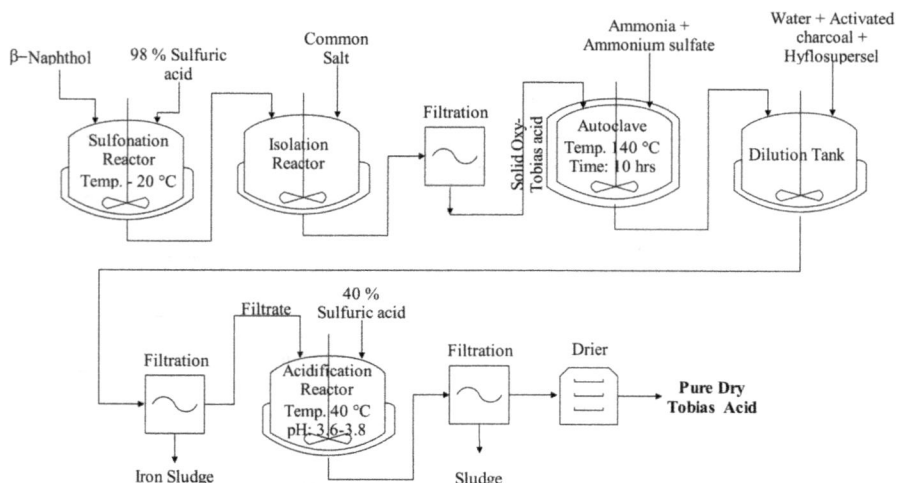

Figure 7.26 Flow sheet diagram of production of Tobias acid

7.6.2.5 Process description

The following major steps are involved in the manufacture of Tobias acid.

(A) Sulfonation: 98% sulfuric acid is taken into a jacketed sulfonation reactor having a stirring facility. It is cooled upto –10 to –12°C using brine in the jacket. Solid β-naphthol is charged slowly for 1 hour with continuous stirring and circulation of brine water. After 8 hours of the sulfonation reaction, the reaction mixture is sent to the isolation reactor. Here, common salt is added to precipitate the oxy-tobias acid. It is then filtered to collect the solid product

(B) Amination: Oxy-Tobias acid is taken in an autoclave. Also, ammonia and ammonium sulfate are added. Heat the mixture at a temperature of 145–150°C for 10–12 hours to give 2-amino naphthalene. The reaction mixture having 2-amino naphthalene, in liquid form, is transferred into a dilution tank. Water and thereafter, activated charcoal as clarifier and hyflosupersel as filter-aid are added into the tank. After stirring for 15 min, filter the reaction mixture and collect the mother liquor in an acidification tank. Also, wash the residue with water and collected it in this tank.

(C) Acidification: The collected mother liquor containing 2-amino naphthalene is heated upto 40–45°C. Also, 40% sulfuric acid is added as fast as possible to adjust the pH between 3.6–3.8 to precipitate tobias acid. During acidification, sulfur dioxide is evaporated, which is scrubbed using an ammonia scrubber. The scrubber yields liquid ammonium sulfite. The ammonium sulfite is further utilized in amination. Now, Tobias is filtered and dried at 80°C.

7.6.3 Uses

It is used as dye intermediate especially, for manufacturing reactive dyes. It is also useful for the preparation of other dye intermediates like J-acid and amino-iso-J-acid.

7.7 Amino J-acid

It is used as a dye intermediate.

7.7.1 Introduction

Its physical and chemical properties are as per Table 7.6.

7.7.2 Manufacturing process

It is prepared from tobias acid.

7.7.2.1 Raw materials

Tobias acid, 25% Oleum and Water (for hydrolysis)

Table 7.6 Physical & Chemical Properties of Amino-iso-J-acid

Sr. No.	Properties	Particular
1	IUPAC name	2-Amino 5-7 Di-sulphonic Acid
2	Chemical and other names	Amino-iso-J-acid
3	Molecular formula	$C_{10}H_9NO_6S_2$
4	Molecular weight	303 gm/mole
5	Physical description	Light yelllo wish solid powder
6	Solubility	Soluble in dilute alkaline Solution
7	Melting point	Not Available
8	Boiling point	Not Available
9	Flash point	Not Available
10	Density	Not Available

7.7.2.2 Chemical reactions

Tobias acid undergoes sulfonation with 25% oleum at 34–36°C to give 2-naphthylamine-disulfonic acid, and thereafter, at a higher temperature of about 104–106°C to obtain a 1,5,7-trisulfonic acid derivative. This product is hydrolyzed with hot water at 114–116°C to yield the final product, amino-iso-J-acid.

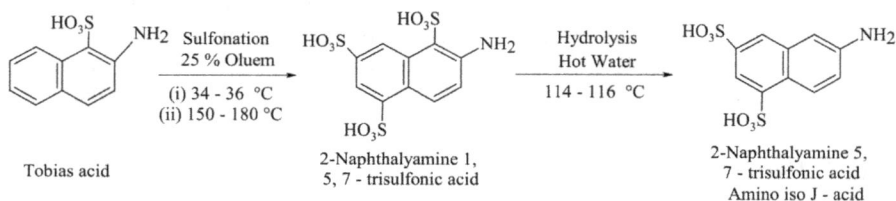

Figure 7.27 Synthesis of Amino J-acid

7.7.2.3 Quantitative requirements

1000 kgs of amino-iso-J-acid requires 900 kgs of tobias acid, 2800 kgs of oleum (25%) having 600 kgs of SO_3 gas and 3500 liters of water (for hydrolysis).

7.7.2.4 Flow sheet diagram

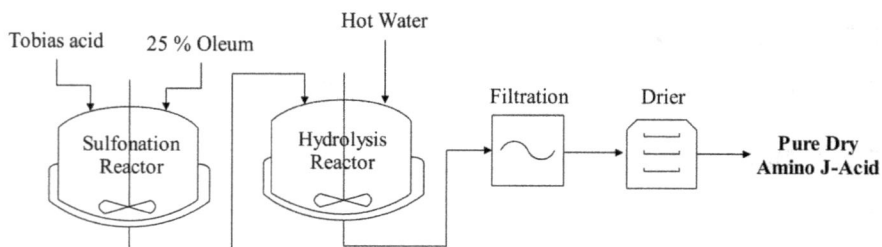

Figure 7.28 Flow sheet diagram of production of Amino J-acid

7.7.2.5 Process description

The following unit processes are conducted for manufacturing amino-iso-J-acid.

(A) Sulfonation: 25% oleum is taken in a jacketed sulfonation reactor equipped with a stirrer. It is cooled upto 18–20°C using cooled water with stirring. Solid Tobias acid is charged slowly upto 1 hour with stirring and circulated cool water to maintain a temperature of 18–20°C. After adding Tobias acid, heat the reaction mixture upto 34–36°C. Then, stir the reaction mixture for 1 hour and maintaina temperature of 34–36°C to obtain a disulfonic acid derivative. To obtainthe trisulfonic derivative, the mixture is further heated upto 104–106°C. Maintain the temperature at 104–106°C for 12 hours with stirring. The sulfonated mixture is ready for hydrolysis.

(B) Hydrolysis and isolation: Hot water at 50°C is taken in a jacketed hydrolysis reactor. Now, the sulfonated mixture is added. Now, steam is inserted into the jacket of the reactor to attaina temperature of 114–116°C. Maintain the temperature with stirring for 5 hr. Then, cool the reaction mixture and filter the product.

7.7.3 Uses

It is used as a dye intermediate (Sun et al., 2012).

7.8 Vat Dyes

7.8.1 Introduction

One of the oldest dyes are vat dyes used to dye textile fibers. It is mainly used for dyeing cellulose and protein fibers dyeing, but it is also useful on cotton, wool and rayon. Rubbing fastness of this dye is not good. These dyes cannot be used directly because they are insoluble in water. However, when reduced to a leuco form, they become soluble in the presence of an alkali and develop an affinity for cellulosic fibers; a leuco compound solution can be applied by dyeing or printing. The original insoluble dye is then built into the fiber's structure after re-oxidation, which is typically accomplished by exposure to air. In order for the dye particles to become firmly set and the shade to fully develop, aggregation or crystallization is caused by a final treatment with hot soap or another detergent. This dyeing process is difficult and costly. But, washing fastness of the vat dye is very good and various shades are available.

7.8.2 Manufacturing process

It is prepared from naphthalene.

7.8.2.1 Raw materials

Naphthalene, Chlorobenzene, Diluted and Concentrated Sulfuric acid, Aluminum chloride, Nitrobenzene, Sodium hydroxide, Potassium hydroxide, Potassium nitrate, Stannous chloride, Chloride gas and Ammonia

7.8.2.2 Chemical reactions

Preheated naphthalene vapor is oxidized using mercury to form phthalic anhydride. After oxidation, it undergoes Friedel & Craft alkylation with chlorobenzene in the presence of aluminum chloride at 100°C and thereafter, it is hydrolyzed with cold dilute sulfuric acid to form 4-chloro-benzoyl 2-benzoic acid, which is then cyclized with sulfuric acid to form 2-chloro anthraquinone at 130°C. Amination of 2-chloro anthraquinone is conducted in an autoclave at a temperature of 200°C and pressure of 1000 atmospheres to give 2-Amino anthraquinone. It is then condensed in the presence of potassium hydroxide/potassium nitrate to form Indanthrene blue R.S. then it further reacts with sulfuric acid/chlorine gas to form Indanthrene G.C.D. Further, 2-amino anthraquinone is condensed with nitrobenzene in the presence of stannous chloride at 220°C to form Indanthrene yellow as per Fig. 7.29.

7.8.2.3 Flow sheet diagram

7.8.2.4 Process description

The following unit processes are conducted for manufacturing vat dyes.

(A) Oxidation: The initial raw material for manufacturing vat dyes is Naphthalene. Preheated naphthalene and mercury (Hg) as an oxidizer are inserted into a fluid bed reactor. Steam is also added into this reactor. The reaction is exothermic and so, the temperature in the reactor is 350–385°C. Naphthalene undergoes oxidation to form phthalic anhydride, which is collected in a steel condenser. Further, water vapor and carbon dioxide are removed from the top of the condenser.

(B) Friedel-Craft alkylation & Hydrolysis: Chlorobenzene is inserted into a steam jacketed Friedel-Craft reactor equipped with an agitator. It is then heated upto 100°C. Anhydrous aluminum chloride and phthalic anhydride are charged into the reactor with stirring and at a constant temperature of 100°C. After 8 hours of reaction, the reaction mixture is cooled down to room temperature. It is then collected from the bottom and transferred into a jacketed hydrolyser with an agitator. Cold diluted sulfuric acid is added into the hydrolyser such that the temperature does not rise upto 100°C. After 2 hours of reaction, 2-(4-chlorobenzoyl) benzoic acid is formed and collected from the bottom of the hydrolyser.

(C) Cyclization: 2-(4-chlorobenzoyl) benzoic acid is taken in a glass lined jacketed cyclizator reactor with an agitator. Concentrated sulfuric acid is added and heated upto 130°C for 2 hours to yield 2-chloro anthraquinone by the cyclization unit process. 2-chloro anthraquinone is collected from the bottom of reactor and is used for further processing.

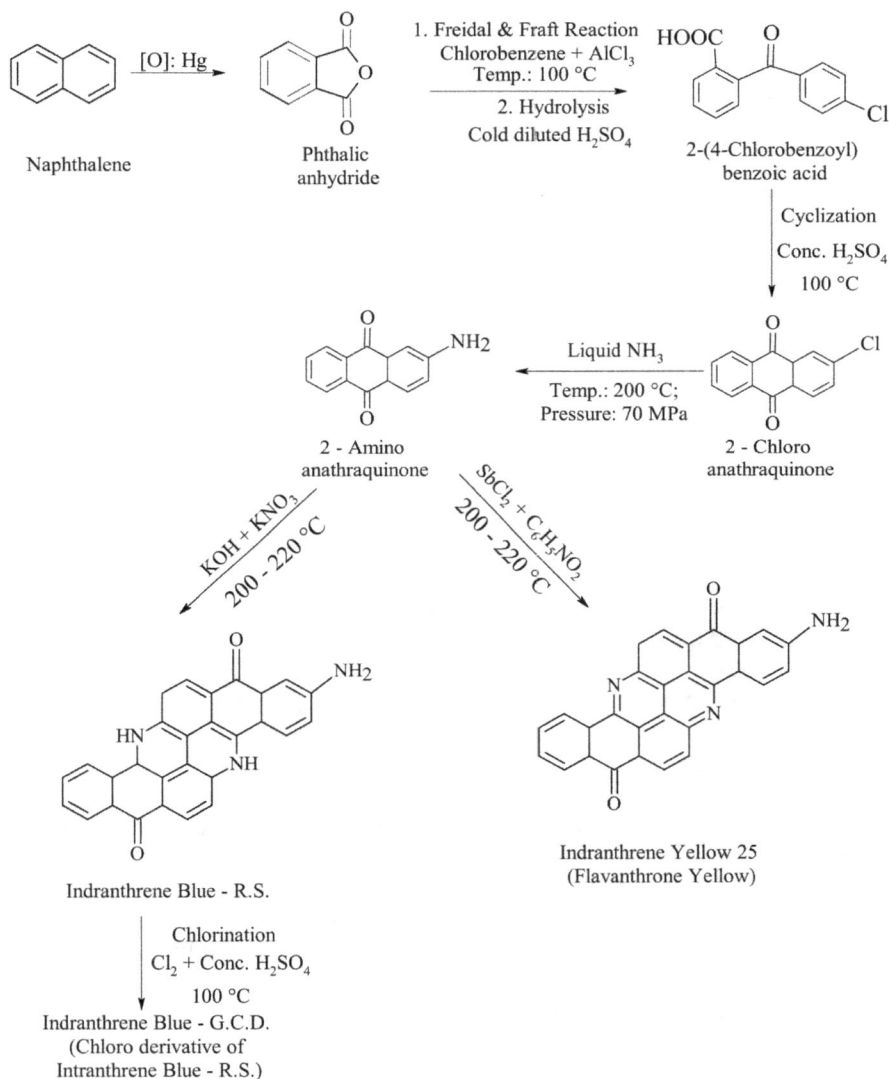

Figure 7.29 Synthesis of Vat dye

(D) Amination: 2-Chloro anthraquinone is taken in an oil jacketed steel autoclave facilitated with an agitatorand a pressure gauge. It is heated upto 200°C. Liquid ammonia is added to the autoclave with stirring at 200°C. The temperature is maintained for 24 hours. The pressure is allowed to rise upto 700 atmospheres using a pressure gauge. Amination of the chloro group is done to yield 2-Amino anthraquinone under these conditions. Now, this 2-Amino anthraquinone is utilized for the preparation of various vat dyes.

(E) Condensation: 2-Amino anthraquinone is transferred into an oil jacketed condensation kettle. Potassium hydroxide and potassium nitrate are added

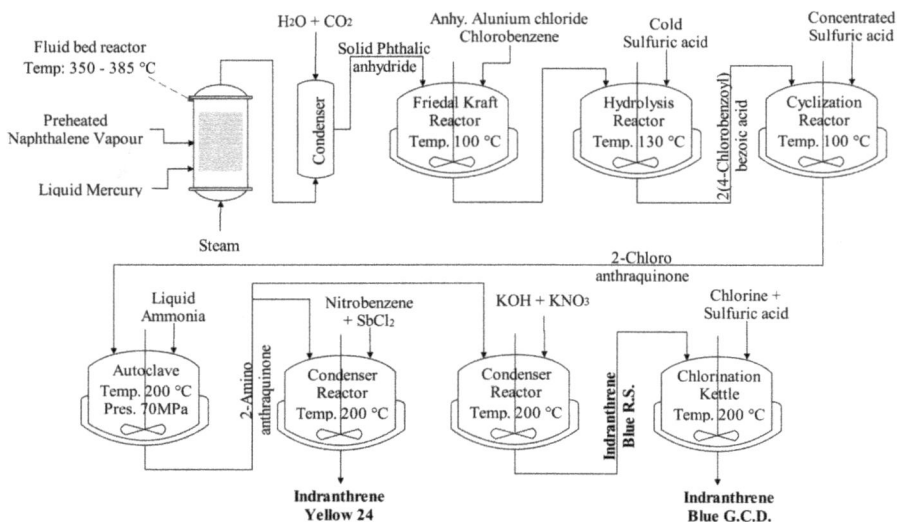

Figure 7.30 Flow sheet diagram of production of Vat dye

with stirring. The reaction kettleis heated upto 200°C. After 2 hours of reaction, Indanthrene Blue R.S. is formed, which is collected from the bottom of the kettle. The Indanthrene Blue R.S. is added to another chlorinator kettle. Chlorine and sulfuric acid are added with stirring in such a manner that the temperature does not rise upto 100°C. After 2 hours, chlorination of Indanthrene Blue R.S. is done to form Indanthrene G.C.D. Further, 2-amino anthraquinone is transferred to a condensation kettle, in which nitrobenzene and stannous chloride are added. The reaction kettle is heated upto 200°C. After 2 hours of reaction, Indanthrene Yellow 24 is formed, which is collected from the bottom of the kettle. A yield of about, 200 kgs of 2-amino anthraquinone and 60 kgs of Indanthrene blue R.S. is obtained by using 148 kgs of phthalic anhydride. Also, a 100% yield of Indanthrene Blue G.C.D. and Indanthrene Yellow 24 is obtained from Indanthrene blue R.S. and 2-amino anthraquinone respectively.

7.8.3 Uses

Vat dyes are utilized as fiber-reactive dyes, direct dyes, and acid dyes for dyeing cotton, wool, and other fibers (Bozic et al., 2006).

7.9 Indigo Dyes

7.9.1 Introduction

This is one of the oldest dyes having enedione ($\overset{O}{\underset{|}{C}}-C=C-\overset{O}{\underset{|}{C}}$) group as the chromophore. Ancient people in India used indigo. It is water insoluble and can't be applied directly on textile material. It is used extensively due to its special colors and very good fastness properties. They are important vat dyes. *Indigofera tinctorie*, a plant, is the source of natural indigo, which was deployed by the Egyptians in 200 BC. In 1897, the first synthetic indigo dye was introduced in the textile industry, entirely replacing the natural substance.

7.9.3 *Manufacturing process*

It is prepared from aniline.

7.9.3.1 *Raw materials*

Aniline, Formaldehyde (Formalin), Sodium oxide, Sodium sulfite, Sodium cyanide, Potassium hydroxide (KOH), Sodium Hydroxide (NaOH), [or Chloroacetic acid, $FeCl_3$, NaOH, KOH], Ammonia, Sodium metal and Air

7.9.3.2 *Chemical reactions*

This method was developed by scientist, Carl Heumann in 1890. In the process of manufacturing Indigo dyes, the first step is to prepared phenylglycine by the individual reaction of aniline with formaldehyde, sodium bisulfite ($NaHSO_3$) and sodium cyanide, followed by alkaline hydrolysis at 100°C. Later on, Heumann modified this process in 1907 due to the usage of hazardous sodium cyanide and increasing the yield of phenylglycine. In the modified method, direct condensation of aniline with chloroacetic acid is conducted in the presence of $FeCl_3$ to give phenylglycine. This phenylglycine is fused at 350°C with sodium hydroxide and sodamide to form indoxyl. Now, two moles of indoxyl undergo air oxidation to yield Indigo dyes.

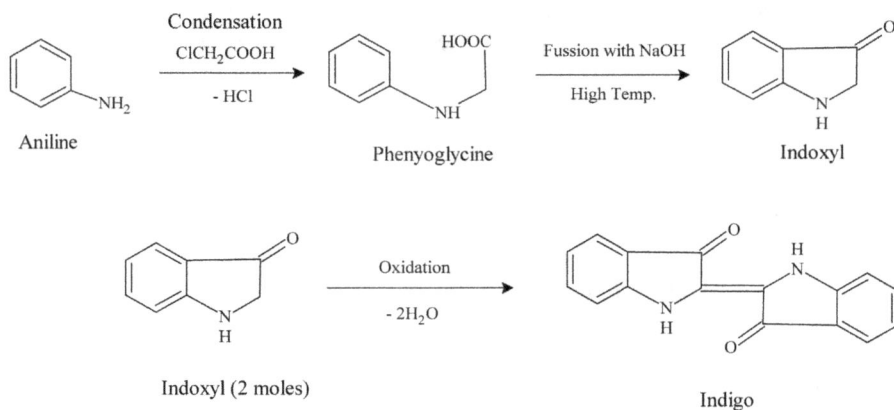

Figure 7.31 Synthesis of Indigo dye

7.9.3.4 Flow sheet diagram

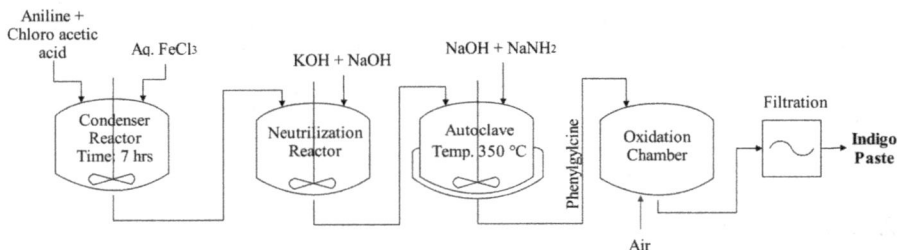

Figure 7.32 Flow sheet diagram of production of Indigo dye

7.9.3.5 Process description

The following major steps are involved in the manufacture of indigo dyes.

(A) Condensation: In this step, aniline, chlorobenzene and aqueous ferric chloride are taken in a reactor having an agitator. The mixture is stirred for several hours. After completion of the condensation reaction, the reaction mixture is collected at the bottom of the reactor and transferred into a neutralization tank. Here, it is neutralized with potassium hydroxide and sodium hydroxide to get phenylglycine.

(B) Fusion: The obtained phenylglycine is inserted into another hot oil jacketed autoclave and the reaction temperature rises upto 230°C. Sodamide, potassium hydroxide and sodium hydroxide are added and a temperature at 230°C is maintained to form indoxyl.

(C) Oxidation: The indoxyl is transferred tothe oxidation chamber, where air is introduced into the bottom of the reactor. The reaction mixture is oxidized to form indigo dyes. The precipitates are filtered, dried and packed. Potassium hydroxide and sodium hydroxide are recovered from the mother liquor. About 80–82% yield of Indigo dyes is achieved using this process.

7.9.3 Uses

They are especially utilized to dye cotton fabrics mainly blue jeans. They are also utilized for dyeing wool and silk. They are also used to manufacture other vat or indigo dyes (Pattanaik et al., 2020).

7.10 C.I. Acid Red - 119

It is an acid dye, prepared by the diazotization process.

7.10.1 Properties

Its main properties are tabulated as per Table 7.7.

Table 7.7 Physical & Chemical Properties of Acid Red - 119

Sr. No.	Properties	Particular
1	IUPAC name	Disodium salt of 3-[[N-ethyl-4-[[4-[(3-sulfonatophenyl) diazenyl] naphthalen-1-yl]diazenyl]anilino]methyl] benzenesulfona
2	Chemical and other names	Acid red – 119, EINECS 274-403-7, 70210-06-9, Benzenesulfonic acid, 3-((ethyl(4-((4-((3-sulfophenyl) azo)-1-naphthalenyl)azo)phenyl)amino) methyl)-
3	Molecular formula	$C_{31}H_{25}N_5Na_2O_6S_2$
4	Molecular weight	673.7 gm/mole
5	Physical description	Red odourless solid powder
6	Solubility	Insoluble in water, soluble in hot water for blue, soluble in ethanol for green light blue.
7	Melting point	Not Available
8	Boiling point	Not Available
9	Flash point	Not Available
10	Density	Not Available

7.10.2 Manufacturing process

Starting material of Acid Red - 119 dye is melanilic acid.

7.10.2.1 Raw materials

Melanilic acid, α-Naphthyl amine, Ethylbenzene aniline sulfonate (EBAMASA), Sulfuric acid, Sodium nitrite, Caustic flask, Hydrochloric acid, Water and Ice

7.10.2.2 Chemical reactions

As per Fig. 7.33, melanilic acid undergoes diazotization with hydrochloric acid and sodium nitrite to give a diazonium salt of the acid. This salt is coupled with α-naphthyl amine in an acidic medium. It is further diazotized using sulfuric acid and coupled with EBAMASA in an alkaline medium.

7.10.2.3 Quantitative requirements

1000 kgs of Acid Red - 119 is obtained using 300 kgs of melanilic acid, 240 kgs of α-naphthyl amine, 460 kgs of EBAMASA, 250 kgs of sulfuric acid, 240 kgs of sodium nitrite, 190 kgs of caustic flask, 270 kgs of hydrochloric acid, 3000 kgs of water and 5000 kgs of ice.

7.10.2.4 Flow sheet diagram

7.10.2.5 Process description

(A) **First diazotization and coupling:** As per the simple flow-sheet, melanilic acid, sodium nitrite and hydrochloric acid are taken in different mild steel tanks and water is added. These solutions are cooled down to 0–5°C using ice. Now, all these solutions are added to the first diazonium reactor which is facilitated with stirring. Maintain the temperature of 0–5°C with stirring for 1 hour till the completion of the diazotization reaction. The diazotizatied

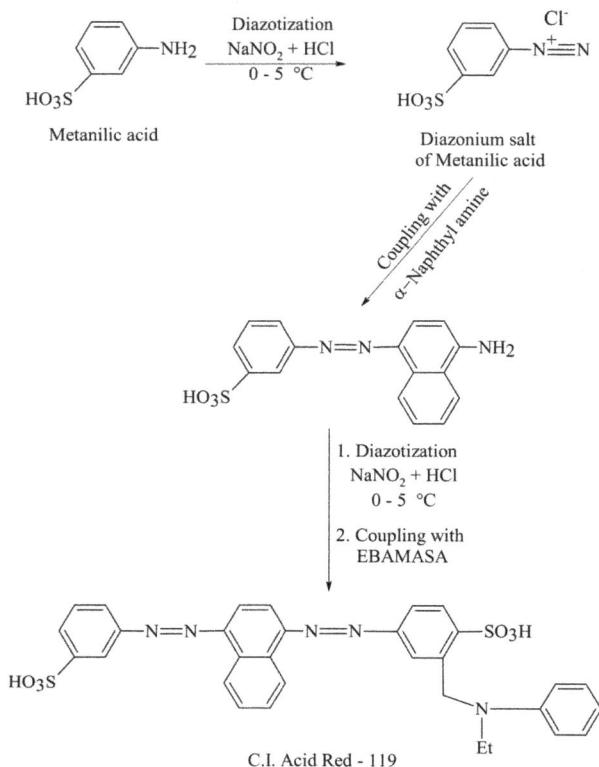

Figure 7.33 Synthesis of C.I. Acid Red - 119

solution is collected from the bottom of the reactor and sent to a coupling reactor with stirring. Now, add α-naphthyl amine and hydrochloric acid solution maintaining a temperature 20–25°C. After completion of the coupling reaction, the reaction mass is collected from the bottom of the reactor and sent to the second diazonium reactor with stirring.

(B) Second diazotization and coupling: The reaction mass is collected in the second diazonium reactor and cooled down to 0–5°C using ice. Also, the solutions of sodium nitrite and sulfuric acid are taken in different mild steel tanks and cooled down to 0–5°C using ice. Thereafter, both solutions are charged into the second diazonium reactor with stirring. Then the temperature is maintained at 0–5°C with stirring for 1 hour till the completion of diazotization. The diazotized solution is collected from the bottom of the reactor and sent to another coupling reactor with stirring. Then, caustic soda and EBAMASA solutions are added to the reactor with stirring at a temperature of 20–25°C. Stir the reaction mass for 3 hours maintaining the temperature at 20–25°C to yield soluble Acid Red - 119. It is precipitated using common salt. It is then filtered using a filter press and dried.

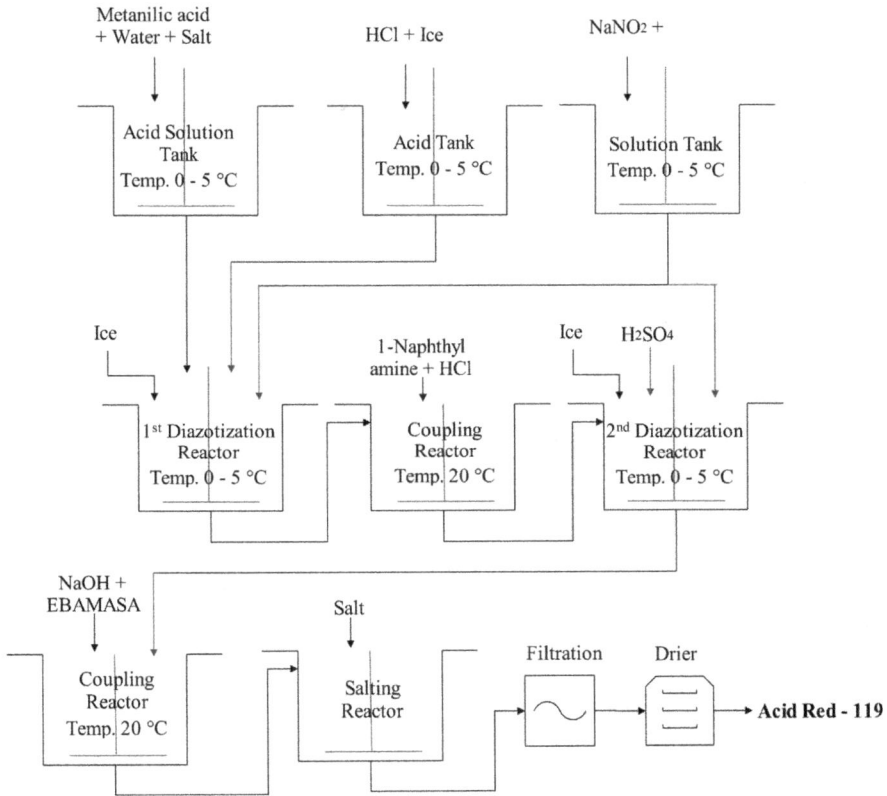

Figure 7.34 Flow sheet diagram of production of C.I. Acid Red - 119

7.10.3 Uses

It is mostly used to color wools, polyamides, and silks. It can also be used to dye leathers.

7.11 Solvent Yellow 163

It is an anthraquinone series dye.

7.11.1 Properties

Its main properties are mentioned in the Table 7.8.

7.11.2 Manufacturing process

It is prepared from 1,8-dichloro anthraquinone and thiophenol.

Table 7.8 Physical & Chemical Properties of Solvent Yellow – 163

Sr. No.	Properties	Particular
1	IUPAC name	1,8-Bis(phenylthio)anthraquinone; 1,8-Bis(phenylthio)-9,10-anthracenedione
2	Chemical and other names	C.I. Solvent Yellow – 163, C.I. 58840, Amaplast Yellow GHS, Navimplast Yellow GHS, Polysolve Yellow, Waxoline Yellow 5RP-FW, Transparent Yellow GHS, Transparent Plastic Yellow 105, Oil Yellow GHS
3	Molecular formula	$C_{26}H_{16}O_2S_2$
4	Molecular weight	424.5 gm/mole
5	Physical description	Yellow odourless solid powder
6	Solubility	Soluble in various organic solvents like methanol, ethano, acetone, dichloromethanol, methylbenzene, etc.
7	Melting point	187°C
8	Boiling point	Not Available
9	Flash point	Not Available
10	Density	0.64 gm/cm³ at 27°C

7.11.2.1 Raw materials

1,8-Dichloro anthraquinone, Thiophenol, Isobutanol and Potassium carbonate

7.11.2.2 Chemical reactions

As per Fig. 7.35, 1,8-Dichloro anthraquinone and thiophenol undergo a condensation reaction in the presence of isobutanol and potassium carbonate to give solvent yellow 163.

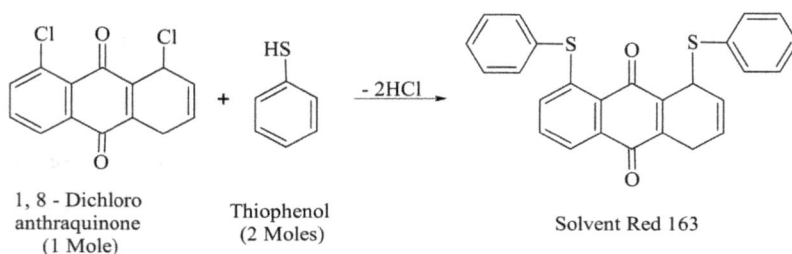

1, 8 - Dichloro anthraquinone (1 Mole)   Thiophenol (2 Moles)   Solvent Red 163

Figure 7.35 Synthesis of Solvent Yellow 163

7.11.2.3 Quantitative requirements

1000 kgs of solvent yellow 163 is obtained using 670 kgs of 1,8-dichloro anthraquinone, 545 kgs of thiophenol, 1500 kgs of isobutanol and 350 kgs of potassium carbonate.

7.11.2.4 Flow sheet diagram

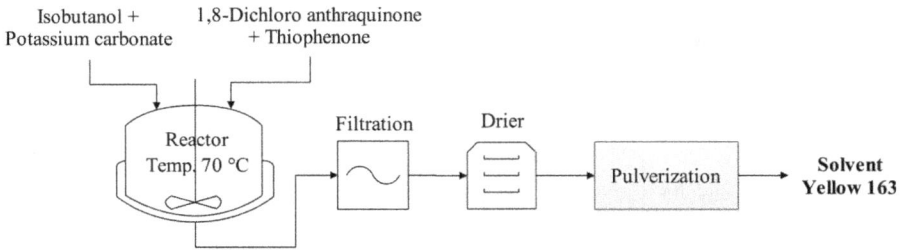

Figure 7.36 Flow sheet diagram of production of Solvent Yellow 163

7.11.2.5 Process description

As per the flow-sheet, insert isobutanol as the solvent and potassium carbonate as the solute in a jacketed stainless steel reactor with a stirring facility. Start stirring and then, add 1,8-dichloro anthraquinone and thiophenol in it. Heat the reaction mixture upto 70°C. Maintain the reaction temperature at 70°C with stirring for 3 hours. Cool the reaction mixture to room temperature. Collect the reaction mixture from the bottom of the reactor and subject it to filtration. Dry the wet cake at 75–80°C and pulverize it.

7.11.3 Uses

It is utilized to color plastics, polyester raw materials, waxes, lubricants, fuels, gasolines, candles, paints, and printing inks (Clark, 2011).

7.12 Classification of Pigments

A pigment is a dry insoluble substance, usually pulverized and crushed, and transformed into paints or inks when suspended in a liquid medium. Pigments are classified as follows.

7.12.1 Inorganic pigments

They are of the mineral-earth variety, but they are typically synthetic materials or metallic oxides. Mineral-earth pigments are relatively straightforward and contain colorful components. The washing, drying, crushing, and mixing procedures make up the preparation process, which is also simple. These pigments are further classified into two types:

7.12.1.1 White pigment

White or concealing pigments are often referred to as pigments that give coatings light-scattering qualities. Due to their very high index of refraction, they act by scattering all solar wavelengths, causing them to appear white to the human eye. They are known as hidden pigments because sunlight scattering lessens the likelihood that light will pass through a colored coating and reach the substrate. When a paint layer is thick enough and contains enough light-scattering pigment,

Table 7.9 Composition, Properties and Uses of White Pigments

Sr. No.	Name of Pigment	Composition	Characteristics	Applications
1	Titanium dioxide	Ti FeO$_3$ and TiO$_2$ (TiO$_2$) iliminite + rutile	Highest index of refraction (2.76), High opacity and hiding power, High Oil absorbing capacity, Spreading power is almost double than that of white lead, No tendency of chalking	In paints, paper and textiles, In other industries
2	White lead PbCO$_3$ + Pb(OH)$_2$	PbCO$_3$: 68.9% Pb(OH)$_2$: 31.1%	Easily applied, High covering power, Toxic in nature, Yellows badly on exposure to atmosphere, Soluble in alkali and paints	In manufacture of paints
3	Sublimed white Lead (Basic sulphate)	PbSO$_4$: 75% PbO: 20% ZnO: 5%	High specific gravity and refractive index, Slow chalking out of the film producing a rough surface	In manufacture of paints
4	Zinc oxide	ZnO: 100%	Brilliantly white having excellent texture, Causes no discoloration even in be contact with CO$_2$ gas, More durable in combination with white lead	It is opaque to UV light and thus protects from UV, Chalking can be prevented
5	Lithopone (ZnS+BaSO$_4$)	ZnS = 28–30% BaSO$_4$: 72–70%	Extremely fine and cheap pigment, Good hiding power, Not as durable as white lead and zinc oxide	Widely used for cold water paints, Traffic paints, In floor covering and oil cloth industry

it truly becomes opaque and conceals the substrate. This class of pigments contributes brightness and opacity, making them one of the most widely used pigments for coatings. The composition, characterizations and applications are mentioned in Table 7.9.

The rutile kind of titanium oxide (TiO$_2$) crystals are the most popular white pigments. Rutile is the most effective white pigment now on the market because it has the highest index of refraction (2.76) compared to any substance that can be produced in pigment form at a reasonable price. Anatase, is a different type of TiO$_2$ crystal frequently used in coatings, however because of its lower index of refraction (2.55), it is a less effective optical pigment. Additionally, coatings made of surface-treated rutile TiO$_2$ are more resistant to exposure than coatings made of comparable anatase pigments. In the so-called trade sales industry, which includes the retail, architectural, and contractor segments, TiO$_2$ pigments are used in extremely high volumes. Light, pastel, and white coatings due to their application, drives up the demand for TiO$_2$.

7.12.1.2 Colored pigments

There are several organic and inorganic color pigments that enable paint users to produce films of practically every color in the visible spectrum. Colored pigments work by absorbing certain visible light wavelengths while transmitting or dispersing the others. The common colored pigments are follows.

(A) Blue pigment: Since ancient times, blue pigments have been widely utilized for both the bulk coloring of polished, unglazed ceramics as well as the surface ornamentation of several classes of pottery that vary in style. This work deals with an exception with an antimony insertion into the host lattice of SnO_2 proposed as an alternative to this system. The conventional sources of blue in currently known ceramic colors contain Cobalt. The composition, characterizations and applications are mentioned in Table 7.10.

Table 7.10 Composition, Properties and Uses of Blue Pigments

Sr. No.	Name of Pigment	Composition	Characteristics	Applications
1	Ultramarine Blue	White – $Na_2Al_3Si_3SO_{12}$ Green – $Na_5Al_3Si_3S_2O_{12}$ Blue – $Na_2Al_3Si_3S_3O_{12}$	Silicate skeleton have a potential influence on the color, Color is due to the fast that S present is in the form of polysulfide	For making Cotton and linen fabrics
2	Cobalt Blue	Co_3O_4: 30–35% Al_2O_3: 65–70%	Very expensive and are not used in paints for ordinary purposes	For making blue inks, Carbon papers and ribbons
3	Phthalocyanine blue	$C_{32}H_{16}N_8$	Higher light fastness properties, Good tinting strength and covering power, Resistance to the effects of alkalies and acids	In manufacture of paints
4	Ferrocyanide blue (Prussian blue)	$C_{18}Fe_7N_{18}$	easily made, Cheap, Nontoxic and Intensely colored	In manufacture of paints

(B) Red pigments: The details of the main red pigments are tabulated in Table 7.11.

Table 7.11 Composition, Properties and Uses of Red Pigments

Sr. No.	Name of Pigment	Composition	Characteristics	Applications
1	Red Lead (Pb_3O_4)	Pb_3O_4 + PbO	Bright-red powder with high specific, Excellent covering power, Inhibits corrosion	For primary coat on structural steel, In imparting red colour to the glass for making bangles
2	Synthetic Iron	Fe_3O_4	Has dark brilliant color, High covering power and tinting strength	Widely used in domestic paints, enamels, floor and paints
3	Basic lead chromate	$[PbCrO_4 \cdot Pb(OH)_2]$		

(C) Yellow pigments: Detailed descriptions of yellow pigments are as follows.

Table 7.12 Composition, Properties and Uses of Yellow Pigments

Sr. No.	Name of Pigment	Composition	Characteristics	Applications
1	Ochre	Naturally occurring yellow Fe_2O_3	Fast to light and inert to chemical action	In paint industry
2	Chrome yellow	-	Great opacity, High brilliance, High hiding power, High tinting strength	In making yellow paints
3	Zinc yellow	$4ZnOK_2O_4CrO_3$ $\cdot 3H_2O$	Excellent corrosion inhibiting effect, High covering power and tinting strength	In making paints and as a priming watt for steel and aluminum

(D) Green pigments: The compositions, characterizations and applications of green pigments are mentioned in Table 7.13.

Table 7.13 Composition, Properties and Uses of Green Pigments

Sr. No.	Name of Pigment	Composition	Characteristics	Applications
1	Chrome Green	Cr_2O_3	High Power of oil absorption It has disadvantages such as lack of brilliancy and opacity	As green pigments
2	Chromium oxide (Guignet's Green)	$[Cr_2O(OH)_4]$	Have high covering power, High corrosion inhibition capacity	As Paint for metal surface, As fast non-fading green for washable distempers
3	Phthalocyanine Green	$C_{32}H_3Cl_{13}CuN_8$ to $C_{32}HCl_{15}CuN_8$	Highly stable, Resistant to alkali, acids, solvents, heat, and UV radiation.	Used in inks, coatings, and many plastics

(E) Black pigments: Detail descriptions of black pigments are as follows.

Table 7.14 Composition, Properties and Uses of Black Pigments

Sr. No.	Name of Pigment	Composition	Characteristics	Applications
1	Natural Black Oxide	Fe_2O_3:94–95%	. Oil absorption power is 10–15 kg of linseed oil per 100 kg for priming metal of Pigment	In making paints
2	Precipitated Black		High hiding power	In cement Iron oxide emulsions and water paints
3	Carbon Black/ Furnace Black		Good tinting strength, Not affected by light, acids and alkalies	Used in making water proof paints, Increases life of paints
4	Lamp Black		Good tinting strength, Resistant to high temperature	In making black pigments

7.12.2 *Organic pigments*

In nature, organic colors are uncommon. This justifies their bulk chemical synthesis. They contain carbon, which is negligibly toxic and poses no significant environmental risk. Pitch and petroleum distillates are examples of raw materials that can be converted into insoluble precipitates. Historically, bulk colorants have been made from organic pigments. Such pigments are utilized in surface coatings including paints and inks, synthetic textiles, and plastics. These days, advanced applications like photo-reproduction, opto-electronic displays, and optical data storage use organic pigments. Organic pigments are generally categorized into five types:

7.12.2.1 Monoazo pigments

This class of compounds was discovered in Germany in 1909. Monoazo pigments are usually yellow colored, also known as Arylide yellow or Hansa yellow. They are a family of organic compounds used as pigments. They are mostly used as industrial colorants for plastics, building paints and inks. They are also utilized in artistic oil paints, acrylics and watercolors. Normally, this pigment is yellow and yellow-green in color. Diarylide pigments are organic pigments that are similar. Overall, hazardous pigments have been partially replaced by these pigments in the market. This compound is prepared by the diazotization of aniline or a substituted primary aniline, and thereafter coupling with acetoacetanilide or its derivatives.

7.12.2.2 Diazo pigments

They cover a range of various colors including yellow, orange, brown, red and violet. Diazo colors, generally, have low solubility in organic solvents and fair to good light fastness. They are used for printing inks, plastics, and paints. They are made by combining acetoacetic arylides (diarylide yellows) or pyrazolones (disazopyrazolones) with di- and tetra-substituted diaminodiphenyls as diazonium salts. By diazotizing aromatic amines and coupling them with bisacetoacetic arylides, the second group of pigments, known as bisacetoacetic arylide pigments, is created.

7.12.2.3 Phthalocyanine pigments

An organic substance with the formula $(C_8H_4N_2) \cdot 4H_2$ is phthalocyanine. It falls within the category of an aromatic macrocyclic chemical. The compound is merely of theoretical or specialized interest, but its metal complexes are valuable as dyes, pigments, and catalysts. So, they are also considered as inorganic pigments.

7.12.2.4 Quinacridone pigments

This type of pigment has a basic quinacridone structure, which is a linear system of five anellated rings and molecular formula $C_{20}H_{12}N_2O_2$. This pigment performs largely like pthalocyanine pigments. It possesses exceptional solvent resistance, light and weather fastness, and migration resistance. High-end industrial coatings including automotive finishes, plastics, and specialized printing inks employ it. Commercially available unsubstituted transquinacridone pigments are created by modifying reddish violet beta and red gamma crystals. The 2,9-dimethyl derivative of the substituted quinacridone pigment, which provides a crisp bluish red tone and outstanding fastness qualities, is one of the more significant substituted quinacridone pigments (Gurses et al., 2016; Oyarzum, 2016).

7.13 Titanium Dioxide

Titanium dioxide (class: inorganic white) is the most generally used white pigment due to its brightness and really high index of refraction, during which it has been surpassed only by a couple of other materials. It is also found in most red-colored candy.

7.13.1 Introduction

The following are the main properties of titanium dioxide.

Table 7.15 Physical & Chemical Properties of Titanium dioxide

Sr. No.	Properties	Particular
1	IUPAC name	Dioxotitanium
2	Chemical and other names	Titanium dioxide, Titanium white, Titania, Rutile, Anatase, Brookite
3	Molecular formula	TiO_2
4	Molecular weight	79.8 gm/mole
5	Physical description	Odorless tasteless white powder
6	Solubility	Insoluble in water and organic solvents.
7	Melting point	1855°C
8	Boiling point	2500–3000°C
9	Flash point	Not Available
10	Density	4.23 gm/cm³ at 27°C

Further, TiO_2 has two crystalline forms: (1) Anatase (2) Rutile (more stable), but only Rulite is used in paints. Rutile is prepared by heating Anatase.

$$Anatase \longrightarrow Rutile$$

It is one of most useful pigments. It is used in exterior paint enamels and lacquers.

7.13.2 Manufacturing process

TiO_2 is manufactured by two methods (1) Sulfate process and (2) Chloride process. Ilmenite, a less expensive indigenous mineral, is used as a raw material in the sulfate process. Since it is made from imported ore material, the chloride process is more expensive.

7.13.2.1 Sulfate process

(A) Raw materials: Ilmenite ore ($FeTiO_3$) and Concentrated Sulfuric acid

(B) Chemical reactions

$$FeTiO_3 + 2H_2SO_4 \longrightarrow FeSO_4 + TiOSO_4 + 2H_2O \text{ (Digestion)}$$

$$TiOSO_4 + (n+1)\,H_2O \longrightarrow TiO_2 \cdot nH_2O + H_2SO_4 \text{ (Hydrolysis)}$$

$$TiO_2 \cdot nH_2O \longrightarrow TiO_2 + nH_2O \text{ (Calcination)}$$

(C) Flow sheet diagram

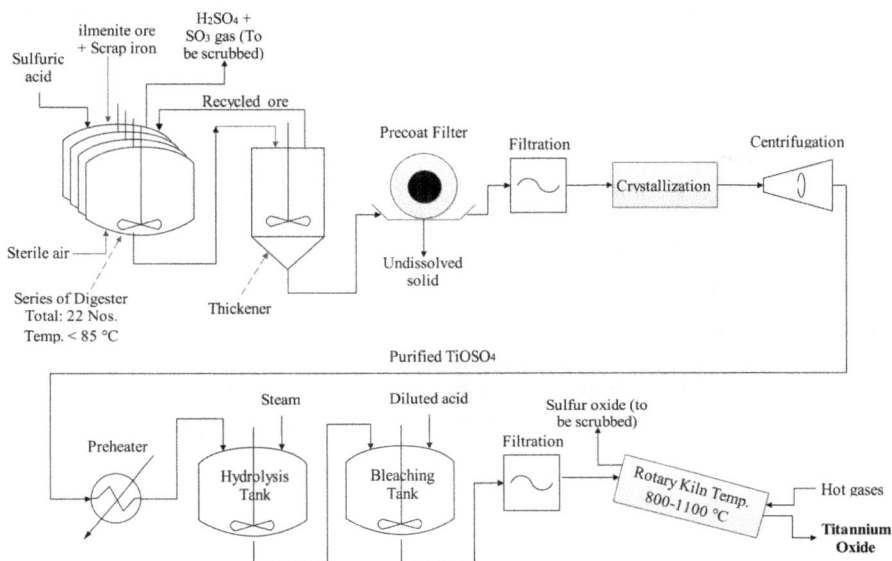

Figure 7.37 Flow sheet diagram of production of Titanium oxide by Sulfate process

(D) Process description: The following steps are involved in manufacturing titanium dioxide.

(a) Digestion: In the first step, diluted sulfuric acid is inserted into a double-paddle screw conveyor digestion tank. Thereafter, a mixture of ilmenite and water is continuously fed into the acid. Also, iron scrape is also charged as a reducing agnt. After completion of reaction time in 45–55 minutes, powdery cake is prepared. After a relatively short dwelling time (< 1 hour), a crumbly cake is produced. There have been several continuous digestion techniques suggested. This reaction is extremely violent and results in the entrainment of enormous volumes of water vapor, which is then absorbed by water-based scrubbing in a scrubber. The tank's bottom is where the digestion mixture is collected, and all undissolved solid material must be removed as completely as possible from the solution. Therefore, it is transferred into a thickener. Here, preliminary settling, followed by filtration of the sediment with a rotary vacuum drum filter or a filter press is conducted and the ore is recycled. The remaining undissolved solid is removed by a precoat filter. The filtrate is further filtered and crystallize. It then undergoes centrifugation and ferrous sulfate is removed to yield pure titanyl sulfate ($TiOSO_4$).

(b) Hydrolysis: The titanyl sulfate is heated in a preheater and transferred into the hydrolysis tank. Steam is blown into this tank. After completion of hydrolysis, the reaction mixture is collected from the bottom and transferred into the bleaching tank. Here, bleaching is carried out using diluted acid or water. Thereafter, it undergoes filtration, and the spent acid filtrate is recycled.

(c) **Calcination:** Calcination is carried out in directly heated rotary kilns heated by a counterclockwise flow of gas or oil. The material needs to be dried for roughly two-thirds of the residence time (7–20 hours overall). Sulfur trioxide is driven out above 500°C, since at higher temperatures, it partially breaks down into sulfur dioxide and oxygen. Depending on the type of pigment, throughput, and temperature profile of the kiln, the product can achieve a maximum temperature of 800–1100°C. The clinker can be directly or indirectly air-cooled in drum coolers once it has left the kiln.

7.13.2.2 Chloride process

(A) **Raw materials:** Rutile, Coke and Chlorine gas

(B) **Chemical reactions:** In a reducing environment, the titanium in the raw material is changed into titanium tetrachloride. Calcined petroleum coke is employed as the reducing agent since it has very little ash and produces very little HCl thanks to its low volatile content. The titanium dioxide reacts exothermically as follows:

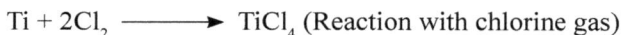

$$TiO_2 + C \longrightarrow Ti + CO_2 \text{ (Reaction with coke)}$$

$$Ti + 2Cl_2 \longrightarrow TiCl_4 \text{ (Reaction with chlorine gas)}$$

This titanium tetrachloride undergoes an oxidation reaction with oxygen or air with a flame to produce fine particles of titanium dioxide and chlorine gas.

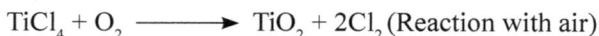

$$TiCl_4 + O_2 \longrightarrow TiO_2 + 2Cl_2 \text{ (Reaction with air)}$$

(C) **Flow sheet diagram**

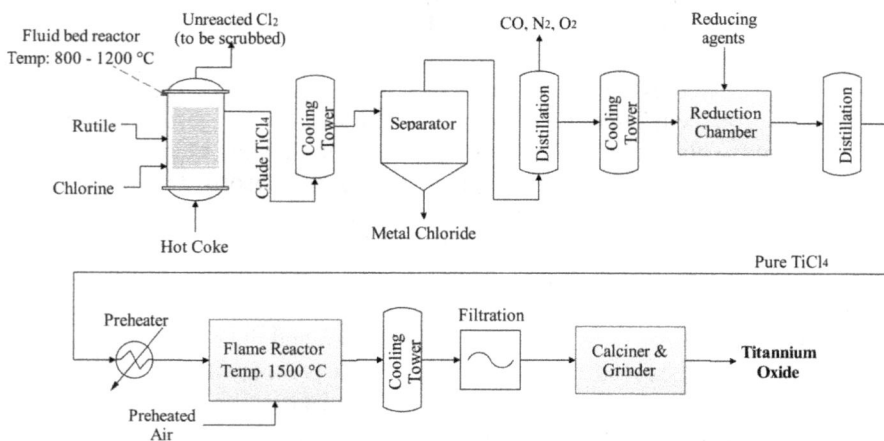

Figure 7.38 Flow sheet diagram of production of Titanium oxide by Chloride process

(D) **Process description:** The following steps are involved in the production of titanium oxide by the chloride process.

(a) **Chlorination of Rutile:** First, titanium raw material-Rutile is crushed into particles that resemble sand in size and petroleum coke is crushed down to a particle size that is five times of the TiO_2. Thereafter, crushed Rutile and TiO_2 are inserted into a brick-lined fluidized-bed reactor. Also, chlorine and oxygen are blown intothe reactor. The raw materials must be as dry as possible to avoid HCl formation. The reaction is exothermic, so, the temperature is maintained at 800–1200°C in the reactor. Gaseous chlorine gas is absorbed by scrubbing it with a caustic soda solution in a scrubber. Due to their low volatility, magnesium chloride and calcium chloride can assemble in the fluidized-bed reactor during the reaction. Zirconium silicate also builds up because, at the utilized temperatures, chlorination occurs very slowly. All the other constituents of the raw materials are volatilized as chlorides in the reaction gases, and thus, crude gaseous titanium chloride having impurities of other metal chlorides, residual dissolved chlorides and exhaust gasesare formed.

(b) **Purification of titanium chloride:** In this stage, the reaction gases are cooled down and metal chloride is separated from $TiCl_4$. Distillation is conducted to remove the exhaust gases (CO_2, CO, O_2 and N_2). The reaction mixture is then cooled and subjected to reduction with metal powders (Fe, Cu, or Sn) to remove dissolved chlorine. Thereafter, it is subjected to distillation to remove solids and purified $TiCl_2$ is formed.

(c) **Preparation and purification of TiO_2:** Purified $TiCl_2$ is vaporized by heating upto 500–1000°C in a preheater and inserted into a flame reactor. Also, air containing oxygen is heated upto > 1000°C using the preheater and blown into the flame reactor. This oxidation reaction is moderately exothermic, so, the temperature is maintained at 1500°C in the reactor. After completion of oxidation, the mixture of gases (Cl_2, O_2, CO_2) and TiO_2 pigment is cooled at room temperature. Using a filtration technique, the pigment is separated from the gas. The gas stream containing chlorine gas is further purified and recycled. It is then calcined and ground.

7.13.4 Uses

The most significant applications include paints and varnishes, paper and plastics, which together make up over 80% of the world's consumption of titanium oxide. An extra 8% of pigment usage is in other items such printing inks, textiles, rubber, cosmetics, and food. The remaining portion is used for other purposes, such as the assembly of technical pure titanium, glass, glass ceramics, electrical ceramics, catalysts, electric conductors, and chemical intermediates (Lakshmanan et al., 2014; Parrino and Palmisano, 2020; Winkler et al., 2014).

7.14 Ultramarine Blue

Ultramarine Blue (class: Inorganic blue) largely comprises of zeolite-based minerals with trace quantities of polysulfides. It can be found in nature as a close relative of lapis lazuli, which contains the blue cubic mineral lazurite.

7.14.1 Properties

Its main properties are mentioned in the Table 7.16.

Table 7.16 Physical & Chemical Properties of Ultramarine Blue

Sr. No.	Properties	Particular
1	IUPAC name	Hexaaluminium(3+) ion octasodium trisulfide(2-) hexaorthosilicate
2	Chemical and other names	Ultramarine Blue, Azure Blue, C.I. Pigment Blue 29, SC-21283
3	Molecular formula	$Al_6Na_8O_{24}S_3Si_6$
4	Molecular weight	994.5 gm/mole
5	Physical description	Bright blue solid powder
6	Solubility	Insoluble in water and organic solvents
7	Melting point	Above 10000°C
8	Boiling point	Not Available
9	Flash point	Not Available
10	Density	2.35 gm/cm³ at 27°C

7.14.2 Manufacturing process

It is prepared from zeolite.

7.14.2.1 Raw materials

Zeolite, Sodium sulfite and Sulfur

7.14.2.2 Chemical reactions:

It is prepared by heating the zeolite, sodium sulfite and sulfur at 750°C for 48 hours.

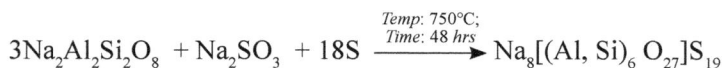

$$3Na_2Al_2Si_2O_8 + Na_2SO_3 + 18S \xrightarrow[\text{Time: 48 hrs}]{\text{Temp: 750°C;}} Na_8[(Al, Si)_6 O_{27}]S_{19}$$

7.14.2.3 Quantitative requirements

1000 kgs of Ultramarine Blue is obtained using 560 kgs of zeolite, 150 kgs of sodium sulfite and 440 kgs of sulfur.

7.14.2.4 Flow sheet diagram

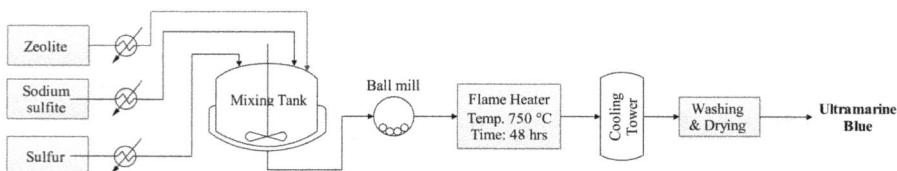

Figure 7.39 Flow sheet diagram for production of Ultramarine Blue

7.14.2.5 Process description

As per a simple flow-sheet, zeolite, sodium sulfide and sulfur are dried individually in a dryer. Thereafter, dried zeolite, anhydrous sodium sulfide and sulfur are mixed properly in a mixing tank. Thereafter, it is ground in a ball mill in the absence of air. The mixture is heated in a leak proof vessel at 750°C for 48 hours. Nitrogen gas is purged to maintain the inert atmosphere in the vessel. After heating, it is allowed to cool down at room temperature. Then it is washed with water, dried and packed.

7.14.3 Uses

Being at the bottom of the rock, ultramarine blue is mostly used for murals, calico printing and paper hangings, and as a remedy for the yellowish tinge that frequently taints items that are supposed to be white, like linenand paper. When washing white clothing, a suspension of synthetic ultramarine (or the chemically distinct prussian blue) may be used as "bluing" or "laundry blue." They are used as colorants and optical brightening agents for paper, textiles, cement, plastics, rubber, inks, paints, cosmetics, and detergents. The natural variant has great chemical stability except when present in acids, making it a highly expensive blue pigment. (Giraldo et al., 2012; Arieli et al., 2004).

7.15 Phthalocyanine Blue BN

The class of phthalocyanine organic pigment known as phthalocyanine Blue BN is a brilliant, crystalline, synthetic blue pigment from the phthalocyanine dye family. It can be found as a copper complex. Usually, paints and dyes use its dazzling blue. It is highly regarded for its exceptional qualities, including light resistance, tinting power, covering ability, and resistance to the effects of alkalis and acids. It resembles a blue powder and is insoluble in most solvents including water.

7.15.1 Properties

Its main properties are mentioned in the table below.

Table 7.17 Physical & Chemical Properties of Phthalocyanine Blue BN

Sr. No.	Properties	Particular
1	IUPAC name	N.A.
2	Chemical and other names	Phthalocyanine blue, Copper phthalocyanine, Cuprolinic blue, Monastral Blue B, Aqualine Blue
3	Molecular formula	$C_{32}H_{16}N_8Cu$
4	Molecular weight	576.1 gm/mole
5	Physical description	Bright blue microcrystals with purple luster odourless powder
6	Solubility	Insoluble in water and organic solvents and hydrocarbon; soluble in 98% sulfuric acid
7	Melting point	480°C
8	Boiling point	600°C
9	Flash point	Not Available
10	Density	1.62 gm/cm³ at 27°C

7.15.2 Manufacturing process

It is prepared from phthalic anhydride.

7.15.2.1 Raw materials

Phthalic anhydride, Urea, Ammonium molybdate, Cuprous chloride, O-Nitro toluene (ONT) and Sulfuric acid

7.15.2.2 Chemical reactions

By condensing phthalic anhydride, cuprous chloride, and urea at 130–150°C, using ammonium molybdate as a catalyst and ONT as a solvent, copper phthalocyanine is produced.

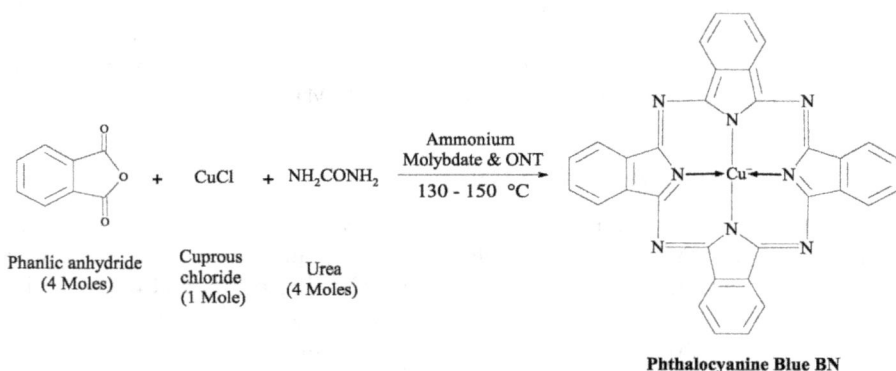

Figure 7.40 Synthesis of Phthalocynime Blue BN

7.15.2.3 Quantitative requirements

1000 kgs of copper phthalocyanine is obtained using 230 kgs of phthalic anhydride, 180 kgs of urea, 1 kg of ammonium molybdate, 50 kgs of cuprous chloride, 700 kgs of ONT and 940 kgs of sulfuric acid (45%).

7.15.2.4 Flow sheet diagram

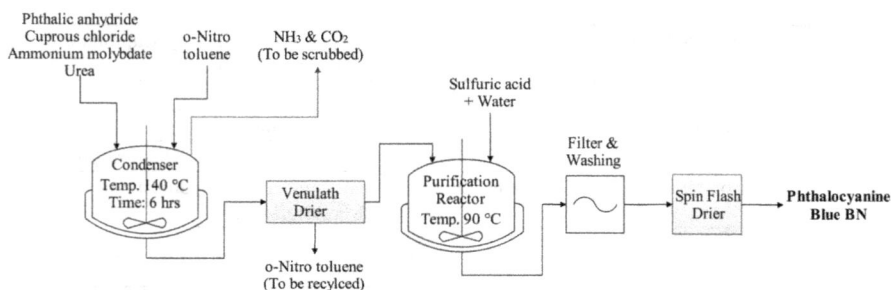

Figure 7.41 Flow sheet diagram for production of Phthalocyanine Blue BN

7.15.2.5 Process description

It is divided into the following parts.

(A) Preparation of Phthalocyanine Blue BN: First o-Nitro toluene as a solvent is charged into a glass line reactor condenser having heating and stirring arrangements. Heat the solvent upto 130–150°C. Also, phthalic anhydride, cuprous chloride, ammonium molybdate and urea are charged into it with stirring. Maintain a reaction temperature of 130–150°C for 6 hours with stirring. Here, a two stage water scrubbing system is used to scrub the liberated ammonia and carbon dioxide gases during the reaction.

(B) Solvent recovery: After condensation, the reaction mass is collected from the bottom of the kettle and transferred to the venulath dryer for solvent recovery. The venulath dryer is a special type of vacuum dryer, which is well known for its ease of operation and robust makeup. Here, ONT is distilled out using a vacuum and recycled. A wet cake of crude copper phthalocyanine blue is obtained after distillation.

(C) Purification of Phthalocyanine Blue BN: The crude wet cake of copper phthalocyanine is transferred into another purification reactor. Also, sulfuric acid and water are charged into it with stirring. Live steam is used to heat the reaction mass upto 90°C. Here, inorganic impurities in the product are dissolved in water and the remaining product remains in solid form. These impurities are removed by filtration. The filtrate is then neutralized with a solution of caustic soda. Following a water wash, the washings are also taken for treatment along with the cake. The spin flash dryer is where the wet cake is taken after that. The final output is a fine powder that is then packaged.

7.15.3 Uses

Phthalocyanine Blue BN may be a water-based pigment paste with a high-dispersed suspension of pigments that have water and pigment auxiliaries, with a pigment blue color index of 15:2. It provides bright colors and stability with good fineness and fastness. The pigment is suitable for use in paints. It is also raw material for other blue pigments like Phthalocyanine Beta Blue, Phthalocyanine Alpha Blue and Pigment Green-7 (Bhoge et al., 2014).

7.16 Pigment Red 122

It is pertaining to the class of Quinacridone organic pigments.

7.16.1 Properties

Its main properties are mentioned in the Table 7.18.

Table 7.18 Physical & Chemical Properties of Pigment Yellow 62

Sr. No.	Properties	Particular
1	IUPAC name	Calcium bis[4-[[1-[[(2-methylphenyl) amino] carbonyl]-2-oxopropyl] azo]-3-nitrobenzene sulphonate]
2	Chemical and other names	Pigment Yellow 62, C.I. 13940, Fast Yellow FW, Pigment Yellow WSR, Irgalite Yellow WSR, Seikafast yellow 1982-5G
3	Molecular formula	$C_{34}H_{30}CaN_8O_{14}S_2$
4	Molecular weight	878.8 gm/mole
5	Physical description	Yellow odorless powder
6	Solubility	Insoluble in water and organic solvents
7	Melting point	Not Available
8	Boiling point	Not Available
9	Flash point	Not Available
10	Density	1.555 gm/cm³ at 27°C

7.16.2 Manufacturing process

The starting materials of Pigment Red 122 are dimethyl succinylosuccinonate (DMSS) and p-methyl aniline.

7.16.2.1 Raw materials

Dimethyl succinylo succinonate (DMSS), p-Methyl aniline (p-Toluidine), Sodium hydroxide, Hydrochloric acid, Phosphorus pentoxide, Phosphoric acid, Dimethyl formamide (DMF), Methanol, Resist salt (R salt) and Water

7.16.2.2 Chemical reactions

As per Fig 7.42, DMSS is condensed with 2 moles of p-toluidine in the presence of methanol to give 2,5-bis-p-tolylamino-cyclohexa-1,4- diene-1 and 4-dicarboxylic acid dimethyl ester at 50°C in 5 hours, which undergoe hydrolysis with sodium hydroxide in the presence of methanol and hydrochloric acid to give 2,5-bis-p-tolylamino-cyclohexa-1,4-diene-1,4-dicarboxylic acid. Thereafter, this compound undergoes cyclization with phosphorus pentoxide and phosphoric acid in the presence of dimethyl formamide and water to give Pigment Red 122.

7.16.2.3 Quantitative requirements

1000 kgs of Pigment Red 122 is obtained using 800 kgs of DMSS, 880 kgs of p-toluidine, 260 kgs of sodium hydroxide, 240 kgs of hydrochloric acid, 1900 kgs of phosphorus pentoxide, 3200 kgs of phosphoric acid, 1000 kgs of DMF, 15400 kgs of methanol, 720 kgs of R salt and 11 tons of water.

7.16.2.4 Flow sheet diagram

7.16.2.5 Process description

It is divided into the following parts.

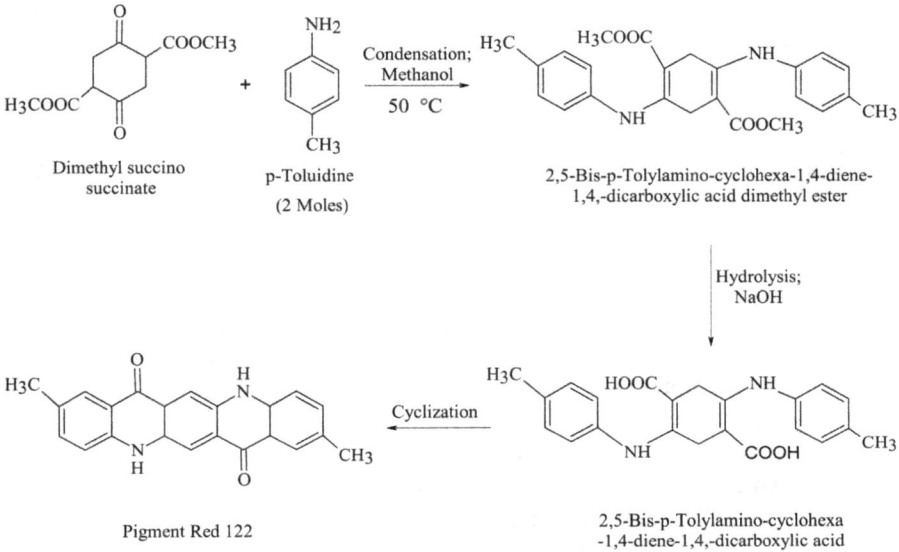

Figure 7.42 Synthesis of Pigment Red 122

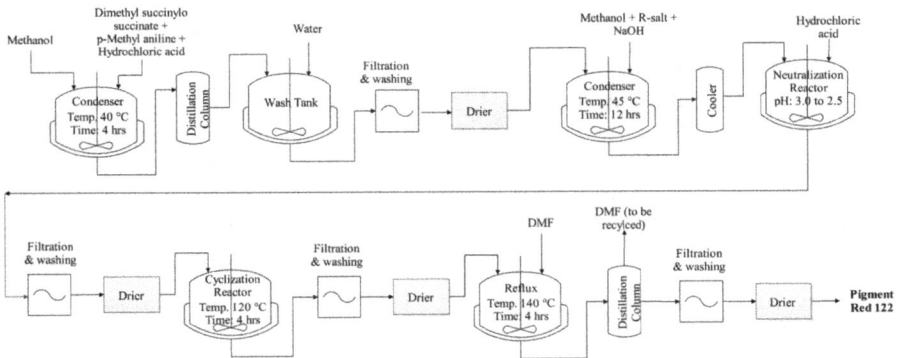

Figure 7.43 Flow sheet diagram for production of Pigment Red 122

(A) Preparation of 2,5-Bis-p-tolylamino-cyclohexa-1,4-diene-1,4-dicarboxylic acid dimethyl ester:

At first, take methanol in a jacketed condensation reactor equipped with a stirring arrangement. Charge dimethyl succinylo succinate and p-methyl aniline into it with stirring. Nitrogen gas is purged to maintain the inert atmosphere in the vessel. The reaction mass is refluxed at 45–50°C. Insert hydrochloric acid with stirring within 15 minutes. The mixture is heated to a reflux temperature of 45–50°C in 4 hours. The reaction mass is collected from the bottom of the reflux and charged into a distillation column. Here, methanol is distilled out and reused. After distillation, the reaction slurry is transferred into another reactor having a heating and stirring facility. Add water and stir for 1 hour at 45–50°C. It is further filtered and washed with water till neutralized. Dry the precipitates of

2,5-bis-p-tolylamino-cyclohexa-1,4- diene-1,4-dicarboxylic acid dimethyl ester at 80°C.

(B) Preparation of 2,5-Bis-p-tolylamino-cyclohexa -1,4-diene-1,4-dicarboxylic acid: Charge the methanol and dry precipitates of 2,5-bis-p-tolylamino-cyclohexa-1,4-diene-1,4-dicarboxylic acid dimethyl ester into a jacketed vessel equipped with an agitator and a reflux condenser. Add R-salt and sodium hydroxide in it. Reflux the reaction mass at 45–50°C with agitation for 12 hours. After reflux, collect the reaction mass from the bottom of the condenser and cool it. The cooled reaction mass is then transferred into a neutralization tank. Hydrochloric acid is slowly added with stirring until a pH of 3.0–2.5 is attained. Precipitates of 2,5-bis-p-tolylamino-cyclohexa -1,4-diene-1,4-dicarboxylic acid are formed. These precipitates are filtered, washed with water and dried at 60°C.

(C) Preparation of Pigment Red 122: Take phosphoric acid having a phosphorus pentoxide content of 85% in a cyclization reactor with a heating and stirring arrangement. Heat the reaction mass upto 120–130°C. Dried 2,5-bis-p-tolylamino-cyclohexa -1,4-diene-1,4-dicarboxylic acid is introduced into it maintaining the temperature at 120–130°C with stirring. The reaction mass is stirred at 120–130°C for 4 hours. After completion of 4 hours, the reaction mass is cooled down to room temperature. It is then filtered and dried to get crude pigment red 122. The filtrate contains 10–30% phosphoric acid, which is recovered and recycled.

(D) Purification of Pigment Red 122: Crude pigment red 122 and dimethyl formamide (DMF) are taken into a closed jacketed vessel equipped with an agitator and a reflux condenser. Reflux the reaction mass upto 140–150°C for 4 hours. The reaction mass is collected from the bottom of the reflux and charge into a distillation column. Here, DMF is distilled out and reused. Add water and filter it. The pigment red 122 is then washed with water and dried.

7.16.3 Uses

Pigment Red 122 is used in letter pressing, gravure, flexo, silk screen inks and synthetic as well as industrial paints. It is used to color various types of plastics like PVC, rubber, master batches and polyolefins (Silbir and Goksungur, 2019).

7.17 Pigment Yellow 62

It is pertaining to a class of monoazo organic pigments. It is available as calcium salt.

7.17.1 Properties

Its main properties are mentioned in the table below.

7.17.2 *Manufacturing process*

It is prepared from o-nitroaniline-p-sulfonic acid.

7.17.2.1 *Raw materials*

o-Nitroaniline-p-sulfonic acid, Hydrochloric acid, Sodium nitrite, Sodium hydroxide, Acetoacetyl-o-toludide, Acetic acid, Calcium chloride and Water

7.17.2.2 *Chemical reactions:*

It is prepared by the diazotization of o-nitroaniline-p-sulfonic acid using hydrochloric acid and sodium nitrite; and thereafter, coupling with acetoacetyl-o-toluidine. The product is heated with calcium chloride at 80–90°C.

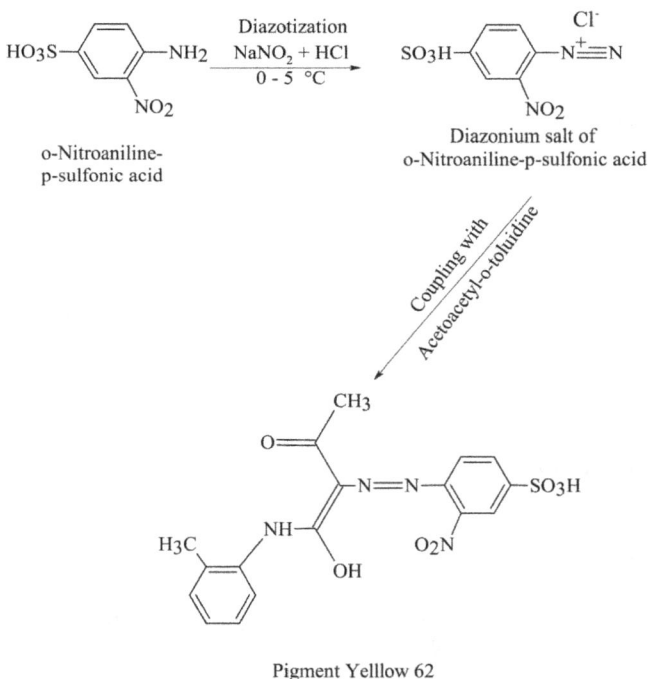

Figure 7.44 Synthesis of Pigment Yellow 62

7.17.2.3 *Quantitative requirements*

1000 kgs of Pigment Yellow 62 is obtained using 570 kgs of o-nitroaniline-p-sulfonic acid, 50 kgs of hydrochloric acid, 325 kgs of sodium nitrite, 200 kgs of sodium hydroxide, 500 kgs of acetoacetyl-o-toluidine, 25 kgs of acetic acid, 500 kgs of calcium chloride and 5000 kgs of water.

7.17.2.4 *Flow sheet diagram*

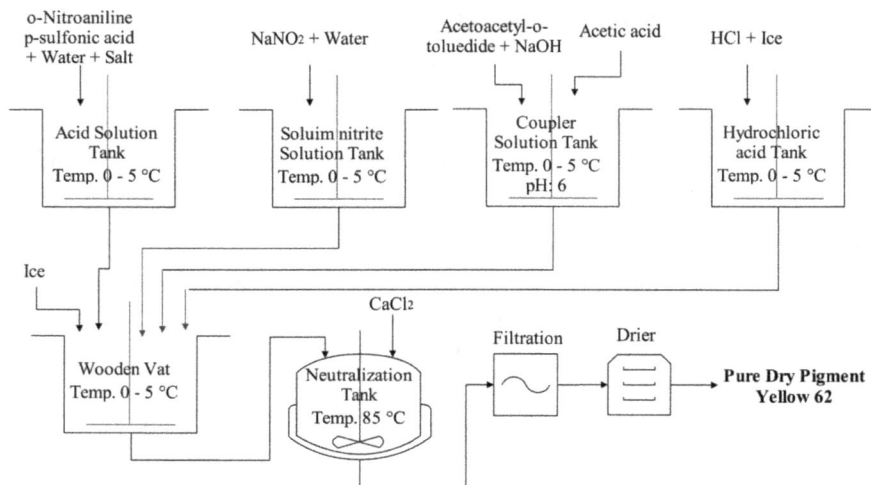

Figure 7.45 Flow sheet diagram for production of Pigment Yellow 62

7.17.2.5 Process description

As per the flow-sheet, water solutions of o-nitroaniline-p-sulfonic acid, hydrochloric acid and sodium nitrite are taken in different tanks and cooled to 0–5°C using ice. All these solutions are mixed with stirring in a diazotization reactor, maintaining a temperature of 0–5°C. Stir the reaction mass for 1 hour while maintaining the temperature at 0–5°C. Meanwhile, dissolve acetoacetyl-o-toludide (coupler) in sodium hydroxide and water in another tank. Stir for 1 hour and cool it down to 0–5°C. The coupler is precipitated with acetic acid and the pH is adjusted to 6.0. Now, slowly add the above prepared diazo solution into the coupler solution. Addition should be such that the temperature is maintained at 0–5°C for a time duration of about 2–3 hours with stirring. After addition, stir the reaction for 30 minutes. After the coupling reaction, transfer the reaction mass into another precipitation tank. Heat the reaction mass upto 80–90°C and add calcium chloride. Maintain the temperature at 80–90°C for 1 hour. Cool the reaction mass down to room temperature. Filter the slurry in the filter press. The wet cake is then dried, pulverized and packed.

7.17.3 Uses

Color Yellow 62 is a pigment powder that is greenish yellow and has superb lighting as well as migration fastness, chemical resistance and heat stability. Its mainly used for coloring PVC, low as well as high density polyethylene, rubber, stationery, industrial paints, water based decorative paints and water based inks (Choi et al., 2010).

References

Arieli, D. Vaughan, D.E.W. and Goldfarb, D. 2004. New Synthesis and Insight into the Structure of Blue Ultramarine Pigments. Journal of the American Chemical Society 126(18): 5776–88. https://doi.org/10.1021/ja0320121

Arnkil, H., Anter, K.F. and Klaren, U.T. 2012. Colour and Light: Concepts and Confusions, Color and Environment. Interim Meeting of the International Colour Association (AIC), Taipei, Taiwan.

Balaban, T.S., Tamiaki, H. and Holzwarth, A.R. 2005. Supermolecular Dye Chemistry, Springer-Verlag Berlin, Heidelberg. https://doi.org/10.1007/b105136

Beal, W. 2008. The Classification of Dyes by their Dyeing Characteristics. Review of Progress in Coloration and Related Topics 72(4): 146–158.

Benkhaya, S., Mrabet, S. and Elharfi, A. 2020. A Review On Classifications, Recent Synthesis And Applications Of Textile Dyes. Inorganic Chemistry Communications 115: 107891. https://doi.org/10.1016/j.inoche.2020.107891

Bhoge, Y., Deshpande, T.D., Patil, V. and Badgujar, N.P. 2014. Synthesis of copper phthalocyanine blue pigment: Comparative evaluations of fusion, solvent and microwave techniques. International Journal of Applied Engineering Research 9(10): 1271–1278.

Bokhari, S.A.R., Khan, R.R.M., Tahir, M.S. and Tahir, N. 2016. Synthesis, Characterization and RP-HPLC Method Development and Validation for Simultaneous Determination of Koch Acid and H Acid. Journal of Chemical Society of Pakistan 38(03): 469–478.

Bozic, M. and Kokol, V. 2006. Vat dyes: Conventional process of dyeing and ecological alternatives, Tekstilec 49(1): 8–15.

Butyrskaya, E.V., Shaposhnik, V.A., Butyrskii, A.M., Merkulova, Y.D., Rozhkova, A.G. et al. 2004. Interpretation of Hypsochromic and Bathochromic Shifts of Vibrational Frequencies of a Cation Exchanger. Chemistry—A European Journal 10(2): 360–70. https://doi.org/10.1134/S106193480710005X

Choi, S., Kwon, O., Kim, N. and Yoon, C. 2010. Preparation of Solvent Soluble Dyes Derived from Diketo-pyrrolo-pyrrole Pigment by Introducing an N-Alkyl group. Bulletin of the Korean Chemical Society 31(4): 1073–1076. https://doi.org/10.5012/bkcs.2010.31.04.1073

Clark, M. 2011. Handbook of Textile and Industrial Dyeing: Principles, Processes and Types of Dyes, Elsevier Science.

Daniel, M. 2012. Color vision and colorimetry: theory and applications, SPIE monograph.

Giraldo, C., Tobon, J.I. and Restrepo, O. 2012. Ultramarine blue pigment: A non-conventional pozzolan, Construction and Building Materials 36: 305–310. https://doi.org/10.1016/j.conbuildmat.2012.04.011

Gregory, P. 2009. Dye and Dye Intermediates. In book: Kirk-Othmer Encyclopedia of Chemical Technology. https://doi.org/10.1002/0471238961.0425051907180507.a01

Gurses, A., Acıkyıldız, M., Gunes, K. and Gurse, M.S. 2016. Dyes and Pigments: Their Structure and Properties 13–29. https://doi.org/10.1007/978-3-319-33892-7_2

Kiernan, J.A. 2001. Classification and naming of dyes, stains and fluorochromes. Biotechnic and Histochemistry 76(5–6): 261–78. DOI: 10.1080/bih.76.5-6.261.278

Lakshmanan, V.I., Bhowmick, A. and Halim, M.A. 2014. Titanium Dioxide - Production, Properties and Applications. In book: Titanium Dioxide: Chemical Properties, Applications and Environmental Effects.

Lye, J. 2002. Colour chemistry. Color Research & Application 27(5): 376–377.

Masuda, Y. 2016. The World of Color Science and Color Chemistry. Shikizai Kyokaishi 89(1): 22–28.

Oyarzum, J.M. 2016. Pigment Processing Physico-Chemical Principles, Vincentz Network.

Parrino, F. and Palmisano, L. 2020. Titanium Dioxide (TiO_2) and Its Applications, Elsevier Science. https://doi.org/10.1016/C2019-0-01050-3

Pattanaik, L., Padhi, S., Hariprasad, P. and Naik, S.N. 2020. Life cycle cost analysis of natural indigo dye production from *Indigoferatinctoria L.* plant biomass: a case study of India, Clean Technologies and Environmental Policy 22(8): 1–16. https://doi.org/10.1007/s10098-020-01914-y

Silbir, S. and Goksungur, M.Y. 2019. Natural Red Pigment Production by *MonascusPurpureus* in Submerged Fermentation Systems Using a Food Industry Waste: Brewer's Spent Grain, Foods 8(5): 161. DOI: 10.3390/foods8050161

Sun, C.-B., Zhang, P.-J., Li, C. and Yu, Y. 2012. Treatment of wastewater from Tobias acid production with supercritical water oxidation, Beijing KejiDaxueXuebao/Journal of University of Science and Technology Beijing 34(10): 1097–1101. https://doi.org/10.5004/dwt.2011.2122

Winkler, J. 2014. Titanium Dioxide Production, Properties and Effective Usage, Vincentz Network.

CHAPTER 8

Drug and Pharmaceutical Industry

8.1 Introduction

A drug is a substance used in the prevention, diagnosis or remedy of a sickness in humans or different animals. As per the WHO (World Health Organization), a drug might also be recognized as any substance or product which is utilized or supposed to be utilized for reworking or exploring physiological structures or pathological states for the gain of the recipient. Also, it is defined as any chemical substance, natural or man-made (usually excluding nutrients, water, or oxygen), that alter the biological structure or functioning when administered and absorbed. The difference between a drug and a pharmaceutical is that a drug is substance that causes a physical or mental change in your body. If it is used therapeutically, then it gets classified as a pharmaceutical (Smith, 2004; Wettermark et al., 2016).

8.2 Classification of Drugs

Drugs are classified based on their application and structure.

8.2.1 Classification based on application or mode of action

Drugs are mainly classified according to their applications as follows.

8.2.1.1 Antibiotic drugs

Antibiotics are compounds of natural, semi-synthetic, or synthetic origin which are used for treatment or prevention of bacterial infections without significant toxicity to the human or animal host. They can also both kill or inhibit the boom of bacteria. Several antibiotics are additionally effective against fungi and protozoans; but they are ineffective against viruses such as the bloodless or influenza virus. Antibiotics fight sure infections and ailments prompted by bacteria. Examples of antibiotics are penicillin, tetracycline, streptomycin, chloramphenicol, and sulfa drugs, or sulfonamides. The preferred synthesis mode of antibiotics is aerobic fermentation, which is carried out in a closed vessel called fermenter. The fermenter is provided with automatic temperature control, pH control, agitation, air sparger, automatic cleaning device and steam sterilization. The raw material for the fermentation process may be of plant origin (e.g., grapes, apples and more), grains (e.g., corn, wheat, barley, rice) or industrial by-products (e.g., molasses, wheat sulfite liquor, whey, lactose) or may be by-products of them eat industry (e.g., gelatin). The important components of the raw materials are carbohydrates, proteins and other nitrogen compounds, phosphates and other salts, vitamins and growth factors. Before being exposed to microbial action, this material is dissolved or crushed and then converted into a liquid medium. Also, the medium is sterilized with hot water. Bacteria, yeast and moulds are developed, propagated and transferred into the inoculum for the culture. Also, nitrogenous compounds play an essential role in fermentation. They serve as vitamins for the boom and metabolic endeavors of yeast in the course of fermentation; and as proteins, they additionally have an impact on the manner and stability of fermentation. This inoculum is required for a contamination free culture and to maintain its product forming capacities. Thereafter, it is filtered, extracted, centrifuged, crystallized and dried.

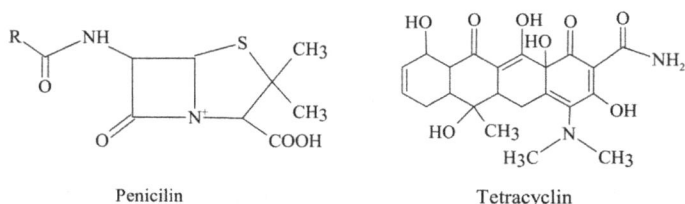

Penicilin   Tetracyclin

Figure 8.1 Examples of Antibiotic drug

8.2.1.2 Anesthetic Drug

An anaesthetic is a substance that induces anaesthesia, a temporary lack of sensation. The opposite of anaesthetics are analgesics (painkillers), which reduce pain without obliterating sensation. These medications are typically given to aid surgeries. Drugs of this kind are used to reduce pain, inhibit reflexes and muscular activity, and ultimately cause unconsciousness. The aforementioned qualities, as well as a broad therapeutic index and the absence of major adverse effects, must be present in the perfect anaesthetic. These medications inhibit or suppress the central nervous system's transmission of neurological signals, making it possible

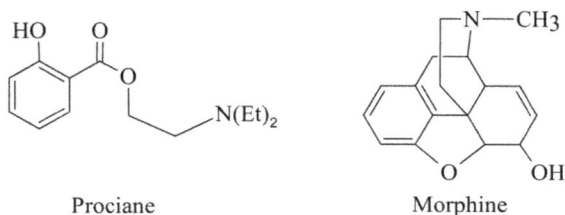

Prociane   Morphine

Figure 8.2 Examples of Anesthetic drug

to perform painless surgical, obstetrical, and diagnostic procedures. Modern anaesthetic therapy involves the use of a wide range of medications.

8.2.1.3 Analgesic drugs

Any medicine used to achieve analgesia, or pain alleviation, is referred to as an analgesic or painkiller. Analgesic medications have a variety of effects on the peripheral and central nervous systems. They differ from anaesthetics, which irreversibly numb the body. There are two categories of analgesics: opioids (substances that resemble morphine), which primarily affect the central nervous system (CNS), and nonopioids (nonsteroidal anti-inflammatory medications, or NSAIDs), which primarily affect the peripheral nervous system. Analgesics include acetaminophen (often referred to as paracetamol in the US), non-steroidal anti-inflammatory drugs (NSAIDs) like salicylates, and opioid medications like morphine and oxycodone.

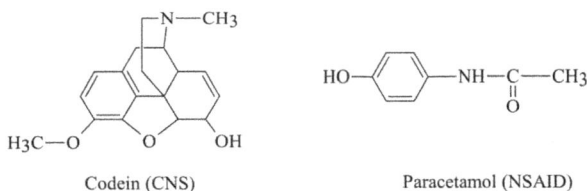

Codein (CNS)   Paracetamol (NSAID)

Figure 8.3 Examples of Analgesic drug

8.2.1.4 Anti-malarial drug

Antimalarial medications are made to either treat or prevent malaria. Of all the diseases brought on by protozoa, malaria is the one that gets carried the furthest. Plasmodia are the malaria-causing organisms. Mefloquine, primaquine, chloroquine, pyrimethamine, amodiaquin, quinine/quinidine, and chloroguanide are antimalarial medications that are now utilized for prophylaxis treatment.

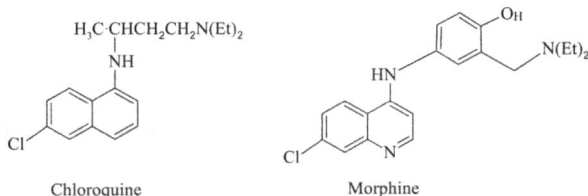

Chloroquine   Morphine

Figure 8.4 Examples of Anti-matarial drug

8.2.1.5 Antihypertensive drugs

Antihypertensive drugs are a class of medications used to treat high blood pressure (high blood pressure). Antihypertensive medication is used to avoid high blood pressure-related problems including myocardial infarctions and strokes. Higher arterial blood pressure is a syndrome that is determined by a variety of causes. The heart rate, blood volume, viscosity, and electrolytic content are the main determinants for determining the arterial blood pressure. This group of medications also includes adrenergic receptor antagonists, calcium channel blockers, ACE (angiotensin-converting enzyme) inhibitors, and diuretics.

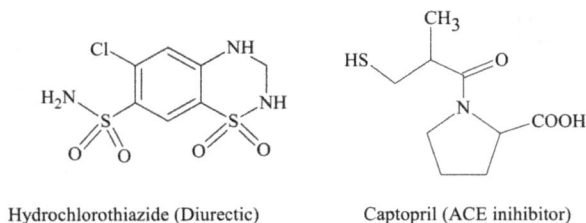

Hydrochlorothiazide (Diurectic)   Captopril (ACE inihibitor)
Figure 8.5 Examples of Anti-hypertentive drug

8.2.1.6 Anti-obesity drugs

Weight-reduction or weight-controlling pharmacological medicines include anti-obesity medications and weight-loss pharmaceuticals. By altering either hunger or calorie absorption, this medication affects weight control, one of the body's basic functions. Dieting and exercise continue to be the main therapeutic techniques for overweight and obese people.

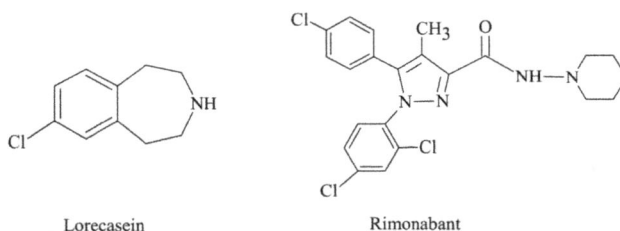

Lorecasein   Rimonabant
Figure 8.6 Examples of Anti-obesity drug

8.2.1.7 Sulfa drugs

Numerous medication classes are based on sulfa medicines or sulfonamides. Synthetic antimicrobial compounds with the sulfonamide group are what the earliest antibacterial sulfonamides were. Some sulfonamides, such as the anticonvulsant sultiame, do not have antibacterial properties. Innovative drug classes known as sulfonylureas and thiazide diuretics are based on sulfonamides, which are used to treat bacteria. The group covers several types of drugs such as child antibacterial drugs, antimicrobials, anti-diabetic agentsand diuretics.

The other drugs based on these applications are antileprotic, gastrointertinal, anti TB psychotropic hormones, vitamins, varodilators, drugs of vegetable origin, vaccines & Sera.

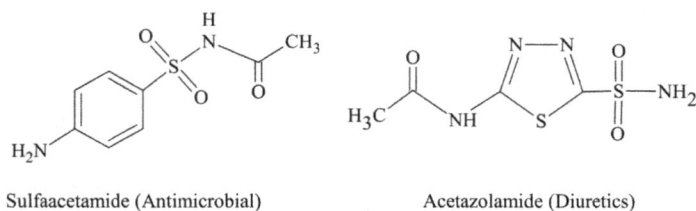

Sulfaacetamide (Antimicrobial)   Acetazolamide (Diuretics)
Figure 8.7 Examples of Sulfa drug

8.2.2 Classification based on structure

This type of classification is according to pharmacophores present in drug molecules. A pharmacophore is a component of a molecular structure that controls a certain biological or pharmacological interaction that a given molecular structure goes through.

8.2.2.1 Drugs based on a substituted benzene ring

There are many substances that are employed as medicinal agents, including benzene rings and other aromatic systems. There are thousands of medications with substituted benzene rings as their central component. These kinds of rings have a variety of roles in pharmaceuticals, from just adding steric bulk to playing a crucial role in the pharmacophore. A variety of sub-parts can be found in substituted benzene rings, such as leukotriene antagonists, arylethanolamines, aryloxypropanolamines, arylsulfonic acid derivatives, arylacetic and arylpropionic acids.

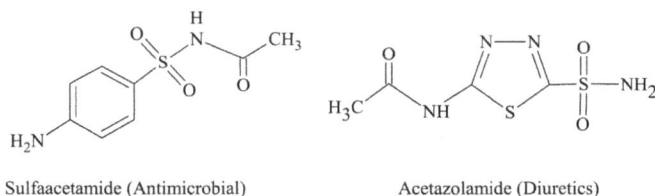

Sulfaacetamide (Antimicrobial)   Acetazolamide (Diuretics)

Figure 8.8 Examples of drug having substituted benzene ring

Norepinephrine is essential for controlling heart rate, blood pressure, and the expansion or contraction of bronchioles. It is a chemical released from the sympathetic nervous system in response to stress. Sulfanilamide is used as an antibacterial agent worldwide.

8.2.2.2 Polycyclic aromatic compounds

These types of compounds contain polycyclic aromatic compounds with two (such as indene, naphthaleneand anthracene) or three rings (i.e., Dibenzocycloheptane, Dibenzocycloheptene, etc.) as the pharmacophore.

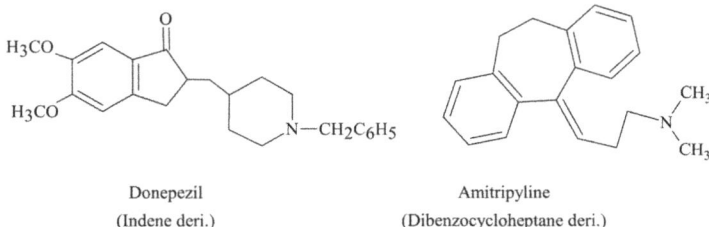

Donepezil   Amitripyline
(Indene deri.)   (Dibenzocycloheptane deri.)

Figure 8.9 Examples of drug having polycyclic aromatic compound

Donepezil is used to maintain functional brain activity in patients with Alzheimer disease. Amitriptyline is used as an antidepressant drug.

8.2.2.3 Drugs based on five-membered heterocyclics

Half of all therapeutic agents contain heterocyclic rings; it may be a five or six or may be seven-membered ring or it may be a heterocyclic ring fused to a benzene ring. Also, they may be compounds having two hetero atoms in their structure. These heterocyclic rings act as a pharmacophore in these compounds. The five-membered heterocyclics include derivatives of furan, pyrrole, thiophene, imidazole, and oxazole.

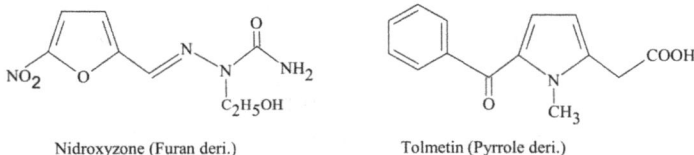

Nidroxyzone (Furan deri.)   Tolmetin (Pyrrole deri.)

Figure 8.10 Examples of drug having five membered heterocycles

Nidroxyzone is an antibacterial agent and tolmetin has anti-inflammatory, analgesic and antipyretic activitities.

8.2.2.4 Drugs based on six-membered heterocycles

The six-membered heterocyclics include derivatives of pyran, pyridine, piperidine, pyridazine, pyrimidine, pyrimidone, pyrazine, piperazine, triazine, etc. The examples are:

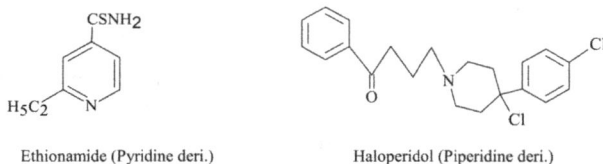

Ethionamide (Pyridine deri.)   Haloperidol (Piperidine deri.)

Figure 8.11 Examples of drug having six membered heterocycles

Ethionamideis an antibiotic utilized in the treatment of tuberculosis. Haloperidol is used primarily to treat schizophrenia and other psychoses. It is a potent antiemetic and is employed in the treatment of intractable hiccups.

8.2.2.5 Five-membered heterocycles fused to a benzene ring

These have a five membered heterocylic compound fused with a benzene ring. The examples are

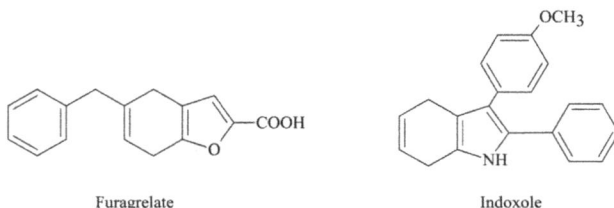

Furagrelate   Indoxole

Figure 8.12 Examples of drug having five membered heterocycles fused to benzene ring

Furagrelate and Indoxole having platelet aggregation inhibitory and anti-inflammatory characteristics respectively.

Other examples of this classification are six-membered and 7 membered heterocyclicrings fused to a benzene ring, heterocyclics fused to two aromatic rings, beta lactam antibiotics, and heterocyclics fused to other heterocyclic rings. (Tripathi, 2014; McEvoy et al., 2004).

8.3 Penicillin

8.3.1 Introduction

Penicillin (PCN or pen), general molecular formula: $C_9H_{11}N_2O_4S$ is a class of antibiotics that includes procaine penicillin, benzathine penicillin, penicillin G (for intravenous use), and penicillin V (for oral use) (intramuscular use). The general structure is mention below, in which group R is different to make a number of varieties of Penicillin. Some Penicillins are mentioned below table, which have different physical and chemical properties.

Table 8.1 Various forms of Penicillin

Name of Penicillin	R =	
Benzyl Penicillin (Penicillin G)	Benzyl	—CH$_2$—R
Ampicillin	α - amino benzyl	—HC—R, NH$_2$
Phenoxylmethyl Penicillin	Phenoxylmethyl	—O—CH$_2$—R
Carbeniciilin	α - carboxy benzyl	—HC—R, COOH

The commercially accessible form of penicillin G (USP grade) with benzyl group R, which is typically the most clinically acceptable, is typically coupled in the salt form with procaine or potassium. This penicillin is a white, finely crystalline powder, having a faint characteristicodor. It is hygroscopic, water soluble and dextro-rotatory. Penicillin is among the first antibiotic medications to be effectively utilized against many bacterial infections caused by staphylococci and streptococci. It is a very old medicine, but widely used today, though many types of bacteria have developed resistance for it following its extensive use. All penicillins are β-lactam antibiotics. First it was invented in 1928 by Scottish scientist Alexander Fleming. It binds to the cell walls of the bacteria, and prevents peptide chains from linking, and lyses them.

8.3.2 Manufacturing process

The industrial production of penicillin was broadly classified into two processes namely, upstream and downstream.

Any technology that contributes to the creation of a product is considered to be upstream. Exploration, development, and production are included in it. Downstream processing is the process of removing and purifying a biotechnology products from fermentation. There is no need to split open the fungal cells because downstream processing is quite simple because penicillin is secreted into the media (to kill other cells). The substance is purified by dissolving and precipitating as a potassium salt because it needs to be very pure because it is used as a therapeutic medical medicine.

8.3.2.1 Raw materials

If used in a concentration of 6%, lactose serves as a very effective carbon supplement. You could also utilize other things, like glucose and sucrose. The procedure uses corn steep liquor (CSL) as a source of nitrogen. As nitrogenous supplements, ammonium sulfate and ammonium acetate can be used. The precipitation of penicillin requires certain potassium, phosphorus, magnesium, sulfur, zinc, and copper salts. Some of these are implemented by corn liquor steeping. Calcium can be added in the form of chalk to counter the natural acidity of CSL. Other raw materials are phosphoric acid, amyl acetate, sodium phosphate and activated charcoal.

8.3.2.2 Flow chart

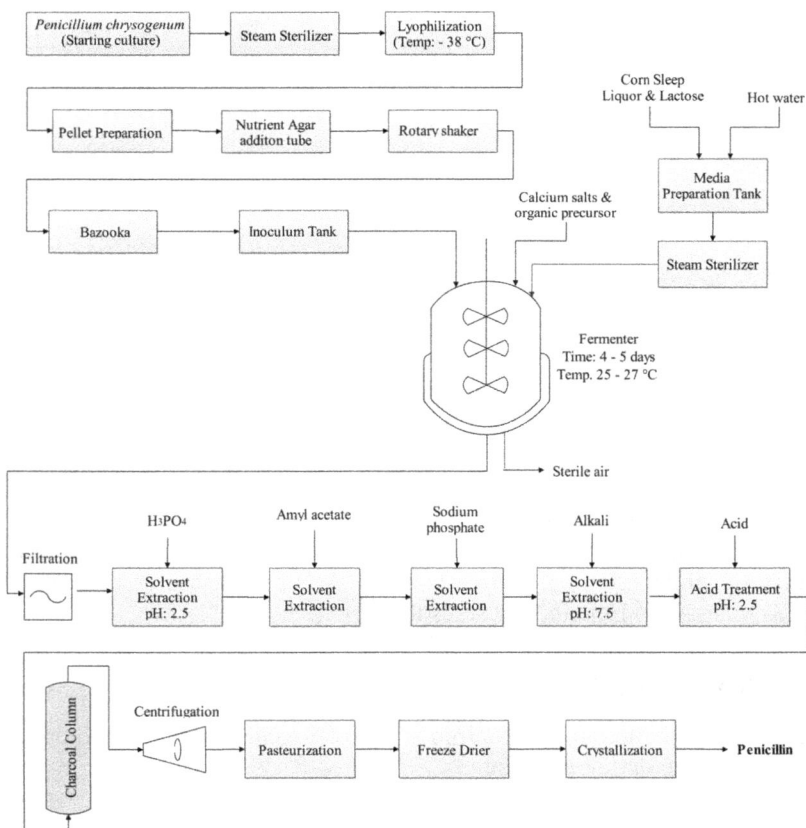

Figure 8.13 Flow sheet diagram of production of Penicillin

8.3.2.3 Process description

As per the flow-diagram, the process is mainly divided into four parts: (A) Raw material preparation, (B) Fermentation, (C) Separation and (D) Purification of penicillin.

(A) Raw material preparation: For culture development and propagation (growth), pure stains of *Penicillium chrysogenum* are sterilized and lyophilized (freeze drying) at −38°C and pellets are prepared (like tablet). Nutrient agar is added for the multiplication of microorganism spores. Then, it is agitated and aerated in a rotary shaker and transferred into an inoculum tank via bazzoka. The culture is agitated with the help of a paddle stirrer in the inoculum tank and transferred into the fermenter. The transfer of culture from the pure stain to the inoculum tank is conducted in a sterile medium.

(B) Fermentation: All the raw materials, i.e., solution of fermentation substrate mainly sucrose found in corn steep liquor and lactose together with minerals and phenylacetic acid, lyophilized *Penicillium chrysogenum* pellets and fermentation medium are sterilized by means of sterile air at 120°C. Thereafter these materials are added into a presterilized fermenter having a paddle type stirrer. Sterile air at a temperature of 25–27°C with continuous stirring is blown at the bottom of the tank for 4–5 days with continuous stirring. Assay of penicillin is continuously measured, until a concentration of 5 to 10 mg/ml is achieved. Suspended solids are removed by continuous filtration at 1–2°C.

(C) Separation: Fermented broth is cooled at room temperature and filtered. Solvent extraction is performed to separate penicillin from the fermented broth, in which the pH is adjusted upto 2.5 with phosphoric acid. Penicillin salt is extracted with amyl acetate, giving a purity of 75 to 80%. The extract is treated with sodium phosphate upto a pH of 7.5 due to enhanced concentration. A second acidic extraction is used to further purify penicillin, and after that, pyrogen-free distilled water is treated with an alkaline salt solution to eliminate any remaining acid.

(D) Purification of penicillin: The purified solution is subjected to charcoal treatment, centrifugation and pasteurization through a biological Seitz filter to remove color and final traces of bacteria and pyrogens. By using vacuum spray drying after freeze drying, this fluid is condensed. By salting out a saturated solution with a neutral salt that contains the cation needed in the final penicillin salt, crystalline penicillin salts can be produced.

8.3.3 Uses

Different types of advanced penicillin such as antistaphylococcal penicillins, aminopenicillins and the antipseudomonal penicillins. Penicillin had been employed to treat many wounded soldiers during World War II (Carmichael and Petrides, 2020).

8.4 Streptomycin

Streptomycin has a wide range of antibacterial activities. Its structure is shown in Fig. 8.14.

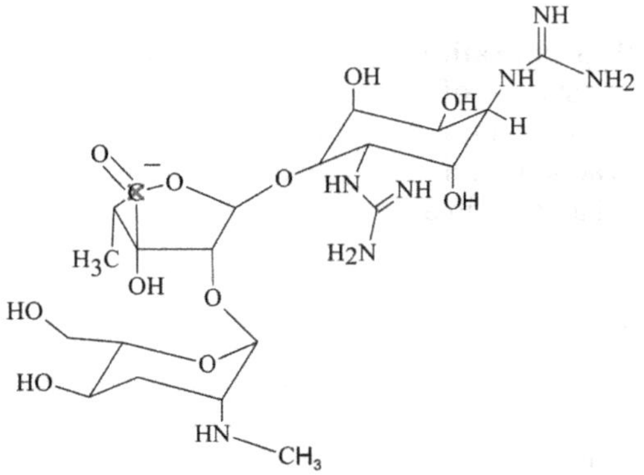

Figure 8.14 Structure of Streptomycin

8.4.1 Introduction

Its properties are as follows.

Table 8.2 Physical and Chemical Properties of Streptomycin

Sr. No.	Properties	Particular
1	IUPAC Name	1-[(1R,2R,3S,4R,5R,6S)-3-carbamimidamido-4-{[(2R,3R,4R,5S)-3-{[(2S,3S,4S,5R,6S)-4,5-dihydroxy-6-(hydroxymethyl)-3-(methylamino)oxan-2-yl]oxy}-4-formyl-4-hydroxy-5-methyloxolan-2-yl]oxy}-2,5,6-trihydroxycyclohexyl]guanidine
2	Chemical and other names	Streptomycin, streptan, streptocol
3	Molecular formula	$C_{21}H_{39}N_7O_{12}$
4	Molecular weight	581.6 gm/mole
5	Physical description	Light yellowish or white powder
6	Solubility	Freely soluble in water but, very slightly soluble in alcohol
7	Melting point	12°C
8	Boiling point	948°C
9	Flash point	527°C
10	Density	1.98 g/cm³

8.4.2 Manufacturing process

It is produced by stains of *Streptomyces griseus*. The fermentation process is carried out in a fermenterat a temperature of 28°C and for 10 days.

8.4.2.1 Raw materials

It is produced by stains of *Streptomyces griseus*. The fermentation process is carried in a fermenter at a temperature of 28°C and for 10 days.

8.4.2.2 Flow chart

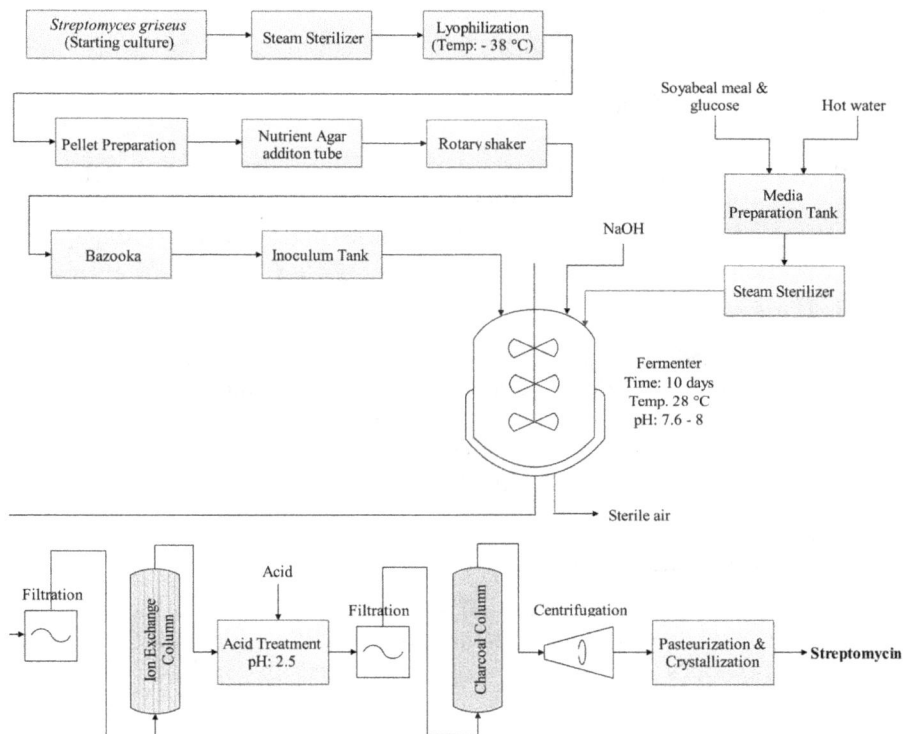

Figure 8.15 Flow sheet diagram of production of Streptomycin

8.4.2.3 Process description

As per flow-diagram, the process is mainly divided into four parts: (A) Raw material preparation, (B) Fermentation, (C) Separation and (D) Purification of Streptomycin

(A) Raw material preparation: For culture development and propagation (growth), pure stains of *Streptomyces griseus are* sterilized and lyophilized (freeze drying) at –38°C and pellets (like tablet) are prepared. Nutrient agar is added for the multiplication of microorganism spores. Then, it is agitated and aerated in a rotary shaker and transferred intothe inoculum tank via the bazzoka. The culture is agitated with the help of a paddle type stirrer in the inoculum tank and transferred into the fermenter. The transfer of the culture from the pure stain to the inoculum tank is conducted in a sterile medium.

(B) Fermentation: All the raw materials, i.e., soyabean meal, glucose, NaCl, lyophilized *Streptomyces griseus* pellets and fermentation medium are sterilized by means of sterile air at 120°C. Thereafter these materials are added into a pre-sterilized fermenter having a paddle stirrer. Sterile air is blown at the bottom of tank for 4–5 days at a temperature of 25–27°C with continuous stirring. In this process, the fermentation tank requires higher agitation and aeration. Here, the pH is increased by liberating ammonia from a microbe. NaOH solution having a pH of 7.6 to 8.0 is used. During this phase, the liberated glucose and ammonia are consumed. Streptomycin accumulates in the medium with an additional production of mycelium.

(C) **Separation:** After fermentation, the broth is filtered and cooled. Cooled fermentation broth undergoes an ion exchange process and acidification to separate streptomycin.

(D) **Purification:** It is thereafter decolorized with activated carbon, crystallized and centrifuged. The desired Streptomycin salt is converted and sterilized through a Seitz filter. Then the product is dried and packed accordingly.

8.4.3 Uses

It is effectively used for the treatment of (i) Staphyloccal skin and eye infections and (ii) G.I. (gastrointestinal) tract infection on oral administration. It also useful in preventing infection in neutropenic patients in conjugation with other antibiotics. It is useful for patients suffering from hepatic coma to suppress the ammonia forming bacteria in the G.I. tract (Chopra et al., 2012).

8.5 Erythromycin

Erythromycin typically has been an oral macrolide antibiotic since the 1950s. By stopping the production of crucial proteins required for the survival of the bacteria, macrolide antibiotics can either slow or even kill sensitive microorganisms. Its structure is shown in Fig. 8.16.

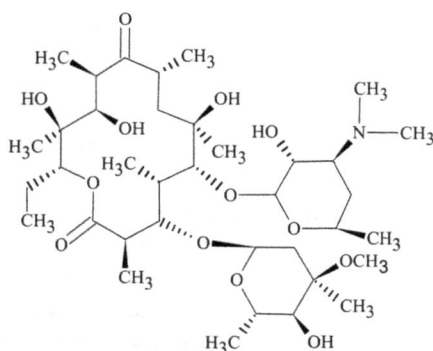

Figure 8.16 Structure of Erythromycin

8.5.1 Introduction

Its main properties are tabulated as per Table 8.3.

8.5.2 Manufacturing process

It is produced by stains of *Streptomyces erythreus*. The fermentation process is carried out in a fermenter at a temperature of 28°C for 72 hours.

Table 8.3 Physical and Chemical Properties of Erythromycin

Sr. No.	Properties	Particular
1	IUPAC Name	(3R,4S,5S,6R,7R,9R,11R,12R,13S,14R)-6- {[(2S,3R,4S,6R)-4-(dimethylamino)-3-hydroxy-6-methyloxan-2-yl]oxy}- 14-ethyl-7,12,13-trihydroxy-4-{[(2R,4R,5S,6S)-5-hydroxy-4-methoxy-4,6-dimethyloxan-2-yl]oxy}-3,5,7,9,11,13-hexamethyl-1-oxacyclotetradecane-2,10-dione
2	Chemical and other names	Erythromycin, Erythromycin A, E-Mycin, Erythromycinum, Abomacetin
3	Molecular formula	$C_{37}H_{67}NO_{13}$
4	Molecular weight	733.9 gm/mole
5	Physical description	White or yellowish odourless powder
6	Solubility	Slightly soluble in water, but readily soluble in dilute HCl
7	Melting point	140°C
8	Boiling point	818°C
9	Flash point	448°C
10	Density	1.226 g/cm³

8.5.2.1 Raw materials

Glucose, Sodium nitrate and *Streptomyces erythreus*

8.5.2.2 Flow chart

8.5.2.3 Process description

As per the flow-diagram, the process is divided into four main parts: (A) Raw material preparation, (B) Fermentation, (C) Separation and (D) Purification of Erythromycin

(A) **Raw material preparation:** For culture development and propagation (growth), pure stains of *Streptomyces erythreusare* sterilized and lyophilized (freeze drying) at –38°C and pellets (like tablet) are prepared. Nutrient agar is added for multiplication of microorganism spores. Then, it is agitated and aerated in a rotary shaker and transferred to the inoculum tank via bazzoka. The culture is agitated with the help of a paddle stirrer in the inoculum tank and transferred to the fermenter. The transfer of culture from a pure stain to the inoculum tank is conducted in a sterile medium.

(B) **Fermentation:** All the raw materials, i.e., soyabean meal, glucose, sodium chloride, lyophilized *Streptomyces griseus* pellets and fermentation medium are sterilized by means of sterile air at 120°C. Thereafter these materials are added in a presterilized fermenter having a paddle stirrer. Sterile air is blown at the bottom of tank for 72 hours at a temperature of 28°C with continuous stirring.

(C) **Separation:** The filtration broth is filtered and cooled. This filtered cooled broth is treated with NaOH and undergoes solvent extraction with amyl acetate. Add sulfuric acid to adjust the pH upto 5. Centrifuge the slime at a pH of 9.5 with NaOH.

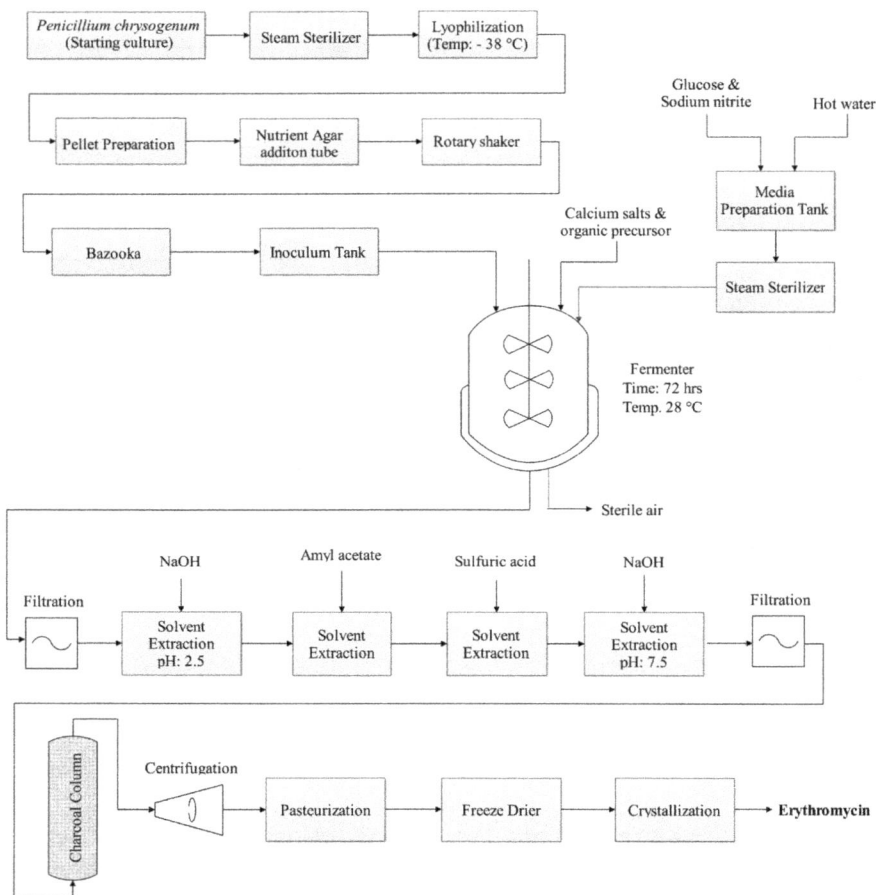

Figure 8.17 Flow sheet diagram of production of Erythromycin

(D) Purification of Erythromycin: It is decolorized using activated charcoal, the intermediate crystals are salted out and centrifuged. Then it is recrystallized and sterilized through a Seitz filter. Then the product is dried and packed accordingly.

8.5.3 Uses

A multitude of bacterial infections can be treated with erythromycin, an antibiotic. This includes syphilis, chlamydia, skin, and respiratory tract infections. It can help prolong delayed stomach emptying. It can be administered orally and intravenously (Minas, 2008; Anisimova and Yarullina, 2018).

Major raw materials including major steps involved in the production of various antibiotics are mentioned in below table.

Table 8.4 Major raw materials and steps involved in the production of antibiotic

Antibiotics	Culture	Raw material		Fermentation condition	Separation and purification
		Carbon source	Nitrogen source		
Penicillin	*Penicillium chrysogenum*	Lactose	Corn steep liquor (CSL)	4 to 5 days, 25–27°C	Solvent extraction, centrifugation, pressurization and freeze drying
Streptomycin	*streptomyces griseus*	Glucose	Soybean meal	4 to 5 days, 25–27°C, with continuous stirring	Solvent extraction, decolorization, crystallization and centrifugation
Erythromycin	*streptomyces erythreus*	Glucose	Sodium nitrate	72 hours, 28°C, with continuous stirring	Solvent extraction, centrifugation, decolorization and crystallization

8.6 Aspirin

Aspirin is one of the drugs that is used the most in the world. It is available in the salicylate form, which is present in plants like myrtle and willow trees. The first NSAID (non-steroidal anti-inflammatory medicine) to be discovered was aspirin. It has interactions with several other medications, such as warfarin and methotrexate. Its structure is as follows.

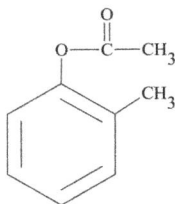

Figure 8.18 Structure of Aspirin

8.6.1 Introduction

Its properties are given below.

Table 8.5 Physical and Chemical Properties of Aspirin

Sr. No.	Properties	Particular
1	IUPAC Name	2-Acetoxybenzoic acid
2	Chemical and other names	Aspirin, Acetylsalicylic acid-ASA, Ecotrin, 2-(Acetyloxy) benzoic acid, Acenterine
3	Molecular formula	$C_9H_8O_4$
4	Molecular weight	180.2 gm/mole
5	Physical description	Colourless to white crystalline odourless powder
6	Solubility	Soluble in acetone, less soluble in water. In solution with alkalis, the hydrolysis proceeds rapidly and the clear solutions formed may consist entirely of acetate and salicylate.
7	Melting point	135°C
8	Boiling point	140°C
9	Flash point	250°C
10	Density	1.40 g/ml

8.6.2 Manufacturing process

It is prepared from Phenol.

8.6.2.1 Raw materials

Phenol, Sodium hydroxide, Carbon dioxide gas, Zinc powder, Zinc sulfate, Acetic anhydride and Activated charcoal

8.6.2.2 Chemical reactions

Phenol is reacted with sodium hydroxide to give Sodium phenolate. Sodium phenolate is isolated and powdered. Then, it is exposed to the action of carbon dioxide under pressure and heated to form sodium salicylate, which undergoes a reaction with sulfuric acid to give salicylic acid. This acid is esterified with acetic anhydride in toluene at 90°C to yield acetyl salicylic acid (Aspirin).

Figure 8.19 Synthesis of Aspirin

8.6.2.3 Quantitative requirements

For the production of 1000 kgs of aspirin, we require 1000 kgs of phenol, 455 kgs of sodium hydroxide, 650 kgs of carbon dioxide, 13 kgs of zinc, 26 kgs of zinc sulfate, 560 kgs of acetic anhydride and 26 kgs of activated charcoal

8.6.2.4 Flow chart

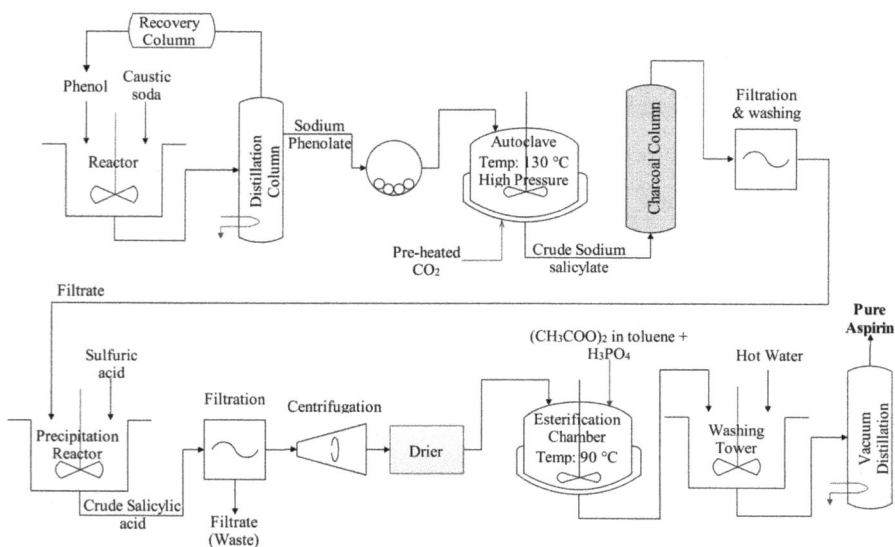

Figure 8.20 Flow sheet diagram for production of Aspirin.

The following major steps are involved in the production of Aspirin.

(A) Preparation of sodium phenolate: As per Fig. 8.20, phenol and caustic soda are fed into a reactor. This reactor has a paddle stirrer. To obtain dry sodium phenolate that has been purified, it is collected from the bottom of the reactor after the completion of the reaction and fed into a vacuum column. The unreacted phenol is recovered and recycled from the recovery column. Dry sodium phenolate is finely powdered using heated ball mill at 130°C. If sodium phenolate is directly used in this reaction, then it is purified, dried and powdered.

(B) Preparation of salicylic acid: Dry powdered sodium phenolate is added to a jacketed stainless steel autoclave with an agitator and pressure gauge. To create sodium salicylate, carbon dioxide gas is delivered into the autoclave's bottom chamber at a temperature of 100°C and a pressure of 7 atmospheres. Activated charcoal is used to decolorize the reaction material. After that, the reaction mixture is moved to a water tank, where it is purified to produce sodium salicylate. This pure sodium salicylate is fed into a stainless steel tank with an anchor type agitator. Sulfuric acid is added with continuous stirring in such way that the temperature does not go upto 50°C to get salicylic acid. Pure solid salicylic acid is obtained by filtration, centrifugation and drying.

(C) Preparation of Aspirin: The dried salicylic acid is taken in a stainless steel jacketed reactor with an agitation facility. The reaction mass is heated to 90°C by adding acetic anhydride while stirring continuously in a reactor containing toluene and phosphoric acid as the catalyst. To produce acetyl salicylic acid, keep the temperature at 90°C for 20 hours (Aspirin). The reaction mass is removed from the bottom of the reactor once the reaction is complete and charged into a washing chamber made of cast iron or stainless steel with an agitator. The chamber is agitated as hot water is introduced. The reaction mixture is collected at the bottom and vacuum distillation is used to extract pure aspirin.

8.6.3 Uses

It is a salicylate drug that is frequently prescribed to relieve inflammation, fever, and pain. Additionally, it has an antiplatelet action by preventing platelets from adhering together and from covering blood vessel walls that have been injured. It is used in the long-term, at modest doses, to help those who are at a high risk of blood clot development to avoid heart attacksand strokes. To lower the chance of suffering another heart attack or losing cardiac tissue, low dosages of aspirin may be administered right away after a heart attack. Certain cancers, particularly colorectal cancer, may be prevented by aspirin (Rainsford, 2004; Kandeh et al., 2020; Patel et al., 2013).

*****Note:** Salicylic acid undergoes esterification with methanol in the presence of a catalyst and sulfuric acid to form methyl salicylate.

8.7 Insulin

Insulin is a polypeptide hormone that controls various bodily functions, including the vascular system and metabolism. Insulin is derived from the Latin insula, meaning "island," and is made in the pancreatic islets of Langerhans. It keeps the blood sugar level stable. It enables the transfer of sugar from the blood to the cells. The symptoms and complications of diabetes are brought on when the body is unable to create enough insulin to move the sugar into the cells, leading to excessive blood sugar levels and insufficient sugar levels in the cells. The primary structure of insulin was explained by Sanger and co-scientists in 1945 to 1953. This is composed of component A and B, two polypeptide chains. While subunit B has 30 amino acids, subunit A only has 21 amino acids (total amino acids: 51). All the 51 amino acids are of different types. These chains are connected by two disulfide bridges. The molecular weight varies from 12,000 to 36000 Daltons. Insulin is a peptide hormone generated by beta cells in the pancreas and by Brockmann bodies in some teleost fish. Insulin is derived from the Latin word insula, which means island.

8.7.1 Introduction

Its main properties are mentioned in the following table.

Table 8.6 Physical and Chemical Properties of Insulin

Sr. No.	Properties	Particular
1	IUPAC Name	L-phenylalanyl-L-valyl-L-asparagyl-L-glutaminyl-L-histidyl-L-leucyl-L-cysteinyl-glycyl-L-seryl-L-histidyl-L-leucyl-L-valyl-L-alpha-glutamyl-L-alanyl-L-leucyl-L-tyrosyl-L-leucyl-L-valyl-L-cysteinyl-glycyl-L-alpha-glutamyl-L-arginyl-glycyl-L-phenylalanyl-L-phenylalanyl-L-tyrosyl-L-threonyl-L-prolyl-L-lysyl-L-alanine (7->7'),(19->20')-bis(disulfide) with glycyl-L-isoleucyl-L-valyl-L-alpha-glutamyl-L-glutaminyl-L-cysteinyl-L-cysteinyl-L-alanyl-L-seryl-L-valyl-L-cysteinyl-L-seryl-L-leucyl-L-tyrosyl-L-glutaminyl-L-leucyl-L-alpha-glutamyl-L-asparagyl-L-tyrosyl-L-cysteinyl-L-asparagine (6'->11')-disulfide
2	Chemical and other names	Insulin, AN-18331
3	Molecular formula	$C_{254}H_{377}N_{65}O_{75}S_6$
4	Molecular weight	5733.5 g/mol
5	Physical description	White or almost white crystalline powder
6	Solubility	It is slightly soluble in water, but practically insoluble in alcohol, chloroform and ether. Also it is soluble in dilute solution of mineral acids and with degradation in solutions of alkali hydroxide.
7	Melting point	233°C
8	Boiling point	Not available
9	Flash point	Not available
10	Density	1.09 g/cm³

8.7.2 Manufacturing process

It is prepared from the glands of beef and pork.

8.7.2.1 Raw materials

Glands of beef and pork, Acid, Ethanol and Ammonia

8.7.2.2 Flow chart

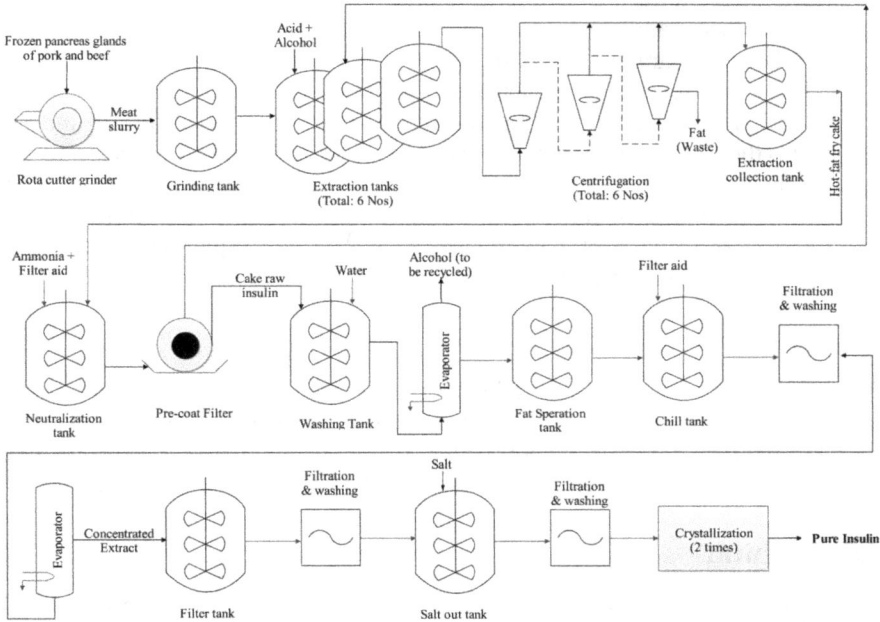

Figure 8.21 Flow sheet diagram of production of Insulin

8.7.2.3 Process description

Following major steps are involved in manufacturing of insulin.

(A) Preparation of raw insulin: In this process, pancreatic glands of beef and pork are refrigerated at –20°C and cut using a roto cut grinder. The cut meat slurry is extracted first with acid and then, ethanol in extraction tanks. A total of six continuous extraction tanks are utilized for the extraction. Ethanol is added to the extractor, if required. Now, this extracted liquor is centrifuged using six continuous centrifugations. The cake is transferred into a hot-fat fry tank from the sixth extractor, where the fat is separated as a "fried residue". The fat free extract is transferred to the collected extract and then, to the neutralization tank, in which the acid is neutralized with ammonia. Also, a filter aid is added to the neutralization tank to increase the rate of filtration. A continuous filtration process is carried out in a precoat drum filter to obtain the cake containing insulin. The same re-purification process of clear liquor (filtrate), i.e., addition of acid and removal of fat is carried out, where the liquor is re-acidified with acid.

(B) Purification of insulin: The cake is subjected to a continuous washing process using water in a washing tank. The liquor is transferred into a 1st stage evaporator to separate the remaining fat. A filter aid is sent and added to the chill tank where. Now, the liquor is filtered and subjected to a 2nd evaporator. The concentrated extract is filtered and then delivered to a salting out tank from this evaporator. The insulin is precipitated by the addition of the desired salt and subjected to filtration. It is then crystallized twice to purify the insulin.

Another manufacturing process of insulin is that the two chains, A and part B are separately synthesized from two gene splicing or recombinant DNA. This process is conducted by a fermentation process using *bacterium E. coli*. The reduction-reoxidation mechanism is then used to join the two chains through disulfide bonds. An oxidizing agent is applied (a substance that results in oxidation or the transfer of an electron). Human insulin is combined with an oxidizing agent, and then its structure and purity are examined using a variety of techniques, including high performance liquid chromatography, X-ray crystallography, gel filtration and amino acid sequencing.

8.7.3 Uses

By encouraging the absorption of glucose from the blood to skeletal muscles and fat tissue and by inducing fat to be stored rather than utilized for energy, it regulates the metabolism of carbohydrates and fats. Insulin also prevents the liver from producing glucose (Saltiel and Pessin, 2007; Kuhlmann and Schmidt, 2014).

8.8 Barbital

Veronal and Medinal were the brand names for the pure acid and sodium salt, respectively, of Barbital (known as Barbitone in other parts of the World). It started with the readily available barbiturate. It served as a sleep aid from 1903 till the middle of 1950.

8.8.1 Introduction

The following table shows the various properties of barbital.

Table 8.7 Physical and Chemical Properties of Barbital

Sr. No.	Properties	Particular
1	IUPAC Name	5,5-diethyl-1,3-diazinane-2,4,6-trione
2	Chemical and other names	Barbital, Barbitone, Diemal, Diethylmalonylurea, Dormileno, Ethylbarbital, Medinal, Veronal
3	Molecular formula	$C_8H_{12}N_2O_3$
4	Molecular weight	184.2 gm/mole
5	Physical description	White needle crystal
6	Solubility	Soluble in water, alcohol, chloroform, ether, acetone, ethyl acetate, alkalies, petroluem ether, acetic acid, etc.
7	Melting point	190°C
8	Boiling point	250°C
9	Flash point	170°C
10	Density	1.29 g/cm²

8.8.2 Manufacturing process

It is prepared from chloroacetic acid.

8.8.2.1 Raw materials

Chloroaceitc acid, Sodium hydroxide, Sodium cyanide, Ethanol, Sulfuric acid, Benzene, Sodium metal, Ethyl bromide and Urea

8.8.2.2 Chemical reactions

It is prepared from chloroacetic acid by different methods. One of the methods is shown below.

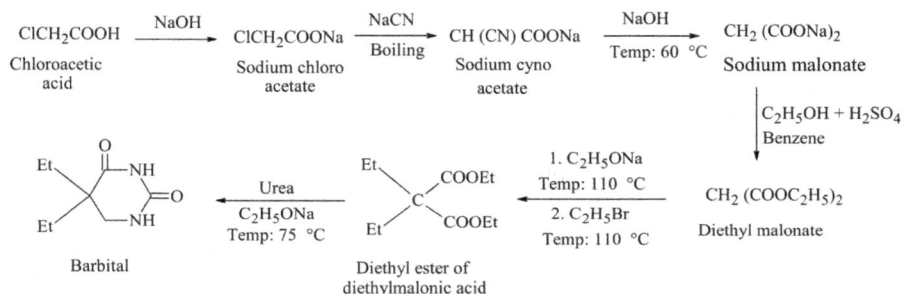

Figure 8.22 Synthesis of Barbital

8.8.2.3 Flow chart

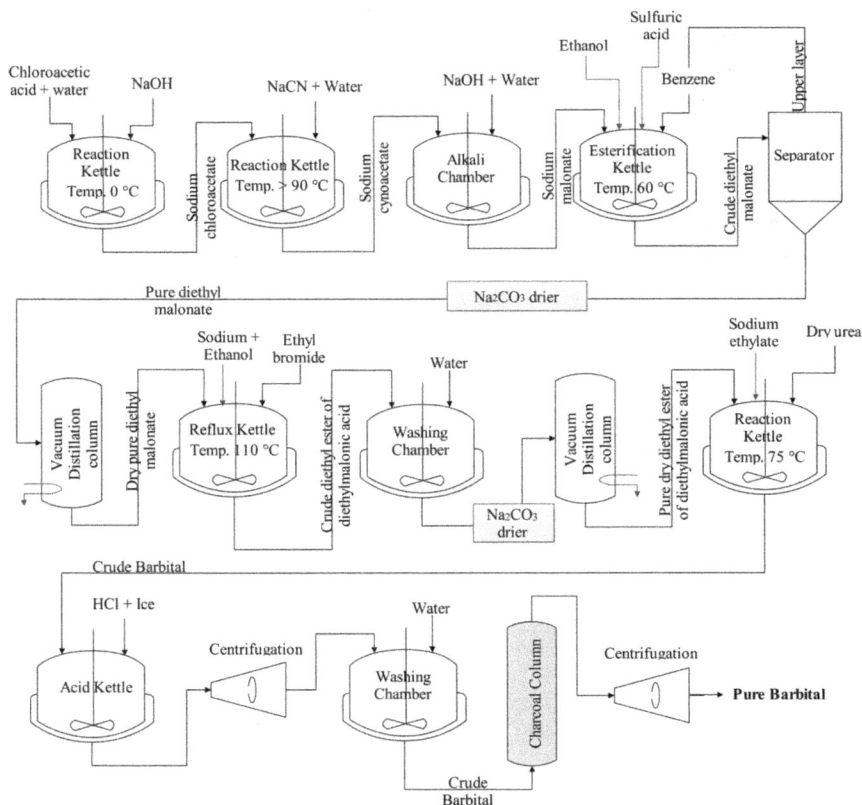

Figure 8.23 Flow sheet diagram of production of Barbital

8.8.2.4 Process description

The following major steps are involved in the production of Barbital.

(A) Preparation of sodium malonate: Chloroacetic acid is taken in a jacketed reactor kettle with an agitator. Dissolve it in the least possible amount of cold water with stirring. While adding a solution of anhydrous sodium hydroxide or sodium carbonate to this mixture, the temperature is maintained at 0°C by utilizing cold water in a jacket. And, thus sodium chloroacetate is formed, collected from the bottom of kettle, and transferred into another jacketed reaction kettle with continuous agitation. A water solution of sodium cyanide is heated to 70°C, and then added to the kettle slowly, with good stirring. The solution is added in such a way that the temperature does not rise above 90°C. The reaction is complete after about 20 minutes and thereafter, the hot solution is transferred into the jacketed alkali chamber. Sodium hydroxide in water is charged into the alkali chamber. This solution is continuously boiled for about 2 hours for the complete evolution of ammonia. After the solution has evaporated, sodium malonate and sodium chloride are combined in the powdered residue.

(B) Preparation of diethyl malonate: Powdered sodium malonate is taken into an esterification kettle. The kettle is facilitated with a jacket and agitation. Ethanol (industrial spirit) and benzene are taken in an esterification kettle with stirring. Sulfuric acid is added; at such a rate that the temperature of this stirred mixture does not exceed 25°C (upto several hours). Thereafter, the temperature is increased upto 60°C using hot water circulation. This temperature is maintained for 8–10 hours for the completion of the reaction and then, cooled to room temperature. The reaction mixture is collected from the bottom and transferred into an extractor. Here, the unreacted benzene is collected from the extractor repeatedly. To obtain pure diethyl malonate, the obtained benzene layers are washed with dilute sodium hydroxide, dried over anhydrous sodium carbonate, and then vacuum-distilled.

(C) Preparation of diethyl malonic ester: First the reflux kettle is washed with water and dried. Water must be kept far away from this reaction; even atmospheric moisture can reduce the yield. Clean sodium metal is taken and dissolved in absolute ethanol in a refluxed kettle with a jacket and an agitator and heated to 60°C, to prepare sodium ethylate solution. With continuous stirring at a temperature of 60°C, diethyl malonate is charged into it. The reaction mass is heated to a temperature of 80°C. Now, ethyl bromide is charge for a time duration of 3–4 hours. After addition of all chemicals, the mass is refluxed for 2–3 hours till its temperature of goes upto 110°C. Thereafter, sodium ethylate and ethyl bromide are added to the reflux kettle at a temperature of 60°C. Again, reflux the mass for 2–3 hours at 110°C. Cool the mass and transfer it to the washing kettle. Charge the water for 30 minutes with brisk agitation, and allow it to settle down to separate into two layers. The crude diethyl malonic ester in the top layer is separated, dried over anhydrous sulfate, and then vacuum-distilled.

(D) Preparation of barbitone (5,5-diethyl malonyl urea): First, dry urea and dry diethyl Malonic ester and malonate are charged and vigorously agitated in the reaction vessel. A heated sodium ethylate solution (clean sodium metal in dry ethanol) is added with stirring. The reaction mixture is heated at 75°C. After some time, the reaction mixture become viscous and the residue left behind is crude barbitone.

(E) Purification of barbitone: Coned hydrochloric acid crushed ice is taken in an acid kettle and dry barbitone powder is charged into this mixture, gradually, with good stirring. The reaction mixture is cooled to a temperature below 0°C. Stir for several hours and centrifuge to get solid barbitone. It is further purified by washing it with water, decolorizing with charcoal and centrifuging under vacuum to get pure diethyl barbituric acid (barbitone).

8.8.3 Uses

Barbital inhibits most metabolic functions when used in large amounts. It can also lead to dependence and is used as a hypnotic and sedative. Barbital is also employed in veterinary medicine to treat depression of the central nervous system.

8.9 Phenobarbital

The World Health Organization advises using phenobarbital, commonly known as phenobarbitone or phenobarb, to treat neurological disorders of the central nervous system.

8.9.1 Introduction

Its main properties are tabulated as follows.

Table 8.8 Physical and Chemical Properties of Pheno Barbital

Sr. No.	Properties	Particular
1	IUPAC Name	5-ethyl-5-phenyl-1,3-diazinane-2,4,6-trione
2	Chemical and other names	Pheno Barbital, Phenylethylbarbituric acid, Phenylbarbital, Phenemal, Phenobarbitone
3	Molecular formula	$C_{12}H_{12}N_2O_3$
4	Molecular weight	232.2 gm/mole
5	Physical description	Odorless white crystalline powder or colorless crystals
6	Solubility	Soluble in alcohol, water and glycerin.
7	Melting point	175°C
8	Boiling point	398°C
9	Flash point	345°C
10	Density	1.23 g/cm²

8.9.2 Manufacturing process

It is prepared from benzyl cyanide.

8.9.2.1 Raw materials

Benzyl cyanide, Sodium ethylate, Diethyl carbonate, Sulfuric acid, Ethyl bromide and Urea

8.9.2.2 Chemical reactions

Diethyl phenyl ethyl malonate is produced by refluxing benzyl cyanide with sodium ethylate, diethyl carbonate, and ethyl bromide. When urea is added to this malonate, a 4-Imino derivative of diethyl phenyl ethyl malonate is prepared. Finally, phenobarbital is produced by acid hydrolysis of this imino derivative.

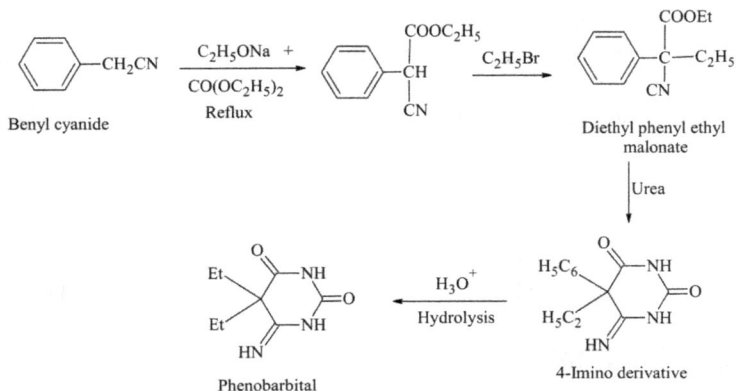

Figure 8.24 Synthesis of Phenobarbital

8.9.2.3 Flow chart

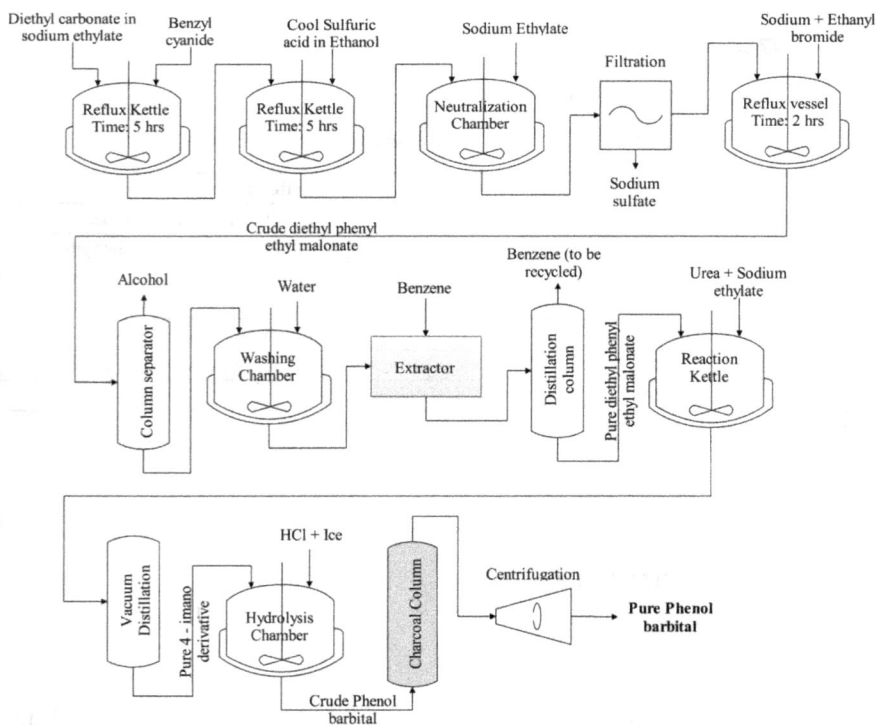

Figure 8.25 Flow sheet diagram of production of Phenobarbital

8.9.2.4 Process description

The following steps are involved in the production of phenobarbital.

(A) **Preparation of diethyl phenyl ethyl malonate:** Clean sodium metal is taken in a reflux kettle with a jacket and an agitator and dissolved in absolute ethanol to prepare sodium ethylate solution. Diethyl carbonate is charged with stirring and dissolved in the sodium ethylate solution. Benzyl cyanide is added drop-wise to the solution and the reaction mixture is refluxed for 5 hours. It is then cooled and collected from the bottom of the kettle and charged into another jacketed reflux kettle. A cooled mixture of sulfuric acid in anhydrous ethanol is then added. This alcoholic solution is refluxed for 5 hours and cooled. The reaction mass is collected from the bottom and transferred into a neutralization chamber. Sodium ethylate is charged with stirring to precipitate sodium sulfate. The mass undergoes filtration to remove sodium sulfate. The filtrate is added to a jacketed reflux kettle containing sodium metal. It is refluxed while adding ethyl bromide drop-wise. Thereafter, the reaction mass is heated for 2 hours. The leftover material is then dissolved in water after the alcohol has been removed by distillation. To obtain pure diethyl phenyl ethyl malonate and benzene, separate the chemical from the water using vacuum distillation and benzene. Recycled benzene is used.

(B) **Preparation of Phenobarbitone:** Pure diethyl phenyl ethyl malonate and dry urea are mixed together in a container with an effective stirring mechanism. Add a clean metallic sodium solution dissolved in dry ethanol to the mixture mentioned above (sodium ethylate). To distil off the ethanol, the reaction mass is slowly cooked for six hours while being stirred. The remaining white powder is put into the hydrolysis chamber once all of the alcohol has been thoroughly distilled out of the combination. In the chamber, condensed hydrochloric acid in water and finely broken ice are added. The mixture is centrifuged and given a charcoal treatment after being held at 0°C till the acid crystallizes to give pure phenobarbitone.

8.9.3 Uses

Phenobarbital, also known as phenobarbitone or phenobarb, is mainly used to control the abnormal electrical activity in the brain that occurs during a seizure. It is also utilized for the short-term treatment of restlessness, anxiety release, tension, and fear (Niazi, 2004).

8.10 Paracetamol

The WHO advises using this medication as the first line of treatment for pain disorders. The American FDA first gave it the go-ahead in 1951.

8.10.1 Properties

IIts main properties are tabulated as per Table 8.9.

Table 8.9 Physical and Chemical Properties of Paracetamol

Sr. No.	Properties	Particular
1	IUPAC Name	N-(4-hydroxyphenyl) acetamide
2	Chemical and other names	Paracetamol, Acetaminophen, Tylenol, Datril, Acamol
3	Molecular formula	$C_8H_9NO_2$
4	Molecular weight	151.2 gm/mole
5	Physical description	Odorless white crystalline solid
6	Solubility	Freely soluble in alcohol, dimethylformamide, ethylene dichloride, acetone, ethyl acetate; Practically insoluble in petroleum ether, pentane, benzene
7	Melting point	170°C
8	Boiling point	> 500°C
9	Flash point	-
10	Density	1.293 g/cm²

8.10.2   Manufacturing process

It is manufactured from p-nitro chlorobenzene.

8.10.2.1 Raw materials

p-Nitro chlorobenzene (PNCB), Sodium hydroxide (Caustic lye), Sulfuric acid, Iron powder, Acetic acid, Acetic anhydride and Activated charcoal

8.10.2.2 Chemical reactions

First, p-Nitro chlorobenzene (PNCB) undergoes hydrolysis with sodium hydroxide at a temperature of 135–140°C and a pressure of 6 atmospheres and thereafter, neutralization with sulfuric acid to get p-Nitro phenol. This p-nitro phenol on reduction with iron and acetic acid under mild acidic condition gives p-aminophenol, which condenses (acelylation) with acetic anhydride to yield paracetamol and acetic acid.

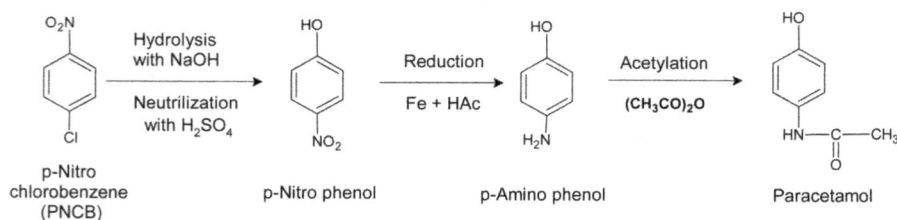

Figure 8.26 Synthesis of Paracetamol

8.10.2.3 Quantitative requirements

For manufacturing 1000 kgs of paracetamol, we require 1250 kgs of PNCB, 1275 kgs of caustic lye (48%), 370 kgs of sulfuric acid, 1035 kgs of iron powder, 580 kgs of acetic acid, 760 kgs of acetic anhydride, 75 kgs of activated charcoal and 7000 liters of water.

8.10.2.4 Flow chart

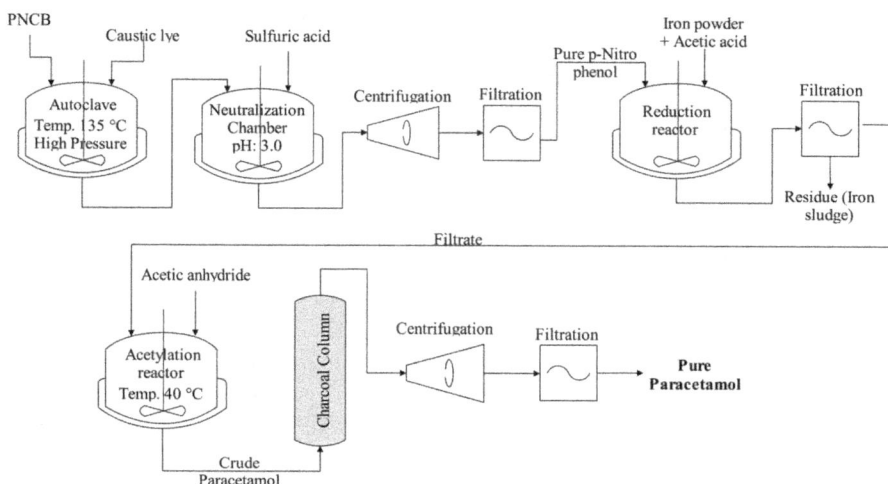

Figure 8.27 Flow sheet diagram of production of Paracetamol

8.10.2.5 Process description

The process is mainly divided into two parts as follows.

(A) Preparation of p-nitro phenol: In the first step of the process, caustic lye solution (48% w/w) is inserted into a hydrolysis autoclave. PNCB is also added into it. Steam is passed into thethe reactor jacket to raise the temperature upto 135–140°C and the pressure to 6 atmospheres. These conditions are maintained in the autoclave for 8 hours. The reaction liquor is cooled and collected from the bottom of the autoclave. This liquor is transferred into a neutralization reactor. Sulfuric acid is slowly added into the reactor to adjust the pH to about 3. The reaction mass is centrifuged and filtered to separate p-nitro phenol.

(B) Preparation of paracetamol: The obtained p-nitro phenol is charged into a reduction reactor. Iron powder and acetic acid are also added to the reactor. After conversion of p-nitro phenol into water soluble p-amino phenol, the liquid reaction mass is collected from the bottom of the reactor and filtered to remove iron slag and mother liquor sent to the acetylation reactor. The liquor is heated upto 40°C using steam in the reactor jacket. Acetic anhydride is slowly charged into the reactor maintaining the reaction temperature of 40°C. After completion of the reaction, the reaction mass is cooled at room temperature. This mass undergoes activated charcoal treatment, followed by crystallization and filtration.

8.10.3 Uses

Paracetamol is considered as the most common, economical, effective and safe drug used for treatment of fever, muscle and body aches, flu and cold symptoms. Lower doses are also given to pregnant women and children without any side effects (Intratec, 2021).

8.11 Proguanil

It is used as a hydrochloride salt, so, it is referred to as Proguanil hydrochloride. Proguanil, also known as chloroguanide, is recommended by WHO as the most efficient and secure drug desired in a health system.

8.11.1 Properties

Its properties are mentioned below.

Table 8.10 Physical and Chemical Properties of Proguanil

Sr. No.	Properties	Particular
1	IUPAC Name	(1E)-1-[amino-(4-chloroanilino) methylidene]-2-propan-2-ylguanidine
2	Chemical and other names	Chloroguanide, Chlorguanide, Paludrin, Paludrine, 1-Isopropyl-5-(4-chlorophenyl)biguanide, N1-p-Chlorophenyl-N5-isopropylbiguanide
3	Molecular formula	$C_{11}H_{16}ClN_5$
4	Molecular weight	253.7 gm/mole
5	Physical description	Odorless white crystalline solid
6	Solubility	Soluble in ethanol (95%), slightly soluble in water, practically insoluble in chloroform and in ether
7	Melting point	129°C
8	Boiling point	290°C
9	Flash point	190°C
10	Density	1.30 g/cm²

8.11.2 Manufacturing process

The starting materials for the manufacture of proguanil are isopropyl amine and dicyandiamide.

8.11.2.1 Raw materials

Isopropyl amine, Dicyandiamide, n-Butanol, p-Chloroaniline hydrochloride, Isopropanol and Water

8.11.2.2 Chemical reactions

Isopropyl amine and dicyandiamide are refluxed for 3 hours in the presence of butanol to give Isopropyl dicyandiamide. Thereafter, this product is again refluxed with p-chloroaniline hydrochloride for 4 hours to yield proguanil. It is purified using recrystallization by isopropanol.

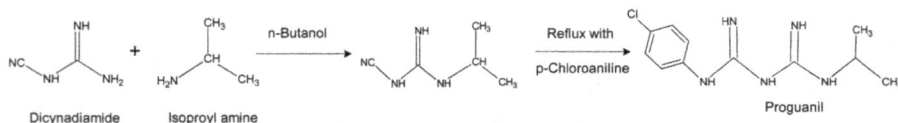

Figure 8.28 Synthesis of Proguanil

8.11.2.3 Quantitative requirements

For manufacturing 1000 kgs of proguanil, we require 227 kgs of isopropyl amine, 800 kgs of dicyanodiamide, 770 kgs of butanol, 630 kgs of p-chloroaniline hydrochloride, 765 kgs of isopropanol and 765 kgs of water.

8.11.2.4 Flow chart

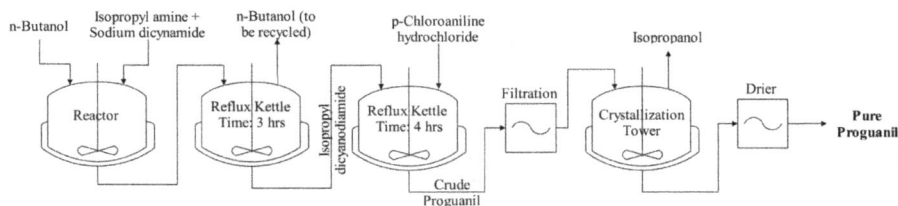

Figure 8.29 Flow sheet diagram of production of Proguanil

8.11.2.5 Process description

The process is divided into the following parts.

(A) Preparation of isopropyl dicyanodiamide: As per the flow chart, butanol is inserted into a glass lined jacketed reactor. Also, isopropyl amine and sodium dicyanamide are charged into a reactor. The reaction mass is refluxed for 3 hours using steam in the reactor jacket. After production of isopropyl dicyanodiamide, the reaction mass is collected from the bottom and fed into a distillation unit, where butanol is distilled out and reused. The remaining liquid mass containing isopropyl dicyanodiamide undergoes further processing.

(B) Preparation of proguanil: p-chloroaniline hydrochloride in water is charged into a glass lined jacketed reactor. The obtained isopropyl dicyanodiamide is also added in it. The reaction mass is refluxed for 4 hours to form solid proguanil. Then it is cooled and filtered. The solid product is sent to a glass lined recrystallization tower. Here, isopropanol is added and the reaction mass is heated. The recovered isopropanol is recycled. The final product is dried, thereafter.

8.11.3 Uses

Proguanil is generally combined with chloroquine or atovaquone and used to treat and prevent malaria in adults and children. The majority of other multi-drug resistant strains of *P. falciparum* can be successfully treated with it as well.

8.12 Diphenhydramine (Benadryl)

Diphenhydramine, 1st Generation Antiemetic Agent, is mainly used to treat allergies.

8.12.1 Properties

Its properties are mentioned in the Table 8.11.

Table 8.11 Physical and Chemical Properties of Diphenylhydramine

Sr. No.	Properties	Particular
1	IUPAC Name	2-benzhydryloxy-N,N-dimethylethanamine
2	Chemical and other names	2-Diphenylmethoxy-N,N-dimethylethylamine, Allerdryl, Benadryl, Benhydramin, Benylin, Benzhydramine
3	Molecular formula	$C_{17}H_{21}NO$
4	Molecular weight	255.4 gm/mole
5	Physical description	Odorless white crystalline solid
6	Solubility	Freely soluble in methanol and ethanol and less soluble in DMSO
7	Melting point	162°C
8	Boiling point	150°C
9	Flash point	9°C
10	Density	1.10 g/cm²

8.12.2 Manufacturing process

Initially, diphenhydramine was prepared by George Rieveschlhad and thereafter, in 1946, it became commercially available. Now-a-days, it is prepared from diphenylmethane.

8.12.2.1 Raw materials

Diphenylmethane, Liquid bromine, Chlorobenzene, Dimethylamino ethanol, Caustic soda, Toluene, Hydrochloric acid, Activated charcoal and Water

8.12.2.2 Chemical reactions

Diphenylmethane is reacted with liquid bromine to give diphenylbromomethane, which is condensed with dimethylamino ethanol in the presence chlorobenzene (as a solvent) to yield diphenphydramine. Thereafter, it is treated with hydrochloric acid to give its salt.

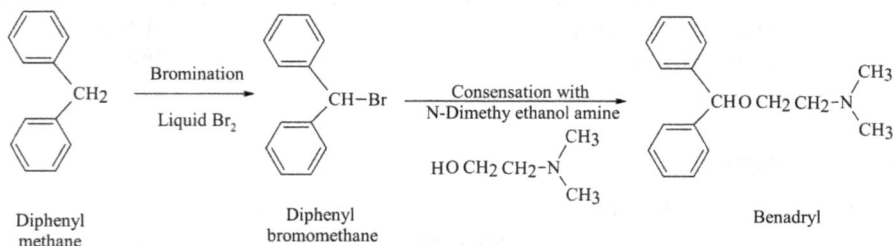

Figure 8.30 Synthesis of Benadryl

8.12.2.3 Quantitative requirements

For manufacturing 1000 kgs of diphenhydramine, we required 830 kgs of diphenylmethane, 800 kgs of liquid bromine, 800 kgs of chlorobenzene, 475 kgs of dimethylamino ethanol, 300 kgs of caustic soda flakes, 300 kgs of hydrochloric acid, 20 kgs of activated charcoal and 1000 kgs of water.

8.12.2.4 Flow chart

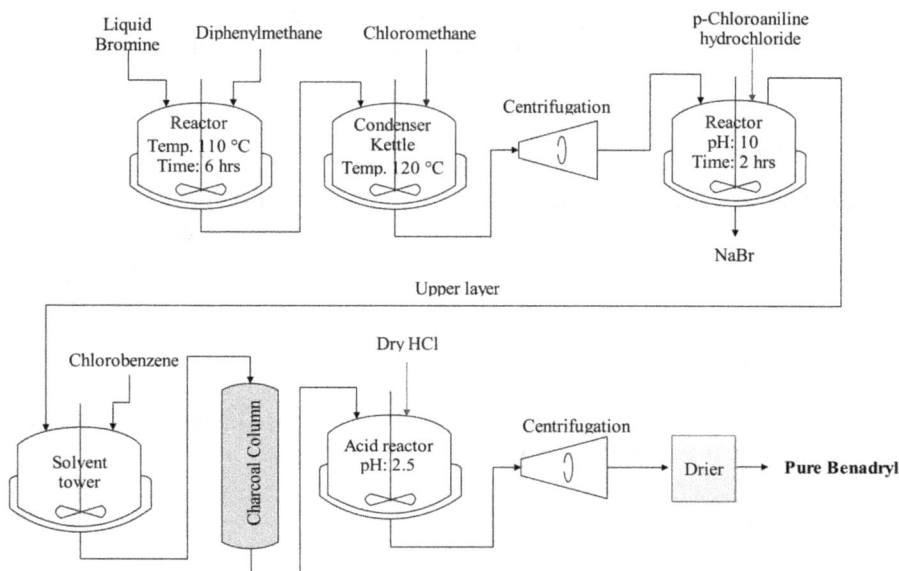

Figure 8.31 Flow sheet diagram of production of Benadryl

8.12.2.5 Process description

The process is divided into two main parts.

(A) Preparation of diphenphydramine: In the first step, liquid diphenylmethane is inserted into a glass lined jacketed reactor having a stirring facility. It is heated upto 110°C with steam in the reactor jacket. Liquid bromine is added at 110–115°C for 5 to 6 hours. The reaction temperature is maintained at 115°C for 1 hour. Excess hydrogen bromide is removed by purging nitrogen gas in it. Also, it maintains an inert layer during the reaction. Now, add chloromethane to the reactor and heat the reaction mass upto 120°C. Charge dimethylamino ethanol slowly into the reactor maintaining a temperature of 120–125°C for a time span of 4–5 hours. Maintain the reaction temperature of 125°C for 2 hours. Cool down the reaction to room temperature. Centrifuge the reaction mass to recover chloromethane from the mother liquor.

(B) Preparation and purification of Diphenphydramine hydrochloride: Wet material and water are added to the jacketed stainless steel reactor, facilitated with stirring. Heat the solution to dissolve the wet material. Now, adjust the pH to10 with caustic lye solution. Stir the reaction mass for 2 hours. Now, the reaction mass is allowed to settle for 8 hours. Here, sodium bromide is removed from the bottom as a byproduct. The upper liquid layer containing the product is collected and chlorobenzene is added to it. The reaction mass then goes to the activated charcoal column. The liquid reaction mass is collected from the bottom of the charcoal column and sent into another glass lined reactor. Dry hydrochloric acid is purged into it to adjust the pH in the range of 2.5 to 3.0. Stir for 1 hour. Centrifuge the material to collect the final product diphenphydramine hydrochloride. Dry the product using a fluid bed or tray drier. Chloromethane is recovered from the mother liquor by centrifugation.

8.12.3 Uses

It is used under various trade names like Benadryl, Genahist, Naramin, Sominex Unisom. It is used to treat allergies, hay fever, motion sickness, and the common cold. It is also used to prevent nausea, vomiting and vigilance (Disouza et al., 2016).

8.13 Cetirizine (Cetirizine)

Cetirizine is a potent second-generation antihistamine used in various forms such as tablet, syrup, injection, cream, gel, ointment and liquid.

8.13.1 Properties

Its main properties are tabulated as follows.

Table 8.12 Physical and Chemical Properties of Cetrizine

Sr. No.	Properties	Particular
1	IUPAC Name	2-[2-[4-[(4-chlorophenyl)-phenylmethyl]piperazin-1-yl]ethoxy] acetic acid
2	Chemical and other names	(2-(4-((4-Chlorophenyl)phenylmethyl)-1-piperazinyl)ethoxy) acetic Acid, Ceti-Puren, Alerlisin, Cetiderm, Cetalerg
3	Molecular formula	$C_{21}H_{25}ClN_2O_3$
4	Molecular weight	388.9 gm/mole
5	Physical description	White crystalline solid
6	Solubility	Freely soluble in water and alcohol, and sparingly soluble in chloroform.
7	Melting point	112°C
8	Boiling point	540°C
9	Flash point	-
10	Density	1.20 g/cm^2

8.13.2 Manufacturing process

Starting materials for cetirizine are 4-Chloro benzhydralpiperazine and 2-Chloroethanol.

8.13.2.1 Raw materials

p-Chlorobenzhydryl chloride, Anhydrous Piperazine, 2-Chloroethanol, Caustic soda flask, Triethyleamine, Toluene, Acetone, Methylene Dichloride (MDC) or Dichloromethane, Sodium monochloroacetate, Dimethylformamide, potassium hydroxide, Hydrochloric acid and Water

8.13.2.2 Chemical reactions

p-Chlorobenzhydryl chloride is reacted with anhydrous piperazine in the presence of toluene in a caustic soda flask at 40°C for 5 hrs to give 4-chloro benzhydralpiperazine. This derivative and 2-chloroethanol are reacted in the presence of triethylamine in toluene for 2 hours to give 2-{4-[(4-chlorophenyl) (phenyl)methyl]piperazin-1-yl}ethanol, which is reacted with sodium monochloroacetate in the presence of dimethyl formamide and potassium hydroxide to give cetirizine base. This cetirizine base further reacts with hydrochloric acid and acetone to yield the final product, cetirizine dihydrochloride.

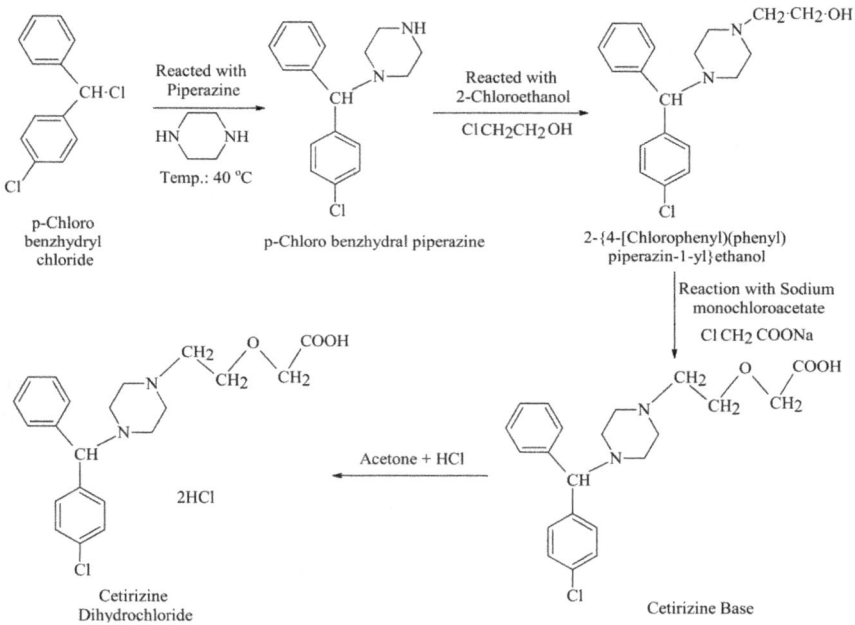

Figure 8.32 Synthesis of Cetirizine dihydrochloride

8.13.2.3 Quantitative requirements:

For manufacturing 1000 kgs of cetirizine, we required 700 kgs of p-chlorobenzhydryl chloride, 270 kgs of anhydrous piperazine, 230 kgs of 2-chloroethanol, 400 kgs of caustic soda, 270 kgs of triethyleamine, 5000 kgs of toluene, 400 kgs of DCM, 210 kgs of MCA, 3000 kgs of DMF, 120 kgs of potassium hydroxide, 6000 kgs of acetone, 1000 kgs of hydrochloric acid and 6000 kgs of water.

8.13.2.4 Flow chart

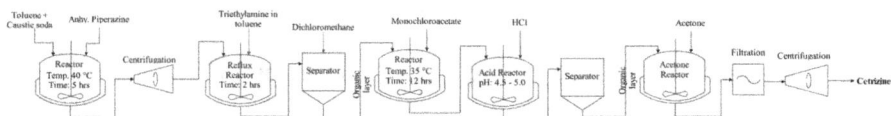

Figure 8.33 Flow sheet diagram of production of Cetirizine dihydrochloride

8.13.2.5 Process description

(A) Preparation of p-chlorobenzhydralpiperazine: First of all, dry the jacketed reactor having a stirring facility. Charge the toluene and caustic soda flask in it. Heat the mass upto 40°C. Add anhydrous piperazine slowly with stirring. Maintain the reaction temperature of 40°C for 5 hours. Cool the reaction mass and centrifuge to collect wet 4-chloro benzhydral piperazine.

(B) Preparation of 2-{4-[(4-chlorophenyl)(phenyl)methyl]piperazin-1-yl} ethanol: Wet 4-chloro benzhydral piperazine and water are charged into a jacketed stainless steel reactor havinga stirring facility. Triethylamine in toluene is also added into it. Reflux the reaction mixture with stirring for 2 hours. Cool the reaction mass at room temperature. Distill out toluene and reuse it. Further, cool to room temperature. For layer separation, charge dichloromethane and stir for 30 minutes. Settle out for 2 hours to separate water and the organic layer. Collect the organic product layer after discharging the water layer.

p-Chlorobenzhydryl chloride is reacted with anhydrous piperazine in the presence of toluene and caustic soda flask at 40°C for 5 hours to give 4-chloro benzhydralpiperazine. This derivative and 2-chloroethanol are reacted in the presence of triethylamine in toluene for 2 hours to give 2-{4-[(4-chlorophenyl) (phenyl)methyl]piperazin-1-yl}ethanol, which is reacted with sodium monochloroacetate in the presence of dimethylformamide and potassium hydroxide to give cetirizine base. This cetirizine base further reacts with hydrochloric acid and acetone to yield cetirizine dichloride as the final product.

(C) Preparation of cetirizine dichloride: Charge the organic layer containing 2-{4-[(4-chlorophenyl)(phenyl)methyl]piperazin-1-yl}ethanol into a jacketed stainless steel reactor with stirring. Charge the monochloroacetate intothe dimethylformamide and potassium hydroxide solution. The reaction mass is stirred for 10–12 hours at a temperature of 33–35°C. After the completion of the reaction, adjust the pH from 4.5 to 5.0 using hydrochloric acid. Stir, settle and separate the two layers. Collect the organic layer and discard the aqueous layer for effluent treatment. Charge the Acetone and filterit througha sparkler filter to remove impurities fromthe filtrate. Now, charge the filtrate into a glass line reactor. Purge it with hydrochloric acid gas and adjust the pH from 1.0 to 1.2. Centrifuge it to yield the final product, cetirizine dichloride. Acetone is recovered from the mother liquor and reused.

8.13.3 Uses

Various brand names for it, including Alatrol, Alerid, Alzene, Cetirin, Cetzine, Histazine, Humex, Letizen, Reactine, Razene, Triz, Zetop, Zirtec, Zirtek, Zodac, Zyllergy, Zynor, Zyrlek, and Zyrtec, are used as antihistamines in diverse contexts. It is used to treat allergy symptoms like itchy, watery eyes and running noses, sneezing, and hives. Its action mechanism is based on selective antagonism of the histamine H1 receptor (Reiter et al., 2012).

8.14 Ciprofloxacin

Ciprofloxacin was introduced in 1987. Like Proguanil, it is also recommended by WHO as the most efficient and protected drug desirable in a health system. It is available as a hydrochloride salt.

8.14.1 Properties

Its main properties are tabulated as per Table 8.13.

Table 8.13 Physical and Chemical Properties of Ciprofloxacin

Sr. No.	Properties	Particular
1	IUPAC Name	1-cyclopropyl-6-fluoro-4-oxo-7-piperazin-1-ylquinoline-3-carboxylic acid
2	Chemical and other names	Ciprofloxacin, Ciprinol, Ciprobay , Ciproxan, Cipromycin
3	Molecular formula	$C_{17}H_{18}FN_3O_3$
4	Molecular weight	331.3 gm/mole
5	Physical description	Faint to light yellow crystalline powder
6	Solubility	Practically insoluble in water, very slightly soluble in dehydrated alcohol and in dichloromethane, and soluble in dilute acetic acid
7	Melting point	255°C
8	Boiling point	590 °C
9	Flash point	298 °C
10	Density	1.50 g/cm²

8.14.2 Manufacturing process

The starting material for manufacturing ciprofloxacin is 2,4-dichloro-5-fluoro acetophenone.

8.14.2.1 Raw materials

2,4-Dichloro-5-fluoro acetophenone, Dimethyl carbonate (DMC), Toluene, Sodium hydride, Dimethyl sulfate, Dimethylformamide, Cyclopropyl amine, Acetic acid, Sodium hydroxide, Hydrochloric acid, Ammonium hydroxide, Methanol, Piperazine, Activated carbon, Hyflo and EDTA

8.14.2.2 Chemical reactions

First, 2,4-Dichloro-5-fluoro acetophenone, dimethyl carbonate, sodium hydride, dimethyl sulfate, dimethylformamide, cyclopropyl amine and acetic acid in the presence of toulene (solvent) are reacted to give Methyl-3-cyclopropylamine-2-(2,4-Dichloro 6-fluoro benzoyl acrylate). This acylate is further cyclized with sodium hydroxide and hydrochloric acid to give Q acid. This acid is reacted with piperazine in the presence of methanol and ammonium hydroxide solvent to give

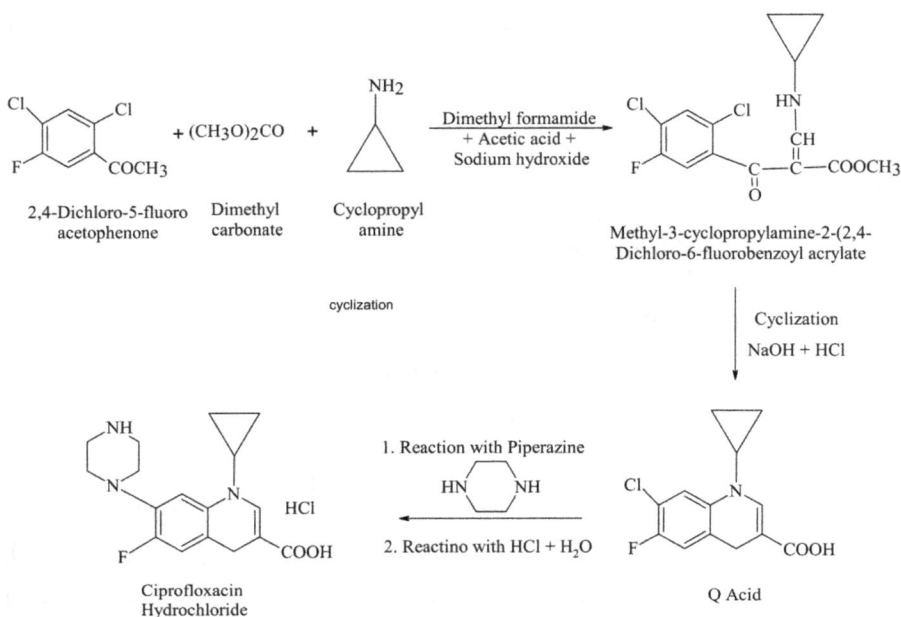

Figure 8.34 Synthesis of Ciprofloxacin hydrochloride

ciprofloxacin base. It is reacted with hydrochloric acid and water to yield the final product, ciprofloxacin hydrochloride.

8.14.2.3 Quantitative requirements

For manufacturing 1000 kgs ciprofloxacin, we require 761 kgs of 2,4-dichloro-5-fluoro acetophenone, 645 kgs of dimethyl carbonate, 300 kgs of sodium hydride, 632 kgs of dimethyl sulfate, 366 kgs of dimethyl formamide, 210 kgs of cyclopropyl amine, 436 kgs of acetic acid, 7100 kgs of toluene, 354 kgs of sodium hydroxide, 1400 kgs of hydrochloric acid (35%), 500 kgs of ammonium hydroxide (25%), 3800 kgs of methanol, 65 kgs of piperazine, 32 kgs of activated carbon, 10 kgs of hyflo and 2 kgs of EDTA.

8.14.2.4 Flow chart

8.14.2.5 Process description

(A) Preparation of Acrylate: First, a stainless steel reactor facilitated with a jacket and stirrer is heated with steam to remove the moisture content upto 0.1%. Then, toluene is charged into it. Nitrogen gas is then purged into it to maintain an inert atmosphere during the process. Now, sodium hydride and dimethyl carbonate are added. This mixture is heated upto 55°C using steam in the reactor jacket. Add 4-dichloro-5-fluoro acetophenone into the toluene slowly for a period of 4 hrs at 60–65°C. Maintain the temperature at 60–65°C for 4 hours. Cool the reaction down to 15–20°C using cooling water. Acetic Acid and dimethylformamide are charged within 2 hours at

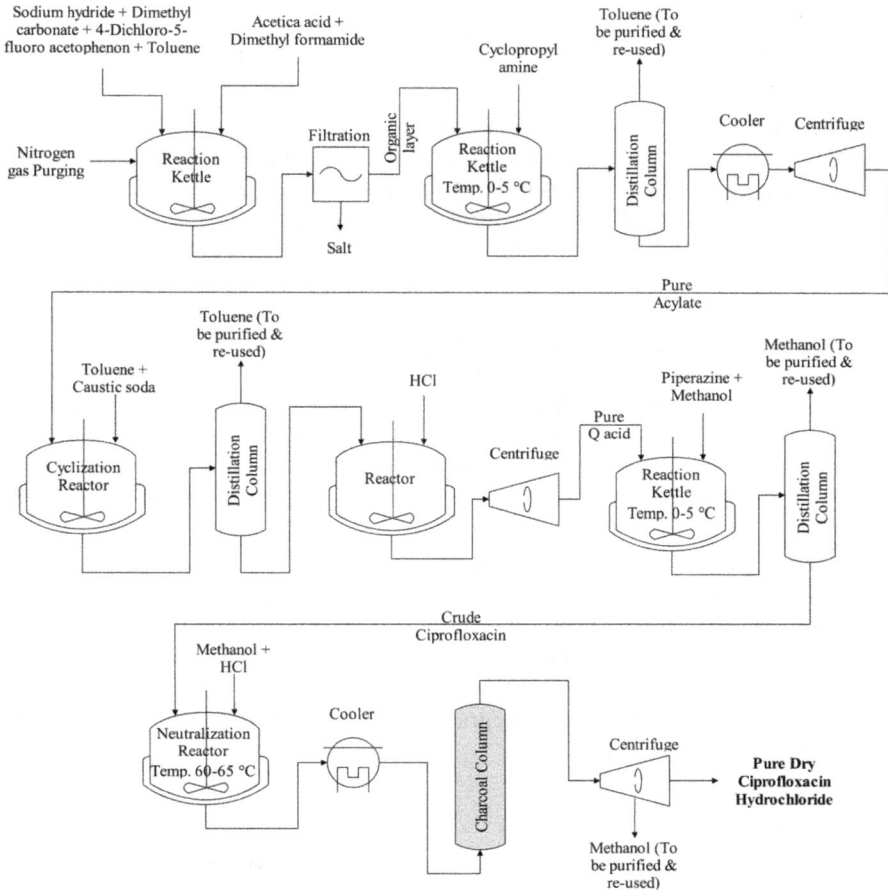

Figure 8.35 Flow sheet diagram of production of Ciprofloxacin hydrochloride

15–20°C. Maintain the temperature of 20°C for 30 minutes. The reaction mass is filtered to separate salt & the organic layer. Now, the organic layer is cooled to 0–5°C. Charge the cyclopropyl amine slowly into the reactor by keeping the temperature at 0–5°C. Maintain the temperature at 0–5°C for 4 hours. After completion of the reaction, the reaction mass is collected from the bottom of the reactor and sent to the distillation unit. Here, toluene is distilled out and reused. Then, the reaction mass is cooled to 5°C. Then it is centrifuged to get solid acylate. Wash the solid acylate with chilled toluene.

(B) **Preparation of Q-acid:** Solid wet acylate and toluene are charged into another jacketed stainless steel reactor having a stirring facility. Heat the mixture to the refluxing temperature and remove the water azeotropically. Now, caustic soda flask is added slowly with stirring. The reaction mass is stirred for 2 hours. After completion of the reaction, the reaction mass is collected from the bottom of the reactor and sent to a distillation unit.

Here, toluene is distilled out and reused. Charge caustic soda lye solution with stirring and heat the mass upto 85–95°C. Maintain the temperature of 85–95°C for 4 hours. Thereafter, add hydrochloric acid to adjust the pH from 3.0to 3.5. Centrifuge the mass to collect Q-acid.

(C) Preparation of ciprofloxacin base: Initially, the solvent, methanol is taken into a jacketed stainless steel reactor having a stirring facility. Add piperazine and Q-acid into it. Heat the mass upto 120–125°C. The reaction mass is maintained at 125–130°C for 15 hours. After the completion of the reaction, the reaction mass is collected from the bottom and sent to a distillation unit. Here, methanol is distilled out and reused. The pH was kept neutral (6.9 to 7.2) by using hydrochloric acid and centrifuging the material and washing with hot water. Crude ciprofloxacin and water are charged into a purification tank. Here, adjust the pH from 4.2 to 4.5 using acetic acid. Add activated charcoal, EDTA and hyflo with stirring. Stir the reaction mass for 30 minutes. The reaction mass is subjected to filtration and its pH is adjustd in the neutral range (6.9 to 7.2) by using ammonia solution at 55 to 60°C. Then the mass is centrifuged at 55–60°C and washed with hot water to get pure ciprofloxacin base.

(D) Preparation of ciprofloxacin hydrochloride: Pure ciprofloxin base and methanol are charged into a jacketed reactor at room temperature. Then the pH is adjusted from 2.0 to 2.5 by using hydrochloric acid. Heat the reaction mass to 60–65°C. Maintain the temperature at 60–65°C for 3 hours. The mass is cooled to 10–15°C. The mass is centrifuged to separate methanol and the technically acceptable grade of ciprofloxacin hydrochloride. Here, methanol is recycled.

8.14.3 Uses

It is available in different trade names such as Ciloxan, Neofloxin, Cipro, Cipro XR, and Cetraxal and; administrated through oral, intravenous, intratympanic, ophthalmic, and otic pathways. A variety of bacterial infections, including joint, skin, respiratory, and urine infections are treated using antibiotic ciprofloxacin. It is used to stop the growth of bacteria (Armstrong et al., 2021; Bhattacharya and Prajapati, 2017).

8.15 Pregabalin

This drug is developed by Parke-Davis, as a successor to gabapentin and available under the brand name, Pfizer. Pregabalin is a gamma-Amino butyric acid analog-anticonvulsant and analgesic used for neuropathic pain.

8.15.1 Properties

Its main properties are tabulated below.

Table 8.14 Physical and Chemical Properties of Pregabalin

Sr. No.	Properties	Particular
1	IUPAC Name	(3S)-3-(aminomethyl)-5-methylhexanoic acid
2	Chemical and other names	(S)-3-(Aminomethyl)-5-methylhexanoic acid. Pregabalin, Lyrica, 3-isobutyl GABA
3	Molecular formula	$C_8H_{17}NO_2$
4	Molecular weight	159.2 gm/mole
5	Physical description	White to off-white crystalline solid
6	Solubility	Freely soluble in water and both basic and acidic solutions
7	Melting point	194°C
8	Boiling point	298°C
9	Flash point	118°C
10	Density	1.01 g/cm²

8.15.2 Manufacturing process

The starting material for manufacturing pregabalin is 1 3-(carbamoylmethyl)-5-methylhexanoic acid.

8.15.2.1 Raw materials

1,3-(Carbamoylmethyl)-5-methylhexanoic acid, 1-Phenyl ethyl amine (PEA), Chloroform, Methanol, Water, Sodium hydroxide, Concentrated Hydrochloric acid, Sulfuric acid, Chlorine gas, n-Butanol, Toulene and p-Toulene sulfonic acid

8.15.2.2 Chemical reactions

As per Fig. 8.30, isovaleraldehyde and cyanoacetamide undego condensation at 100–110°C for 12 hours in the presence of water and piperidine, followed by heating with concentrated hydrochloric acid and toluene for 2 hours to give Isobutyl glutaric acid. This acid reacts with urea in the presence of sodium hydroxide at 145–150°C for 5 hours to yield 3-(carbamoylmethyl)-5-methyl hexanoic acid. This acid is converted into (R)-3-(carbamoylmethyl)-5-methyl hexanoic acid by heating it in the presence of chloroform and methanol at 50–55°C, followed by treatment with sodium hydroxide solution in water and concentrated hydrochloric acid. Now, this product further reacts with sodium hypochloride and chlorine gas in the presence of sodium hydroxide to give crude (3S)-3-(aminomethyl)-5-methylhexanoic acid (Pregabalin). The unwanted isomer is removed by treating it with sulfuric acid. Finally, it is purified with n-butanol and water.

8.15.2.3 Quantitative requirements

For manufacturing 1000 kgs of pregabalin, we require 750 kgs of isovaleraldehyde, 980 kgs of cyanoacetamide, 500 kgs of piperidine, 710 kgs of 1-phenyl ethyl amine, 27900 kgs of chloroform, 350 kgs of methanol, 39000 kgs of water, 4300 kgs of sodium hydroxide, 3700 kgs of concentrated hydrochloric acid, 3500 kgs of sulfuric acid, 450 kgs of chlorine gas, 5750 kgs of n-butanol, 4700 kgs of toluene and 100 kgs of p-toulene sulfonic acid.

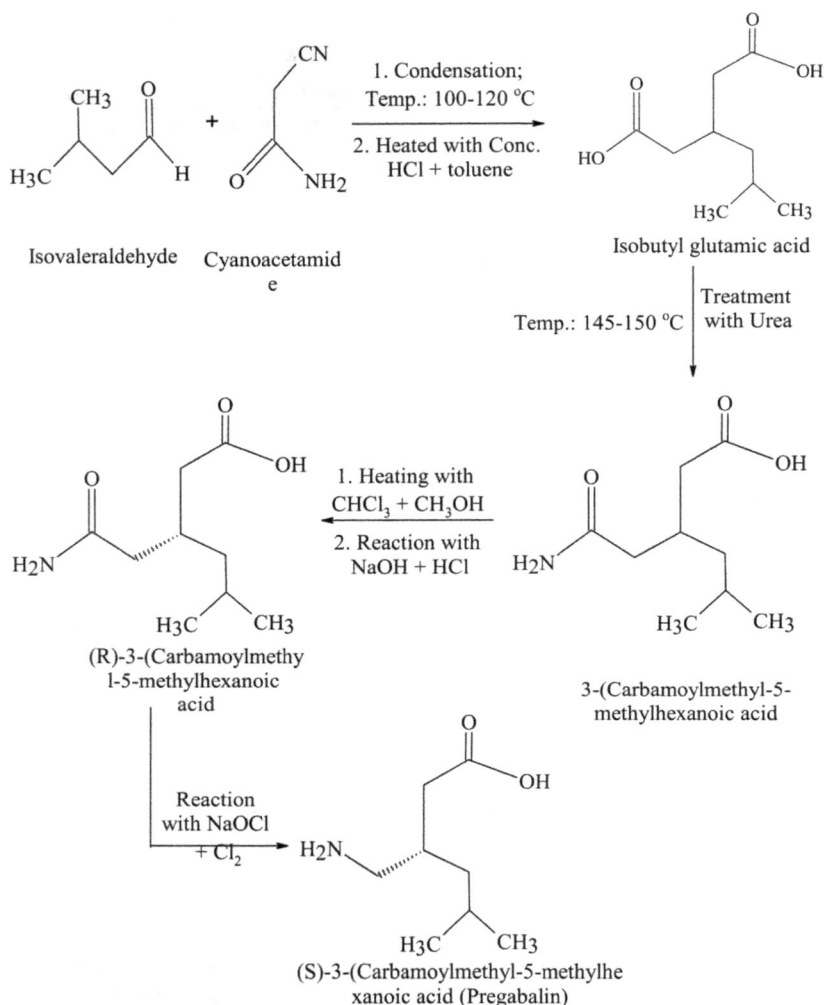

Figure 8.36 Synthesis of Pregabalin

8.15.2.4 Flow chart

8.15.2.5 Process description

It is divided into the following steps.

Step 1: Preparation of (3R)-3-(carbamoylmethyl)-5-methyl hexanoic acid phenyl ethyl amine salt: As per the flow-chart, first water and piperidine are charged into a stainless steel jacketed reactor having a stirring facility. The mixture is heated upto 100–110°C with stirring. Now, isovaleraldehyde and cyanoacetamide are charged slowly into the reactor with stirring. The reaction mixture is refluxed at 100–110°C for 12 hours. After completion of the reaction, the mass is collected from the reactor and fed into another jacketed reactor having a stirring

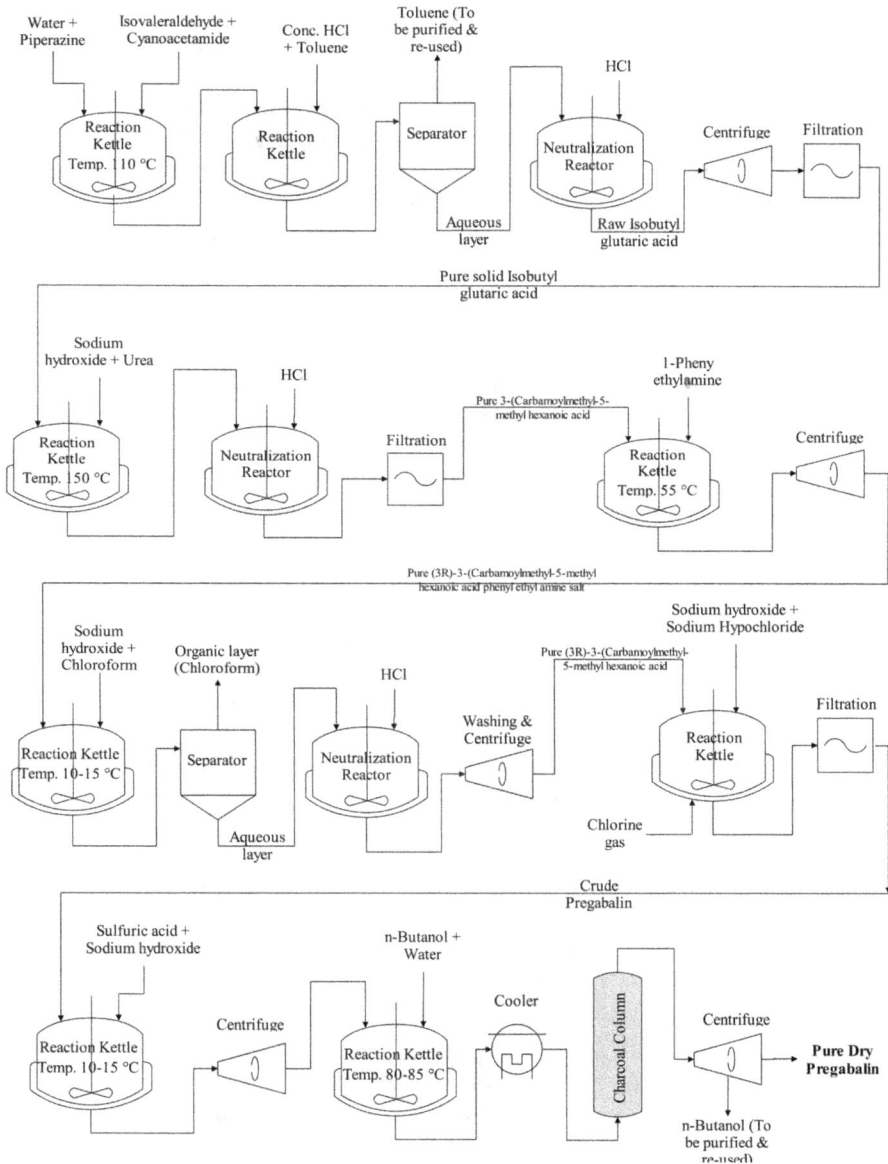

Figure 8.37 Flow sheet diagram of production of Pregabalin

facility. Concentrated HCl and toluene are charged with heating and stirring. The mixture is refluxed for 2 hours. After completion of the reaction, the mass is moved into a separator, where the top layer, which contains the organic layer, is separated from the watery layer. In order to obtain raw isobutyl glutaric acid, this aqueous layer is neutralized with HCl. It is then centrifuged and filtered to get pure solid isobutyl glutaric acid. Now, sodium hydroxide solution is taken in another stainless steel jacketed reactor having a stirring facility. Also, urea and solid isobutyl glutaric acid are charged into the reactor. The reaction mass is heated at 145–150°C. A temperature of 145–150°C is maintained for 5 hours to get 3-(carbamoylmethyl)-5-methyl hexanoic acid. The reaction mass is transferred into neutralization, where 3-(carbamoylmethyl)-5-methyl hexanoic acid is precipitated using HCl and collected by filtration. Now, chloroform and methanol are charged into a jacketed stainless steel reactor having a stirring facility. 3-(carbamoymethyl)-5-methyl hexanoic acid is also added to it. The temperature of the reaction mass is raised upto 50–55°C using hot water circulation in the reactor jacket. Now, 1-phenyl ethylamine is also added to the above slurry slowly at 50–55°C. A temperature of 50–55°C is maintained at 30 minutes with stirring. After the completion of the reaction, the seeding agent (3R)-(–)-3-(carbamoylmethyl)-5-methyl-hexanoic acid is added into the reaction mass. Collect the reaction mass from the bottom of the reactor, cooled down to room temperature and centrifuge the material [(3R)-3-(carbamoylmethyl)-5-methyl hexanoic acid phenyl ethyl amine salt].

Step 2: Preparation of (3R)-3-(carbamoylmethyl)-5-methyl hexanoic acid: Add water and wet cake of (3R)-3-(carbamoylmethyl)-5-methylhexanoic acid phenyl ethyl amine salt into a jacketed stainless steel reactor having a stirring facility. Cool the reaction mass down to 10–15°C using cooling water in the jacket. Adjust the mass pH to 10.0–11.0 using sodium hydroxide solution. Maintain the temperature for 30 minutes with stirring. Charge chloroform into the reactor. Maintain the reaction for 30 min with stirring. Allow the mass to settle for 30 min and separate the upper chloroform layer. Now, take the aqueous layer from the bottom and slowly adjust the pH to 2.0–3.0 using concentrated hydrochloric acid. Collect the reaction mass from the bottom of the reactor and wash the material with water. Centrifuge the reaction mass to get pure solid (3R)-3-(carbamoylmethyl)-5-methyl hexanoic acid.

Step 3: Preparation of (3S)-3-(aminomethyl)-5-methylhexanoic acid (crude pregabalin): Take sodium hydroxide solution into a jacketed stainless steel reactor having a stirring facility. Cool the solution down to 0–5°C. Chlorine gas is purged from the bottom of reactor. After addition of the gas, (3R)-3-(carbamoylmethyl)-5-methylhexanoic acid is added into reactor. A temperature of 0–5°C is maintained for 30 minutes. Now, sodium hypochlorite solution is fed slowly at a temperature of 0–15°C. The reaction mass is stirred for 2 hours at 15–20°C. Then, the temperature is slowly increased upto 50–55°C and maintained for 45 minutes. Cool the mass and filter it through a sparkler filter to collect the raw crude Pregabalin.

Step 5: Purification of pregabalin: Crude pregabalin is taken into a mixing tank. Adjust the reaction mass pH to 3.5 by adding 50% sulfuric acid solution at 30–35°C. Adjust the reaction mass pH from 7.0 to 7.5 using

50% NaOH solution at 25–25°C. Maintain the mass at 20–25°C for 1 hour. This material is collected from the bottom of the mixing tank and centrifuged. Insert n-butanol and crude pregabalin into a jacketed reactor, facilitated with stirring. Add water in it. Heat the reaction mass to 80–85°C. Maintain the temperature at 80–85°C for 30 minutes. Cool the mass slowly to 8–12°C. Pass the reaction mass into a charcoal column. Finally, centrifuge the pure pregabalin.

8.15.3 Uses

It is available with different trade names like LyricaAvertzGabalin, Gabamax, Gb-Lin, Maxgalin, Nova, Nova Plus, Nurlin, Prebaxe, Pregalin M, Pregastar M, Regab-50 and Zylin. It is used to treat symptoms of pain created by nervous damage due to diabetes or shingles. Injuries to the spinal cord that produce nerve pain are also treated with it (Abdel-Magid and Caron, 2206).

References

Abdel-Magid, A.F. and Caron, S. 2006. Fundamentals of Early Clinical Drug Development: From Synthesis Design to Formulation, Willey Publication.

Anisimova, E. and Yarullina, D. 2018. Characterization of Erythromycin and Tetracycline Resistance in Lactobacillus fermentum Strains. International Journal of Microbiology (3–4): 1–9. doi: 10.1155/2018/3912326

Armstrong, C., Miyai, Y., Formosa, A. and Thomas, D. 2021. On-Demand Continuous Manufacturing of Ciprofloxacin in Portable Plug-and-Play Factories: Development of a Highly Efficient Synthesis for Ciprofloxacin. Organic Process Research & Development 25(7): 1524–1533. https://doi.org/10.1021/acs.oprd.1c00118

Bhattacharya, S.A. and Prajapati, B.G. 2017. Formulation, design and development of ciprofloxacin hydrochloride floating bioadhesive tablets. e-Journal of Surface Science and Nanotechnology.

Carmichael, D. and Petrides, D. 2020. Penicillin V Production via Fermentation—Process Modeling and Techno-Economic Assessment (TEA) with SuperPro Designer.

Chopra, M., Kaur, P., Bernela, M. and Thakur, R. 2012. Synthesis And Optimization of Streptomycin Loaded Chitosan-Alginate Nanoparticles. International Journal of Scientific & Technology Research 1(10): 31–34

Disouza, J.I., Rustomjee, M. and Patravale, V.B. 2016. Pharmaceutical Product Development Insights Into Pharmaceutical Processes. Management and Regulatory Affairs, CRC Press. https://doi.org/10.1201/b19579

Intratec, 2021. Paracetamol Production From Nitrobenzene Report, Intratec Solutions LLC.

Kandeh, A.M., Adjivon, A.F. and Iyekowa, O. 2020. Biological Activities and Industrial Production of Aspirin (Acetylsalicylic Acid), Global Scientific Journal 8(10).

Kuhlmann, M. and Schmidt, A. 2014. Production and manufacturing of biosimilarinsulins: implications for patients, physicians, and health care systems. Biosimilars, 45.https://doi.org/10.2147/BS.S36043

McEvoy, G.K., Jane, M. and Kathy, L. 2004. AHFS Drug Information, American Society Health-System Pharmacists, 2004.

Minas, W. 2008. Production of Erythromycin with Saccharopolysporaerythraea, In: Methods in Biotechnology. Microbial Processes and Products 18. DOI: 10.1385/1-59259-847-1:065

Niazi, S.K. 2004. Handbook of Pharmaceutical Manufacturing Formulations, Liquid Products, 3.

Patel, R.P., Patel, K.P., Modi, K.A. and Pathak, C.J. 2013. Formulation, development and evaluation of injectable formulation of Aspirin, Drugs and Therapy Studies 3(1): 2. https://doi.org/10.4081/dts.2013.e2

Rainsford, K.D. 2004. Aspirin and Related Drugs, CRC Press.

Reiter, J., Trinka, P., Bartha, F.L. and Pongo, L. 2012. New Manufacturing Procedure of Cetirizine. Organic Process Research & Development 16(7): 1279–1282. https://doi.org/10.1021/op300009y

Saltiel, A.R. and Pessin, J.E. 2007. Mechanisms of insulin action, Landes Bioscience. Springer Science+Business Media.

Smith, H.J. 2004. Hywel Williams, Smith and Williams' Introduction to the Principles of Drug Design and Action, CRC Press. https://doi.org/10.1201/9780203304150

Tripathi, K.D. 2014. Pharmacological Classification of Drugs with Doses and Preparations, Jaypee Brothers Medical Publishers.

Wettermark, B., Elseviers, M.M., Almarsdottir, A.B. and Andersen, M. 2016. Introduction to drug utilization research. In book: Drug Utilization Research 1–12.

CHAPTER 9

Fermentation Industries

9.1 Introduction

Historically, the term "fermentation" (Latin for "to boil") referred to the degradation of food, which is frequently accompanied by the development of gas. One of the earliest instances is the fermentation of carbohydrates by yeast into alcohols and carbon dioxide. The word "fermentation" is now used to describe changes caused by microbes. Gas evolutions are not a crucial factor. We observe several intricate chemical processes in our daily lives that are induced by the activities of living things. Examples include the souring and curdling of milk, the dissolution of meals, the production of indigo dye from the compound indicated in food, the curing of tobacco, the creation of benzaldehyde or oil of bitter almonds from the amygdaline found in almond seeds and the fermentation of fruit juice into wines. All these processes, together are referred to as fermentation processes, involving the breakdown of complex organic materials into simpler components and the production of the method-appropriate enzymes by living organisms. Weizmann invented a fermentation method to turn grain into acetone and n-butanol during World War I. Between 1920 and 1940, citric and gluconic acids were formerly successfully created, but today, they are prepared economically by the petrochemical sector. The development of antibiotics like penicillin during World War II paved the way for major breakthroughs in microbiological process control technology that are still in use today. Numerous fermentation techniques compete directly with pure chemical synthesis. The production of ethanol, acetone, butanol, and other alcoholic beverages through fermentation has mostly been replaced by synthetic alternatives. Chemical transformations occur during fermentation under controlled conditions. Some crucial procedures include oxidation (turning alcohol into acetic acid, sucrose into citric acid, and dextrose into gluconic acid), reduction (turning aldehyde to alcohol and sulfur to sulfurdioxide), hydrolysis (turning starch and sucrose into glucose and fructose), and esterification (Hexose phosphate from hexose & phosphoric acid).

Primitive people discovered how to make fermented drinks, such as beer, wine and liquors, and have practiced this skill for thousands of years. Malt beverages, fermented wines, and distilled liquors make up the three categories of alcoholic beverages. Wine is created by the activity of yeast on fruit sugar, while distilled liquors are fermented liquors that are then distilled to increase the alcohol concentration. Beer requires malted (germinated) grain to make the carbs ferment able, while wine is created by the action of yeast on grape sugar. The primary raw ingredients are the carbohydrates found in grains and fruits. The variety of grains and fruits used varies from one country or beverage to another. Russia ferments potatoes to create vodka by distillation (type of wine). The cereals, including corn, barley, rice, and grapes are raw ingredients for fermentation.

There are mainly three types of fermentation processes: (1) Batch (2) Fed-batch and (3) Continuous. Batch and fed-batch processes are quite common, but, continuous operations are rarely used. In a batch process, a batch of the culture medium in a fermenter is inoculated with a microorganism. The fermentation process occurs for a certain duration (is called "fermentation time") and the product is harvested. The typical fermentation time is extends upto 4–5 days. In fed-batch fermentations, a sterile culture medium is added either continuously or periodically to the inoculated fermentation batch. The volume of the fermenting broth increases with each addition of the medium, and it is achieved in the batch time duration. In continuous fermentations, the sterile medium is charged continuously into a fermenter. The fermented product is continuously withdrawn so that the volume of the fermenter remains constant. After attaining a certain microbial population concentration by adding batch cultures and feeding, the continuous fermentations begin. In some continuous fermentations, a small part of the received culture may be recycled, to continuously inoculate the sterile feed medium entering the fermenter. Industrial fermentation is almost a batch process, in which, the culture is first grown in one vessel. The growth is transferred to another vessel containing the nutrient media so that a teeming population of the desired microbe is obtained. The cultivation of microbes is performed by shaking the flask containing them on a shaker. The growth from the vessel is then transferred to a seed fermenter containing the seed medium. The cell or spore suspension which is transferred is termed inoculum and the operation is called inoculation. Once the seed is ready, it is transferred to the production fermenter, a big vessel, containing the production media (or raw materials to be fermented). The vessel is provided with an agitator for mixing the contents. The temperature is kept steady by the circulation of hot and cold water through the vessel coils; in addition the vessel has inlets for air, inoculation and introduction of feeds. Valves for sampling and withdrawal of fermented broth are also fitted in it. During fermentation, various parameters such as pH (a measure of acidity or alkalinity), oxygen level, rate of air flow, and vessel pressure are controlled. Thus, individual sockets for mounting the appropriate probes are also provided. After the fermentation is over the products are isolated. Microbes, which play an important part in the fermentation process, include various types of bacteria, yeast and molds. These microbes depend on the raw materials to be fermented. Raw materials to be fermented may be either sugary materials (molasses, whey, glucose, sucrose), or starchy materials (wheat, rice, maize, potato, cassava, corn) or cellulosic material (wood, agricultural wastes). The nutrition of microorganisms depends on carbon and nitrogen sources, minerals, oxygen, vitamins and other necessary growth factors. A combination of these materials constitutes their growth media. Sugars such as glucose, lactose and sucrose are common carbon sources. Cheaper carbon sources include starch, corn steep liquor, molasses. Soyabean meal, cotton seed meal and peanut meal serve as organic nitrogen sources. Inorganic nitrogen sources are ammonium salts or nitrates. In general, the mineral requirement is met by salts of sodium, potassium, magnesium, phosphates and chlorides (Cinar et al., 2003; Vogel and Haber, 2007; Mcneil, 2008; Chen, 2013).

Some chemicals are prepared by industrial fermentation processes, which are as follows.

9.2 Industrial Alcohol/Ethyl alcohol (Ethanol – Purity – 95.6%)

Chemically, it is 95% ethyl alcohol and 5% water. Industrial alcohol, known as denatured alcohol, is an outgrowth of alcoholic beverages.

9.2.1 Properties

Table 9.1 Physical and Chemical Properties of Industrial alcohol (Ethanol)

Sr. No.	Properties	Particular
1	IUPAC Name	Ethanol
2	Chemical and other names	Ethyl alcohol, Methylcarbinol, Grain alcohol, Ethyl hydroxide
3	Molecular formula	C_2H_6O
4	Molecular weight	46.1 gm/mole
5	Physical description	Clear colorless liquid with a characteristic vinous odor and pungent taste
6	Solubility	Miscible with ethyl ether, acetone, chloroform; soluble in benzene
7	Melting point	−112°C
8	Boiling point	78.2°C
9	Flash point	16.6°C
10	Density	0.791 g/ml at 20°C

Its main properties are as follows.
Industrial alcohol is manufactured by the following processes.

1. Fermentation process
   (A) From sucrose substrate
   (B) From waste sulfite substrate of paper mills
   (C) From starch substrate

2. Petroleum process
   (A) Catalytic hydration of ethylene
   (B) Esterification and hydrolysis of ethylene

(C) Oxidation of petroleum

Chemically, it is produced by hydrating ethylene (C_2H_4), although in underdeveloped nations, microbial fermentation using readily available, inexpensive staples is favored. Molasses, whey, glucose, sucrose, and other sugary materials, as well as starchy materials (wheat, rice, maize, potato, cassava, and corn) and cellulose materials, are the basic ingredients used in the fermentation process (wood, agricultural wastes). The most promising raw material for the fermentation of alcohol, particularly gasohol, is thought to be corn. The United States generates the majority of the world's ethanol from corn, making it the country with the highest production of the fuel. Table 2 represents the total quantity of ethanol produced by different countries/regions and their respective raw materials.

Table 9.2 Production of Industrial alcohol from various country/region

Sr. No.	Country/Region	Raw materials	Total Production (Millions of U.S. liquid gallons per year - 2016)
1	USA	Corn	15,330
2	Brazil	Sugarcane	7,295
3	European Union	cereals and beet	1,377
4	China	Corn and sweet potato	845
5	Canada	Corn	436
6	Thailand	Cassava and molasses	322
7	Argentina	Corn	264
8	India	Molasses	225
9	Rest of World	-	26,094

9.2.2 Manufacturing process from corn:

9.2.2.1 Raw materials

Dehull and dehydrated corn, Water

9.2.2.2 Chemical reactions

Reaction for Monosaccharides Production:

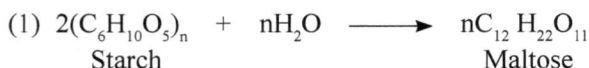

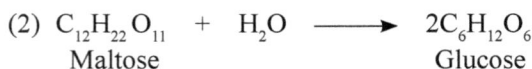

(1) $2(C_6H_{10}O_5)_n$ + nH_2O $\longrightarrow$ $nC_{12}H_{22}O_{11}$
      Starch   Maltose

(2) $C_{12}H_{22}O_{11}$ + H_2O $\longrightarrow$ $2C_6H_{12}O_6$
      Maltose   Glucose

Fermentation Reaction:

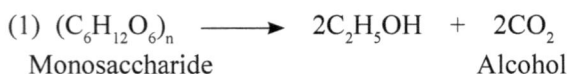

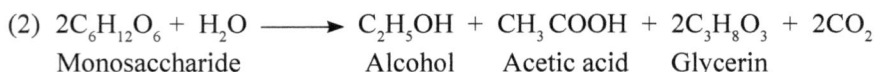

(1) $(C_6H_{12}O_6)_n$ $\longrightarrow$ $2C_2H_5OH$ + $2CO_2$
      Monosaccharide   Alcohol

(2) $2C_6H_{12}O_6 + H_2O$ $\longrightarrow$ C_2H_5OH + CH_3COOH + $2C_3H_8O_3$ + $2CO_2$
      Monosaccharide   Alcohol   Acetic acid   Glycerin

9.2.2.3 Quantitative requirements

For production of 1000 liters ethyl alcohol, we require 2500 kgs of corn.

9.2.2.4 Flow sheet diagram

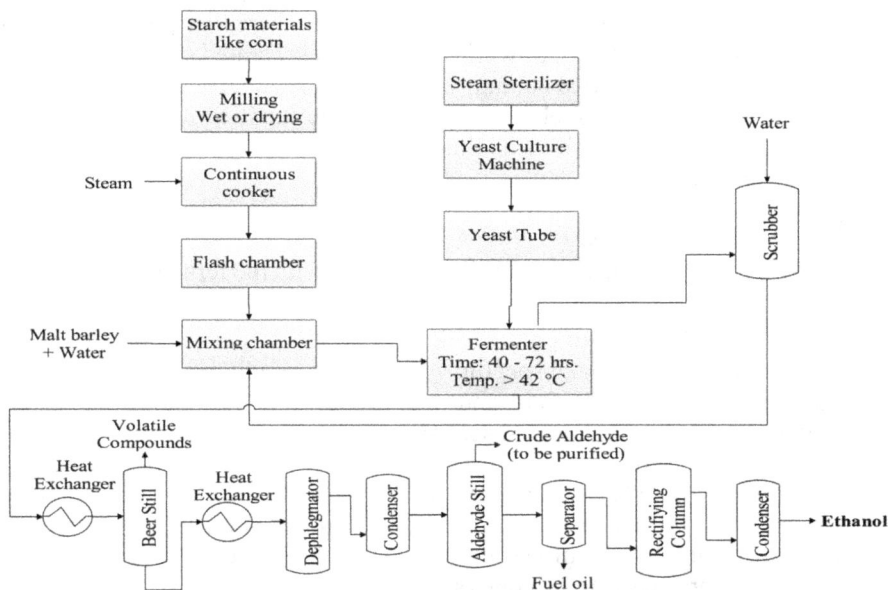

Figure 9.1 Flow sheet diagram of production of Ethanol from starchy materials

9.2.2.5 Process description

This process is conducted in three main steps:

(1) **Preparation of raw materials:** The flow sheet diagram in Fig. 9.1 shows the various operations involved in the production of alcohol, milled and corked under a pressure of 0.6 to 1.0 atmosphere and a high temperature upto 175°C by means of steam in a continuous cooker to gelatinize the ground grainso that the starch can be transformed into fermentable sugars by barley malt amylases. The mash is transferred into a flash chamber to bring the temperature down to 60°C in the flash chamber. The cooked grain is delivered to a mixer, where it is combined with water and malted barley. In the brief time spent in the converter, the starch is hydrolyzed to approximately 70% maltose and 30% dextrin. 5% percent of the total mixer is sterilized under pressure and cooled down. On another side, selected yeast is grown in the mixer. Thereafter the yeast is transferred to the fermenter.

(2) **Fermentation:** The mixer and yeast undergo fermentation reactions in the fermenter. Here, sulfuric acid is employed to adjust the pH from 4.8 to 5.0. As this process is exothermic, cooling water is circulated in the cooling coils to ensure that the maximum temperature does not exceed 32°C. The fermentation time is altered from 40 to 72 hrs. Carbon dioxide is liberated from the fermenter, which is absorbed by water in the scrubber.

(3) **Purification:** After the fermentation reaction, liquor called beer having 5.5 to 11% alcohol is pumped into the beer still through a heat exchanger. Heat exchanged removes proteins, residue sugars and vitamins, called stillage from beer, which is utilized as a constituent of animal feed. Beer still removed the volatile compounds. Further, beer passes through a heat exchanger, partial condenser, dephlegmator, and thereafter gets transferred into the aldehyde column, in which aldehyde is separated out. The liquor is transferred into a rectified column via a separator. The role of separator is to remove higher boiling fuel oil from liquor. The rectified column increases the concentration of alcohol and finally 95–96% of alcohol is collected from condensation.

9.2.2.5 Major engineering problems

The following engineering problems are associated with the fermentation of corn for the production of ethanol.

1. Collection and storage of corn.
2. Maintenance of sterile and specific yeast culture conditions.
3. Continuous sugary dilution and distillation require additional space, equipment and operating costs, so, batch processes are more preferable than continuous process.
4. The process is also associated with waste disposal problems. It is uneconomical to concentrate the cattle feed if the value of the biological oxygen demand (BOD) of wastewater increases. So, during waste disposal, trickling filters, activated sludge or anaerobic digestion must be conducted before discharging to water run-off.
5. For fuel economy in the series of distillations preheat exchangers are used.

9.2.3 Manufacturing process from molasses:

Molasses is created through the processing of sugar cane in sugar industries. Molasses contains 50 to 55 percent sucrose, which has the molecular formula $C_{12}H_{22}O_{11}$. This source compound is utilized to make ethyl alcohol. Molasses can be used to produce ethanol in the forms of absolute and rectified spirits.

9.2.3.1 Raw materials

Molasses, Sulfuric acid and Ammonium sulfate

9.2.3.2 Chemical reactions

Fermentation reactions:

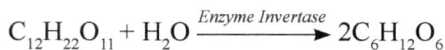

$$C_{12}H_{22}O_{11} + H_2O \xrightarrow{\text{Enzyme Invertase}} 2C_6H_{12}O_6$$

$$C_6H_{12}O_6 \xrightarrow{\text{Enzyme Zymase}} 2C_2H_5OH + 2CO_2; \Delta H = -130.5kJ$$

Side reaction:

$$2C_6H_{12}O_6 + H_2O \longrightarrow ROH + RCHO$$

9.2.3.3 Quantitative requirements

For the production of 1 tonne of Ethanol, we required 5.6 tons of molasses, 27 kgs of sulfuric acid and 2.5 kgs of ammonium sulfate.

9.2.3.4 Flow sheet diagram

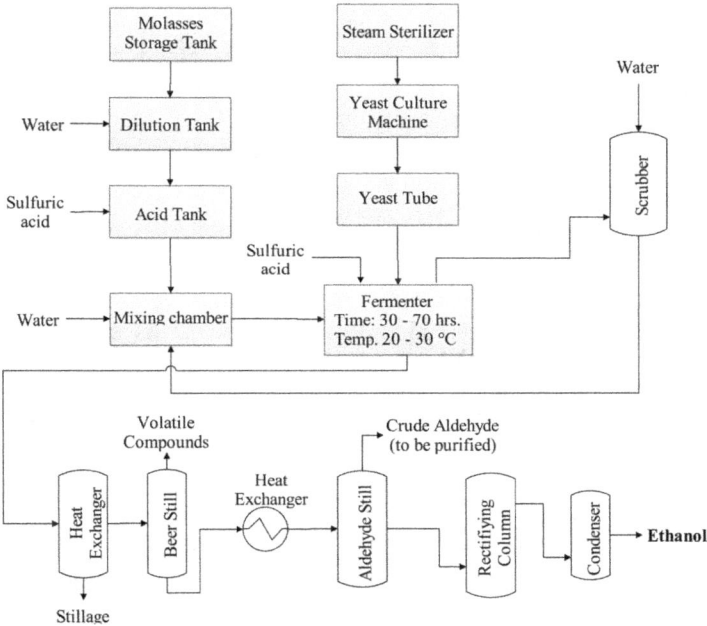

Figure 9.2 Flow sheet diagram of production of Ethanol from molasses

9.2.3.5 Process description

This process is conducted in three main steps:

(1) Preparation of raw materials: As per Fig. 9.2, molasses is stored in a storage tank and it is transferred into a dilution tank, where the sugar is diluted with water to get a concentration range of 10–15%. With the use of sulfuric acid, the pH is kept between 4 and 5. On the opposite side, to create a yeast culture, yeast is sterilized under pressure, cooled, and transported to a tank. Ammonium and magnesium phosphate or sulfate nutrition is offered. The yeast benefits from the acidic environment by producing the catalytic enzymes invertase and zymase. So, the pH is maintained at about 4–5 using sulfuric acid.

(2) Fermentation: The diluted molasses and yeast undergo a fermentation reaction in the fermenter. Here, sulfuric acid is employed to adjust the pH from 4.8 to 5.0. As this process is exothermic, cooling water is circulated in the cooling coils to ensure that the maximum temperature does not exceed 32°C. The fermentation time is altered from 30 to 70 hrs at 20–30°C. Carbon dioxide is liberated from the fermenter, which is absorbed by water in the scrubber. The water scrubber is used to scrub carbon dioxide with ethanol. The fermented product consists of 8–10% ethanol, water and aldehyde.

(3) **Purification:** After the fermentation reaction, liquor called beer having 8 to 10% alcohol is pumped into the beer still through the heat exchanger. The heat exchanged removes proteins, residue sugars and vitamins, called stillage from the beer, which is utilized as a constituent of animal feed. The beer still removes the volatile compounds. The beer stillage containing 50% ethanol and 50% aldehyde, is transferred to the aldehyde still. Aldehydes from the top, and a mixture of ethanol and water from the middle and bottom streams emerge from stills at various points along the column. To create a product called rectified spirit, which contains 95% ethanol, the middle stream is put into a rectification column.

9.2.4 Uses

Industrial alcohol, referred to as ethyl alcohol, is a development of alcoholic beverages, although it has gained prominence due to its economic benefits as a solvent, (first choice is water) and for synthesis of other chemicals like acetaldehydes, esters, ethanoic acid, ethylene dibromide, glycols and local anesthetics. In medicine, it is injected into neurons and ganglia as a painkiller and used as a solvent. It is crucial to a variety of industries like vinegar, liquid soaps, detergents, hand sanitizers, coatings (lacquers), printing inks, adhesives and liquid dish washing detergents. Now-a-days, it is used as a biofuel in engines. It is also used to prepare 99% concentrated ethanol. By distilling a 95% azeotrope with a third ingredient a minimum constant-boiling combination boiling at a temperature lower than that of 95% alcohol, is made. Due to its low toxicity and capacity to dissolve non-polar molecules, highly concentrated ethanol is frequently employed in both industry and science for synthetic chemical processes and solvents. Additionally, anhydrous ethanol is employed as an antiseptic to cleanse the skin prior to injections, frequently in conjunction with iodine (Roozbehani et al., 2013; Schwietzke et al., 2009; Lima et al., 2000).

9.3 Acetic Acid

The second-simplest carboxylic acid is acetic acid (after formic acid). It has a carboxyl group with a methyl group attached to it. When pure acetic acid is cooled, a solid resembling ice forms. So, it also goes by the name glacial acetic acid. Ethanoic acid makes up between 3 and 9% of the total volume of vinegar, making it the main ingredient besides water. Although concentrated ethanoic acid is caustic and may harm the skin, it is categorized as a weak acid since it only partially dissociates in solution.

9.3.1 Properties

Its main properties are as per Table 9.3.

Table 9.3 Physical and Chemical Properties of Acetic acid

Sr. No.	Properties	Particular
1	IUPAC Name	Acetic acid
2	Chemical and other names	Glacial Acetic Acid, Vinegar, Ethanoic acid, Ethylic acid, Acetasol, Methanecarboxylic acid
3	Molecular formula	$C_2H_4O_2$
4	Molecular weight	60.1 gm/mole
5	Physical description	Clear colorless liquid with a vinegar odor and pungent taste.
6	Solubility	Miscible with ethanol, ethyl ether, acetone, benzene; soluble in carbon tetrachloride, carbon disulfide
7	Melting point	16.6°C
8	Boiling point	117.9°C
9	Flash point	40°C
10	Density	1.05 g/ml at 20°C

Industrially it is manufactured by synthetic as well as bacterial fermentation. About 75% of ethanoic acid produced in the industry is formed by the carbonylation of methanol. Other chemical routes are the oxidation of acetaldehyde and ethylene respectively. Even though the fermentation technique contributes only to 10% of the global output, it is nonetheless crucial for the manufacturing of vinegar because many food purity rules demand that vinegar used in foods be biologically derived. The fermentation route consists of aerobic oxidation of acetaldehyde and ethanol and anaerobic fermentation of sucrose, in which aerobic oxidation of ethanol is most effective.

9.3.2 Manufacturing process by air-oxidation of ethanol

This process is important because now-a-days a batch process is used to prepare vinegar with 15% acetic acid content in 24 hrs, while a continuous process is used to prepare vinegar of 20% acetic acid content within 60 hrs. In this process, ethanol solution undergoes fermentation using *Bacterium aceti* with air, to yield acetic acid. But the disadvantage of this process is that sometimes over-oxidation of ethanol leads to the liberation of carbon dioxide which is undesirable. Also, it requires a number of days to produce purified acetic acid.

9.3.2.1 Raw materials
Ethanol and air

9.3.2.2 Chemical reactions

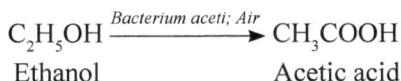

$$C_2H_5OH \xrightarrow{\textit{Bacterium aceti; Air}} CH_3COOH$$
$$\text{Ethanol} \qquad\qquad\qquad\qquad \text{Acetic acid}$$

9.3.2.3 Quantitative requirements
For production of 1 tonne of acetic acid, we require 7770 kgs of ethanol and 530 kgs of air (oxygen).

9.3.2.4 Flow sheet diagram

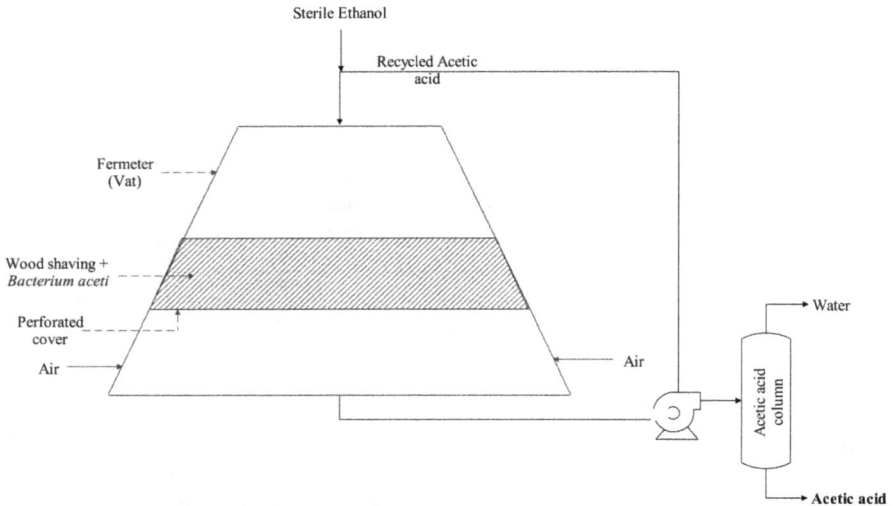

Figure 9.3 Flow sheet diagram of production of Acetic acid from Ethanol

9.3.2.5 Process description

First, an autoclave is used to sterilize the diluted alcohol solution before it is added to the fermenter. The fermenter is a large wooden container called vat, which is filled with wood shavings. These wood shavings are already impregnated with *Bacterium aceti*. Also, a perforated cover with holes for air insertion from the bottom is also arranged in the fermenter. Now, ethanol is introduced at the top of the fermenter and allowed to trickle down the shavings. Also, air (oxygen) flow is maintained at the bottom. Here, acetic acid and heat are formed in the fermenter. The generated heat helps to maintain the reaction temperature at 35°C, which is more preferable for the growth and activity of bacteria. The liquid leaving from the bottom is further introduced at the top of next fermenter by a pump. Nitrogen is the unreacted part of the air during fermentation. Air as an inexpensive oxidizing agent yielding acetic acid as the liquid product with a large amount Nitrogen after fermentation. So, after achieving the desirable concentration, the resulting stream undergoes scrubbing using cooled water to remove Nitrogen from the top. And, the bottom stream containing acetic acid and water is sent to the acetic acid column to separate water from acetic acid. Vinegar of 15% concentration and 70% acetic acid is manufactured in 24 hrs and 10 days respectively.

9.3.2.6 Major engineering problems

(A) Bacteria become dormant if the concentration of ethyl alcohol is greater than 15%.
(B) Regulation of air supply is essential.
(C) The alcohol flow is controlled in such a way that the temperature never rises above 35°C, which is the ideal temperature for bacterial development.

9.3.3 Manufacturing process by anaerobic fermentation of sucrose

Without the use of oxygen, anaerobic bacteria (like *Clostridium thermoaceticum*) can directly convert carbohydrates to acetic acid digest sucrose. This procedure has the benefit of lowering the cost of raw materials by up to 60%. Anaerobic processes have drawbacks since they produce low productivities because microorganisms are inhibited at a low pH. Therefore, research is concentrated on creating bacterial strains with a greater pH tolerance in order to increase the productivity of acetic acid. The final fermentation broth has low levels of acetic acid, making it difficult to separate or purify it because traditional separation techniques, including distillation, are not viable at such low levels. Also, the mechanism of anaerobic fermentation of sugar is quite complex.

9.3.3.1 Raw material

Sucrose (corn, sugar cane and corn stover) and Nitrogen

9.3.3.2 Chemical reactions

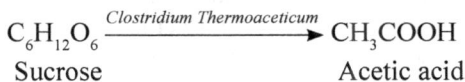

$$C_6H_{12}O_6 \xrightarrow{\textit{Clostridium Thermoaceticum}} CH_3COOH$$

Sucrose   Acetic acid

9.3.3.3 Flow sheet diagram

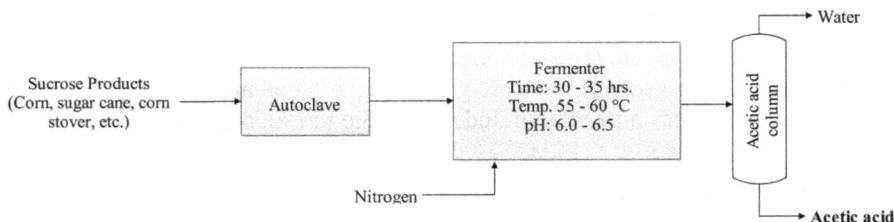

Figure 9.4 Flow sheet diagram of production of Acetic acid from Sucrose

9.3.3.3 Process description

A dilute solution of raw materials is first sterilized in an autoclave and then charged into the fermenter. Bacteria *Clostridium Thermoaceticum* is added to the fermenter. Also, nitrogen is bubbled out from the bottom of the fermenter to maintain an inert atmosphere duringanaerobic fermentation. A temperature of 55–60°C and a pH of 6.0–6.5 is maintained for 30–35 hrs. The resulting stream undergoes scrubbing using cooled water to remove Nitrogen from the top. The bottom stream containing acetic acid and water is sent to an acetic acid column to separate water from the acetic acid. Vinegar of 15% concentration and 70% acetic acid is manufactured in 24 hrs and 10 days respectively.

9.3.5 Uses

Various uses of acetic acid are as follows.

1. It is used as a primary solvent in numerous industrial processes.
2. It is used for the production of insecticides, wood glues, soft drink bottles, rubbers and plastics, rayon fibers, synthetic fibers, and textile inks and dyes.
3. It serves as a stop bath in the development of photographic films.
4. It is also used as household vinegar to clean indoor climbing holds of chalk.
5. It is utilized as a solvent for re-crystallization to purify organic compounds.
6. It is used in testing blood in clinical laboratories.
7. It is also used in the film industry.
8. Acetic acid, sometimes referred to as vinegar, is employed as a medicine to treat a variety of ailments. It is frequently used as an eardrop to treat infections of the auditory meatus. It can be utilized as an ear wick. As a liquid it is used to flush the bladders having a urinary catheter in an effort to prevent infections or blockage.
9. It is used as a food additive (Nayak, 2022; Li et al., 2014).

9.4 Beer

The third most common beverage overall, after water and tea, is beer, one of the oldest and most popular alcoholic beverages in the world. Beer is typically made from malted barley, although it can also be made from wheat, maize (corn), and rice.

9.4.1 Introduction

Beer and related products are alcoholic drinks with a low alcohol concentration (2–7%) that are brewed from a variety of cereals. Hops are typically added to provide a more or less bitter flavor and to manage the subsequent fermentation.

9.4.2 Manufacturing process

The cereals utilized are malted barley, oats, and maize as well as malt adjuncts like flaked rice, oats, and millet. Wheat is used in Germany, whereas rice and millet are used in China. Yeast, brewing sugars, and syrups (such as glucose or cone sugar) round out the raw components. The most crucial grain for brewing beer is barley, which is partially germinated to create malt. Instead of producing their own finished malt, the majority of brewers buy it from wholesalers.

9.4.2.1 Raw materials

Barley is the main raw material for preparing beer. Other raw materials are yeast, hops and water.

9.4.2.2 Quantitative requirements

The production of 1 ton of beer, requires 250 kgs of malt barely.

9.4.2.3 Flow sheet diagram

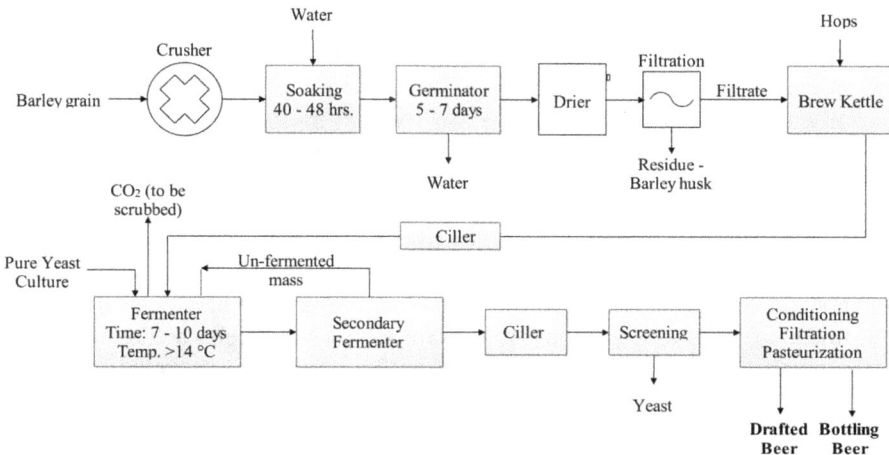

Figure 9.5 Flow sheet diagram of production of Beer

9.4.2.5 Process description

The process is conducted in four steps as follows.

(1) Malting: In the first step of steeping, barely grains are soaked in water for a period of 40–48 hrs, so as to increase the kernels' moisture content to around 45%, allowing for regulated germination. The water is drained off and the grains germinate for 5–7 days, until the desired embryo growth has been achieved, which is continuously monitored. Barley is grown under controlled moisture and temperature conditions, and carbon dioxide availability. Maximum enzyme production, minimal enzymatic activity, and plant growth are the goals of the germination process. Some of the barley starch is subsequently converted to maltoseand dextrins. This explains why the final malt, which resembles typical barley grains, tastes sweet. Kiln drying at 75–80°C stops germination and stabilizes malted barley at the conclusion of germination. It gives the malt, such as light malt, dark malt, amber malt, and black patent malt, flavor, aroma, and color. For the subsequent mashing process, the malt that is thus produced serves as a supply of carbs and enzymes.

(2) Brewing: After malting, brewing is carried out where milling of malted barley is conducted with the addition of hot water. Here, malted barley undergoes careful cracking retaining large pieces of husk and shattering endosperms at a controlled temperature for enzymatic action. The enzymes are allowed to degrade the starches to maltose and proteins into amino acids that the yeast can utilize during fermentation. The sweet wort during the brewing process, is then subjected to filtration to remove barley husks as a by-product. The remaining sweet wort (filtrate) is transferred to another kettle, where the hop is added and boiled under pressure at 105°C for 1 hr. During heating, it extracts flavors (bitter acids) and aromatic compounds from the hop. Thereafter, the spent hop is removed. The hopped wort is cooled down and subjected to fermentation. Brewing serves to extract hop flavors and aromatic compounds, converts starch to maltose, proteins to amino acids, and sterilizes solutions of maltose and hop flavors by enzymatic conversion.

(3) **Fermentation and purification:** The brewer's strain of either a bottom or a top fermenting yeast is used to ferment the hopped wort. The strain you choose will affect the flavor of your beer because of the quantitatively insignificant metabolites it produces. The fermentation process, which lasts 7 to 10 days at a temperature of 14°C and yields CO_2 and ethanol, allows CO_2 to escape through blow-off tubes with their ends submerged in a water tank, creating an air-lock that keeps oxygen out of the fermentation tanks. Secondary fermentation is conducted to remove aldehydes and other undesirable products. It is cooled down and yeast cells are removed by screening. If contaminants are found in the filtrate, it is recycled. The beer is carbonated, filtered and pasteurized, thereafter, bottled as normal or draft beer (not proper pasteurization).

9.4.3 Uses

Beer is an alcoholic drink. Its uses are as follows.

1. Beer is used to prevent disorders of the digestive and cardiovascular systems, such as atherosclerosis, coronary failure, attacks, and strokes. It is known to stop blood clots.
2. Several B vitamins are present in beer (B1, B2, B6 and B12).
3. Beer is also used to prevent Alzheimer's disease, gallstones, type 2 diabetes, heart disease in people with type 2 diabetes, kidney stones, prostate cancer, breast cancer, and other cancers. It also strengthens bones and prevents gallstones.
4. Some people drink beer to boost their digestion, appetite, and breast milk production.
5. It is used to strengthen the bones (Pires and Branyik, 2015; Rodman et al., 2015).

9.5 Wine

A wine (Vinum in Latin) is a fermented alcoholic beverage created from grapes, often *Vitis vinifera*, without the addition of sugars, acids, enzymes, water, or other nutrients (from the Latin vinum).

9.5.1 Introduction

Since at least a few thousand years ago, grape juice has been fermented to make wine. The product's flavor, bouquet, and aroma vary depending on the grape, soil, and sun, which is significantly responsible for its quality. The color is mostly influenced by the type of grapes used and whether the skins are removed prior to fermentation. Wines are divided into categories such as natural (alcohol content

7 to 24%), fortified (alcohol content 14 to 30%), sweet or dry, still, or sparkling. Alcohol or brandy are added to fortified wines. Some of the sugars in the sweet wines are still unfermented.

9.5.2 *Manufacturing process*

It is prepared by the fermentation of grapes.

9.5.2.1 *Raw materials*
Red or black grapes, Sulphurous acid (H_2SO_3) and Bentonite

9.5.2.2 *Quantitative requirement*
For the production of 1 ton of wine, we require 1000 kgs of grapes.

9.5.2.3 *Flow sheet diagram*

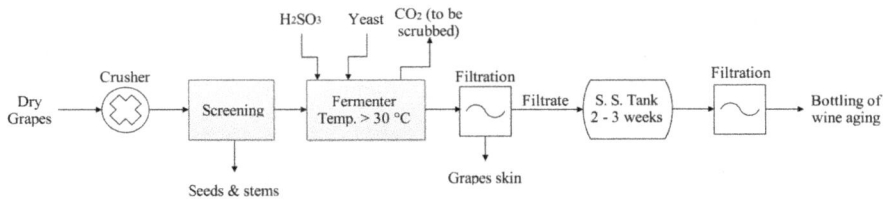

Figure 9.6 Flow sheet diagram of production of Wine

9.5.2.4 *Process description*
Dry red wine, red or black grapes are conveyed to the crusher where the leaves and stems are removed, and the grapes are crushed. The resulting pulp is pumped into a fermentation tank (having a capacity 11000 to 38000 liters and cooling coils) where sulphurous acid is added to maintain an acidic pH; 3 to 5% of the juice's volume of an active culture of chosen and cultured yeast is added. The temperature rises throughout the fermentation process and managed by the cooling coils to keep it below 30°C. A grating float in the tank helps to partially block the carbon dioxide evolved from being transported from the stems and seeds to the top. Graph wines undergo pressing after fermentation to separate the wine from the grape skins. The wine is then settled in enormous stainless steel or vertical oak tanks. Wines and heavier-bodied white wines are aged in small oak barrels after settling. Some wines are filtered in order to stabilize and clarify them during barrel aging and before bottling. Wines that are finished are bottled. In a bottle, wine can mature longer.

9.5.3 *Uses*

It reduces the risk of heart-attack and type-2 diabetes (Claus, 2019; Alba-Lois and Segal-Kischinevzky, 2010; Mathew et al., 2017).

9.6 Liquors

A sizable component of the market for alcoholic beverages is made up of distilled beverages (spirits) that range in alcohol content from 35 to 50 percent. They are resistant to microbial deterioration because of the high ethanol level in them. They are split into two major groupings. Beverages that are a mainstay in the final product and have a significant impact on sensory quality belong to one group. Included in this category are brandies, which are distilled from fermented grapes or other fruit juices, whiskey, which is made from cereals, and rum, which is made from sugarcane juice or molasses. Products like vodka and gin, which have a distilled alcohol base that has been treated to provide certain flavor qualities, are included in the second group. Although quite intoxicating, liquors lack the health-promoting ingredients found in fermented beer because distillation removes vitamins, amino acids, carbohydrates, and other ingredients.

9.6.1 *Properties*

Liqueurs are often highly sweet; they don't age for very long after the ingredients are combined, though they may rest during manufacture to let the flavors meld.

9.6.2 *Manufacturing process*

They are prepared by the distillation and aging of wine or grain or husk depending upon the liqueur type.

9.6.2.1 Raw materials

Raw materials for distilled liqueurs are wine, grain or husk.

9.6.2.2 Flow sheet diagram

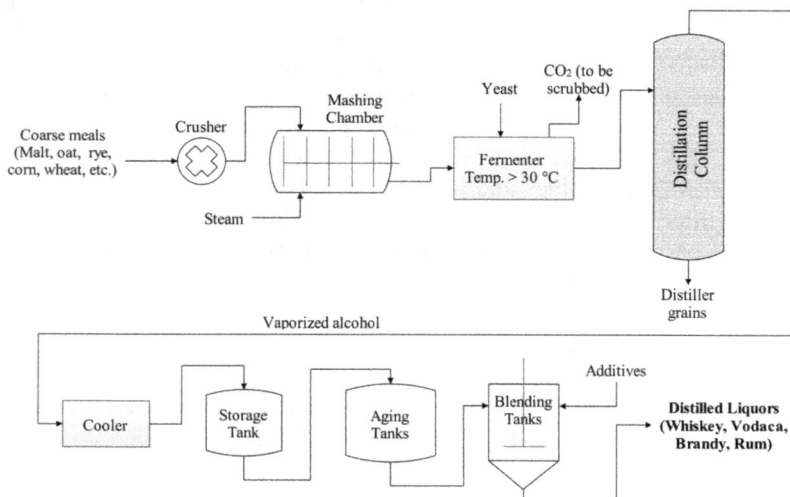

Figure 9.7 Flow sheet diagram of production of Distilled liquors

9.6.2.3 Process description

Since different fermented items are distilled and aged to produce distilled liquor, such as brandy, different whiskies and gin, there is no specific method described for making distilled liquor. Brandy is made by distilling wine or the marc, which is the residue left over after racking or filtering. Bourbon whiskey can be made by distilling a grain mixture that contains at least 51% maize, then aging it. Similar to this, for rye whiskey to be crushed and fermented, the grain must contain 51% rye. Scotch whiskey's characteristic flavor comes from the peat-dried barley used in its production. Equipments like stills are made of copper or stainless steel in modern distilleries. According to the legislation, bourbon or rye whiskey must be aged for at least one to five years in charred, new, white oak barrels with a white wood content of 65 to 70 percent or more. The charred white oak imparts color and other ingredients to the whiskey as it ages. Whiskey production requires the fermentation of whole grains so that the husks and germs (which carry the corn oil) are suspended in the beer still's liquor. Slop, or stillage, is the name for this discharge liquor.

There are many intricacies involved in making various spirit beverages, however as an illustration, the procedure for a spirit made using cereal is as follows: A coarse meal is made from the raw material. The procedure removes starch from the raw material by dissolving the protective shell. The sugar created from the starch is combined with purified water and boiled. A mash results from this. The inclusion of yeast in the fermentation tank causes the sugar to be transformed to alcohol and carbon dioxide. When yeast is added to sugar, it multiplies, creating carbon dioxide that bubbles up and a mixture of alcohol, particles, and congeners, or the flavors that give each drink its distinct flavor. Heat is applied to the congeners, water, grain particles, and alcohol. The water, the grain fragments, and some of the congeners remain in the boiling kettle after the alcohol has first vaporized. The alcohol is then chilled or condensed, producing crystal-clear droplets of distilled spirits. Ageing and blending are two extra stages that are frequently used to create various distilled spirits. Certain distilled spirits (including rum, brandy, and whiskey) are aged by being let to grow old in wooden casks, where they progressively acquire a unique flavor, aroma, and color. Additionally, certain spirits undergo a blending procedure in which two or more spirits belonging to the same category are mixed. Since the blended spirit belongs to the same specialized category as its components, this method differs from mixing.

9.6.3 Uses

Liqueurs are consumed as alcoholic beverages (Vazquez et al., 2005).

9.7 n-Butanol

One of the earliest fermentation techniques used to produce a chemical for human use is the preparation of n-butanol.

9.7.1 Properties

Its main properties are as follows.

Table 9.4 Physical and Chemical Properties of n-Butanol

Sr. No.	Properties	Particular
1	IUPAC Name	Butan-1-ol
2	Chemical and other names	1-Butanol, n-Butanol, Butyl alcohol, Propylmethanol, Methylolpropane
3	Molecular formula	$C_4H_{10}O$
4	Molecular weight	74.1 gm/mole
5	Physical description	Colourless liquid with characteristic odour
6	Solubility	Soluble in acetone; miscible with ethanol and ethyl ether
7	Melting point	−89.9°C
8	Boiling point	117.9°C
9	Flash point	29°C
10	Density	0.81 g/ml at 20°C

9.7.2 Manufacturing process

Weizmann, a scientist, formed a procedure that uses the fermentation of grains that contain starch to produce acetone and butyl alcohol. Selective bacterial fermentation of carbohydrates-containing materials like molasses and grains results in the production of butyl alcohol. Also, it gives acetone and ethyl alcohol as byproducts.

9.7.2.1 Raw materials

Starch (carbohydrates) containing materials such as molasses and grains are selective raw materials for butyl alcohol.

9.7.2.2 Chemical reactions

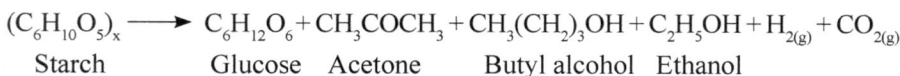

$$(C_6H_{10}O_5)_x \longrightarrow C_6H_{12}O_6 + CH_3COCH_3 + CH_3(CH_2)_3OH + C_2H_5OH + H_{2(g)} + CO_{2(g)}$$

Starch   Glucose   Acetone   Butyl alcohol   Ethanol

Three chemicals, i.e., Acetone, Butanol and Ethanol are prepared in this fermentation process simultaneously; so, this fermentation is known as Acetone-Butanol-Ethanol (ABE) fermentation.

9.7.2.4 Flow sheet diagram

9.7.2.5 Process description

As per the flow-sheet diagram, molasses is diluted with water upto a concentration of approximately 5% sugar, thereafter, sterilized, cooled to 30°C and then pumped to the fermenter. Yeast is also added to it and a temperature of 37°C is maintained. During the fermentation of molasses, CO_2 and H_2 gases are evolved. After 36–48 hrs of the fermentation process, beer containing 1.5 to 2.5% mixed solvents is taken out from the bottom of the fermenter and transferred to a distillation

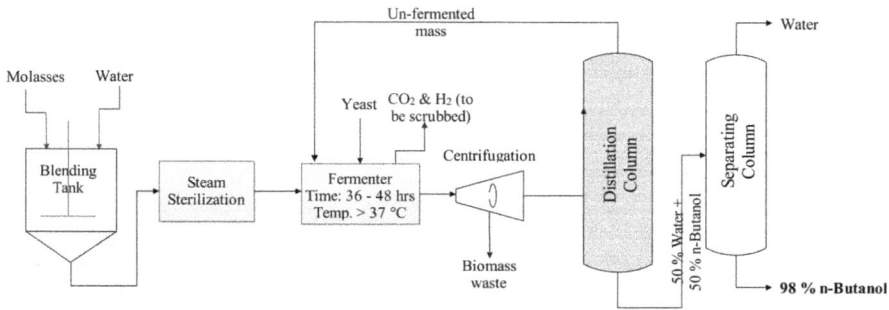

Figure 9.8 Flow sheet diagram of production of Butanol from molasses

column. In this column, water and ethanol are removed and, the butanol section, which includes around 50% H_2O, is removed from the middle of the column and put into the separator, where 98% butanol is produced.

The major problem faced during the manufacturing of Butanol using the fermentation process of molasses is a low acetone-butanol-ethanol (ABE) yield, low reactor productivity, use of diluted sugar solutions, and unfavorable product recovery all contributing to product inhibition. Maximum solvent concentrations of 25–30 gm/l are infrequently reached during the fermentation process. The crucial step in this fermentation process is hence simultaneous recovery. Low ABE yield also has another issue, i.e., it yields 0.3–0.45 gm/gm with the release of the rest of the carbon in form of carbon dioxide gas. In addition, hydrogen gas is also liberated, which results in a loss of energy. A low butanol yield adversely affect its cost-price. When batch fermentation is employed, it is also associated with low reactor productivity. We are getting a productivity of 0.50 l/hr, so, a continuous immobilized cell reactor or cell recycled reactors are preferable. In this reactor, a productivity as high as 6.5 to 15.8 l/hr has been achieved. A lower cost is another significant factor considered during the pricing of butanol.

9.7.3 Uses

Butyl alcohol can be used as a solvent to make butyl acetate, glycol ester, and ester, among other products. Additionally, it is utilized in the formation of Amine-resin and plasticizers. A good variety of formulated products, including paints and coatings, cosmetics and fragrances, adhesives and inks, and even food tastes and extracts, use butanol as an intermediary to create high-value resins and specialty solvents. Numerous products and processes employ n-butanol, including plastics and polymers, lubricants, synthetic rubber, fire retardants, and brake fluids. (Sauer et al., 2016; Ndaba et al., 2015; Sun et al., 2009).

9.8 Citric Acid

A crucial metabolic intermediary is citric acid. Citrus fruits contain this acidic chemical. Because they may chelate calcium, the salts of citric acid (citrates) can be used as anticoagulants.

9.8.1 *(Introduction) properties*

Its main properties are as follows.

Table 9.5 Physical and Chemical Properties of Citric acid

Sr. No.	Properties	Particular
1	IUPAC Name	2-Hydroxypropane-1,2,3-tricarboxylic acid
2	Chemical and other names	Citric acid, Hydrocerol A, 2-Hydroxytricarballylic acid, Aciletten, Citretten
3	Molecular formula	$C_6H_8O_7$
4	Molecular weight	192.1 gm/mole
5	Physical description	White or colourless, odourless, crystalline solid, having a strongly acid taste
6	Solubility	Soluble in water, ethanol, ether
7	Melting point	153.0°C
8	Boiling point	310.0°C
9	Flash point	100.0°C
10	Density	1.66 g/ml at 20°C

9.8.2 *Manufacturing process*

Around 1826, Italian lemons were transported to England and used to make the first commercially successful citrus acid (lemons contain 7–9% citric acid). Up until 1919, when the principal technique utilizing *Aspergillus niger* commenced in Belgium, juice remained the commercial source of acid. Grimoux and Adams first produced acid from glycerol and then from symmetrical dicloroacetone. Since then, more synthetic material-based ways have been reported, but chemical methods have so far failed to be competitive due to their high cost and low yield. So, the fermentation process is more preferable than other chemical methods. It is prepared by the fermentation of molasses.

9.8.2.1 *Raw materials*

Raw material for citric acid is beet molasses containing 48–50% sugar by the weight of sugar from cane-molasses. Other raw materials are water, sulfuric acid and phosphorous, potassium and nitrogen in the form of acid or salts.

9.8.2.2 *Chemical reactions*

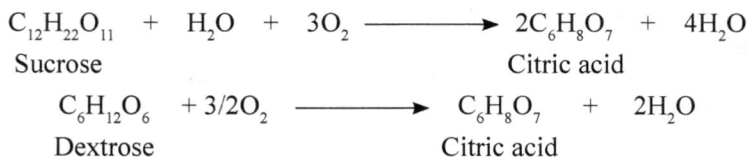

$$C_{12}H_{22}O_{11} \ + \ H_2O \ + \ 3O_2 \longrightarrow 2C_6H_8O_7 \ + \ 4H_2O$$

Sucrose   Citric acid

$$C_6H_{12}O_6 \ + 3/2O_2 \longrightarrow C_6H_8O_7 \ + \ 2H_2O$$

Dextrose   Citric acid

9.8.2.3 *Flow sheet diagram*

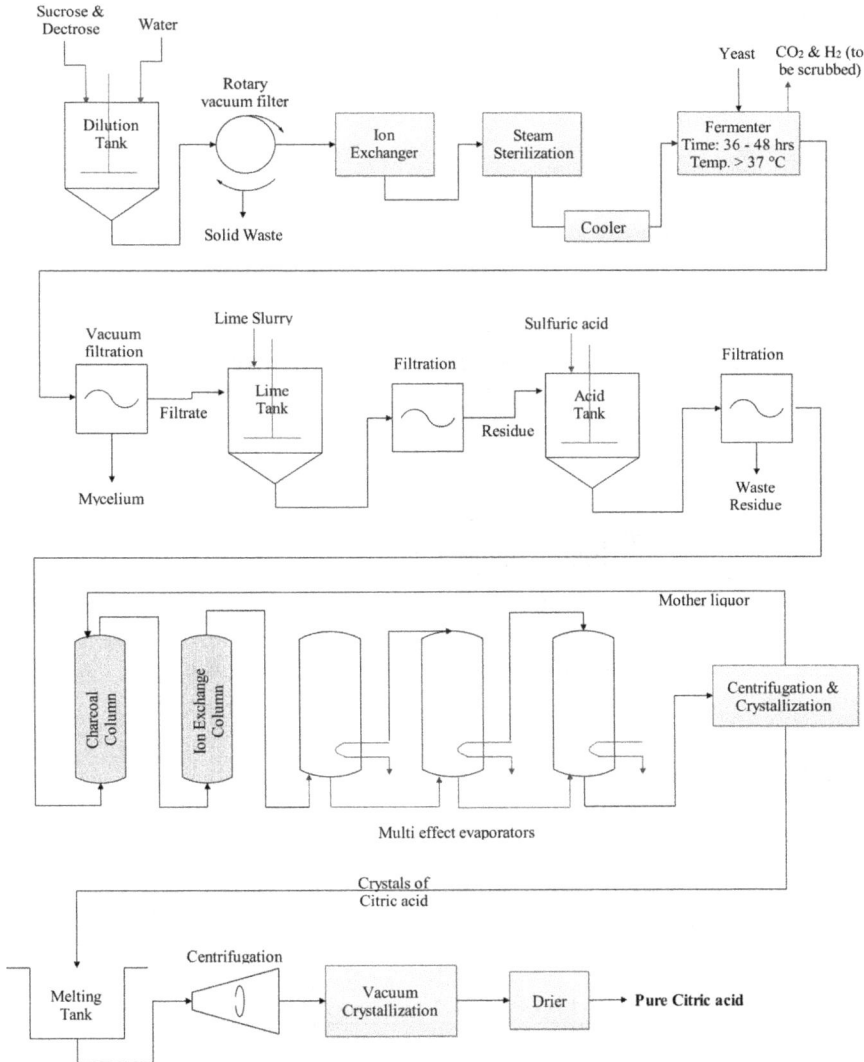

Figure 9.9 Flow sheet diagram of production of Citric acid

9.8.2.4 Process description

As per the flow-sheet diagram, the molasses is diluted with water, rotary vacuum filtered, ion-exchanged, sterilized, and cooled to 30°C and pumped to the fermenter. Yeast is also added to it and a temperature of 37°C is maintained. After 36–48 hrs of the fermentation process, the liquor is filtered to remove mycelium as a by-product and the filtrate is precipitated by adding lime. The resulting calcium oxalate and liquid are heated to 95°C, filtered by vacuum, and then acidified with H_2SO_4 to produce calcium-citrate. Citric acid is present in the filtrate, which is then decolored by charcoal treatment, ion exchanged, and concentrated by a multi-effect evaporator. Thereafter, it is centrifuged and crystallized to obtained citric acid crystals and mother liquor. Mother liquor is recycled through charcoal treatment. Re-purified citric acid is obtained by re-melting, centrifugation and crystallization, and drying to get pure citric acid.

9.8.3 Uses

Its main application is as an acidulant in jams, jellies, and other foods such as carbonated beverages. It is used extensively in the pharmaceutical industry, especially in the production of citrates and effervescent salts. It is also utilized as an ion-sequesting agent buffer, acetyl tributyl citrate, a vinyl resin plasticizer (Kristainsen et al., 1998; Yalcin et al., 2010).

*Other fermentation products are Penicillin, Streptomycin and Erithromycin, which are mentioned in previous chapter.

10.9 Lactic acid

Swedish chemist Carl Wilhelm Scheele discovered this substance for the first time in sour milk in 1780; thereafter it was manufactured using the fermentation process and utilized in different industries by French scientist Fremy in 1881. It is an optically active compound and has two isomers, i.e., D-Lactic acid and L-Lactic acid. Naturally, it is available in equal amounts, so it is known as DL-Lactic acid. Furthermore, it is regarded as one of the most significant hydroxycarboxylic acids.

10.9.1 (Introduction) properties

Its main properties are as follows.

Table 9.6 Physical and Chemical Properties of Lactic acid

Sr. No.	Properties	Particular
1	IUPAC Name	2-Hydroxypropanoic acid
2	Chemical and other names	2-Hydroxypropionic acid, 2-Hydroxypropanoic acid 2-Hydroxypropionic acid, Lactic acid
3	Molecular formula	$C_3H_6O_3$
4	Molecular weight	90.04 gm/mole
5	Physical description	Viscous, colorless to yellow liquid or colorless to yellow crystals
6	Solubility	soluble in water, ethanol, diethyl ether, and other organic solvents
7	Melting point	16.8°C
8	Boiling point	122°C
9	Flash point	110.0°C
10	Density	1.25 g/ml at 20°C

10.9.2 Manufacturing process

Lactic acid is manufactured either by the fermentation of carbohydrates or by chemical synthesis from acetaldehyde.

(1) **Fermentation of carbohydrates:** Lactic acid is manufactured by anaerobic fermentation of carbohydrates through pyruvic acid using *Lactobacillus helveticus*.

(2) **From acetaldehyde:** Acetaldehyde is reacted with hydrogen cyanide to yield cyanohydrin, which on undertone acid hydrolysis gives lactic acid.

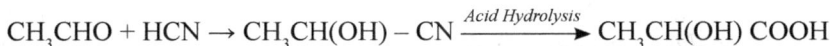

$$CH_3CHO + HCN \rightarrow CH_3CH(OH) - CN \xrightarrow{Acid\ Hydrolysis} CH_3CH(OH)\ COOH$$

Here, we are discussing anaerobic fermentation of carbohydrates due to a cheap process and sustainable development.

10.9.2.1 Raw materials

Carbohydrates containing materials such sugarcane and sugar beet having 45–50% sucrose are used for production of lactic acid. Sometime, molasses is used, but its productivity is very low, due to a low carbohydrate content. Other raw materials are *Lactobacillus helveticus*, sulfuric acid, hydrated lime and water.

10.9.2.2 Chemical reactions

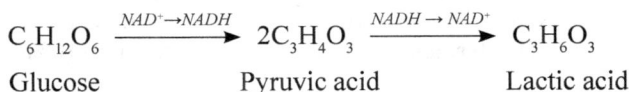

$$C_6H_{12}O_6 \xrightarrow{NAD^+ \rightarrow NADH} 2C_3H_4O_3 \xrightarrow{NADH \rightarrow NAD^+} C_3H_6O_3$$

$$\text{Glucose} \qquad\qquad \text{Pyruvic acid} \qquad\qquad \text{Lactic acid}$$

NAD - Nicotinamide adenine dinucleotide

NADH - Nicotinamide adenine dinucleotide Hydrogen

10.9.2.3 Flow chart

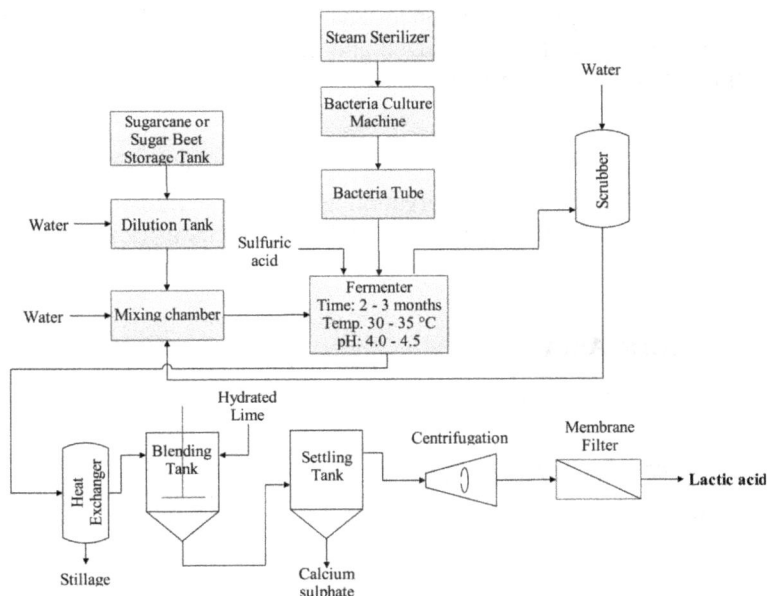

Figure 9.10 Flow sheet diagram of production of Lactic acid from Glucose

10.9.2.4 Process description

This process is conducted in three main steps in three:

(1) **Preparation of raw materials:** As per Fig. 9.10, sugarcane and/or sugar beet are stored in a storage tank and transferred into a dilution tank where the sugar content is diluted with water to provide a range of 10–15%. Sulfuric acid is used to keep the pH at 4–5. On other side, bacteria namely *Lactobacillus helveticus* is sterilized under pressure, cooled, and transferred intothe yeast culture tank to develop yeast culture. Nutrition of ammonium and magnesium phosphate or sulfate is provided. The acidic condition favors the yeast to produce catalytic enzymes, invertase and zymase, so, the pH is maintained at about 4–5 using sulfuric acid.

(2) **Fermentation:** The diluted molasses, sugarcane or/and sugar beet and bacteria undergo a fermentation reaction in fermenter. Here, sulfuric acid is employed to adjust the pH from 4.0 to 4.5. As this process is exothermic, cooling water is circulated in the cooling coils to ensure that the temperature is maintained at 30–35°C. The fermentation duration is altered from 40 to 2–3 hrs. Carbon dioxide is liberated from fermenter, which is absorbed in the scrubber water. A water scrubber is used to scrub carbon dioxide alongwith ethanol. The fermentation product consists of 8–10% ethanol, water and aldehyde.

(3) **Purification:** After the fermentation reaction, liquor called beer having 15 to 20% lactic acid is pumped into a beer still through a heat exchanger. The heat exchanged removes the protein, residue sugar and vitamin, called stillage from the beer, which is utilized as a constituent of the animal feed. Thereafter, lactic acid is treated with hydrated lime to remove excess sulfuric acid. It then undergoes separation to separate calcium sulfate. The upper part containing lactic acid is centrifuged and given a special type of membrane treatment to remove by-products.

10.9.3 Uses

Lactic acid is a common food preservation and curing ingredient. Additionally, it is used in the pharmaceutical, cosmetic and chemical industries (Paramithistis, 2017; Krishna et al., 2018).

10.10 Succinic Acid

It is a dicarboxylic acid derivate and first isolated from amber in 1546.

10.10.1 (Introduction) properties

Its main properties are as follows.

Table 9.7 Physical and Chemical Properties of Succinic acid

Sr. No.	Properties	Particular
1	IUPAC Name	Butanedioic acid
2	Chemical and other names	Amber acid, Succinic acid, 1,2-ethanedicarboxylic acid, 1,2-ethanedicarboxylic acid
3	Molecular formula	$C_4H_6O_4$
4	Molecular weight	118.09 gm/mole
5	Physical description	Colourless or white, odourless crystals
6	Solubility	Soluble in water, ethanol, ethyl ether, acetone, methanol
7	Melting point	188°C
8	Boiling point	235°C
9	Flash point	160°C
10	Density	1.572 g/ml at 20°C

10.10.2 Manufacturing process

It is produced by different methods like hydrogenation of maleic acid, oxidation of 1,4-butanediol, and carbonylation of ethylene glycol. But it is produced by anaerobic fermentation of carbohydrates industrially.

10.10.2.1 Raw materials

Corn or molasses or sugarcane or beet sugar having carbohydrates are used for fermentation. Also, other raw materials are *Actinobacillus succinogenes*, sulfuric acid, calcium hydroxide and water. Sometime, agro-wastes are used to reduce the cost of succinic acid.

10.10.2.2 Chemical reactions

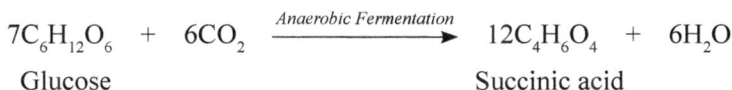

$$7C_6H_{12}O_6 \ + \ 6CO_2 \ \xrightarrow{\textit{Anaerobic Fermentation}} \ 12C_4H_6O_4 \ + \ 6H_2O$$

$$\text{Glucose} \qquad\qquad\qquad\qquad\qquad \text{Succinic acid}$$

10.10.2.3 Flow chart

10.10.2.4 Process description

This process is conducted mainly in three steps:

(1) **Preparation of raw materials:** As per Fig. 9.11, molasses, sugarcane or/ and sugar beet is stored in a storage tank and it is transferred into a dilution tank, where it is diluted with water to obtain a sugar concentration of around 10–15%. The pH is maintained at about 4–5 using sulfuric acid. On other side, bacteria namely *Actinobacillus succinogenes* is sterilized under pressure, cooled, and transferred into the yeast culture tank to develop it. Nutrients such as ammonium and magnesium phosphate or sulfate are provided. The acidic condition favors the yeast to produce catalytic enzymes, invertase and zymase, so, the pH is maintained at about 4–5 using sulfuric acid.

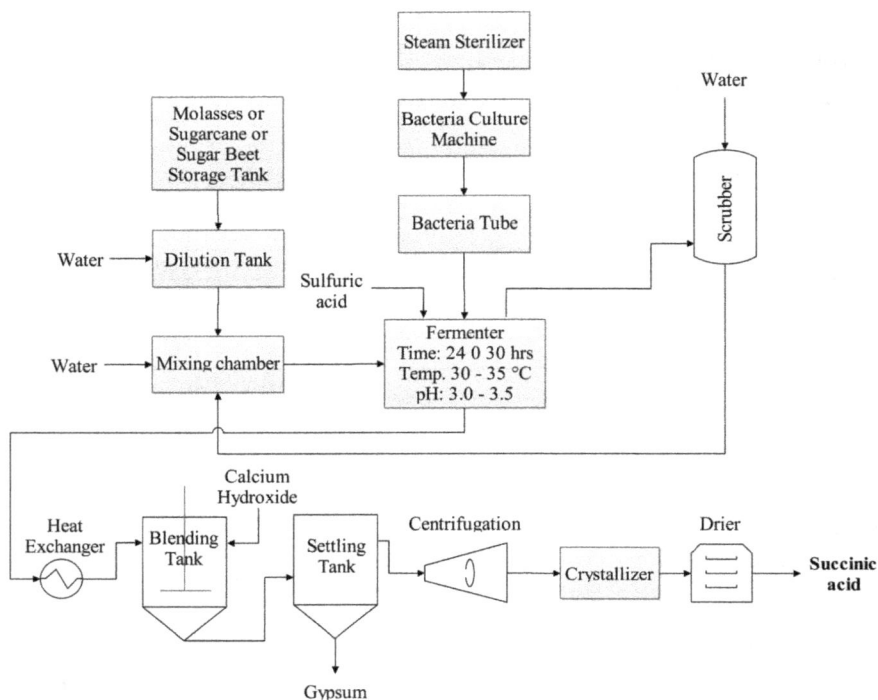

Figure 9.11 Flow sheet diagram of production of Succinic acid from Glucose

(2) **Fermentation:** The diluted molasses, sugarcane or/and sugar beet and bacteria undergo a fermentation reaction in the fermenter. The initial pH is adjusted from 4.0 to 4.5 with sulfuric acid. The fermentation is exothermic, so a cooling facility (via cooling coils) is provided to ensure that the maximum temperature does not exceed 35°C. The residence time is 2–3 months and the temperature is maintained at 30–35°C. Carbon dioxide is liberated from the fermenter, which is absorbed in the scrubber water. Here, the by-product CO_2 contains some ethanol due to vapor liquid evaporation and can be recovered by water scrubbing.

(3) **Purification:** After the fermentation reaction, liquor called beer having 15 to 20% lactic acid is pumped into a beer still through a heat exchanger. The heat exchanged removes the proteins, residue sugars and vitamins, called stillage from beer, which is utilized as a constituent of animal feed. Thereafter, lactic acid is treated with hydrated lime to remove excess sulfuric acid. It then undergoes separation to separate calcium sulfate. The upper part containing lactic acid is centrifuged and given a special type of membrane treatment to remove by-products.

10.10.3 Uses

It is used as a green solvent or lubricant in cosmetics, foods, dyes and perfumes. It is also useful in, plasticizers, metal treatment, lacquers, photographic chemicals and coatings. It is utilized in the manufacture of medicines for sedatives, antispasmers, antiplegm, antiphogistic, anrhoers, contraceptives and cancer (Ferone et al., 2018; Stylianou et al., 2021).

References

Alba-Lois, L. and Segal-Kischinevzky C. 2010. Yeast Fermentation and the Making of Beer and Wine.

Chen, H. 2013. Modern Solid State Fermentation: Theory and Practice, Springer Netherlands. https://doi.org/10.1007/978-94-007-6043-1

Cinar, A., Parulekar, S.J., Undey, C. and Birol, G. 2003. Batch fermentation: modeling, monitoring, and control, Marcel Dekker Publication.

Claus, H. 2019. Wine Fermentation, Fermentation 5(1): 19.

Ferone, M., Raganati, F., Olivieri, G., Salatino, P. and Marzocchella, A et al. 2019. Continuous Succinic Acid Fermentation by *Actinobacillus Succinogenes*: Assessment of Growth and Succinic Acid Production Kinetics. Applied Biochemistry and Biotechnology 187(1). https://doi.org/10.1186/s13068-018-1143-7

Krishna, B.S., Nikhilesh, G.S.S., Tarun, B., Narayana S.K.V. and Gopinadh, R. et al. 2018. Industrial production of lactic acid and its applications. International Journal of Biotech Research 1(1): 42–54.

Kristiansen, B., Linden, J. and Mattey, M. 1998. Citric Acid Biotechnology, Taylor & Francis.

Li, Y., He, D., Niu, D. and Zhao, Y. 2014. Acetic acid production from food wastes using yeast and acetic acid bacteria micro-aerobic fermentation, Bioprocess and Biosystems Engineering 38(5). https://doi.org/10.1007/s00449-014-1329-8

Lima, U.A., Basso, L.C. and Amorim, H.V. 2000. Ethanol production, Industrial Biotechnology 3: 16–17.

Mathew, B., Mohammed S.S.D., David E.S. and Ugboko, H. 2017. Production of Wine from Fermentation of Grape (*Vitisvinifera*) and Sweet Orange (*Citrus seninsis*) Juice using Saccharomyces cerevisiae Isolated from Palm Wine, International Journal of Current Microbiology and Applied Sciences 6(1): 868–881. http://dx.doi.org/10.20546/ijcmas.2017.601.103

McNeil, B. 2008. Linda Harvey, Practical Fermentation Technology, Willey. DOI: 10.1002/9780470725306

Nayak, J. 2022. A Green Process for Acetic Acid Production, 7th International Conference on Chemical, Ecology and Environmental Sciences, Thailand.

Ndaba, B., Chiyanzu, I. and Marx, I. 2015. N-Butanol derived from biochemical and chemical routes: A review, Biotechnology Reports, 8. https://doi.org/10.1016/j.btre.2015.08.001

Paramithiotis, S. 2017. Lactic acid fermentation of fruits and vegetables. In: Food biology series, CRC Press.

Pires, E. and Branyik, T. 2015. Biochemistry of Beer Fermentation, Springer International Publishing. https://doi.org/10.1007/978-3-319-15189-2

Rodman, A., Jones, H. and Gerogiorgis, D.I. 2015. Process optimisation of beer fermentation through dynamic simulation, 35th Congress of the European Brewery Convention (EBC) Porto, Portugal.

Roozbehani, B., Mirdrikvand, M., Moqadam, S.I. and Roshan, A.C. 2013. Synthetic ethanol production in the Middle East: A way to make environmentally friendly fuels. Chemistry and Technology of Fuels and Oils 49(2).

Sauer, M. 2016. Industrial production of acetone and butanol by fermentation-100 years later, FEMS Microbiology Letters 363(13).

Schwietzke, S., Kim, Y., Ximenes, E. and Mosier, N.S. 2009. Ethanol Production from Maize, In book: Molecular Genetic Approaches to Maize Improvement 347–364. https://doi.org/10.1007/978-3-540-68922-5_23

Stylianou, E., Pateraki, C., Ladakis, D. and Vlysid, A. 2021. Optimization of fermentation medium for succinic acid production using *Basfiasucciniciproducens*. Environmental Technology & Innovation 24: 101914. https://doi.org/10.1016/j.eti.2021.101914

Sun, Z., Fitzgerald, L., Mukherjee, S.S. and Liu, S. 2009. N-Butanol Production via Fermentation of Sugar Maple Wood Extract Hydrolysate, AIChE Annual Meeting.

Vazquez, M.J. Garrote, G., Alonso, J.L., Dominguez, H. and Parajo, J.C. et al. 2005. Refining of autohydrolysis liquors for manufacturing xylooligosaccharides: Evaluation of operational strategies. Bioresource Technology 96(8): 889–96. https://doi.org/10.1016/j.biortech.2004.08.013

Vogel, H.C. and Haber, C.C. 2007. Fermentation and Biochemical Engineering Handbook, 3rd Edition. https://doi.org/10.1016/C2011-0-05779-4

Yalcin, S.K., Bozdemir, M.T. and Ozbas, Z.Y. 2010. Citric acid production by yeasts: Fermentation conditions, processoptimization and strain improvement. In Current Research, Technology and Education Topics in Applied Microbiology and Microbial Biotechnology 1374–1382.

CHAPTER 10

Agrochemical Industries

10.1 Introduction

The argochemical industry is mainly based on the two following types of industries.

10.1.1 Fertilizer industries

A fertilizer is described as any material of natural or synthetic origin (except liming materials), organic, or inorganic, that is applied to soils or to plant tissues (often leaves) to give one or more plant nutrients necessary for the growth of plants. Fertilizers help plants develop more quickly. This objective is achieved in two methods, the common one being nutrient-rich additions. There are three major nutrients, namely nitrogen, phosphorus, and potassium (N, P & K), which are necessary for different crops. Inland-grown raw materials are typically obtained for nitrogenous fertilizers, which act to increase the effectiveness of the soil by altering its water retention and aeration. These nutrients aid plants in the ways listed below.

(1) **Nitrogen:** It is a crucial ingredient for many different plant processes and one of the essential building blocks of chlorophyll. Nitrogen stimulates the growth of vegetation, especially the leaves, stems, and branches. Nitrogen may be quickly removed from the soil by plants, whereas the opposing elements takes longer. If a plant has fading leaves or appears small and has poor growth, the soil may be deficient in nitrogen. You may find out which element might even be missing from your soil by using a soil test kit. Test kits can be bought at the top home and garden stores.

(2) **Phosphorus:** When the pH of the soil is between 5.5 and 7, phosphorus is most readily available to plants. It stimulates the growth of roots, seeds, and flowers. The pH measures how acidic or alkaline the soil is and can range from 0 to 15, with 7 being regarded as neutral. Phosphorus, unlike nitrogen, has a tendency to remain in the soil; it should only be added for new growth. Like carrots, root crops consume a lot of phosphorus, especially in the early stages of growth.

(3) **Potassium:** Potassium encourages the growth of buds, roots, and fruit ripening. It improves tolerance for heat, cold, and drought as well as disease resistance. All plants require these components to grow, especially during periods of fluctuating weather. Although it has a propensity to remain in the soil, potassium is strongly used by plants that grow vegetables, thus it should be supplemented when needed, as determined by a soil test.

(4) **Additional Nutrients:** In addition to nitrogen, phosphorous, and potassium, fertilizers frequently contain trace levels of other nutrients. Calcium is used to build the cell walls of plants (Ca). A crucial element in photosynthesis and the creation of chlorophyll is iron (Fe). Magnesium (Mg) helps plants develop and repair themselves as well as create molecules that contain chlorophyll. Proteins are required for plant growth and, as a result, the maturation of fruits and seeds. Sulfur (S) aids in this process.

Two categories serve as the main divisions for fertilizers. Straight fertilizers provide one nutrient (say, N, P, or K). To create fertilizers with nitrogen, ammonia is needed (NH_3). All other nitrogen fertilizers, such as urea ($CO(NH_2)_2$) and anhydrous nitrates, need this ammonia as a feedstock (NH_4NO_3). All phosphate fertilizers, including potash, which is a mixture of potassium minerals used to make potassium fertilizers, are derived from minerals that contain the anion PO_4^{3-}. Compound fertilizers contain N, P, and K (Nielsson, 2018; Park, 2001).

10.1.2 Pesticide industries

Pesticides are described by the Food and Agriculture Organization (FAO) as "any substance or mixture of drugs intended for preventing, destroying, or controlling any pest, including human or disease vectors, unwanted plants or animal species, causing damage during or otherwise interfering with the assembly, processing, storage, transport, or sale of food, agricultural commodities, wood and wood products, or substances which will be administered. The term comprises compounds designed to be employed as phytohormones, defoliants, desiccants, or agents to thin or stop fruits from falling from trees too soon. Additionally, they are used as fertilizers sprayed on crops before or after harvesting to protect the produce from deterioration during storage and transportation. Herbicides, insecticides (which may include insect growth regulators, termiticides and more), nematicides, molluscicides, pesticides, avicides, rodenticides, bactericides, insectifuges, animal repellents, antimicrobials, fungicides, disinfectants (antimicrobials), and sanitizers are all included under the umbrella term "pesticide". Different methods of action are used by insecticides. Some insecticides cause nervous system disruption, while others may harm their exoskeletons, repel them, or exert other forms of control. Insecticides are frequently used in industrial, public health, and agricultural settings as well as in residential and commercial settings (e.g., control of roaches and termites). The three insecticides that are most frequently employed are carbamates, pyrethroids, and organophosphates. Herbicides are chemicals used to manage undesirable plants. They are also frequently referred to as weed-killers. The use of selective herbicides in agricultural industries allows for the control of particular weed species while causing little to no damage to the intended crop. While non-selective herbicides (commonly referred to as "total weed-killers") are frequently used to clear ground wastes, industrial and construction sites, railways, and railway embankments because they kill all plant material with which they come into contact. Other significant distinctions besides selective/non-selective included (Harris, 2000; Maolcsy et al., 1988; Ollinger and Fernandez-Cornejo, 1998).

10.2 NPK Fertilizers

10.2.1 Introduction

Straight fertilizers include Potash (potassium chloride) (MOP), Calcium ammonium nitrate (CAN), Ammonium nitrate (AN), Ammonium sulfate (AS), Urea, Single superphosphate (SSP), Triple superphosphate (TSP), and Mono-ammonium phosphate (MAP). Di-ammonium phosphate (DAP) is a well-known product produced using a precise process. There are an infinite number of N/P/K-ratios and numerous production processes, making compound or complex fertilizers like NPK more difficult to define. The product name "NPK" is typically followed by three numbers to indicate the percent of N, P_2O_5, and K_2O which the merchandise contains, for example, 24-6-12 shows that this particular grade contains 24% N (nitrogen compounds), 6% P_2O_5, and 12% K_2O (potassium compounds). In addition, the fertilizer may contain micro nutrients, magnesium, boron, sulfur, and other elements. The standard nutrient content (N + P_2O_5 + K_2O) will typically be between 40 and 60%.

10.2.2 Manufacturing process

Four different processes are available for manufacturing NPK fertilizers:

1. Nitrophosphate-based NPK fertilizers (mixed acid route)
2. Ammonium phosphate/ammonium nitrate-based NPK fertilizers
3. Nitrophosphate-based NPK fertilizers (ODDA-route)
4. Mechanical blending of single or multi-nutrient components

We study the first process, i.e., Mixed acid process with phosphate rock digestion, because, it has the advantage of a broad range of formulations. This process is very flexible and produces grades with varying degrees of phosphate water solubility. It gives a high quality product and medium labor and investment costs. The process is able to use cheap raw materials such as phosphate rock. But, this process has a complex gas scrubbing process.

10.2.2.1 Raw materials

Phosphate rock, Nitric acid, Ammonia, Phosphoric acid, Sulfuric acid and Potash

10.2.2.2 Quantitative requirements

The quantities of raw materials are based on the N, P and K ratio.

10.2.2.3 Flow sheet diagram

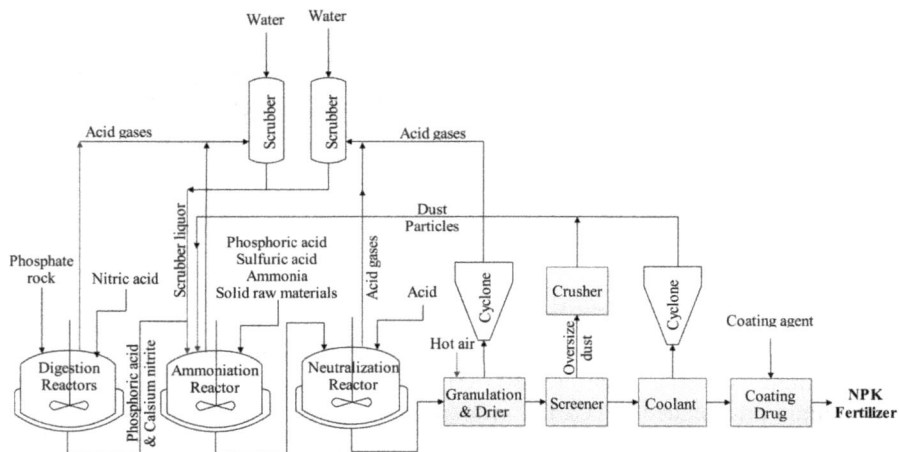

Figure 10.1 Flow sheet diagram of production of NPK fertilizer

10.2.2.4 Process description

The first step, in the digesting reactor, nitric acid and phosphate rock are charged to produce a solution of phosphoric acid and calcium nitrate. Acid gases including nitrogen and fluorine oxides, as well as others, are produced during the digestion process depending on the type of phosphate rock. After digestion, the reaction mass is transferred to the jacketed ammoniation reactor. Also, ammonia, phosphoric acid, sulfuric acid, all the solid materials like potash, ammonium phosphates, superphosphates, ammonium sulfate and compounds containing potassium and magnesium, are added. The reaction is exothermic, so, cooling water is passed through the jacket. The ammoniation mass is collected from the bottom and charged into a neutralization reactor. Here, the pH of the reaction mass is adjusted upto arange of 5 to 6 by adding acid. After neutralization, the reaction mass (fertilizer) goes to the granulator where granulation of the reaction mass is conducted. Thereafter, this granulated fertilizer is dried using hot air. During the drying process, the upward stream is inserted into a cyclone, where gases and dust particles are separated. Gases coming from the digester, ammoniator, neutralizer and cyclone are scrubbed in a water scrubber. The scrubbed water is recycled into ammoniation reactor. After drying, screening is performed. Here, oversized particles are separated from regular particles. Oversized particles are crushed using a crusher to make dust particles. Thereafter, fertilizer cooling is performed, where also dust particles are separated. Dust particles coming from the cyclone, crusher and cooler are recycled into the ammoniation tower. Finally, regular sized cooled fertilizer particles are coated using a mixture of organic agent and inorganic powder as the coating agent in the coating drug.

10.2.3 Uses

Fertilizers are used to help plants grow faster and more abundantly. Adequate amounts of Potassium can increase stress conditions on plants during drought conditions. Potassium is also accountable for producing quality crops. Potassium plays a vital role in the breakdown of carbohydrates, giving energy to plants. Potassium is important in the translocation of minerals (Udom and Nengibenwari, 2022; Nurudeen, 2015).

10.3 Urea

The first Dutch scientist Herman Boerhaave invented Urea from urine in 1727 but this discovery is typically credited to the French chemist Hilaire Rouelle.

10.3.1 Properties

Its main properties are as follows.

Table 10.1 Physical & Chemical Properties of Urea

Sr. No.	Properties	Particular
1	IUPAC name	Urea
2	Chemical and other names	Carbamide, Carmol, Basodexan, Carbonyldiamide, Ureophil, Carbamimidic acid, Ureaphil
3	Molecular formula	CH_4N_2O
4	Molecular weight	60.05 gm/mole
5	Physical description	Colourless to white odorless crystalline powder
6	Solubility	Soluble in water, methanol, glycerol; Insoluble in chloroform, ether
7	Melting point	132°C
8	Boiling point	200°C
9	Flash point	-
10	Density	1.323 gm/cm³ at 27°C

10.3.2 Manufacturing process

Firstly, urea was synthesized in 1727 by Herman Boerhaave from urine. Thereafter, the German chemist Friedrich Wohler obtained urea artificially by treating silver cyanate with ammonium chloride in 1828.

$$AgNCO + NH_4Cl \rightarrow (NH_2)_2CO + AgCl$$

The Bosch-Meiser urea process, named after its inventors, was created in 1922 and is an industrial method for producing urea by ammonium carbonate decomposition.

10.3.2.1 Raw materials

Ammonia and Carbon dioxide

10.3.2.2 Chemical reactions

First, ammonium carbonate is manufactured from carbon dioxide and ammonia under high pressure and temperature. The reaction is highly exothermic.

$$\leftrightharpoons 2NH_3 + CO_2 \leftrightharpoons H_2NCOONH_4; \Delta H = -155 \text{ MJ/kg·mole}$$

Formed ammonium carbonate is dehydrated to give urea and water. This reaction is reversible and highly endothermic.

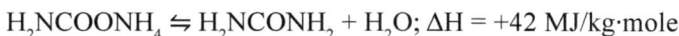

$$H_2NCOONH_4 \leftrightharpoons H_2NCONH_2 + H_2O; \Delta H = +42 \text{ MJ/kg·mole}$$

10.3.2.3 Quantitative requirements

1000 kgs of urea is obtained using 1150 kgs of ammonia and 1470 kgs of carbon dioxide.

10.3.2.4 Flow sheet diagram

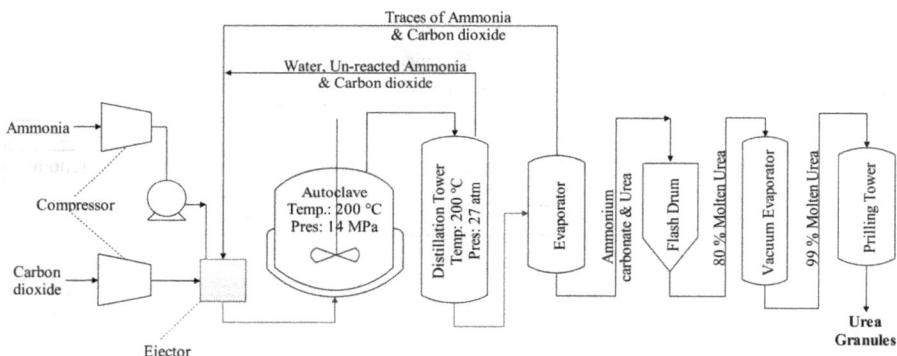

Figure 10.2 Flow sheet diagram of production of Urea

10.3.2.5 Process description

As per the flow-sheet diagram, liquid ammonia is compressed using a compressor and pumped into an autoclave using various multiple pumps to maintain the pressure. Thereafter, carbon dioxide is also compressed and transferred into the autoclave. Both raw materials enter the autoclave from the bottom. The autoclave is made of stainless steel having a jacket, which is lined with a film of oxides to protect it from corrosion. A catalyst bed is placed on the inner side of the autoclave structure. Now, a pressure of 40 atmospheres and temperature of about 180–200°C is maintained for 2 hours. After 2 hours of reaction, a plug flow operation takes place and ammonium carbonate, un-reacted ammonia, carbon dioxide, water and decomposed ammonium carbonate are removed from the top of the autoclave. The reaction mixture is fed into a special flash-evaporator having a temperature of 140°C and pressure of 27 atmospheres. This evaporator includes a gas-liquid separator (distillation tower) and condenser, which separates the un-reacted ammonia, decomposed ammonium carbonate, carbon dioxide and water from ammonium carbonate and urea; and recycled. Now, the aqueous mixture of ammonium carbonate (major part) and urea (minor part) is heated by means of steam and inserted into a flash drum. Here, decomposition of ammonium carbonate takes place to form urea and water. Traces of ammonia and carbon dioxide are recycled. The solution is fed to a vacuum evaporator for concentrating the slurry at 135°C and 60 cm Hg. The concentrated molten slurry is collected from the bottom and passed from the top of the Prilling tower. Here, the tower rotates and sprinkles the slurry and cooled air is passed into the bottom. All the moisture is removed from the bottom of the tower as the urea forms granules. These granules are sent by conveyors to the bagging section.

10.3.2.6 Major engineering problems

(A) **Carbamate decomposition and recycle:** The production of urea can be carried out using a variety of procedures. The recycling design is the primary distinction between competing methods. Since conversion only reaches 40–50% per pass, unreacted off gases must be recycled or profitably employed some where else. Due to corrosion and the production of solid carbamate in compressors, recompression of off gases is almost impossible.

(B) **Production of granular urea:** One more issue is the formation of biurets. Vacuum evaporation of urea from 80% to about 99%, spraying to air cool and solidification must be done immediately above the melting point of urea with a minimum residence time in the range of several seconds.

$$2NH_2CONH_2 \leftrightarrow NH_2CONHCONH_2 + NH_3$$

(C) **Heat dissipation in the autoclave:** The exothermic heat of reaction can be eliminated by coils or wall cooling.

(D) **Corrosion:** This has been the main factor holding back the development of the NH_3-CO_2 process. With haste alloy C, titanium, stainless steel (321 SS), and aluminum alloys are used in other parts of the plant; expensive silver or tantalum liners are used in the autoclaves. To reduce the severe corrosion rates, it is preferable to have minimum pressure, temperature, and excess NH_3 conditions. In the autoclave, chromium steel is frequently used under these circumstances.

10.3.3 Uses

1. Urea is a solid nitrogenous fertilizer with the highest nitrogen content that is currently in use, accounting for around 90% of the world's fertilizer production. Therefore, it has rock bottom transportation costs.
2. Some skin creams, moisturizers, and hair conditioners contain it as an ingredient.
3. Making urea nitrate, a strong explosive used in industry and some improvised explosive devices, is possible using urea (Yahya, 2018; Baboo, 2022; Khan, 2016).

10.4 Ammonium Nitrate

The primary nitrogenous fertilizer that accounts for all of the nitrogen consumed globally is ammonium nitrate. It is easier for crops to access than urea.

10.4.1 Properties

Its main properties are as follows.

Table 10.2 Physical & Chemical Properties of Ammonium nitrate

Sr. No.	Properties	Particular
1	IUPAC name	Azanium nitrate
2	Chemical and other names	Ammonium nitrate, Nitram, Ammonium nitricum, mmonium saltpeter, Nitrate of ammonia
3	Molecular formula	$H_4N_2O_3$
4	Molecular weight	80.43 gm/mole
5	Physical description	Colourless to whilte colored odorless powder
6	Solubility	Soluble in water, alcohol, ketone, ammonia
7	Melting point	170°C
8	Boiling point	210°C
9	Flash point	°C
10	Density	1.72 gm/cm³ at 27°C

10.4.2 Manufacturing process

It is prepared from anhydrous ammonia and nitric acid worldwide. Ammonium nitrate can also be made via a metathesis reaction, in which ammonium chloride is reacted with silver nitrate to give ammonium nitrate.

10.4.2.1 Raw materials

Anhydrous ammonia and Nitric acid

10.4.2.2 Chemical reactions

A Neutralization reaction occurs between alkaline ammonia and acidic nitric acid to give ammonium nitrate salt.

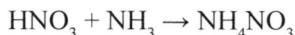

$$HNO_3 + NH_3 \rightarrow NH_4NO_3$$

10.4.2.3 Quantitative requirements

1000 kgs of ammonium nitrate is obtained using 250 kgs of anhydrous ammonia and 900 kgsof nitric acid.

10.4.2.4 Flow sheet diagram

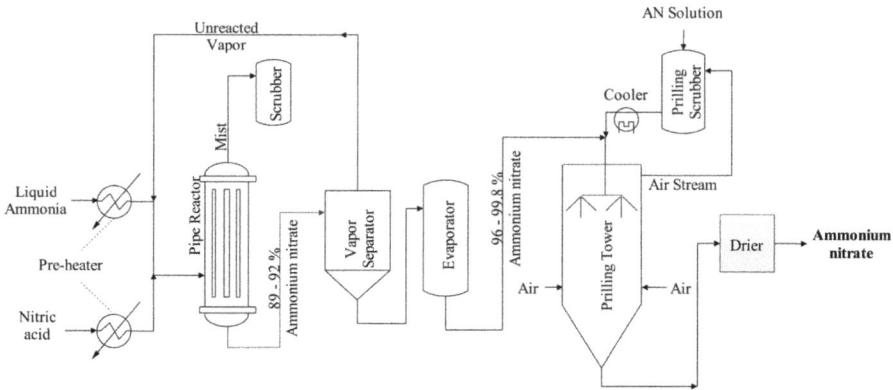

Figure 10.3 Flow sheet diagram of production of Ammonium nitrite

10.4.2.5 Process description

It is divided into two main parts.

(A) Preparation of ammonium nitrate solution (Wet Process): Firstly, liquid ammonia is superheated in a heater to convert ammonia vapor and pipe-fed to a neutralization reactor. The stoichiometric quantity of nitric acid (55% concentration wt/wt) is also heated and added to the reactor. The elevated pressure of 4 atmospheres and pH 3–6 is maintained in the reactor to get ammonium nitrate in solution form. The pH of the ammonium nitrate solution obtained in the reactor controls the feed of ammonia gas to the reactor. The reaction is exothermic, so, a temperature below 105°C is maintained by cooling water. The generated steam is further utilized for evaporation purposes. Here, mist is removed from the process stream using a water scrubber. Thereafter, the reaction mass is transferred into a vapor separator, in which vapor is separated and transferred into a reactor and recycled. The solution containing 85–92% ammonium nitrate is collected from the bottom of separator and sent to an evaporator. The clean steam is used in the shell of the evaporator to increase the concentration of ammonium nitrate in the solution. The concentrated ammonium nitrate solution (about 96–99.8%) leaves the bottom of the evaporator and pumped to the rilling tower.

(B) Preparation of solid ammonium nitrate (Dry Processes): In the prilling tower, liquid concentrated ammonium nitrate solution is sprayed using a spray nozzle or prilling bucket, which falls by gravity within a tall tower (prilling tower) where it cools down and solidified against a counter-current air stream, converting it into solid ammonium nitrate prills. Thereafter it is collected on the belt conveyors located at the bottom of the prilling tower. The temperature of solid ammonium nitrate at the bottom of the prilling tower is around 80–110°C depending upon the product cycle. Here, the air stream leaving the top of the prilling tower is scrubbed and cooled in the prilling scrubber by means of ammonium nitrate solution, which is recycled and most of the ammonium nitrate contained in the air is recovered. The ammonium nitrate scrubber solution is cooled with cooling water and recycled into the prilling tower. Thereafter, it is dried using hot air, cooled down to room temperature, coated with an inert material (usually, kaolin, limestone or dolomite), if required and packed.

10.4.3 Uses

1. Commonly ammonium nitrate is utilized as a nitrogen-release fertilizer and is a very significant chemical in the agricultural industry. It is a source of nitrogen for plants.
2. It is used for manufacturing explosives. It's actually the most significant component of an explosive called nitrate heating oil.
3. It is also utilized in certain industries, construction settings and as a food preservative in cooling bags (Ochoa-Gonzalez et al., 2020; Kirova-Yordanova, 2014).

10.5 Ammonium Sulfate

It is an inorganic salt with diverse economic uses. The most common usage is as a soil fertilizer. Due to the quick increase in urea, ammonium nitrate, and less nitrogen (21%) use, it now contributes to a relatively tiny portion of the world's total nitrogen fertilizer consumption. Ammonium sulfate has several key benefits, including low hygroscopicity, good physical qualities (when made correctly), chemical stability, and strong agronomic performance.

10.5.1 Properties

Its main properties are as follows.

Table 10.3 Physical & Chemical Properties of Ammonium sulphate

Sr. No.	Properties	Particular
1	IUPAC name	Diazanium sulfate
2	Chemical and other names	Ammonium Sulfate, Diammonium sulfate, Mascagnite, Actamaster
3	Molecular formula	$H_8N_2O_4S$
4	Molecular weight	132.13 gm/mole
5	Physical description	Brownish gray to white colored odorless crystals
6	Solubility	Freely soluble in water, insoluble in ethanol, ether
7	Melting point	280°C
8	Boiling point	-
9	Flash point	-
10	Density	1.77 gm/cm³ at 27°C

10.5.2 *Manufacturing process*

Ammonium sulfate is naturally present in volcanic fumaroles as the rare mineral mascagnite as a result of coal burns on some dumps. Commercially, it is made by mixing sulfuric acid with ammonia, which is frequently a byproduct of coke furnaces. It is also manufactured from gypsum ($CaSO_4 \cdot 2H_2O$). Gypsum that has been finely split is added to an ammonium carbonate solution. Ammonium sulfate remains in the solution as calcium carbonate precipitates as a solid.

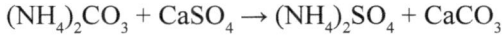

$$(NH_4)_2CO_3 + CaSO_4 \rightarrow (NH_4)_2SO_4 + CaCO_3$$

10.5.2.1 *Raw materials*
Coke oven gas (ammonia) and Concentrated Sulfuric acid

10.5.2.2 *Chemical reactions*
It is prepared by treating ammonia (coke oven gas-by-product from coke ovens) with concentrated sulfuric acid.

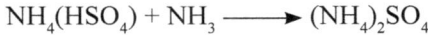

$$NH_3 + H_2SO_4 \longrightarrow NH_4(HSO_4)$$

$$NH_4(HSO_4) + NH_3 \longrightarrow (NH_4)_2SO_4$$

10.5.2.3 *Quantitative requirements*
1000 kgs of ammonium sulfate is obtained using 400 kgs of coke oven gas and 600 kgs of concentrated sulfuric acid.

10.5.2.4 *Flow sheet diagram*

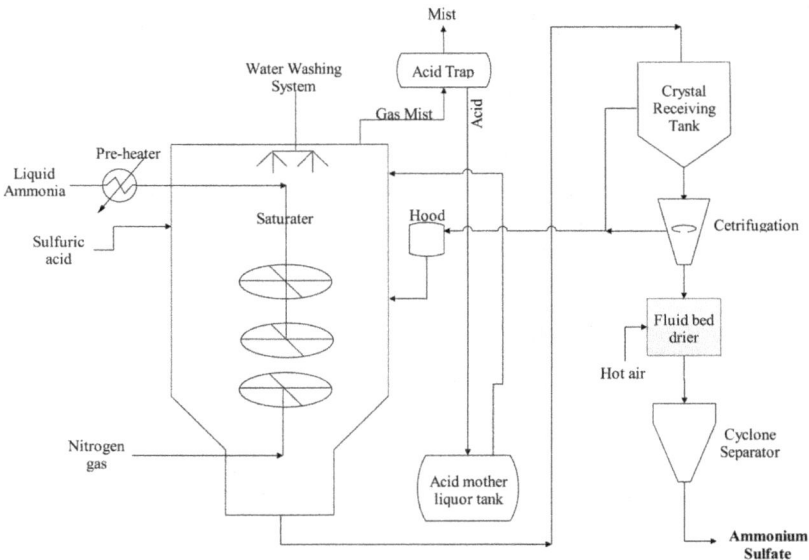

Figure 10.4 Flow sheet diagram of production of Ammonium sulfate

10.5.2.5 Process description

Coke oven gas (liquid ammonia) is preheated and evaporated at a temperature of 60–70°C in an evaporator using low pressure steam at 16 atmospheres and preheated. Stoichiometric quantities of preheated gaseous ammonia and concentrated sulfuric acid are introduced to the saturator, in such way that preheated gaseous ammonia is bubbled into the sulfuric acid. The saturator is a cylindrical vessel with a conical bottom having a bubbler with a duct in the middle attached to it of bubbler and duct is at the middle of it. The duct, containing a hood at the bottom is provided with a vane like arrangement. Also, it is provided with a ring type system to insert nitrogen gas at the bottom. Also, it is facilitated with a water washing system at the top. The saturator is continuously maintained with liquor in an acid bath that comprises of 4–5% sulfuric acid. The bubbler hood, which is submerged in the bath, is where the coke oven gas enters. The gas rises through the mother liquor. During this interval, the ammonia contained in the gas interacts with the sulfuric acid in the liquid to generate solid ammonium sulfate, which settles at the bottom of the saturator. At 6–7 kg/cm², pure nitrogen is purged into the saturator through N_2 rings. Purging nitrogen promotes crystal formation. To keep the saturator's acidity at a constant level, pure sulfuric acid (98%) is fed into the device. To gather the ammonium sulfate, water is sprayed at the saturator's wall. The mist and acid are separated in the acid trap. The gas mist gathers at the top of the saturator. The acid is recycled in the saturator, while the mist is removed as a stack after proper treatment. Now, the ammonium sulfate containing slurry is collected from the bottom of the saturator and transferred into a settling tank having a conical shape. The top of the settling tank contains liquor, which is recycled into the saturator. The bottoms containing ammonium sulfate are further centrifuged. Thereafter, ammonium sulphate is dried using hot air at 120–150°C in a fluid bed drier. After drying, it is fed into a cyclone separator to separate finely divided ammonium sulfate and stored.

10.5.3 Uses

1. Primarily it is used as a crop fertilizer and in lawns.
2. It is used in vaccine preparations and as a water purifier for pharmaceuticals.
3. It is also used in the chemical, wood pulp, textile and pharmaceutical industries (Hanlocon and Zhao, 2015; Kandil and Cheira, 2016).

10.6 Potassium sulfate

The second-largest potassium based chemical is potassium sulfate, which is mostly used as a fertilizer. 44% of it is potassium.

10.6.1 Properties

Its main properties are as per Table 10.4.

Table 10.4 Physical & Chemical Properties of Potassium sulphate

Sr. No.	Properties	Particular
1	IUPAC name	Dipotassium sulfate
2	Chemical and other names	Potassium sulfate, Arcanum duplicatum, Sal Polychrestum
3	Molecular formula	K_2O_4S
4	Molecular weight	174.25 gm/mole
5	Physical description	Colourless or white crystals or crystalline odorless powder
6	Solubility	Soluble in water, ammonia; Insoluble in ethanol, ether, acetone
7	Melting point	1067°C
8	Boiling point	1689°C
9	Flash point	-
10	Density	2.660 gm/cm³ at 27°C

10.6.2 Manufacturing process

Various processes are used to manufacture potassium sulfate: (1) Decantation Process or Langbeinite Process, in which Langbeinite (ore of potash) is treated with concentrated solution of potassium chloride, (2) Glascrite Process, in which sodium sulfate is reacted with potassium chloride (3) Mannheim Process, in which sulfuric acid reacts with potassium chloride or hydroxide (4) Recovery from Kainite using float floatation and (5) Hargreaves Process, in which sulfur dioxide is reacted with air (O_2), water vapor (H_2O) and hot potassium chloride. Among the above processes, the Mannheim process is the most feasible process industrially. Originally, the Mannheim process was originally developed from sodium sulfate production by reacting sodium chloride with sulfuric acid. But sodium chloride is replaced by potassium hydroxide.

10.6.2.1 Raw materials

Potassium hydroxide and 50% Sulfuric acid

10.6.2.2 Chemical reactions

Potassium sulfate is prepared by reacting potassium hydroxide with 50% sulfuric acid.

$$2KOH + H_2SO_4 \longrightarrow K_2(SO_4) + 2H_2O$$

10.6.2.3 Quantitative requirement

1000 kgs of potassium sulfate is obtained using 1150 kgs of potassium hydroxide and 804 kgs of 50% sulfuric acid.

10.6.2.4 Flow sheet diagram

10.6.2.5 Process description

As per Fig. 10.5, water is taken in a vessel containing an agitator. Solid potassium hydroxide is added to it to make a 50% solution. Now, 50% potassium hydroxide solution is slowly added to a brick lined reactor, in which 50% sulfuric acid

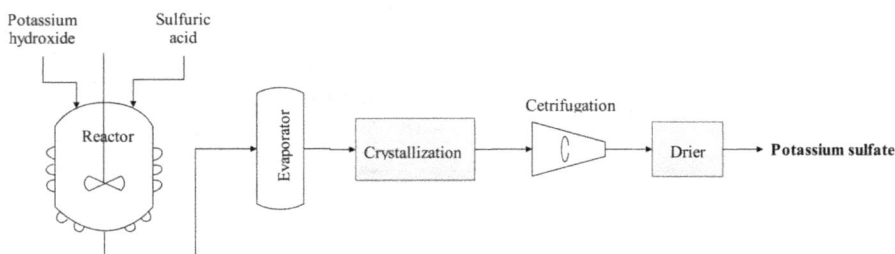

Figure 10.5 Flow sheet diagram of production of Potassium sulfate

is already present to yield potassium sulfate. The reaction gets automatically started. After competition of the reaction, the solution is transferred into an evaporator. Here, potassium sulfate solution is concentrated by means of hot air. Thereafter, it is crystallized, centrifuged and dried.

10.6.3 Uses

1. 53% of potassium and 17% of sulfur, both of which are needed for good plant growth, are present in potassium sulfate. Potassium provides several advantages for plants, including promoting root growth, increasing their yield and quality, balancing their metabolism to promote protein synthesis and energizing them for healthy growth, and enhancing their immune systems.
2. Glass manufacturing occasionally uses it as well.
3. Potassium sulfate is additionally utilized in artillery propellant charges as a flash suppressor. Muzzle flash, flareback, and blast over pressure are all reduced. It is also used in pigments, detergents, cleaning agents and drilling fluids (Grzmil and Kic, 2005; Bantang and Camacho et al., 2020).

10.7 Potassium nitrate

It appears naturally as the mineral niter. It's a source of nitrogen, hence its name is derived from it. One of the many nitrogen-containing compounds collectively referred to as saltpetre or saltpetre is potassium nitrate.

10.7.1 Properties

Its main properties are as per Table 10.5.

10.7.2 Manufacturing process

Different methods are available for manufacturing potassium nitrate, which are mentioned below.

1. Reaction between sodium nitrate and potassium chloride
2. Reaction between ammonium nitrate and potassium chloride

Table 10.5 Physical & Chemical Properties of Potassium nitrite

Sr. No.	Properties	Particular
1	IUPAC name	Potassium nitrite
2	Chemical and other names	Potassium nitrite
3	Molecular formula	KNO_2
4	Molecular weight	85.10 gm/mole
5	Physical description	Yellowish white crystalline odorless solid
6	Solubility	Soluble in water; Insoluble in ethanol, ether, acetone
7	Melting point	441°C
8	Boiling point	537°C
9	Flash point	-
10	Density	1.915 gm/cm³ at 27°C

3. Reaction between molten salt [Mixture of sodium nitrate (54.5% by wt) and potassium nitrate (44.5% by wt)] and potassium chloride.
4. Reaction between potassium chloride and nitric acid
5. Production of potassium nitrate by ion exchange process (continuous process)
6. Production of potassium nitrate by reacting calcium nitrate and potassium chloride (continuous process)
7. Production of potassium nitrate by the reaction of aluminum nitrate and potassium chloride

Out these methods, it is prepared by the double displacement reaction between sodium nitrate and potassium chloride on an industrial scale.

10.7.2.1 Raw materials

Sodium nitrate and Potassium chloride

10.7.2.2 Chemical reactions

It is prepared by reacting sodium nitrate with potassium chloride.

$$NaNO_3 + KCl \longrightarrow NaCl + KNO_3$$

10.7.2.3 Flow sheet diagram

10.7.2.4 Process description

In the first step, water is taken in a jacketed reactor with a stirring facility. Now, solid sodium nitrate is added with stirring to make the solution. After addition, heat the sodium nitrate solution upto 70–75°C by mean of steam in the jacket. Now, solid potassium chloride is charged into the reactor with stirring and maintaining a temperature of 70–75°C. By adding, KCl crystals in the reactor they change to NaCl crystals, which allows the hot potassium nitrate solution to pass through them. Some water is charged to avoid further deposition of NaCl as the solution is cooled. Thereafter, it is crystallized, centrifuged and dried.

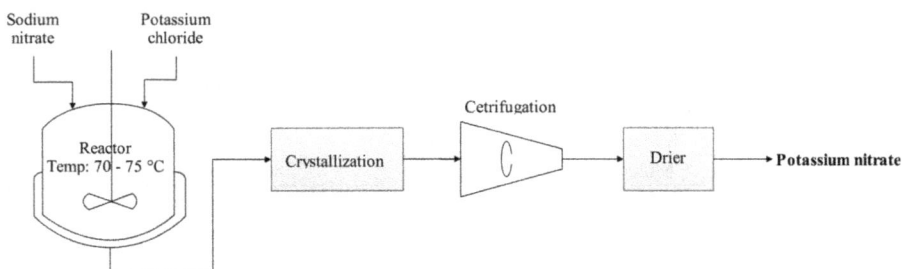

Figure 10.6 Flow sheet diagram of production of Ammonium nitrite

10.7.3 Uses

1. Primary it is used as a fertilizer being a source of potassium and nitrogen.
2. It is used for the production of nitric acid.
3. It is used as a food preservative, oxidizer and electrolyte in a salt bridge.
4. It is also utilized in pharmaceutical industries (Joshi et al., 2014; Jurisova et al., 2013).

10.8 Triple Super Phosphate (TSP)

It is an inorganic compound, also known as monocalcium phosphate.

10.8.1 Properties

Its main properties are as follows.

Table 10.6 Physical & Chemical Properties of Triple Super Phosphate (TSP)

Sr. No.	Properties	Particular
1	IUPAC name	Calcium phosphoric acid
2	Chemical and other names	Triple super phosphate, Single superphosphate, Enriched superphosphate
3	Molecular formula	$CaH_6O_8P_2^{+2}$
4	Molecular weight	236.06 gm/mole
5	Physical description	Grey or brown colored granular grain odorless color
6	Solubility	Soluble in water; Insoluble in ethanol, ether, acetone
7	Melting point	109°C
8	Boiling point	203°C
9	Flash point	-
10	Density	2.220 gm/cm³ at 27°C

10.8.2 Manufacturing process

Two processes are available for manufacturing the triple super phosphate. In the first process calcium hydroxide is treated with phosphoric acid. In another process it is manufactured from phosphate rock, which is more feasible. In this process, phosphate rock is digested with phosphoric acid to give triple super phosphate.

Sometimes, phosphoric acid is also manufactured by reacting phosphate rock with sulfuric acid. Here, at this stage, gypsum is a by-product which is removed.

10.8.2.1 Raw materials

Phosphate rock and Sulfuric acid

10.8.2.2 Chemical reactions

TSP is prepared by reacting phosphate rock and sulfuric acid.

$$Ca_5FO_{12}P_3 + 7H_2SO_4 + H_2O \longrightarrow 3CaH_4(PO_4)_2 \cdot H_2O + 2HF + 7CaSO_4$$

Side Reaction:

$$4HF + 3SiO_2 + H_2O \longrightarrow SiO_2 \cdot H_2O + 2H_2SiF_6$$

10.8.2.3 Quantitative requirements

1000 kgs of triple super phosphate is obtained using 560 kgs of phosphate rock and 360 kgs of 98% sulfuric acid.

10.8.2.4 Flow sheet diagram

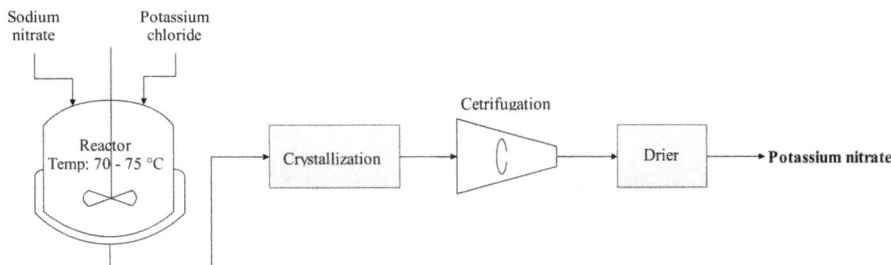

Figure 10.7 Flow sheet diagram of production of Triple Super phosphate

10.8.2.5 Process description

In this process, rock phosphate is crushed using a crusher. This crushed rock is fed into a two stage reactor (digester). Rock phosphate is reacted with sulfuric acid to yield phosphoric acid, which is further digested with rock phosphate to give triple super phosphate. As phosphate rock contains some silicon dioxide (SiO_2), Fluorosilicic acid (H_2SiF_6), nitrosyl chloride (NOCl) and other gases are generated during the reaction, which are scrubbed with water in a scrubber. This scrubbed water (fluoro silicate acid liquor) is recycled into the reactor. Now, the reaction mass is collected from the bottom of the reactor and subjected to filtration to remove insoluble gypsum ($CaSO_4$). The reaction mass containing TSP is transferred into an evaporator. Here, the solution is concentrated by means of hot air. Thereafter, it is crystallized, centrifuged and dried.

10.8.3 Uses

1. Inorganic nutrients including calcium and potassium are included in Triple Super Phosphate fertilizer, which is used to replenish vital soil elements for agricultural purposes.
2. Additionally, it is employed in the food business as a leavening agent, which makes baked items rise (Hakan et al., 2012; Oueslati, 2015).

10.9 DDT

Organochlorine chemical dichlorodiphenyltrichloroethane, also known as DDT, was initially created as an insecticide but later gained notoriety for its adverse effects on the environment. The insecticidal properties of DDT, which were first produced in 1874, were discovered in 1939 by Swiss chemist Paul Hermann Muller. DDT was utilized in the last half of world war II to regulate malaria and typhus among civilians and troops. It causes neurons in insects to open their sodium ion channels and fire spontaneously, which leads to spasms and eventually death. DDT and other insecticides do not affect insects with specific sodium channel gene alterations. Additionally, DDT resistance is provided by up-regulation of genes encoding cytochrome P450 in some insect species because more of these enzymes speed up the conversion of the toxin into inactive metabolites.

10.9.1 Properties

Its main properties are as follows.

Table 10.7 Physical & Chemical Properties of DDT

Sr. No.	Properties	Particular
1	IUPAC name	1-chloro-4-[2,2,2-trichloro-1-(4-chlorophenyl) ethyl] benzene
2	Chemical and other names	DDT, 4,4' DDT, Clofenotane, Chlorophenothane, Dicophane, Chlorphenotoxum, Parachlorocidum
3	Molecular formula	$C_{14}H_9Cl_5$
4	Molecular weight	354.48 gm/mole
5	Physical description	Colorless crystals or off-white powder with a slight, aromatic odor
6	Solubility	Soluble in acetone, ether, benzene, carbon tetrachloride, kerosene, dioxane, pyridine.
7	Melting point	109°C
8	Boiling point	260°C
9	Flash point	75°C
10	Density	16. 01 gm/cm^3 at 27°C

10.9.2 Manufacturing process

Starting materials for preparing DDT are ethanol, chlorine gas and sulfuric acid.

10.9.2.1 Raw materials

Ethanol, Chlorine gas, Sulfuric acid, Monochlorobenzene (MCB), Oleum and Lime

10.9.2.2 Chemical reactions

First, Chloral is prepared by chlorination of ethanol with chlorine gas in the presence of 98% sulfuric acid. Thereafter, Chloral is reacted with two moles of MCB in the presence of fuming sulfuric acid (oleum) as a catalyst to yield DDT.

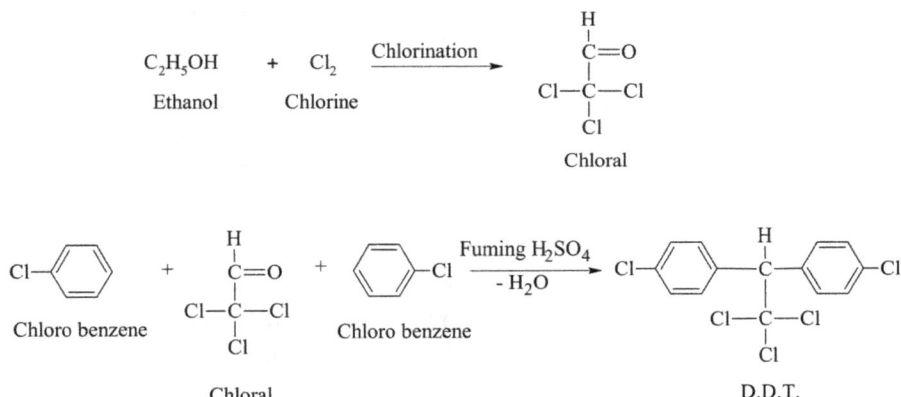

Figure 10.8 Synthesis of DDT

10.9.2.3 Flow sheet diagram

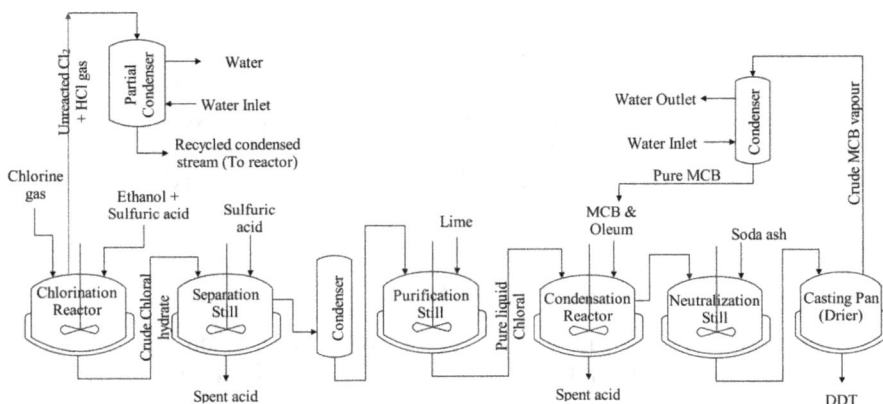

Figure 10.9 Flow sheet diagram of production of DDT

10.9.2.4 Process description

In this process, ethanol and sulfuric acid are taken in a jacketed chlorination reactor with a stirring facility. Dry chlorine is absorbed into ethanol at room temperature to form chloral hydrate and HCl gas. HCl and the remaining chlorine gas undergo partial condensation using a partial condenser for liquidation. The uncondensed gases are scrubbed using water. Condensed liquid from the partial condenser is recycled into the reactor. The reaction slurry is collected from the bottom of reactor and transferred into a separation still facilitated with a jacket. In the still, waste acid is separated in the presence of sulfuric acid which acts as a desiccant at a high temperature and pressure. Also, chloral hydrate is converted into chloral. This waste acid is collected from the bottom and a gas stream having chloral is accumulated on the top of the reactor. The chloral is converted into liquid chloral using a water condenser. The chloral is treated with lime to remove dissolved acidic impurities. Purified liquid chloral is taken into a condensation reactor facilitated with a jacket and a stirrer. Also, Monochlorobenzene and fuming sulfuric acid (Oleum) are added into the reactor with stirring. Water is circulated in the jacket to control the temperature. After the reaction, the reaction mass settles down to separate the organic and spent acid layer. The spent acid is withdrawn from the bottom. The organic layer containing DDT and un-reacted MCB is neutralized with soda ash. Now, the neutralized mass is collected from the bottom of the reactor and transferred into a casting pan (drier). Here, vapor having MCB is condensed using a water condenser and recycled. Dry DDT powder is collected and packed.

10.9.2.5 Major engineering problems
1. In the condensation reaction, the room temperature must be controlled using water to avoid the sulfonation reaction of MCB using oleum at a higher temperature.
2. Composition of DDT is avoided while DDT is come out in contact with iron.

10.9.3 Uses

It is a well-known organochloride known for its insecticidal properties and environmental impacts. In addition to xylene solutions or petroleum distillates, emulsifiable concentrates, water-wettable powders, granules, aerosols, smoke candles, and charges for vaporizers and lotions there are several formulations that contain DDT. Its substitute Methoxy chloral [Bis(methoxy phenyl) trichloroethane] is more useful than DDT. It is more efficient and less toxic than BHC (Turusov et al., 2002).

10.10 Hexachlorobenzene (BHC)

Hexachlorobenzene, also known as per chlorobenzene, is an organochloride. Previously it was a fungicide used for mainly for wheat seed treatment to suppress fungal disease.

10.10.1 Properties

Its main properties are as per Table 10.8.

Table 10.8 Physical & Chemical Properties of HBC

Sr. No.	Properties	Particular
1	IUPAC name	1, 2, 3, 4, 5,6 -hexachlorobenzene
2	Chemical and other names	Hexachlorobenzene, HBC, Perchlorobenzene, Sanocide, Hexachlorbenzol, Amatin
3	Molecular formula	C_6Cl_6
4	Molecular weight	284.77 gm/mole
5	Physical description	White to colorless odorless crystalline
6	Solubility	Soluble in cold alcohol, benzene, chloroform, and ether
7	Melting point	442°C
8	Boiling point	325°C
9	Flash point	242°C
10	Density	2.044 gm/cm³ at 27°C

10.10.2 Manufacturing process

It is prepared from benzene and chlorine.

10.10.2.1 Raw materials

Benzene and Chlorine gas

10.10.2.2 Chemical reactions

Dry benzene undergoes chlorination with chlorine gas in the presence of Ultraviolet rays to yield BHC.

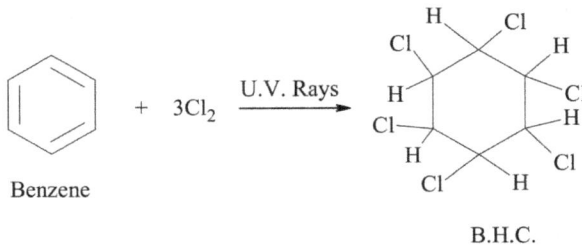

Figure 10.10 Synthesis of BHC

10.10.2.3 Flow sheet diagram

10.10.2.4 Process description

As per the flow-sheet, benzene and water are inserted into a chlorination reactor facilitated with a jacket and stirring. The solution is heated upto 70–75°C. Now, chlorine is inserted through the solution of benzene and water with stirring. The temperature is maintained at 70–75°C for two hours to form BHC. The remaining chlorine and HCl gas are absorbed in the scrubber water. After 2 hours, the mixture containing BHC, chlorobenzene and unreacted benzene is collected

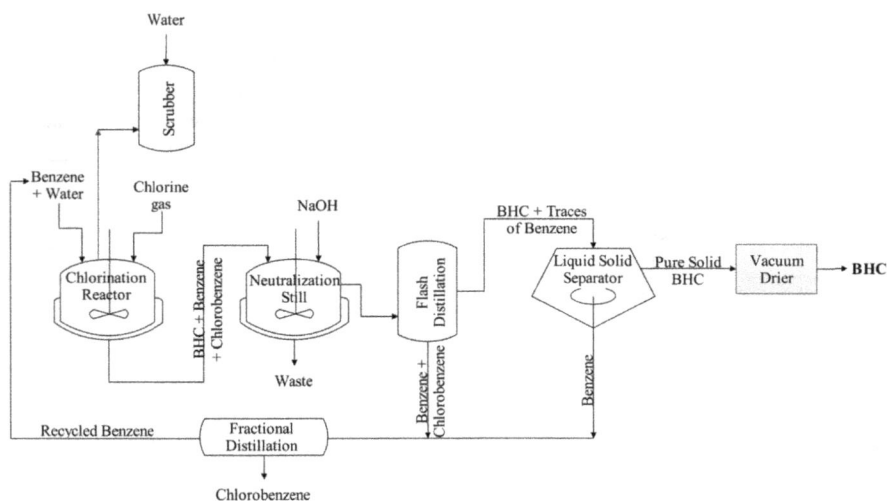

Figure 10.11 Flow sheet diagram of production of BHC

from the bottom of the reactor and transferred into a neutralization vessel. Here, the reaction mass gets neutralized using NaOH. This neutralized mass is distilled out in a flash distillation column using steam to separate out chlorobenzene and unreacted benzene from BHC. The remaining reaction mixture contains BHC and a trace of benzene, which undergoes solid-liquid separation to separate out liquid benzene and solid BHC. Solid dry BHC is purified by vacuum drying. Here, unreacted benzene and chlorobenzene are subjected to fractional distillation to separate out benzene which is recycled.

10.10.2.5 Major engineering problems

Due to chlorine usage and hydrochloric acid emission in this process, corrosion may be a major problem during the entire process.

10.10.3 Uses

1. Hexachlorobenzene is mainly used as a fungicide to control wheat bunt and other fungus on grains.
2. It is most often used to make other compounds, but has also been used as a wood preservative.
3. It is also used as an additive in explosives, as well as a part of the production process for rubber, aluminum, and dyes (Kumar et al., 2013).

10.11 2,4-Dichlorophenoxyacetic acid (2,4-D)

Dichlorophenoxyacetic acid, known as 2,4-D, is one of the oldest and most commonly accessible herbicides in the World, commercially available since 1945. Numerous commercial lawn herbicide blends contain it.

10.11.1 Properties

Its main properties are as follows.

Table 10.9 Physical & Chemical Properties of 2, 4-D

Sr. No.	Properties	Particular
1	IUPAC name	2-(2, 4-dichlorophenoxy) acetic acid
2	Chemical and other names	2, 4 – D, Hedonal, Agrotect, Tributon
3	Molecular formula	$C_8H_6Cl_2O_3$
4	Molecular weight	221.03 gm/mole
5	Physical description	White to yellow, crystalline, odorless powder
6	Solubility	Soluble in water, alcohol, ether, ketone, benzene
7	Melting point	-
8	Boiling point	139°C
9	Flash point	-
10	Density	1.42 gm/cm³ at 27°C

10.11.2 Manufacturing process

Starting material for 2,4-D is phenol.

10.11.2.1 Raw materials

Phenol, Iron, Chlorine, Monochloroacetic acid (MCA), Sodium hydroxide and Hydrochloric acid

10.11.2.2 Chemical reactions

The chlorination of Phenol with chlorine gas in the presence of iron at 50–60°C yields 2,4-dichloro phenol, which undergoes a reaction with MCA to form 2,4-dichlorophenoxyacetic acid (2,4-D). It is neutralized with sodium hydroxide to yield a sodium salt of 2,4-D.

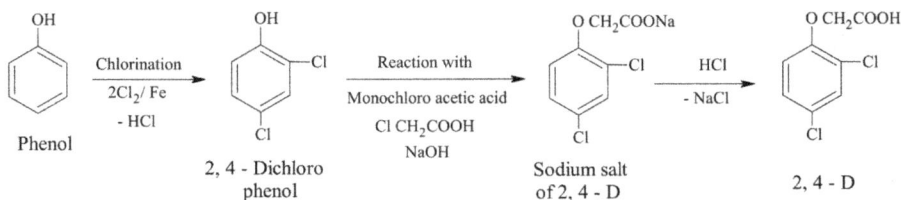

Figure 10.12 Synthesis of 2,4-D

First Dichlorophenol and Monochloroacetic acid react with NaOH and thereafter the products formed also react with each other as follows.

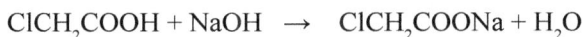

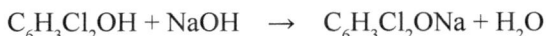

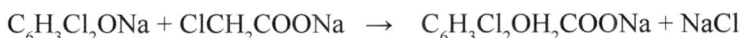

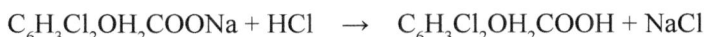

$$ClCH_2COOH + NaOH \rightarrow ClCH_2COONa + H_2O$$

$$C_6H_3Cl_2OH + NaOH \rightarrow C_6H_3Cl_2ONa + H_2O$$

$$C_6H_3Cl_2ONa + ClCH_2COONa \rightarrow C_6H_3Cl_2OH_2COONa + NaCl$$

$$C_6H_3Cl_2OH_2COONa + HCl \rightarrow C_6H_3Cl_2OH_2COOH + NaCl$$

10.11.2.3 Quantitative requirements

1000 kgs of 2,4-D is obtained using 530 kgs of phenol, 700 kgs of chlorine, 350 kgs of iron, 700 kgs of MCB, 600 kgs of sodium hydroxide and 2500 kgs of hydrochloric acid.

10.11.2.4 Flow sheet diagram

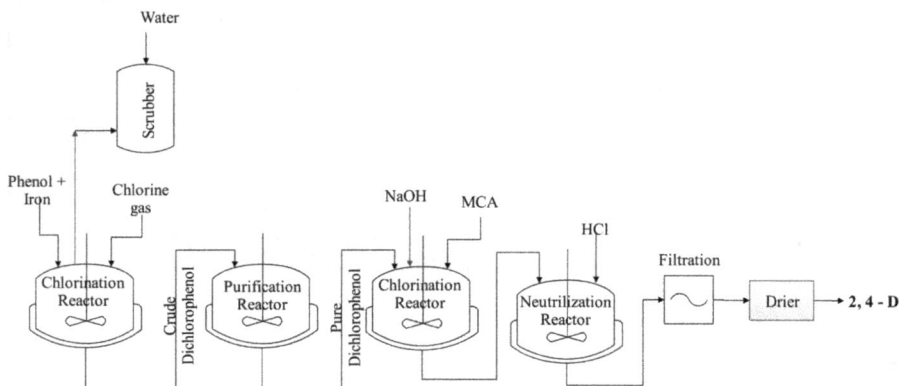

Figure 10.13 Flow sheet diagram of production of 2,4-D

10.11.2.5 Process description

In the first step, Phenol, chlorine gas and iron are introduced into the chlorination chamber. The reaction mass is heated upto 50–60°C with stirring. The remaining chlorine gas is absorbed into the scrubber using water. 2,4-Dichloro phenol (DCP) is purified and inserted into a jacketed agitated reactor. Sodium hydroxide solution is charged into the reactor with stirring to form sodium dichlorophenoxide. Thereafter, Monochloroacetic acid is charged with stirring. The reaction mass is heated upto 70–80°C. This temperature is maintained for 6–8 hours to complete the reaction. The reaction mass is collected from the bottom of the reactor and transferred into a neutralization tank. 30% hydrochloric acid solution is added into the tank and the acidic pH is adjusted. The product, 2,4-D comes out as a precipitate, which is filtered out, washed with water and dried in a dryer.

10.11.3 Uses

It has nine different types of isomers like alpha, beta, gamma, delta, in which the gamma-isomer (γ-isomer), is a 1,000 times more efficient insecticide. The market name of γ-BHC is Gamexin (Burns and Swaen, 2012).

10.12 2,4,5-Trichlorophenoxyacetic Acid (2,4,5-T)

2,4,5-Trichlorophenoxyacetic acid (commonly known as 2,4,5-T), a synthetic auxin, a chlorophenoxy acetic acid herbicide used to defoliate broad-leafed plants. It was created in the late 1940s and found extensive use in the agriculture sector.

10.12.1 Properties

Its main properties are as follows.

Table 10.10 Physical & Chemical Properties of 2, 4, 5-T

Sr. No.	Properties	Particular
1	IUPAC name	2-(2, 4, 5-trichlorophenoxy) acetic acid
2	Chemical and other names	2, 4, 5 – T, Fortex, Trioxon
3	Molecular formula	$C_8H_5Cl_3O_3$
4	Molecular weight	255.47 gm/mole
5	Physical description	Colorless to tan odorless crystalline solid
6	Solubility	Soluble in water, alcohol, ether, n-heptane, toluene, and xylene
7	Melting point	138°C
8	Boiling point	-
9	Flash point	-
10	Density	1.803 gm/cm³ at 27°C

10.12.2 Manufacturing process

The starting material for 2,4,5-T is 1,2,4,5-tetrachloro benzene.

10.12.2.1 Raw materials

1,2,4,5-Tetrachloro benzene, Sodium hydroxide, Monochloroacetic acid (MCAA) and Hydrochloric acid

10.12.2.2 Chemical reactions

1,2,4,5-Tetrachloro benzene undergoes alkaline hydrolysis with dilute sodium hydroxide to form 2,4,5-trichlorophenol, which is condensed with monochloroacetic acid in the presence of sodium hydroxide to yield a sodium salt of 2,4,5-T. It is neutralized with hydrochloric acid to yield 2,4,5-T.

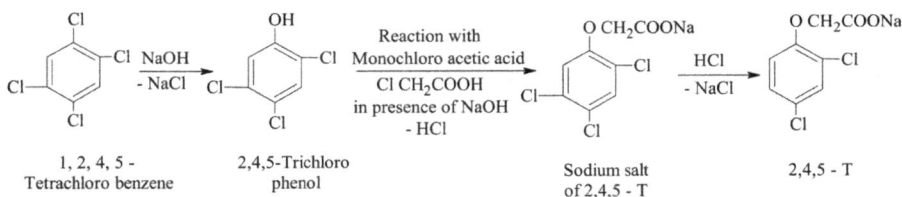

Figure 10.14 Synthesis of 2,4,5-T

10.12.2.3 Flow sheet diagram

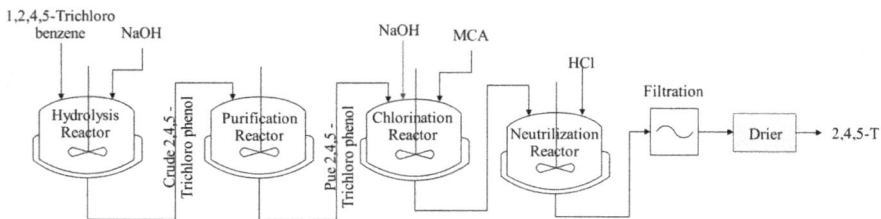

Figure 10.15 Flow sheet diagram of production of 2,4,5-T

10.12.2.4 Process description

At first, a solution of sodium hydroxide in water and 1,2,4,5-tetrachloro benzene are taken into an agitated jacketed reactor. The reaction mass is heated till the reaction is complete to yield 2,4,5-trichlorophenol. After completion of the reaction, crude 2,4,5-trichlorophenol is collected from the bottom of the reactor and transferred into a purification chamber. The product is purified using methoxy dichlorophenol in aqueous chromic acid and thereafter, separated by washing. The purified 2,4,5-trichlorophenol is taken into another jacketed agitated reactor. Monochloroacetic acid and sodium hydroxide solution are also charged into it. The reaction mass is heated upto 140°C with stirring to form a sodium salt of 2,4,5-T. Here, the temperature during this step must be controlled carefully, because if it rises above 160°C a side reaction occurs, and tetrachlorodioxin is produced. After completion of the reaction, the mass is collected from the bottom of the reactor and transferred into a neutralization reactor. Hydrochloric acid is added till a pH of 2–4 is attained. The product is filtered and dried.

10.12.3 Uses

It is an important herbicide because it is capable of killing weeds without affecting the crops. It is very cheap and effective at a low concentration. It is available as ammonium or alkaline ester salt in the market (Huisman and Smit, 2010).

10.13 Aldrin

Until the 1990s, aldrin, organochlorine insecticide, was extensively utilized.

10.13.1 Properties

Its main properties are as per Table 10.11.

Table 10.11 Physical & Chemical Properties of Aldrin

Sr. No.	Properties	Particular
1	IUPAC name	1,2,3,4,10,10-Hexachloro-1,4,4a,5,8,8a-hexahydro-1,4-endo,exo-5,8-dimethano-naphthalene
2	Chemical and other names	Aldrin, HHDN, Aldrex, Octalen, Seedrin
3	Molecular formula	$C_{12}H_8Cl_6$
4	Molecular weight	364.89 gm/mole
5	Physical description	White crystalline odorless solid
6	Solubility	Soluble in water, petroleum, acetone, benzene and xylene
7	Melting point	105°C
8	Boiling point	145°C
9	Flash point	-
10	Density	1.70 gm/cm³ at 27°C

10.13.2 Manufacturing process

The starting material for manufacturing Aldrin is n-Pentane.

10.13.2.1 Raw materials

n-Pentane, Chlorine gas, Anhydrous Sulfuric acid, Acetylene and Cyclopentadiene

10.13.2.2 Chemical reactions

The first step is to manufacture hexachlorocyclopentadiene (HCCP) by the chlorination of n-pentane with chlorine gas in the presence of anhydrous sulfuric acid. It is prepared by the reaction of hexachlorocyclopentadiene, acetylene and cyclopentadiene. In the second step, first acetylene and cyclopentadiene undergoea reaction at 160–180°C to yield aldrin.

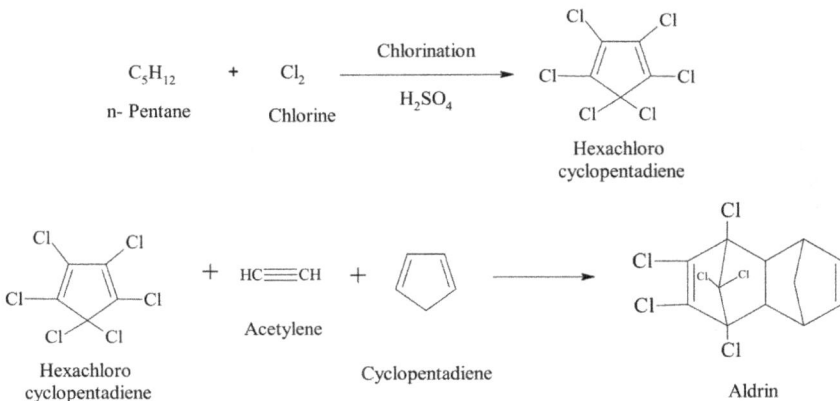

Figure 10.16 Synthesis of Aldrin

10.13.2.3 Flow sheet diagram

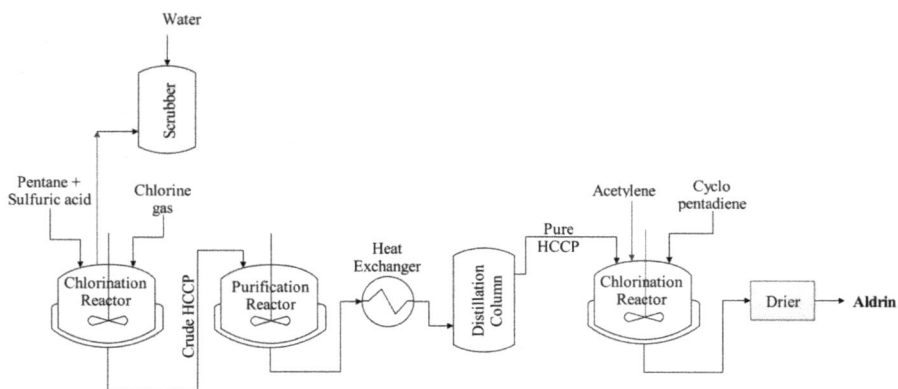

Figure 10.17 Flow sheet diagram of production of Aldrin

10.13.2.4 Process description

As per flow-sheet, anhydrous sulfuric acid is taken into an agitated and jacketed chlorination reactor. Also, n-pentane and chlorine gas are charged into it with stirring. The reaction mass is heated till the completion of the reaction. The remaining chlorine gas is absorbed in the scrubber using water. The crude HCCP is collected from the bottom of the reactor and passed through a heat exchanger and distillation column to obtained purified HCCP. Now, cyclopentadiene and acetylene are taken into another agitated and jacketed reactor. The reaction mass is heated upto 160–180°C for 2 hours to give norbornadiene. Thereafter, purified HCCP is charged into a reactor with stirring at 160–180°C to yield Aldrin. The product is filtered and dried.

10.13.3 Uses

Pesticide Aldrin is used to control insects that live in the soil, including termites, corn root worms, wire worms, rice water weevils, and grasshoppers. It has a long history of use in protecting crops like corn and potatoes, and it works well to keep termites away from wooden constructions. Aldrin is readily metabolized to dieldrin by both plants and animals. Aldrin residues are consequently infrequently and sparingly detected in foods and animals. It is particularly resistant to leaking into groundwater and strongly adheres to soil particles. It is prohibited in many nations due to its toxicity.

10.14 Pretilachlor

Pretilachlor is one of recognized synthetic herbicides.

10.14.1 Properties

Table 10.12 Physical & Chemical Properties of Pretilachlor

Sr. No.	Properties	Particular
1	IUPAC name	2-Chloro-N-(2,6-diethylphenyl)-N-(2-propoxyethyl) acetamide
2	Chemical and other names	2-Chloro-2',6'-diethyl-N-(2-propoxyethyl) acetanilide, Pretilachlor, Solnet
3	Molecular formula	$C_{17}H_{26}ClNO_2$
4	Molecular weight	311.85 gm/mole
5	Physical description	Pale yellow to colorless oily liquid with sweet odor
6	Solubility	Miscible with water, diethyl ether, acetone, benzene, ethanol, ethyl acetate, hexane
7	Melting point	25°C
8	Boiling point	135°C
9	Flash point	-
10	Density	1.052 gm/cm³ at 27°C

10.14.2 Manufacturing process

Starting material for pretilachlor is 2,6-diethyl aniline.

10.14.2.1 Raw materials

2,6-Diethyl aniline (2,6-DEA), Propoxy ethyl chloride, Sodium hydroxide, Chloroacetyl chloride (CAC), Sodium bicarbonate and Toluene

10.14.2.2 Chemical reactions

Firstly, 2,6-DEA, propoxy ethyl chloride and sodium hydroxide are reacted at 185–190°C to give intermediate 2,6-diethyl-N-(2-propoxyethyl) aniline, which is reacted with CAC at 50–55°C in the presence of toluene to give pretilachlor.

Figure 10.18 Synthesis of Pretilachlor

10.14.2.3 Quantitative requirements

1000 kgs of pretilachlor is obtained using 1150 kgs of 2,6-DEA, 440 kgs of propoxy ethyl chloride, 350 kgs of sodium hydroxide, 400 kgs of CAC, 300 kgs of sodium bicarbonate and 1800 kgs of toluene.

10.14.2.4 Flow sheet diagram

10.14.2.5 Process description

In the first step, an excessive amount of 2,6-DEA is taken in a jacketed reactor having stirrer alignment. The material is heated upto 185–190°C using hot oil

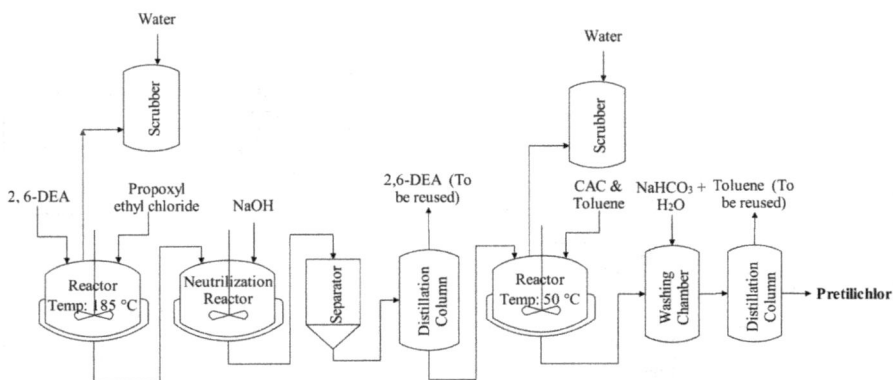

Figure 10.19 Flow sheet diagram of production of Pretilachlor

circulation in the jacket. Then propoxy ethyl chloride is charged into it with stirring. The temperature is maintained at 185–190°C with stirring. During the reaction, hydrochloric acid is formed, which is scrubbed in the scrubber. After 2 hrs, the reaction mass is cooled down to room temperature. The cooled reaction mass is collected from the bottom of the reactor and sent to a neutralization reactor. Sodium hydroxide solution is added to this reactor, until the pH turns alkaline (10–12). Here, 2,6-DEA is regenerated from its hydrochloride. The reaction mass settles down for 2-4 hours for layer separation. The aqueous layer is discarded and the organic layer (containing unreacted 2,6-DEA and intermediate 2,6-diethyl-N-(2-propoxyethyl) aniline) is sent to a distillation tower. In this tower, 2,6-DEA is distilled off and recycled. The remaining pure intermediate 2,6-diethyl-N-(2-propoxyethyl) aniline is sent to another reactor equipped with a jacket and a stirrer. Charge the CAC and toluene into it with stirring. Start heating upto 50–55°C for 4 hours to give final product, Pretilachlor. During the reaction, HCl gas evolved is scrubbed in water to get 30% HCl by-product. After completion of the reaction, the mass is collected from the bottom and sent to a washing reactor. The final product is washed with sodium bicarbonate solution followed by water. The reaction mass is sent to a distillation column, where toluene is distilled off to give pure pretilachlor.

10.14.3 Uses

1. It is a selective herbicide that quickly penetrates the hypocotyls, mesocotyls, and coleoptiles of plants as well as, to a lesser extent, the roots of weeds that are germinated.
2. Pretilachlor is active against primary annual grasses, broad-leaved weeds and sedges in transplanted and seed rice.
3. It is also utilized either for a pre-planting or post-emergence application (Jong et al., 1997; Chen et al., 2020).

10.15 Atrazine

Atrazine is one of the most popular and divisive herbicides in the world because it effectively controls a wide range of weeds, resulting in high crop yields, low treatment costs, application flexibility, little chance of crop damage and drift. As the second in a line of 1,3,5-triazines, it was created in the Geigy laboratories in 1958.

10.15.1 Properties

Table 10.13 Physical & Chemical Properties of Atrazine

Sr. No.	Properties	Particular
1	IUPAC name	6-Chloro-4-N-ethyl-2-N-propan-2-yl-1,3,5-triazine-2,4-diamine
2	Chemical and other names	Atrazine, Gesamprim, Chromozin, Aktikon, Atranex, Fenamin
3	Molecular formula	$C_8H_{14}ClN_5$
4	Molecular weight	215.65 gm/mole
5	Physical description	Colorless or white, odorless, crystalline powder
6	Solubility	Soluble in water, alcohol, ether, ester, DMSO, chloroform, pentane
7	Melting point	173°C
8	Boiling point	-
9	Flash point	-
10	Density	1.220 gm/cm³ at 27°C

10.15.2 Manufacturing process

The starting material for manufacturing Atrazine is cyanuric chloride.

10.15.2.1 Raw materials

Cyanuric chloride, Isopropyl amine, Toluene, Caustic soda, Hydrochloric acid and Monoethyl amine (MEA)

10.15.2.2 Chemical reactions

Cyanuric chloride and isopropyl amine react in the presence of toluene at 15–25°C, followed by treatment with MEA to yield atrazine.

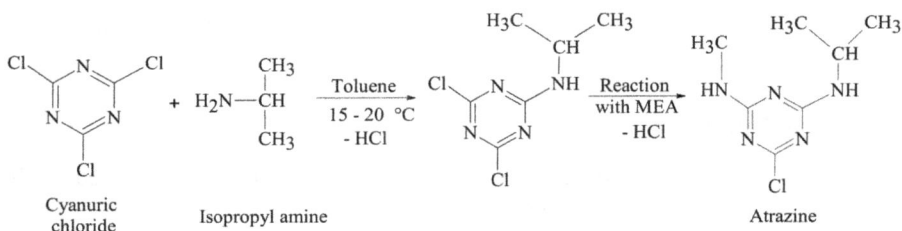

Figure 10.20 Synthesis of Atrazine

10.15.2.3 Quantitative requirements

1000 kgs of atrazine is obtained using 910 kgs of cyanuric chloride, 410 kgs of isopropyl amine, 6000 kgs of toluene, 700 kgs of caustic soda, 250 kgs of MEA and 180 kgs of hydrochloric acid.

10.15.2.4 Flow sheet diagram

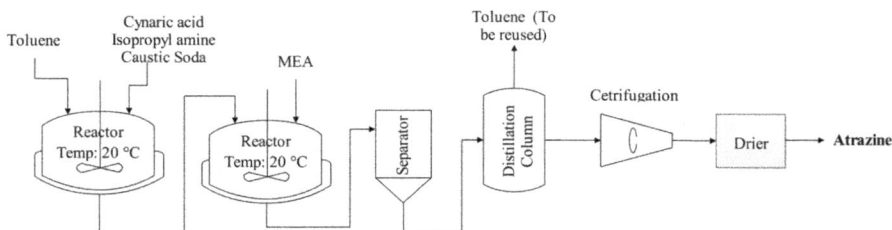

Figure 10.21 Flow sheet diagram of production of Atrazine

10.15.2.5 Process description

Toluene is taken in a jacketed reactor equipped with a stirrer as a solvent. Cyanuric chloride, isopropyl amine and caustic soda solution are charged into it with stirring. The reaction mixture is cooled down to 15–25°C. This temperature is maintained for one hour with stirring. After one hour the reaction mass is collected from the bottom of the reactor and transferred into another jacketed reactor equipped with a stirrer. Now, MEA is added and the same temperature is maintained for another hour with stirring. After completion of the reaction, the reaction mass is allowed to settle for 2–4 hours for layer separation. The aqueous layer is discarded and the organic layer is sent to a distillation tower. Here, toluene is distilled out and recycled. The remaining mass containing the final product is centrifuged and dried.

10.15.3 Uses

A significant pesticide for the management of a wide range of broad leaf weeds and grasses in crops is atrazine. It plays a part in the fight against weeds that are immune to other herbicides, which poses a serious threat to the agricultural sector, and its long-lasting effects in the soil provide stronger control over a wider range of time.

Additionally, it is used on triazine-tolerant canola, sorghum, maize, sugarcane, lupins, pine, and eucalyptus plantations to prevent pre- and post-emergence broadleaf and grassy weeds. (Wen et al., Motheo et al., 2017).

10.16 Bifenthrin

Initially, pyrethroid pesticide is employed to affect the red imported ant's systemanervosum. It is extremely poisonous to aquatic life.

10.16.1 Properties

Table 10.14 Physical & Chemical Properties of Bifenthrin

Sr. No.	Properties	Particular
1	IUPAC name	(2-Methyl-3-phenylphenyl)methyl (1R,3R)-3-[(Z)-2-chloro-3,3,3-trifluoroprop-1-enyl]-2,2-dimethyl cyclopropane-1-carboxylate
2	Chemical and other names	Empower, Fanfare, Biphenthrin, Kappa-bifenthrin
3	Molecular formula	$C_{23}H_{22}ClF_3O_2$
4	Molecular weight	422.87 gm/mole
5	Physical description	Off-white to pale tan waxy solid with a very faint slightly sweet odor
6	Solubility	Soluble in water, methylene chloride, chloroform, acetone, ether, toluene
7	Melting point	69°C
8	Boiling point	-
9	Flash point	165°C
10	Density	1.26 gm/cm³ at 27°C

10.16.2 Manufacturing process

The starting material for manufacturing bifenthrin is 2-methyl 3-biphenylmethanol.

10.16.2.1 Raw materials

2-Methyl-3-biphenyl methanol, Hydrochloric acid, n-Hexane, Cyhylothric Acid, Potassium carbonate and Sodium carbonate

10.16.2.2 Chemical reactions

2-Methyl-3-Biphenylmethanol is chlorinated with hydrochloric acid solution in the presence of n-hexane as a solvent to give BPC (2-biphenyl methyl chloride), which is further reacted with cyhylothric acid and potassium carbonate to yield bifenthrin.

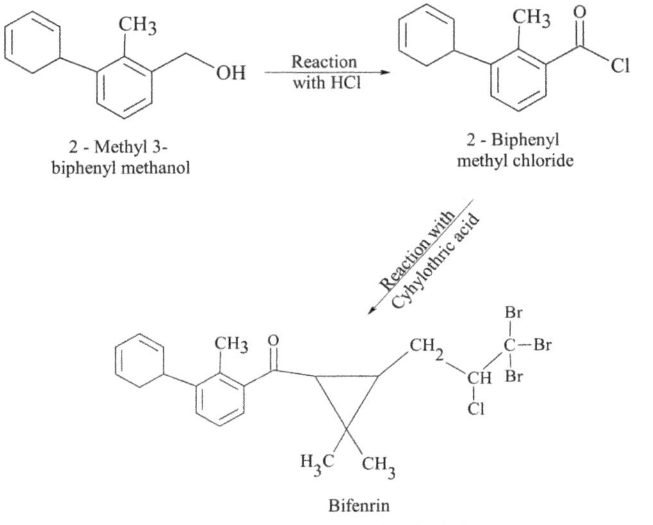

Figure 10.22 Synthesis of Bifenthrin

10.16.2.3 Quantitative requirements

1000 kgs of bifenthrin is obtained using 510 kgs of 2-methyl 3-biphenyl methanol, 500 kgs of hydrochloric acid, 1700 kgs of hexane, 650 kgs of cyhylothric acid, 500 kgs of potassium carbonate and 160 kgs of sodium carbonate.

10.16.2.4 Flow sheet diagram

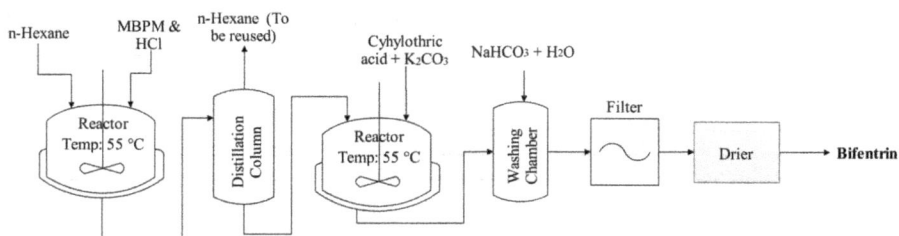

Figure 10.23 Flow sheet diagram of production of Bifenthrin

10.16.2.5 Process description

As per Fig. 10.23, n-Hexane is taken in a reactor equipped with jacket and stirrer. Charge MBPM and hydrochloric acid solution in it with stirring. Heat the reaction mass upto 50–55°C. Maintain this temperature for 4 hours with stirring. After completion of the reaction, the reaction mass containing BPC and hexane is collected from the bottom and sent to a distillation unit. Here, hexane is distilled off and recycled. The reaction mass having BPC is charged into another jacketed reactor having a stirrer. The cyhylothric acid and potassium carbonate are charged into it with stirring. The reaction mass is heated upto 50–55°C. This temperature is maintained for 8 hours with stirring to yield bifenthrin. The reaction mass is collected from the bottom and transferred into a washing unit. Here, the reaction mass is washed with sodium carbonate and thereafter, water. It is then filtered and dried.

10.16.3 Uses

1. When dealing with red imported fire ants, bifenthrin is frequently used.
2. It is effective against many different insects, including termites, aphids, worms, ants, gnats, moths, beetles, earwigs, midges, spiders, ticks, yellow jackets, maggots, and thrips.
3. It is primarily utilized in homes, nurseries, and orchards.
4. It is heavily applied to some crops, such as maize, in the agricultural sector. Bifenthrin is used to treat about 70% of the hops and raspberries grown in the United States (Nazir and Iqbal, 2020; Kougard et al., 2002).

10.17 Diafenthiuron

Diafenthiuron was introduced in 1990 by Ciba-Geigy as an insecticide and acaricide. It is a chemical compound from the group of thioureas.

10.17.1 Properties

Table 10.15 Physical & Chemical Properties of Diafenthiuron

Sr. No.	Properties	Particular
1	IUPAC name	1-Tert-butyl-3-[4-phenoxy-2,6-di(propan-2-yl) phenyl] thiourea
2	Chemical and other names	Diafenthiuron, Pegasus, 3-(2,6-Diisopropyl-4-phenoxyphenyl) -1-tert-butylthiourea
3	Molecular formula	$C_{23}H_{32}N_2OS$
4	Molecular weight	384.58 gm/mole
5	Physical description	Colorless odorless solid
6	Solubility	Insoluble in water; soluble in organic solvents like acetone, ether, toluene
7	Melting point	147°C
8	Boiling point	449°C
9	Flash point	225°C
10	Density	1.11 gm/cm³ at 27°C

10.17.2 Manufacturing process

Previously, it was prepared by the reaction of phenol with 2,6-diisopropyl-4-chloroaniline, thiophosgene and tert-butylamine. Now-a-days it is prepared from 4-phenoxy-2,6-diisopropylaniline.

10.17.2.1 Raw materials

4-Phenoxy-2,6-diisopropylaniline (PIPA), Sodium thiocyanate, Toluene, Hydrochloric acid, ter-Butylamine, and Acetonitrile

10.17.2.2 Chemical reactions

As per Fig. 10.24, PIPA is reacted with sodium thiocyanate at a temperature of 90°C in the presence of hydrochloric acid in solvent toluene to obtain N-[4-phenoxy 2,6-di(propan-2-yl) phenyl] thiourea. After the reaction, the reaction mass is filtered, washed and dried. This thiourea derivative condensation reaction with tertiary butylamine takes place in solvent acetonitrile at 140°C to get Diafenthiuron.

10.17.2.3 Quantitative requirements

1000 kgs of diafenthiuron is obtained using 875 kgs of PIPA, 300 kgs of sodium thiocyanate, 2000 kgs of toluene, 450 kgs of 30% hydrochloric acid, 220 kgs of t-butylamine and 1500 kgs of acetonitrile.

10.17.2.4 Flow sheet diagram

10.17.2.5 Process description

As per the flow sheet diagram, 30% HCl solution and toluene are charged into a stainless steel jacketed reactor facilitated with a stirrer. The solution is heated upto 90°C using steam in the jacket. Charge PIPA and sodium thiocyanate into it with

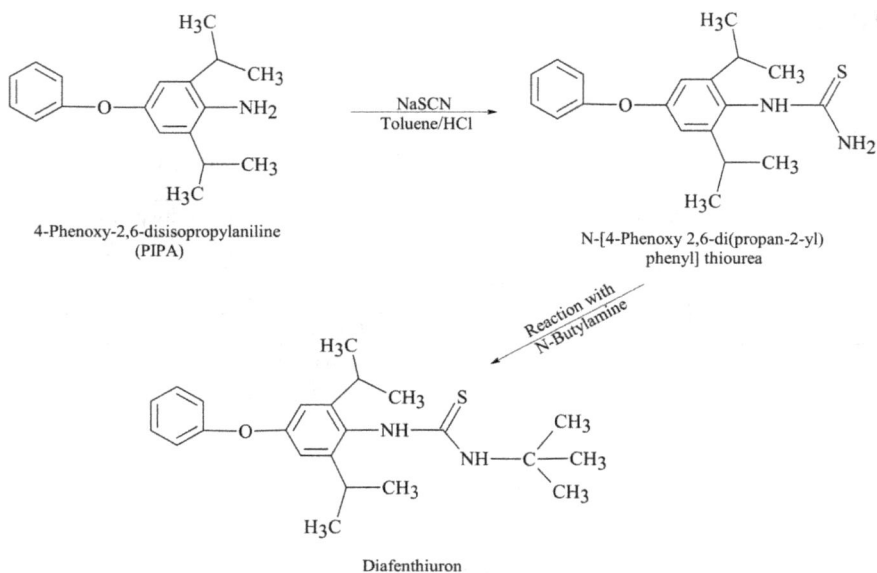

4-Phenoxy-2,6-disisopropylaniline
(PIPA)

N-[4-Phenoxy 2,6-di(propan-2-yl)
phenyl] thiourea

Reaction with
N-Butylamine

Diafenthiuron

Figure 10.24 Synthesis of Diafenthiuron

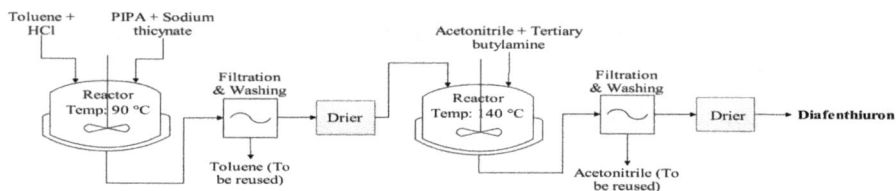

Figure 10.25 Flow sheet diagram of production of Diafenthiuron

stirring. After 2 hours, filter the solution, wash the precipitate of N-[4-phenoxy 2,6-di(propan-2-yl) phenyl] thiourea and dry it. Toluene is recovered from the filtrate and reused. Take acetonitrile in another jacketed reactor with a stirring facility. Charge thiourea derivative and tertiary butylamine into the reactor. The reaction mass is heated upto 140°C using hot oil in the jacket. Maintain the temperature of 140°C with stirring for 4–5 hours to obtain diafenthiuron. Thereafter, the reaction mass is filtered, washed and dried. Acetonitrile is recovered from the filtrate and reused.

10.17.3 Uses

1. Diafenthiuron is an insecticide and an acaricide without regulatory approval to be used in Europe.
2. It is mostly used to prevent and eradicate pests and mites in cotton, soybeans, ornamental plants, fruit trees, and other crops.
3. It can also be used to control other pests of cruciferous plants, such as the diamondback moth and the red spider on apple and citrus trees (Aravind et al., 2015; Schareina et al., 2008).

10.18 Clomazone

A herbicide used in agriculture, clomazone is also the main component of the "Command" and "Commence" products. First it was registered by the USEPA on March 8, 1993 and was commercialized by FMC Corporation. It belongs to the isoxazolidinone family and is bonded to a 2-chlorobenzyl group. It is applied either pre- or post-emergence and comes in the form of an emulsifiable concentrate and a microencapsulated flowable granule (5% clomazone).

10.18.1 Properties

Table 10.16 Physical & Chemical Properties of Clomazone

Sr. No.	Properties	Particular
1	IUPAC name	2-[(2-Chlorophenyl)methyl]-4,4-dimethyl-1,2-oxazolidin-3-one
2	Chemical and other names	Clomazone, Command, Dimethazone, Gamit
3	Molecular formula	$C_{12}H_{14}ClNO_2$
4	Molecular weight	239.69 gm/mole
5	Physical description	Clear colorless to light brown viscous liquid
6	Solubility	Miscible in water, acetone, chloroform, cyclohexanone, methanol, toluene, heptane, DMF
7	Melting point	25°C
8	Boiling point	275°C
9	Flash point	75°C
10	Density	1.192 gm/cm³ at 27°C

10.18.2 Manufacturing process

The starting material for manufacturing clomazone is 3-chloro-2,2-dimethylpropanoyl chloride.

10.18.2.1 Raw materials

3-Chloro-2,2-dimethylpropanoyl chloride (3-CPC), Hydroxylamine hydrochloride, Caustic soda, o-Chlorobenzylchloride, Hydrochloric acid and Sodium carbonate

10.18.2.2 Chemical reactions

As per Fig. 10.26, 3-CPC and Hydroxyl amine hydrochloride react at a pH of 8–10 using caustic solution to yield 3-chloro-N-hydroxy 2,2-dimethyl propanamide, followed by a reaction with caustic lye for 4-5 hours to get 4,4-dimethyl isoxazolidin-3-one. This product undergoes a condensation reaction with o-chloro benzyl chloride for 5–6 hours to yield clomazone isomer. This isomer is reacted with HCl and subsequently, sodium carbonate to give clomazone.

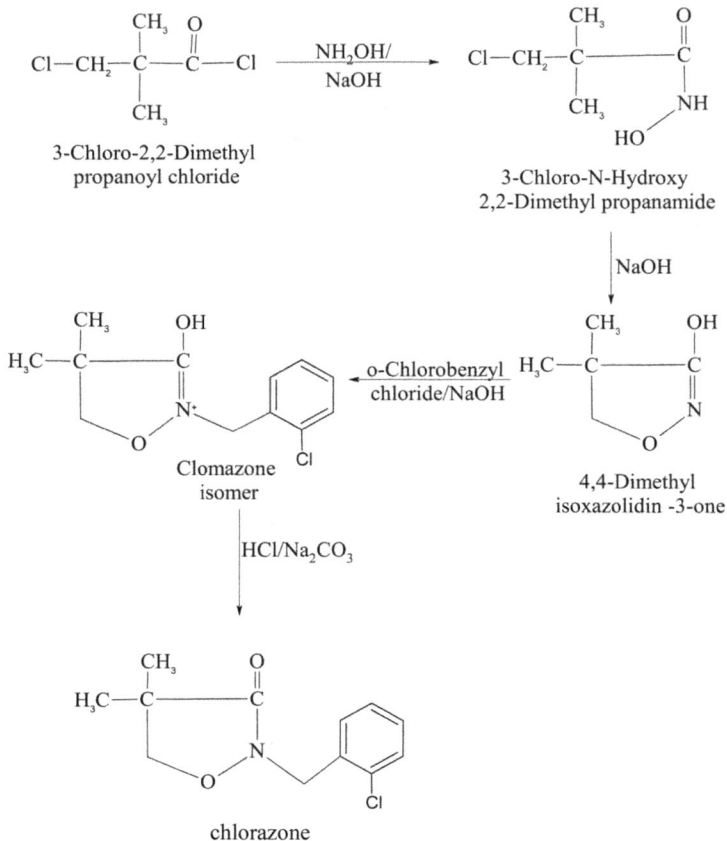

Figure 10.26 Synthesis of Clomazone

10.18.2.3 Quantitative requirements

1000 kgs of chlomazone is obtained using 850 kgs of 3-CPC, 460 kgs of hydroxyl amine hydrochloride, 800 kgs of caustic soda, 700 kgs of o-chloro benzyl chloride, 400 kgs of HCl and 50 kgs of sodium carbonate.

10.18.2.4 Flow sheet diagram

10.18.2.5 Process description

Take water, hydroxylamine and HCl in a continuously stirred reactor. Adjust the pH upto 7–8 using caustic lye. After the addition of caustic lye, charge 3-CPC and caustic lye simultaneously slowly to prepare solid 3-chloro-N-hydroxy 2,2-dimethyl propanamide. The reaction mass is collected from the bottom of the reactor and subjected to filtration. Then it is washed and dried. Charge the water and dry 3-chloro-N-hydroxy2,2-dimethyl propanamide into a caustic reactor. Adjust the pH of the reaction mass upto 7–8 with caustic lye along with stirring. Maintain the pH for 4–5 hours with stirring to get liquid 4,4-Dimethyl

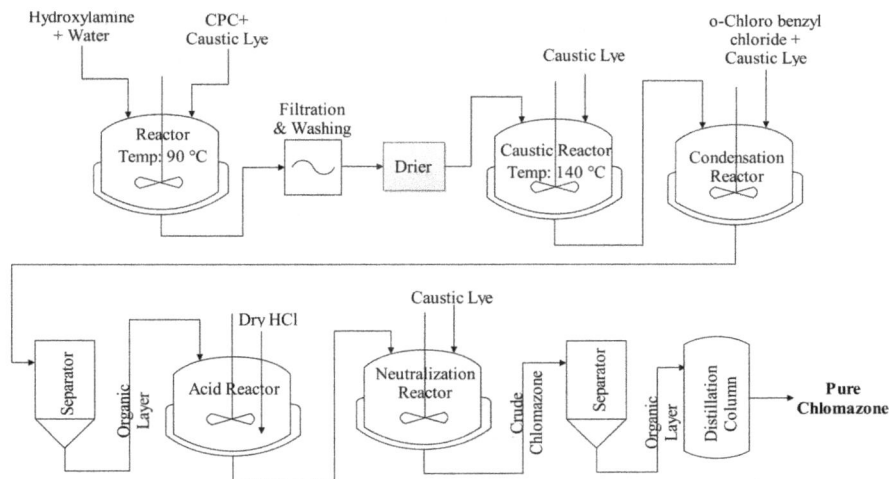

Figure 10.27 Flow sheet diagram of production of Clomazone

isoxazolidin-3-one. Collect the liquid reaction mass from the bottom of the reactor and charge it into another condensation jacketed reactor. Add o-chloro benzyl chloride and caustic lye at a temperature of 45–50°C with stirring. Maintain the temperature of 45–50°C with stirring for 5–6 hours. Cool the reaction mass upto room temperature and collect it from the bottom of the reactor. The reaction mass is transferred into a separator to separate the organic and aqueous layers. The lower organic layer is transferred into another acid tank and the aqueous layer is discarded. Dry hydrochloric acid gas is passed into the reactor from the bottom and maintained for 4–5 hours. The reaction mass is collected from the bottom of the reactor and transferred into a jacketed neutralization reactor equipped with a stirrer. Add sodium carbonate and caustic lye into it. Heat the reaction mass upto 70–80°C. Maintain the temperature for 30 minutes to get crude Clomazone. Cool the reaction mass down to room temperature and collect it from the bottom of the reactor. The reaction mass is transferred into a separator, where the organic and aqueous layers are separated, Then, the aqueous layer is rejected and the organic layer undergoes distillation to get pure liquid Clomazone.

10.18.3 Uses

To control annual grass and broad leaf weeds, clomazone is an extensive herbicide used on fields of rice, peas, pumpkins, soybeans, sweet potatoes, winter squash, cotton, tobacco, and fallow wheat (Ferhatoglu and Barrett, 2006).

10.19 Lambda Cyhalothrin

The U.S. has registered the insecticide lambda-cyhalothrin in 1988 with the Environmental Protection Agency (EPA). Lambda Cyhalothrin is an organic pesticide from the group of pyrethroids. Lambda cyhalothrin is available as an emulsifiable concentrate, wettable powder, or ULV liquid. It is compatible with the majority of other insecticides and fungicides and is frequently combined with buprofezin, pirimicarb, dimethoate, or tetramethrin.

10.19.1 Properties

Table 10.17 Physical & Chemical Properties of Lambda Cyhalothrin

Sr. No.	Properties	Particular
1	IUPAC name	[cyano-(3-phenoxyphenyl)methyl] (1R,3R)-3-[(Z)-2-chloro-3,3,3-trifluoroprop-1-enyl]-2,2-dimethyl cyclopropane-1-carboxylate
2	Chemical and other names	Lambda Cyhalothrin,α-Cyano-3-phenoxybenzyl 3-(2-chloro-3,3,3-trifluoroprop-1-enyl)-2,2-dimethyl cyclopropanecarboxylate, Icon, Karate, Matador
3	Molecular formula	$C_{23}H_{19}ClF_3NO_3$
4	Molecular weight	449.85 gm/mole
5	Physical description	Colorless odorless solid
6	Solubility	Miscible in water, acetone, chloroform
7	Melting point	49°C
8	Boiling point	499°C
9	Flash point	-
10	Density	1.331 gm/cm³ at 27°C

10.19.2 Manufacturing process

The starting materials for manufacturing Lambda Cyhalothrin are Lambda Cyhalothric acid and Thionyl chloride.

10.19.2.1 Raw materials

Lambda Cyhalothric acid, Thionyl chloride, m-Phenoxybenzaldehyde, Sodium cyanide, Caustic lye, n-Hexane, Sodium hypochloride, Isopropyl alcohol and Diisopropyl amine

10.19.2.2 Chemical reactions

As per Fig. 10.28, Lambda cyhalothric acid is chlorinated with thionyl chloride as the chlorinating agent using hexane at 40–45°C to get lambda cyhalothric acid chloride. This chloride product is condensed with m-phenoxybenzaldehyde and sodium cyanide at 20–25°C using n-hexane to form cyhalothrin. Finally, cyhalothrin is epimerized at –5 to 0°C in isopropyl alcohol and diisopropylamine to give lambda cyhalothrin technical.

10.19.2.3 Quantitative requirements

1000 kgs of Lambda cyhalothrin is obtained using 540 kgs of Lambda cyhalothric acid, 270 kgs of thionyl chloride, 440 kgs of m-phenoxybenzaldehyde, 110 kgs of sodium cyanide, 50 kgs of caustic lye, 1200 kgs of n-hexane,75 kgs of sodium hypochloride, 1000 kgs of isopropyl alcohol and100 kgs of diisopropyl amine.

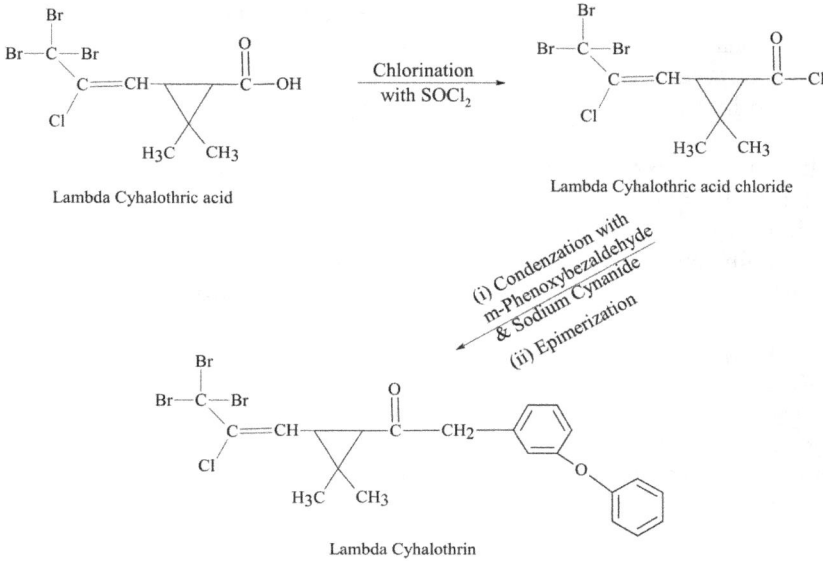

Figure 10.28 Synthesis of Lambda Cyhalothrin

10.19.2.4 Flow sheet diagram

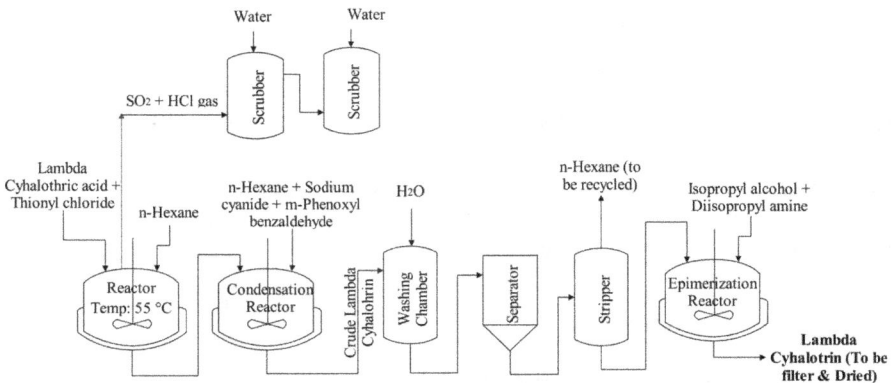

Figure 10.29 Flow sheet diagram of production of Lambda Cyhalothrin

10.19.2.5 Process description

First, n-hexane is taken as a solvent into a jacketed reactor with a stirrer. Start heating and stirring. Heat the solvent upto 40–45°C. Charge the lambda cyhalothric Acid and thionyl chloride slowly with stirring. Maintain the temperature of 40–45°C with stirring for 2–4 hours to obtain 2-chlorotrifluoropropenyl acid chloride. Sulfur dioxide and hydrogen chloride gas which are evolved are scrubbed in a two stage scrubber system. Collect the reaction mass from the bottom of the reactor and transferred into another condensation reactor. Add m-phenoxybenzaldehyde, sodium cyanide and n-hexane to it. During the addition of chemicals, maintain the temperature of 20–25°C with stirring to get crude lambda cyhalothrin. In this process n-hexane is employed as a solvent along with a phase transfer catalyst. The reaction mass is transferred into a water washing tower, where crude lambda cyhalothrin is washed with water. This mass is transferred into a separator to separate the organic and aqueous layers. The aqueous layer is detoxified by sodium hypochloride. The organic layer containing cyhalothrin and n-hexane undergoes stripping to recover n-hexane, which is recycled. Finally, cyhalothrin is transferred into an epimerization reactor. Here, the reaction mass is cooled down to –5 to 0°C with stirring. Charge the isopropyl alcohol and diisopropyl amine into it obtain pure lambda cyhalothrin. The final product is filtered and dried. Isopropyl alcohol is recovered from the filtrate and recycled.

10.19.3 Uses

Since it remains effective for longer periods of time than pyrethrin, it is frequently chosen as an active component in insecticides. Aphids and butterfly larvae are among the pests it frequently controls. It is used with various types of crops, including cotton, grains, hops, ornaments, potatoes, and vegetables. Insects including cockroaches, mosquitoes, ticks, and flies that could serve as disease vectors can be controlled using lambda-cyhalothrin for structural pest management or public health purposes (Subbiah and Muraleedharan, 2009; He et al., 2008).

10.20 Hexaconazole

Hexaconazole may be a broad-spectrum systemic triazole fungicide used for the control of numerous fungi mainly Ascomycetes and Basidiomycetes. It controls fungal diseases fast and effectively because it is swiftly absorbed and translocated throughout the leaf and plant system. It has a phytotonic impact and enhances the plant's visual traits, yield, and product quality.

10.20.1 Properties

Table 10.18 Physical & Chemical Properties of Hexaconazole

Sr. No.	Properties	Particular
1	IUPAC name	2-(2,4-Dichlorophenyl)-1-(1,2,4-triazol-1-yl) hexan-2-ol
2	Chemical and other names	Hexaconazole, Contaf, Sitara, Chlortriafol, Anvil Liquid
3	Molecular formula	$C_{14}H_{17}Cl_2N_3O$
4	Molecular weight	314.21 gm/mole
5	Physical description	White odorless crystalline powder
6	Solubility	Soluble in water, methanol, glycerol, n-hexane, acetone, ethyl acetate
7	Melting point	112°C
8	Boiling point	490°C
9	Flash point	250°C
10	Density	1.313 gm/cm³ at 27°C

10.20.2 Manufacturing process

It is prepared from 2,4-dichlorovalerophenone.

10.20.2.1 Raw materials

2,4-Dichlorovalerophenone, Dimethyl sulfate, Dimethyl sulfide, Potassium hydroxide, 1, 2, 4-Triazole, Anhydrous Potassium carbonate, N,N-Dimethyl formamide (DMF) and Methanol

10.20.2.2 Chemical reactions

2,4-Dichlorovalerophenone reacts with Dimethyl sulfate at 30–35°C in the presence of dimethyl sulfide and potassium hydroxide to give 2-butyl-2-(2,4 dichloro phenyl) oxirane. This oxirane is further condensed with 1,2,4-Triazole at 85–90°C in the presence of anhydrous Potassium carbonate and DMF to give crude Hexaconazole. It is purified using Methanol.

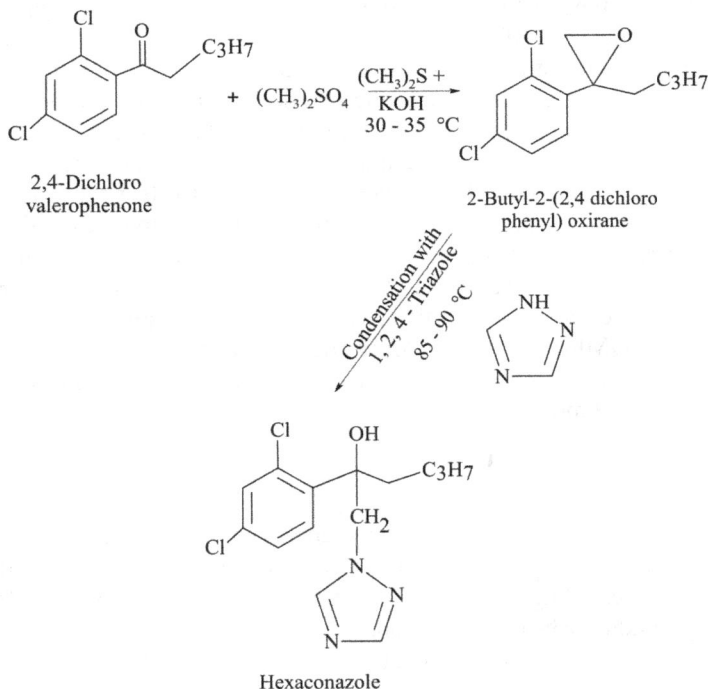

Figure 10.30 Synthesis of Hexaconazole

10.20.2.3 Quantitative requirements

1000 kgs of hexaconazole is obtained using 950 kgs of 2,4-dichlorovalerophenone, 850 kgs of dimethyl sulfate, 1580 kgs of dimethyl sulfide, 920 kgs of potassium hydroxide, 330 kgs of 1,2,4-triazole, 100 kgs of anhydrous potassium carbonate, 3500 kgs of DMF and 985 kgs of methanol.

10.20.2.4 Flow sheet diagram

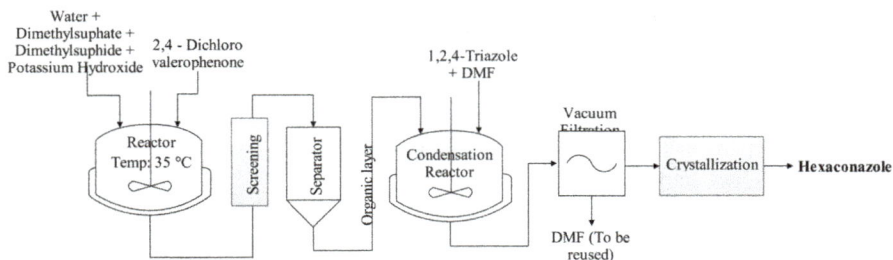

Figure 10.31 Flow sheet diagram of production of Hexaconazole

10.20.2.5 Process description

First, water is charged into a jacketed reactor equipped with a stirrer. Start stirring and charge the Dimethyl sulfate, Dimethyl sulfide and Potassium hydroxide solid into it. Heat the reaction mass upto 30–35°C in hot water in the jacket. After achieving a temperature of 30–35°C, add 2,4-dichloro valerophenone. Maintain the temperature with stirring for 2–3 hours to get oxirane. Charge additional water into it to make slurry. Collect the reaction slurry from the bottom of the reactor and screen it to remove insoluble impurities. After screening, transfer the reaction mass at 65–70°C into a separator to separate the organic and aqueous layers. Discard the aqueous layer from the bottom. Collect the organic layer and transfer it into a condensation reactor having stirring and jacketed facilities. Start the stirring and heating at 85–90°C. After achieving this temperature, 1,2,4-triazole and DMF are added. Maintain the temperature with stirring for 2-3 hours. The reaction mass is cooled and subjected to filtration to get crude solid Hexaconazole. DMF is recovered from the filtrate under vacuum and reused. The product is purified by dissolving methanol and crystallization. Methanol is recovered from the mother liquor from crystallization and reused.

10.20.3 Uses

Hexaconazole is employed mainly for the control of rice sheath blight in China, India, Vietnam and parts of East Asia. Additionally, it is utilized to treat disorders like scab, blast, seath blight, tikka leaf spoil, powdery mildew in various fruits and vegetables including wheat, peas, groundnut, tea, coffee, chilly, tomato, apple, rice, groundnut, grapes, mango, etc. (Han et al., 2013; Singh and Dureja, 2000).

10.21 Malathion

One of the first and most widely used insecticides in the family of organophosphate compounds is malathion, an aliphatic organophosphate that was released in 1950. It was known as carbophos in the USSR, maldison in New Zealand and Australia, and mercaptothion in South Africa.

10.21.1 Properties

Table 10.19 Physical & Chemical Properties of Malathion

Sr. No.	Properties	Particular
1	IUPAC name	Diethyl 2-dimethoxy phosphinothioyl sulfanyl butanedioate
2	Chemical and other names	Malathion, Carbofos, Cythion, Mercaptothion, Prioderm, Phosphothion
3	Molecular formula	$C_{10}H1_9O_6PS_2$
4	Molecular weight	330.3 gm/mole
5	Physical description	Yellow to dark-brown liquid with a garlic-like odor
6	Solubility	Miscible with organic solvents like alcohols, esters, ketones, ethers, aromatic and alkylated aromatic hydrocarbons
7	Melting point	3°C
8	Boiling point	157°C
9	Flash point	> 163°C
10	Density	1.211 gm/cm^3 at 27°C

10.21.2 Manufacturing process

It is prepared from diethyl malonate and o,o-dimethyl dithiophosphate (DDPA).

10.21.2.1 Raw materials

Maleic anhydride, Ethanol, Phosphorus penta sulfide, Methanol, Soda ash, Benzene, Sodium hydroxide and Toluene

10.21.2.2 Chemical reactions

As per Fig. 10.32, Maleic anhydride is esterified with ethanol to yield diethyl maleate (DEM). The obtained DEM is condensed with o,o-dimethyldithiophosphateat 70-80°C to yield Malathion in the presence of benzene as a solvent. o,o-Dimethyl dithiophosphate (DDPA) is manufactured by a reaction with -phosphorus penta sulfide and methanol in the presence of toluene.

10.21.2.3 Quantitative requirements

1000 kgs of malathionis is obtained using 570 kgs of maleic anhydride, 300 kgs of ethanol, 700 kgs of benzene, 560 kgs of toluene, 400 kgs of phosphorus penta sulfide, 240 kgs of methanol, 550 kgs of soda ash and 300 kgs of sodium hydroxide.

10.21.2.4 Flow chart

10.21.2.5 Process description

First, Ethanol and maleic anhydride are taken into an esterification reactor with a stirring facility. The reaction mass is stirred for 2 hours at room temperature. Thereafter, it is transferred from the bottom of the reactor and into a neutralization reactor. Here, soda ash is also added with stirring to get diethylmaleate (DEM). Subsequently, phosphorus penta sulfite, methanol and toluene are inserted into another reactor. It is than filtered, washed and dried to obtain solid DDPA. The filtrate further undergoes recovery of phosphorus penta sulfite. DEM and DDPA are added into a condensation chamber. The reaction mass is stirred for 3 hours at

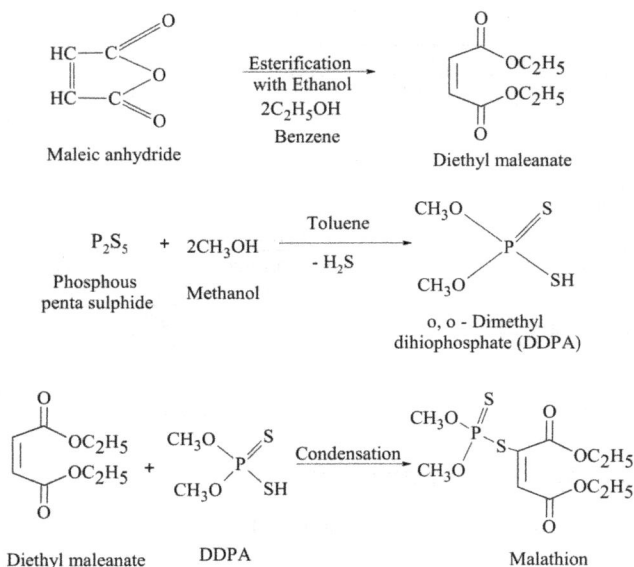

Figure 10.32 Synthesis of Malathion

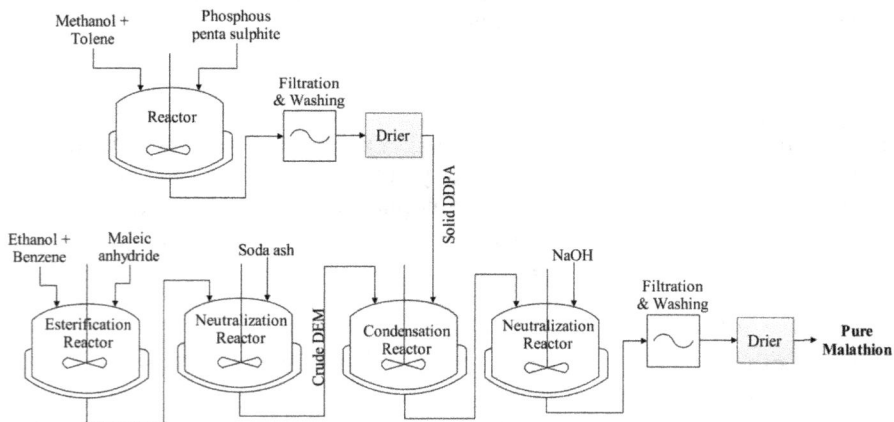

Figure 10.33 Flow sheet diagram of production of Malathion

70–80°C. It is then transferred into the neutralization reactor, where, Malathion is obtained by neutralizing the reaction mass with sodium hydroxide. Finally, it is filtered and dried.

10.21.3 Uses

Alfalfa, apple, apricot, asparagus, avocado, barley, bean (succulent and dry), beets (garden, table, and sugar), blackberry, blueberry, boysenberry, broccoli, cabbage, carrot and cauliflower are just a few of the crops that use Malathion. Additionally, it is utilized by homeowners for ornamental fruit trees, ornamental lawns, and ornamental turf (Velkoska-Markovska and Petanovska-Ilievska, 2020; Hussein and Shawki, 2017).

References

Aravind, J., Samiayyan K. and Kuttalam, S. 2015. Physical and Biological Compatability of Insecticide diafenthiuron 50wp (NS) With Agrochemicals and also to test the Phytotoxicity study against Cardamom Pests. IOSR Journal of Agriculture and Veterinary Science 8(9): 64–67.

Baboo, P. 2022. A case study on Urea Plant, Fertilizer Research.

Bantang, J.P.O. and Camacho, D.H. 2020. Green Production of Potassium Sulfate by Hydrothermal Carbonization of Carrageenan. Asian Journal of Chemistry 32(12): 3105–3108.

Burns, C. and Swaen, G. 2012. Review of 2,4-dichlorophenoxyacetic acid (2,4-D) biomonitoring and epidemiology. Critical Reviews in Toxicology 42(9): 768–86. https://doi.org/10.3109/10408444.2012.710576

Chen, H., Liu, X., Wang, H. and Wu, S. 2020. Polyurea microencapsulate suspension: An efficient carrier for enhanced herbicidal activity of pretilachlor and reducing its side effects. Journal of Hazardous Materials 402: 123744. https://doi.org/10.1016/j.jhazmat.2020.123744

Ferhatoglu, Y. and Barrett, M. 2006. Studies of clomazone mode of action. Pesticide Biochemistry and Physiology 85(1): 7–14. https://doi.org/10.1016/j.pestbp.2005.10.002

Grzmil, B.U. and Kic, B. 2005. Single-stage process for manufacturing of potassium sulphate from sodium sulphate. 32nd International Conference of the Slovak Academy of Sciences 59.

Hadlocon, L.J.S. and Zhao, L.W. 2015. Production of ammonium sulfate fertilizer using acid spray wet scrubbers. Agricultural Engineering International: The CIGR e-journal 41–51.

Hakam, A., Khouloud, M. and Zeroual, Y. 2012. Manufacturing of Superphosphates SSP & TSP from Down Stream Phosphates. Procedia Engineering 46: 154–158. https://doi.org/10.1016/j.proeng.2012.09.459

Han, J., Jiang, J., Su, H. and Sun, M. 2013. Bioactivity, toxicity and dissipation of hexaconazole enantiomers. Chemosphere 93(10). doi: 10.1016/j.chemosphere.2013.09.052.

Harris, J. 2000. Chemical pesticide markets, health risks and residues. Biopesticides Series 1, CABI Publication.

He, L.M., Troiano, J. and Wang, A. 2008. Environmental chemistry, ecotoxicity, and fate of lambda-cyhalothrin. Reviews of Environmental Contamination and Toxicology 195: 71–91. https://doi.org/10.1007/978-0-387-77030-7_3

Hougard, J.M., Zaim, M. and Guillet, P. 2002. Bifenthrin: A Useful Pyrethroid Insecticide for Treatment of Mosquito Nets. Journal of Medical Entomology 39(3): 526–33. DOI:10.1603/0022-2585-39.3.526

Huisman, H.O. and Smit, A. 2010. A new synthesis of 2,4,5-trichlorophenoxyacetic acid (2,4,5-T). Compendium of Chemical Works of the Netherlands 74(2).

Hussein, S.M. and Shawki, M.E.G. 2017. Degradation of malathion in aqueous solutions using advanced oxidation processes and chemical oxidation. Direct Research Journal of Agriculture and Food Science 5: 174–185.

Jong, G., Swaen, G. and Slangen, J. 1997. Mortality of Workers Exposed to Dieldrin and Aldrin: a Retrospective Cohort Study, Occupational and Environmental Medicine 54(10): 702–707. http://dx.doi.org/10.1136/oem.54.10.702

Joshi, C.S., Shukla, M.R., Patel, K. and Joshi, J. 2014. Environmentally and Economically Feasibility Manufacturing Process of Potassium Nitrate for Small Scale Industries: A Review, International Letters of Chemistry Physics and Astronomy 41: 88–99. https://doi.org/10.18052/www.scipress.com/ILCPA.41.88

Jurisova, J., Fellner, P., Danielik, V. and Lencses, M. 2013. Preparation of potassium nitrate from potassium chloride and magnesium nitrate in a laboratory scale using industrial raw materials. ActaChimicaSlovaca 6(1). https://doi.org/10.2478/acs-2013-0003

Kandil, A.T. and Cheira, M.F. 2016. Ammonium sulfate preparation from phosphogypsum waste. Journal of Radiation Research and Applied Sciences 10(1): 24–33. https://doi.org/10.1016/j.jrras.2016.11.001

Khan, A.R. 2016. Control of Ammonia and Urea Emissions from Urea Manufacturing Facilities of Petrochemical Industries Company (PIC)-Kuwait. Journal of the Air & Waste Management Association 66(6). https://doi.org/10.1080/10962247.2016.1145154

Kirova-Yordanova, Z. 2014. Energy Analysis of Ammonium Nitrate Production Plants, 2014, Conference: Proceedings of ECOS: 27th International Conference on Efficiency, Costs, Optimization, Simulation and Environmental Impact of Energy Systems, Turku, Finland.

Kumar, M., Kumar, S.D., Devaraj, K. and Kalaichelvan, P.T. 2013. Hexachlorobenzene-sources, remediation and future prospects. International Journal of Current Pharmaceutical Review and Research 05(01).

Matolcsy, G., Nadasy, M. and Andriska, V. 1988. Pesticide Chemistry, In Studies in Environmental Science 32, Elsevier.

Motheo, A.D.J. 2017. Treatment of actual effluents produced in the manufacturing of atrazine by a photo-electrolytic process. Chemosphere 172: 185–192. doi: 10.1016/j.chemosphere.2016.12.154

Nazir, A. and Iqbal, J. 2020. Method validation for bifenthrinemulsifiable concentrate and uncertainty calculation using gas chromatographic approach. Future Journal of Pharmaceutical Sciences 6(1). https://doi.org/10.1186/s43094-020-0022-9

Nielsson. 2018. Manual of Fertilizer Processing. In Fertilizer Science and Technology, Routledge Publication.

Nurudeen, A.R. 2015. Profitable NPK Fertilizer Rates for Maize Production on Ferric Lixisol. Journal of Agricultural Science 7(12): 233.

Ochoa-Gonzalez, O., Coronado-Hernandez, J.R., Macias-Jimenez, M.A. and Conrado, A.R.R. 2020. Quality Improvement in Ammonium Nitrate Production Using Six Sigma Methodology, Lecture Notes in Computer Science 12133: 172–183.

Ollinger, M. and Fernandez-Cornejo, J. 1998. Innovation and Regulation in the Pesticide Industry, Agricultural and Resource Economics Review 27(1): 15–27. https://doi.org/10.1017/S1068280500001660

Oueslati, A.B.L. 2015. Simulation of triple super phosphate (TSP) manufacturing process, Conference: BTM-Trainings, Tunis.

Park, M. 2001. The Fertilizer Industry, Woodhead Publishing.

Schareina, T., Zapf, A. Cotte, A. and Muller, N. 2008. A Practical and Improved Copper-Catalyzed Synthesis of the Central Intermediate of Diafenthiuron and Related Products. Organic Process Research & Development 12(3): 537–539. https://doi.org/10.1021/op700287s

Singh, N. and Dureja, P. 2000. Persistence of hexaconazole, a triazole fungicide in soils, Journal of Environmental Science and Health Part B Pesticides Food Contaminants and Agricultural Wastes 35(5): 549–58. https://doi.org/10.1080/03601230009373291

Subbiah, S. and Muraleedharan, N.N. 2009. Residues of lambda-cyhalothrin in tea, Food and chemical toxicology: An International Journal 47(2): 502–5. doi: 10.1016/j.fct.2008.12.010.

Turusov, V., Rakitsky, V. and Tomatis, L. 2002. Dichlorodiphenyltrichloroethane (DDT): Ubiquity, Persistence, and Risks. Environmental Health Perspectives 11(2): 125–8. https://doi.org/10.1289/ehp.02110125

Udom B.E. and Nengi-benwari, O. 2022. Green manures and NPK fertilizer applications Effect on organic carbon pool, soil physical Properties and cucumber production, Journal of Global Ecology and Environment 15(4): 30–36.

Velkoska-Markovska, L. and Petanovska-Ilievska, B. 2020. Pesticide formulation by high-performance liquid chromatography. Agriculture & Forestry 66(4): 171–181.

Wen, D.,Chen, B. andLiu, B. 2022. An ultrasound/O_3 and UV/O_3 process for atrazine manufacturing wastewater treatment: a multiple scale experimental study, Water Science & Technology 85(1). https://doi.org/10.2166/wst.2021.633

Yahya, N. 2018. Green Urea: For Future Sustainability. In Green Energy and Technology, Springer.

CHAPTER 11

Explosive & Propellant Industries

11.1 Introduction

A substance or mixture that, when exposed to thermal or mechanical shock, quickly oxidizes exothermally to produce products with a significantly increased volume and a sudden release of potential energy is referred to as an explosive. "Power to weight ratio" refers to the amount of power available from a given weight of explosive. Due to the exothermic nature of the explosive reaction, the byproducts are heated to a high temperature and a large quantity of heat is transferred to the surroundings when this exceedingly quick chemical reaction occurs in a small area. This can be utilized for construction or destruction purposes. Explosives are largely used for military purposes and also, utilized for drilling tunnels through mountains, excavating land for dams, seismic prospecting and moving rocks. One of the earliest explosives was gun powder or black powder which was used during World Wars. Now-a-days, lead azide, mercury fulminate, diazodinitrophenol and trinitrotoluene are widely used as explosives.

11.2 Classification of Explosives

Explosives differ widely in their sensitivity and power and are, mainly classified into the following three parts.

11.2.1 Primary explosives

They are extremely sensitive explosives that will detonate in the presence of fire or a mild shock. As a result, they need to be handled with extreme caution. They are mostly utilized in shells and cartridges in modest amounts to begin or initiate explosions. So, they are also referred to as initiating explosives or detonators. Lead azide and diazodinitrophenol (DDNP) are widely used as detonators for highly explosive materials like trinitrotoluene (TNT).

11.2.2 Low explosives or Propellants

They simply burn and do not abruptly burst. Such explosive combustion reactions occur rather slowly, and they burn in layers, moving around 20 cm per second from the outside to the inside of the device. Examples of propellants are gun powder, Nitrocellulose and cordite. Gun powder or blackpoeder is a mixture of Potassium nitrite (75%), charcoal (15%) and sulfur (10%). This mixture is explosive, but no sparking occurs as it acts as an explosive due to redox reactions.

11.2.3 High explosives

They contain more energy than basic explosives. They are, nonetheless, very stable and remarkably resistant to fire and mechanical shocks. A little quantity of primary explosive is often placed in contact with a high intensity explosive. The high intensity explosive's body begins a quick chemical reaction that is subsequently carried out by the primary explosive at a high pace of 1000–1500 m/s. The volume of gases produced especially when high temperatures are manipulated, has a shattering impact. An example of a high intensity explosive is trinitrotoluene (TNT) (Zapata and Garcia-Ruiz, 2021; Dehkharghani, 2005).

11.3 Propellants

A fuel with a high oxygen content or a blend of fuel and oxidant burns with precise control and the evolution of a significant amount of gas constitutes a rocket propellant. When a propellant reacts quickly, a huge volume of hot gases are created, and they quickly escape through a small aperture at supersonic speed. Pushing the gas rearward causes a corresponding and opposing reaction that propels the rocket forward. The performance of the rocket increases as the exhaust velocity rises. The propellant is retained in the combustion chamber of a rocket engine, and generally burns at a high temperature of about 3000°C and generates a high pressure of about 300 kg cm^2. For pyrotechnic effect signaling, carrying a person, launching explosives at an adversary, propelling a spacecraft into orbit, and other purposes, rockets are utilized.

11.3.1 Characteristics of a good propellant

1) It should to burn with a high specific impulse (defined as the amount of thrust produced divided by the rate of propellant).
2) During burning, it should provide low molecular weight yields.
3) It should to burn slowly and steadily (at a desired rate)
4) It should to handle brief ignition delays.
5) It must have a high density.
6) It should to be stable over a greater temperature range.
7) It should not explode under shock, heat, or impact, making it safe to handle and store in normal circumstances.
8) It must be easily ignitable and should burn with a predictable rate.

9) It should not be hydrophobic or corrosive.
10) After ignition, there should be no solid residue.
11) It shouldn't create any hazardous byproducts.
12) Upon burning, it should to produce a high temperature.

11.3.2 Classification of propellants

Propellants are classified into two according to their physical state: (1) Solid and (2) Liquid.

11.3.2.1 Solid propellants

They may be homogeneous or heterogeneous solid. Homogeneous propellants are further classified as single propellants, which have only one propellant like nitrocellulose. In addition a double base propellant is a thoroughly mixed homogeneous mixture of two propellants in the colloidal state. Examples of double base propellants include a plasticizer, which is a mixture of Nitrocellulose (65%), Nitroglycerine (30%), and Petroleum Jelly (5%), and Ballistite, which is a mixture of Nitrocellulose (60%) and Nitroglycerine (40%). Gun powder which is composed of a mixture of potassium nitrite (75%) charcoal (15%) and sulfur (10%), is an example of a heterogeneous propellant. Other common oxidizing agents are KNO_3, $KClO_4$, NH_4ClO_4 incorporating linear or cross-linked polymers like PVC, Polyurethanes or Thiokol rubber. Solid propellants have the advantage of being easily and safely stored, also, handled and transported easily. Engines used for this propellant have a simple design. They have the disadvantageof having a low specific impulse and are difficult to check and calibrate. Also their use is more economical.

11.3.2.2 Liquid propellants

They are in liquid state and may be monopropellant or bipropellant. Monopropellants worked as a fuel and oxidizing agent such as Nitromethane, Hydrogen peroxide, Hydrazine, Ethylene oxide, Propyl nitrate, while bipropellants contain liquid fuels and oxidizing agents. In bipropellant fuels, liquid fuels commonly used are Hydrogen, Hydrazine, Gasoline, Kerosene, Ethanoland Aniline.They are kept separate from the oxidizing agent and injected whenever required. The oxidizers include liquid oxygen, ozone, fuming nitric acidand dinitrogen tetroxide. The liquid propellants have the advantage of having a high impulse compared to solid propellants. They are comparatively less economical to use. Engines using them can be checked and calibrated more easily. Disadvantages of liquid fuels are that they have storing, handling and transporting problems. Their engine design is more complicated (Kubota, 2007; Agrawal, 2010).

11.4 Nitroglycerin

Chemically, it is an organic nitrate compound developed in 1847. Nitroglycerin is an active ingredient in the production of explosives.

11.4.1 Properties

Its main properties are as follows.

Table 11.1 Physical and Chemical Properties of Nitroglycerin

Sr. No.	Properties	Particular
1	IUPAC Name	1,3-Dinitrooxypropan-2-yl nitrate
2	Chemical and other names	Nitroglycerin, Glyceryl trinitrate, Trinitroglycerin, Nitroglycerol, Glycerol trinitrate, Nitroderm
3	Molecular formula	$C_3H_5N_3O_9$
4	Molecular weight	227.1 gm/mole
5	Physical description	Colorless to pale-yellow, viscous liquid or solid
6	Solubility	Slightly soluble in water; soluble in glacial acetic acid, nitrobenzene, pyridine, ethylene bromide, dichloroethylene, ethyl acetate.
7	Melting point	13.5°C
8	Boiling point	250.0°C
9	Flash point	240.0°C
10	Density	1.60 g/ml at 20°C
11	Decomposition temperature	50–60°C

11.4.2 Manufacturing process

Working under Theophile-Jules Pelouze at the University of Turin, the Italian chemist Ascanio Sobrero created the compound for the first time in 1847. Initially referring to his finding as "pyroglycerine", Sobrero vehemently forbade its usage as an explosive. Alfred Nobel experimented with safer ways to handle the hazardous substance after his younger brother Emil Oskar Nobel, Heleneborg, Sweden, and several factory workers were killed in an explosion at the Nobels' armaments factory in 1864. Alfred Nobel later adopted nitroglycerin as a useful explosive for commercial purposes. Glycerol serves as the precursor to Nitroglycerin.

11.4.2.1 Raw materials

Glycerol, Concentrated Nitric acid, Concentrated Sulfuric acid and Sodium carbonate

11.4.2.2 Chemical reactions

As per Fig. 11.1, it is prepared by the nitration of glycerol using a cooled mixture of concentrated nitric acid and concentrated sulfuric acid.

11.4.2.3 Quantitative requirements

1000 kgs of nitroglycerin is obtained using 430 kgs of glycerol, 1000 kgs of conc. nitric acid, 1000 kgs of conc. sulfuric acid and 150 kgs of sodium carbonate.

11.4.2.4 Flow sheet diagram

$$
\begin{array}{l}
\text{CH}_2\text{OH} \\
| \\
\text{CHOH} \\
| \\
\text{CH}_2\text{OH}
\end{array}
\quad + \; 3 \text{ HNO}_3 \;
\xrightarrow[\text{Temp.: 10 °C}]{\substack{\text{Conc. H}_2\text{SO}_4 \\ \text{Nitration}}}
\quad
\begin{array}{l}
\text{CH}_2\text{ONO}_2 \\
| \\
\text{CHONO}_2 \\
| \\
\text{CH}_2\text{ONO}_2
\end{array}
\quad + \quad 3 \text{ H}_2\text{0}
$$

Glyceral   Nitroglycerine

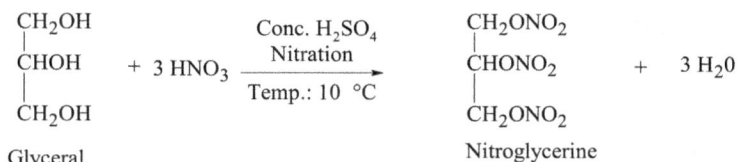

Figure 11.1 Synthesis of Nitroglycerine

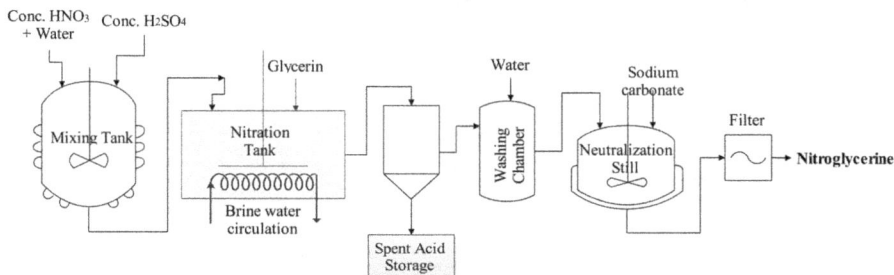

Figure 11.2 Flow sheet diagram of production Nitroglycerine

11.4.2.5 Process description

Concentrated nitric acid, concentrated sulfuric acid and water are poured into a mixed acid tank in a ratio of 40:59:1. This mixture of acids is transferred into a nitration tank equipped with an agitator and steel cooling facilities. Now, 99.9% gylcerol is slowly added to the nitration tank with continuous stirring at a temperature of 5–10°C. The temperature of the nitration reaction is maintained by means of brine water circulation in the tank coils. After completion of the nitration reaction for about 60 to 90 minutes, the reaction mixture containing nitroglycerin and spent acid is allowed to flow through a separating tank. Here, nitroglycerin is carefully separated from the spent acid and sent to a wash tank. Sulfuric acid is recovered from the spent acid. The nitroglycerin is washed with warm water twice. Thereafter, the washed nitroglycerine is neutralized with 2% sodium carbonate solution in a neutralizing tank to remove any remaining acid and filtered.

11.4.3 Uses

1. It has been used as a key component in the production of explosives, primarily dynamite, which is used in the building, demolition, and mining sectors.
2. Since the 1880s, it has been used by the military as an active component and a gelatinizer for nitrocellulose in several solid propellants, such as ballistite and cordite.
3. Reloaders' double-based smokeless gunpowder contains nitroglycerin as a key ingredient. Rifle, pistol, and shotgun reloaders use a wide variety of powder mixtures together with nitrocellulose.
4. For more than 130 years, nitroglycerin has been used therapeutically as a powerful vasodilator to treat cardiac disorders such angina pectoris and chronic heart failure (Jin et al., 2017; Martel et al., 2010).

11.5 Nitrocellulose

The main component of modern gunpowder, nitrocellulose, commonly known as cellulose nitrate, is a mixture of nitric esters of cellulose and a highly combustible substance. It is also used in some lacquers and paints.

11.5.1 (Introduction) properties

Its main properties are as follows.

Table 11.2 Physical and Chemical Properties of Nitrocellulose

Sr. No.	Properties	Particular
1	IUPAC Name	[(2S,3R,4S,5R,6R)-2-[(2R,3R,4S,5R,6S)-4,5-dinitrooxy-2-(nitrooxymethyl)-6-[(2R,3R,4S,5R,6S)-4,5,6-trinitrooxy-2-(nitrooxymethyl)oxan-3-yl]oxyoxan-3-yl]oxy-3,5-dinitrooxy-6-(nitrooxymethyl)oxan-4-yl] nitrate
2	Chemical and other names	Nitrocellulose, Cellulose Nitrate, Collodion, Pyroxylin, Pyroxylinum, Cellulose tetranitrate, Flash paper, Flash cotton, Guncotton, Flash string
3	Molecular formula	$C_{18}H_{21}N_{11}O_{38}$
4	Molecular weight	999.4 gm/mole
5	Physical description	Colorless, or slightly yellow, clear or slightly opalescent, syrupy liquid
6	Solubility	Soluble in acetone and an ether-alcohol mixture
7	Melting point	165.0°C
8	Boiling point	N.A.
9	Flash point	4.4°C
10	Density	0.778 g/ml at 20°C
11	Decomposition temperature	192–209°C

11.5.2 Manufacturing process

The starting material for nitrocellulose is purified cellulose (cotton linters or wood pulp).

11.5.2.1 Raw materials

Cellulose, Concentrated Nitric acid, Concentrated Sulfuric acid, Dilute Caustic soda, Bleaching agent such as Calcium hypochlorite, Dinitrogen tetroxide and Dilute sodium carbonate

11.5.2.2 Chemical reactions

It is prepared by the nitration of purified cellulose (cotton linters or wood pulp) using concentrated nitric and sulfuric acids.

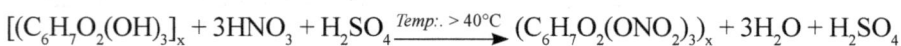

$$[(C_6H_7O_2(OH)_3]_x + 3HNO_3 + H_2SO_4 \xrightarrow{Temp:. > 40°C} (C_6H_7O_2(ONO_2)_3)_x + 3H_2O + H_2SO_4$$

Cellulose   Nitric acid Sulfuric acid   Nitrocellulose

11.5.2.3 Quantitative requirements

1000 kgs of nitrocellulose is obtained using 340 kgs of cotton liner, 460 kgs of conc. nitric acid, 230 kgs of conc. sulfuric acid, 100 kgs of caustic soda, 0.2 kg of sodium carbonate and 8 kgs of dinitrogen tetroxide.

11.5.2.4 Flow sheet diagram

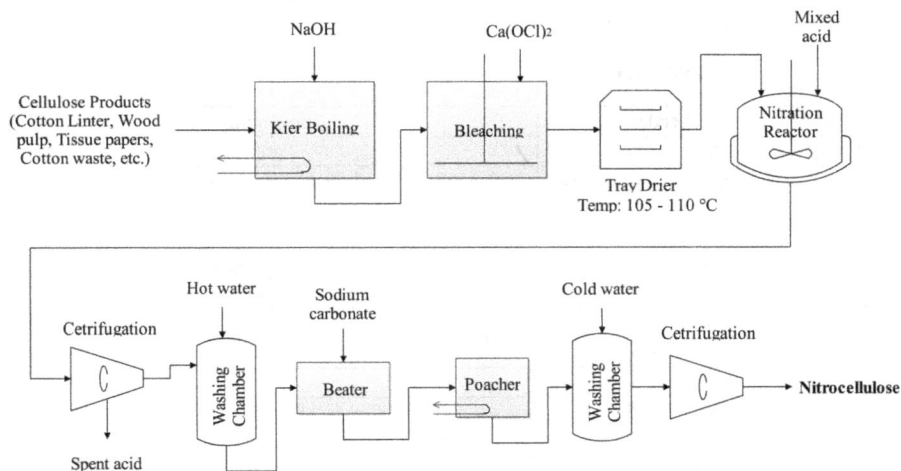

Figure 11.3 Flow sheet diagram of production Nitrocellulose

11.5.2.5 Process description

As per the flow-sheet, cellulosic materials like cotton linters or wood pulp are purified by boiling in Kiers (wooden vessel) with dilute caustic soda solution. It is then bleached with sodium carbonate or calcium hypochlorite with stirring. This cotton is dried in a tray oven at 105–110°C for loosening cotton fibers and transferred into a nitration reactor having a stirring facility. Mixed acid (mixture of 21% HNO_3, 63% H_2SO_4, 15.5% H_2O and 0.5% N_2O_4) is also added in it. After nitration, the nitration mixture is collected from the bottom of the reactor and charged into a centrifuge, where the spent acid is separated from the nitrated cellulose. The spent acid is reused for the nitration process. The nitrated cellulose is washed with hot water to remove the ester. The remaining traces of ester are removed in a beater and poacher, where nitrated cellulose is boiled with dilute sodium carbonate solution. Poached nitrocellulose is washed with cold water and concentrated by centrifugation to get nitrocellulose.

11.5.3 Uses

Nitrocellulose is used primarily as a propellant as are most modern military low-order explosives. It is used to prepare double base homogeneous solid propellants like Plasticizer [mixture of Nitrocellulose (65%), Nitroglycerine (30%) and Petroleum jelly (5%)] and Ballistite [mixture of Nitrocellulose (60%) and Nitroglycerine (40%)]. It is also a major component of smokeless gunpowder (Lavanya et al., 2011).

11.6 Picric Acid

The word "picric", which refers to its bitter flavor, is derived from the Greek word "pikros", which means "bitter". This is one of the most acidic phenols.

11.6.1 *Properties*

Its main properties are as follows.

Table 11.3 Physical and Chemical Properties of Picric acid

Sr. No.	Properties	Particular
1	IUPAC Name	2, 4 ,6 – Trinitrophenol
2	Chemical and other names	Picric acid, Carbazotic acid, Picronitric acid, Melinite, Picral, Picrate
3	Molecular formula	$C_6H_3N_3O_7$
4	Molecular weight	229.1 gm/mole
5	Physical description	Yellow, odorless solid crystal
6	Solubility	Soluble in alcohol, benzene, chloroform and ether
7	Melting point	122°C
8	Boiling point	N.A.
9	Flash point	150.1°C
10	Density	1.76 g/ml at 20°C
11	Decomposition temperature	290–300°C

11.6.2 *Manufacturing process*

First it is manufactured from phenol by an electrophilic substitution reaction, i.e., nitration, but there are various by-products such as phenyl sulfonic acids and dinitrophenols. Recently, it has been manufactured from chlorophenol.

11.6.2.1 *Raw materials*

Monochlorobenzene (MBC), Sulfuric acid, Nitric acid, Caustic lye, Urea and Ice

11.6.2.2 *Chemical reactions*

Chlorobenzene undergos nitration using nitric acid and sulfuric acid at 110–120°C to give 2,4-dinitrochlorobenzene. 2,4-Dinitro phenol is obtained by the alkaline hydrolysis of 2,4-dinitro chlorobenzene at 80–100°C. And, 2,4,6-trinitrophenol also commonly known as "picric acid" is obtained from the nitration of 2,4-dinitrophenol at 40–50°C in presence of excess sulfuric acid as the medium.

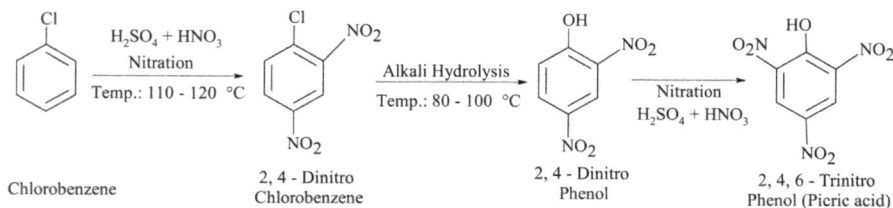

Figure 11.4 Synthesis of Picric acid

11.6.2.3 Quantitative requirements

1000 kgs of Picric acid is obtained using 565 kgs of MCB, 6400 kgs of sulfuric acid (98%), 1900 kgs of sulfuric acid (98%) 1400 kgs of nitric acid (98%), 1200 kgs of caustic lye, 20 kgs of urea and 5000 kgs of ice.

11.6.2.4 Flow sheet diagram

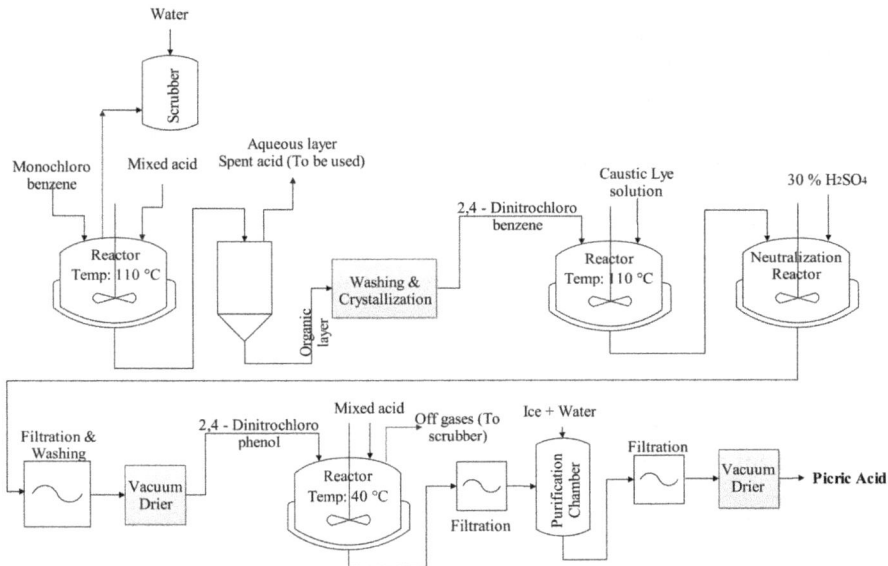

Figure 11.5 Flow sheet diagram of production Picric acid

11.6.2.5 Process description

It's divided into three parts

(A) Preparation of 2,4-Dinitrochloro benzene: In first step, monochloro benzene, 98% sulfuric acid and 98% nitric acid are inserted into a jacketed stainless steel reactor facilitated with stirring. The reaction mass is heated upto 110–120°C. Maintain this temperature for 2 hours with stirring. The remaining gases are removed using a scrubber. The reaction mass is collected from the bottom of the reactor and sent to a layer separation tower. The upper aqueous layer (spent acid) containing 80% sulfuric acid is separated, concentrated and further reused. The lower layer containing 2,4-dinitrochlorobenzene with impurities like 2,5–/2,6-dinitrochloro benzene is collected from the bottom. It is further washed with water to remove any free acidity. Then this organic layer is cooled and crystallized to give pure dinitrochloro benzene.

(B) Preparation of 2,4-dinitro phenol: In this step, pure dinitrochloro benzene and 50% caustic lye solution are inserted into a jacketed stainless steel reactor facilitated with stirring. The reaction mass is heated upto 80–100°C. This temperature is maintained for 3 hrs with stirring. The reaction mass is collected from the bottom of the reactor and sent to a neutralization tank. Here, 30% sulfuric acid is added in it. The reaction mass is further subjected to filtration and given a fresh water wash to remove any traces of free acid and salts. The product, 2,4-dinitro phenol is vacuum dried.

(C) **Preparation of 2,4,6-trinitrophenol:** In the final step, 2,4-dinitro phenol, 98% sulfuric acid and 98% nitric acid are inserted into an anchor type stirring facilitated jacketed stainless steel reactor. The reaction is highly exothermic, so, a temperature of 40–50°C is maintained by circulating cooling water in the jacket. The remaining gases are removed by scrubbing. Excess sulfuric acid is added to avoid the side reactions. This temperature is maintained for 2 hrs with stirring. The reaction mass is collected from the bottom of the reactor and subjected to filtration. The filtrate contains 80–85% sulfuric acid, which is concentrated and further reused. The precipitates are charged into a purification reactor. Ice and water are added to remove the un-reacted acid and other impurities. Thereafter, it is subjected to filtration to collect the final product. It is further vacuum dried and packed.

11.6.3 Uses

It serves as a primary explosive and booster charge to detonate a more insensitive explosive, such as TNT. Additionally, it has been used in dyes, antiseptics, and as a medicine for treating burns (Saleem et al., 2019).

11.7 2,4,6-Trinitro Toluene (TNT)

It is an important explosive.

11.7.1 Properties

Its main properties are as follows.

Table 11.4 Physical and Chemical Properties of 2, 4, 6 – Trinitro toluene

Sr. No.	Properties	Particular
1	IUPAC Name	2-Methyl-1,3,5-trinitrobenzene
2	Chemical and other names	2, 4, 6 – Trinitrotoluene, TNT, s-Trinitrotoluol, Tritol, Tolite, Gradetol
3	Molecular formula	$C_7H_5N_3O_6$
4	Molecular weight	227.1 gm/mole
5	Physical description	Colorless to pale-yellow, odorless solid or crushed flakes
6	Solubility	Soluble in water, chloroform, ethanol, carbon tetrachloride, acetone
7	Melting point	80.1°C
8	Boiling point	240°C
9	Flash point	150.1°C
10	Density	1.654 g/ml at 20°C
11	Decomposition temperature	290–300°C

11.7.2 Manufacturing process

It is manufactured from toluene

11.7.2.1 Raw materials

Toluene, Concentrated Nitric acid, Oleum and Sodium sulfite

11.7.2.2 Chemical reactions

Toluene is put through a two-step nitration procedure with concentrated nitric acid to produce 2,4-dinitro and 2,-mononitro toluene, respectively. To produce 2,4, 6-trinitro toluene (TNT), this 2,4-dinitro toluene is further nitrated in the presence of nitric acid and oleum. The chemical reaction is summarized as below.

11.7.2.3 Quantitative requirements

1000 kgs of TNT is obtained using 540 kgs of toluene, 680 kgs of nitric acid (58%), 370 kgs of nitric acid (98%), 1250 kgs of oleum (20%) and 100 kgs of sodium sulfite.

11.7.2.4 Flow sheet diagram

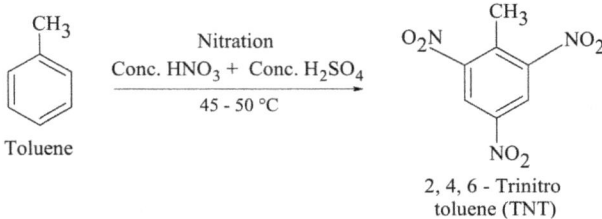

Figure 11.6 Synthesis of TNT

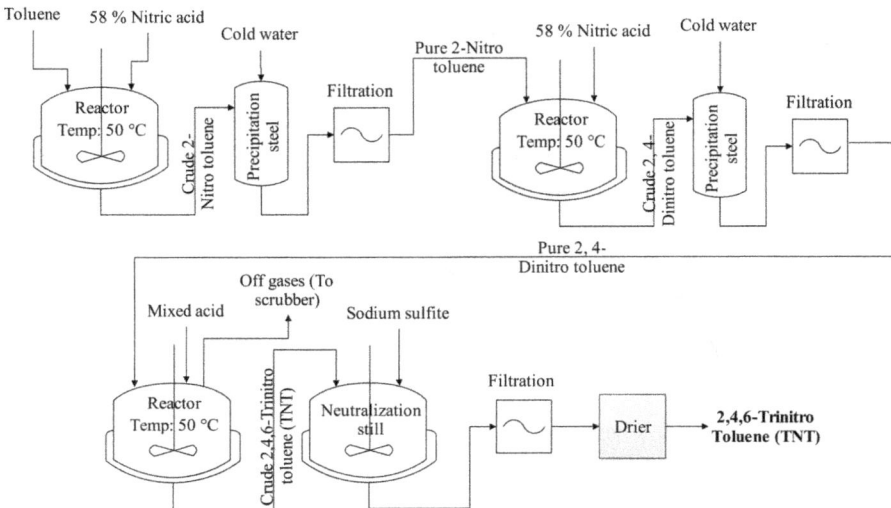

Figure 11.7 Flow sheet diagram of production TNT

11.7.2.5 Process description

To start with toluene and 58% nitric acid are added to a stainless steel reactor with a jacket and stirring facilities. Thereafter, to prevent the oxidation of the methyl group in toluene, 58% nitric acid is charged slowly for 2 hours. The temperature is kept at 45–50°C since this reaction is highly exothermic and poses the risk of a runaway reaction that results in an explosion using cold water in the reactor jacket. The formed 2-nitro toluene is separated out by pouring the reaction mixture into ice-cold water in a steel precipitator followed by filtration. Further, crude 2-nitro toluene and nitric acid are added to a stainless steel reactor with a jacket and stirrer. The same process is carried using 98% nitric acid i.e. the temperature is maintained at 45–50°C for 30 minutes. The formed 2,4-dinitro toluene is separated out by pouring the reaction mixture in ice-cold water in a steel precipitator followed by filtration. This 2,4-dinitro toluene precipitate is charged into a jacketed stainless steel reactor with a stirrer. Also, an anhydrous mixture of nitric acid and oleum is added. The temperature is maintained at 45–50°C for 30 minutes. The emitted gases are scrubbed using cold water. Crude TNT is collected from the bottom of the reactor and sent into a neutralization tank. Here, less stable isomers of TNT and other undesirable reaction products are removed from the pure TNT by treating it with an aqueous sodium sulfite solution. It is then dried and filtered.

11.7.3 Uses

Since TNT is less sensitive to shock and friction than more sensitive explosives like nitroglycerin it has a lower risk of accidental detonation and one of the most widely utilized explosives for military, industrial, and mining uses. It is also blended with other materials for the preparation of a variety of explosives such as Amatex, Amatol, Ammonal, Baratol, Cyclotol, Ednatol, Hexaniteand Minol. (Millar et al., 2011).

11.8 Tetryl

Tetryl, also known as 2,4,6-trinitrophenylmethylnitramine, is a delicate secondary high intensity explosive. Tetryl was first created in 1877 by German scientists Wilhelm Michler and Carl Meyer.

11.8.1 Properties

Its main properties are as follows.

Table 11.5 Physical and Chemical Properties of Tetryl

Sr. No.	Properties	Particular
1	IUPAC Name	N-Methyl-N-(2,4,6-trinitrophenyl)nitramide
2	Chemical and other names	Tetryl, Tetralite, Tetril, N-Methyl-N-picrylnitramine, N-Picryl-N-methylnitraminel, Picrylmethylnitramine
3	Molecular formula	$C_7H_5N_5O_8$
4	Molecular weight	287.1 gm/mole
5	Physical description	Colorless to yellow, crystalline odorless solid
6	Solubility	Soluble in alcohol, ether, glacial acetic acid, benzene
7	Melting point	130°C
8	Boiling point	187°C
9	Flash point	187°C
10	Density	1.73 g/ml at 20°C
11	Decomposition temperature	180–190°C

11.8.2 Manufacturing process

Two methods are available for production of tetryl. Kirk-Othmer in 1980 synthesized it from N,N'-dimethylaniline in 1980. In a second manufacturing process, methylamine and 2,4- or 2,6-dinitrochlorobenzene are combined to create dinitrophenyl methylamine, which is then nitrated to create tetryl (Gibbs and Popolato 1980). The first approach, which is described below, is more practical.

11.8.2.1 Raw materials

N,N'-Dimethylaniline, Concentrated Sulfuric acid, Concentrated Nitric acid and Acetone

11.8.2.2 Chemical reactions

N,N'-Dimethylaniline undergoes nitration in the presence of concentrated sulfuric acid and concentrated nitric acid to give tetryl.

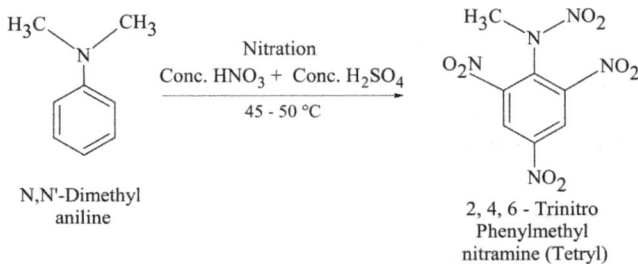

Figure 11.8 Synthesis of Tetryl

11.8.2.3 Flow sheet diagram

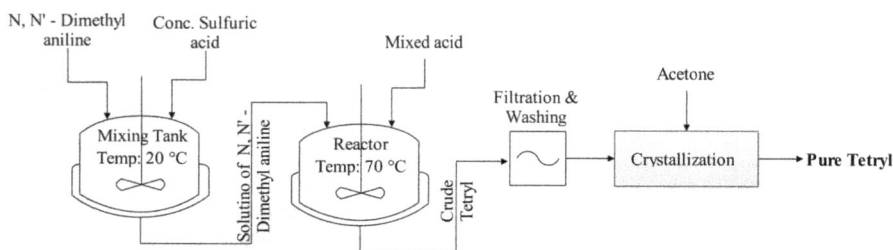

Figure 11.9 Flow sheet diagram of production Tetryl

11.8.2.4 Process description

In this simple process, first solid N,N'-dimethylaniline is charged into a jacketed mixing reactor with a stirrer facility. Now, an excess amount of concentrated sulfuric acid is slowly added to dissolve N,N'-dimethylaniline with stirring. The addition is such that the temperature is maintained at 20–30°C. Also, cooling water is circulated in the jacket. After addition, a solution of N,N'-dimethylaniline is transferred into the jacketed mixing reactor with a stirrer. Now, a mixture of nitric and sulfuric acids is added. The reaction mass is heated upto 60–70°C. This temperature is maintained for 30 minutes with stirring to form solid tetryl. Also, by-products like methyltetryl, ethytetryl, ditetryl and tritetrylare formed during the reaction. After completion of the reaction, the crude product is filtered and washed with water. Thereafter, it is purified by crystallization in acetoneat a temperature of 50°C.

11.8.3 Uses

As it is sensitive to shock and heat like picric acid, it is used for manufacturing good boosters, but it is generally not as safe as TNT or ammonium picrate. As a booster, it's utilized in blasting caps, with primary explosives like mercury fulminate and salt to line it off. It is used to manufacture another explosive called tetrytol (mixture of 70% tetryl and 30% TNT) (Alfaraj et al., 2016).

11.9 PETN

The nitrate ester of pentaerythritol is known as pentaerythritol tetranitrate (PETN), sometimes known as PENT, PENTA, TEN, corpent, penthrite, or-rarely and mostly in German-nitropenta.

11.9.1 Properties

Its main properties are as follows.

Table 11.6 Physical and Chemical Properties of PETN

Sr. No.	Properties	Particular
1	IUPAC Name	[3-Nitrooxy-2,2-bis(nitrooxymethyl)propyl] nitrate
2	Chemical and other names	Pentaerythritol tetranitrate, Nitropentaerythrite Penthrite, Nitropentaerythritol, Nitropenta
3	Molecular formula	$C_5H_8N_4O_{12}$
4	Molecular weight	316.1 gm/mole
5	Physical description	Colourless to white crystals with faint, mild odor
6	Solubility	Soluble in acetone. Sparingly soluble in alcohol, ether
7	Melting point	138°C
8	Boiling point	210°C
9	Flash point	187°C
10	Density	1.773 g/ml at 20°C
11	Decomposition temperature	205–210°C

11.9.2 Manufacturing process

The German explosives company Rheinisch-Westfslische Sprengstoff A.G. of Cologne created and patented pentaerythritol tetranitrate in 1894. The German government patented a modified manufacturing technique, and in 1912, PETN production began. Afterwards the German military used this explosive throughout World War I.

11.9.2.1 Raw materials

Formaldehyde, Concentrated Nitric acid, Acetaldehyde, Calcium hydroxide, Sodium carbonate and Acetone

11.9.2.2 Chemical reactions

Cannizaro reaction is conducted between formaldehyde and acetaldehyde at a temperature of 40–50°C using calcium hydroxide as a catalyst to yield pentaerythritol, which undergoes nitration with nitric acid at 20–25°C to give pentaerythritol tetranitrate.

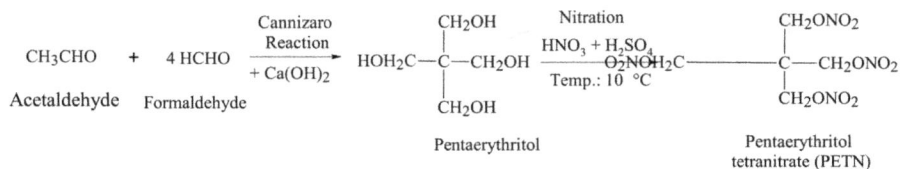

Figure 11.10 Synthesis of PETN

11.9.2.3 Quantitative requirements

1000 kgs of PETN is obtained using 410 kgs of formaldehyde, 200 kgs of acetaldehyde, 320 kgs of calcium hydroxide, 1200 kgs of concentrated nitric acid, 10 kgs of sodium carbonate and 1200 kgs of acetone.

11.9.2.4 Flow sheet diagram

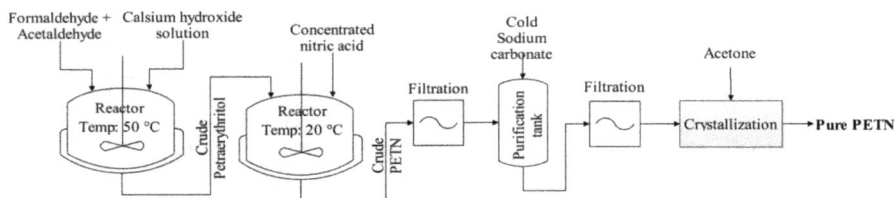

Figure 11.11 Flow sheet diagram of production PETN

11.9.2.5 Process description

In the first step, a 50% aqueous solution of formaldehyde and acetaldehyde is charged into a jacketed reactor having a stirrer. Calcium hydroxide solution is also added. The reaction mass is heated upto 40–50°C. This temperature is maintained for 1 hour to get solid pentaerythritol. Then the precipitates are filtered. Now, crude pentaerythritol is charged into a reactor facilitated with a stirrer and a jacket. Now, concentrated nitric acid is slowly charged into the reactor with stirring. The addition is such that the temperature is maintained at 20–25°C. Also, cooling water is circulated in the jacket to maintain the temperature. After the addition, a temperature of 20–25°C is maintained for 30 minutes with stirring to form crude PETN. Then it is filtered. Thereafter, the obtained precipitates are charged into a purification tank having a cooled sodium carbonate solution. Again, it is filtered and purified by crystallization in acetone at a temperature of 50°C.

11.9.3 Uses

Most commonly, PETN is as an explosive with high brisance. PETN is used in a number of compositions like Semtex (plastic explosive), Pentolite (booster charger), and XTX8003 (extrudable explosive) (Liu, 2019; Tarver et al., 2003).

11.10 Mercury Fulminate

It is a salt of fulminic or paracyanic acid. It is the main explosive that is extremely susceptible to stress, friction, and heat. Thermal decomposition starts at temperatures as low as 100°C, but it continues at a much faster pace as the temperature rises, producing a mixture of relatively stable mercury salts as well as carbon dioxide and nitrogen gas.

11.10.1 Properties

Its main properties are as follows.

Table 11.7 Physical and Chemical Properties of Mercury fulminate

Sr. No.	Properties	Particular
1	IUPAC Name	Mercury (2+); oxidoazaniumylidynemethane
2	Chemical and other names	Mercury fulminate, Mercuric cyanate, Mercury difulminate, Fulminate of mercury, Bis(fulminato)-mercury,
3	Molecular formula	$C_2HgN_2O_2$
4	Molecular weight	284.6 gm/mole
5	Physical description	Grey, pale brown or white crystalline solid
6	Solubility	Soluble in alcohol, ammonium hydroxide, and hot water; slightly soluble in cold water
7	Melting point	160°C
8	Boiling point	357°C
9	Flash point	N.A.
10	Density	4.43 g/cm³ at 20°C
11	Decomposition temperature	100°C

11.10.2 Manufacturing process

Baron Johann Kunkel von Lowenstern, a Swedish-German alchemist, invented mercury fulminate in the 17th century. He produced this lethal explosive using nitric acid and alcohol-treated mercury. This compound's crystal structure wasn't discovered until 2007.

11.10.2.1 Raw materials

Mercury, Concentrated Nitric acid and Ethanol

11.10.2.2 Chemical reactions

It is made by dissolving mercury in sufficient nitric acid, adding ethyl alcohol to the resultant solution, and then heating the mixture until mercury fulminate precipitates.

$$6Hg + 12HNO_3 + 7C_2H_5OH \longrightarrow 6Hg(ONC)_2 + 2CO_2 + 27H_2O$$

Mercury   Nitric acid   Ethanol   Mercury fulminate

11.10.2.3 Flow sheet diagram

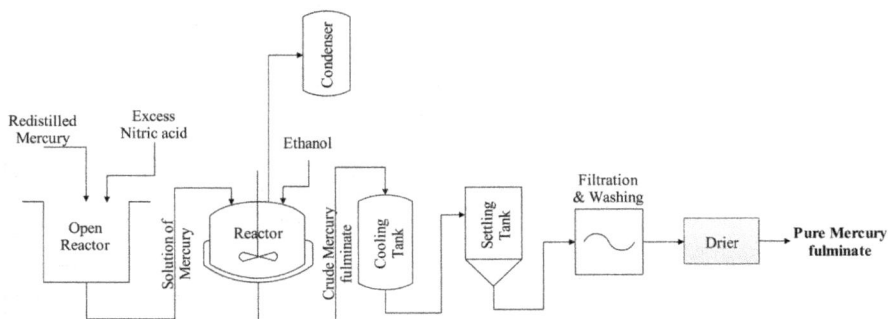

Figure 11.12 Flow sheet diagram of production Mercury fulminate

11.10.2.4 Process description

Redistilled mercury and an excessive amount of nitric acid are added to an open reactor without a stirrer. This reaction mass is allowed to stand overnight until the mercury completely dissolves. Meanwhile ethanol is charged into a jacketed reactor containing a stirrer. The solution of acidic mercury is poured slowly into a reactor. After the addition for a short while, a strong reaction begins, and the liquid boils and white vapors are released. The condenser recovers a majority of these pollutants. Nitric acid is broken down by heat, which causes the vapors to turn brownish-red as the reaction nears its conclusion. Normally, it takes the reaction an hour and a half to produce mercury fulminate. Also, by-products, ethyl nitrate, ethyl nitrite and nitroethane are formed during the reaction. The mercury fulminate settles down by the time the reaction mixture has cooled to room temperature. After filtration, the mercury fulminate crystals are cleaned with cold water to remove any remaining contaminants and acid. Next, the cleaned mercury fulminate is gathered. The acid mother liquor is treated to remove any mercury or mercury salts, neutralized with alkali to recover the alcohol, and evaporated to recover the acid.

11.10.3 Uses

It is widely utilized in mining and is used for dynamite and other explosives. Additionally, it is utilized in the development of canals, roads, railroads, and ammunition (Kurzer, 2000; Wilson, 2007).

11.11 Tetrazene

It is used as an explosive.

11.11.1 Properties

Its main properties are as follows.

Table 11.8 Physical and Chemical Properties of Tetrazene

Sr. No.	Properties	Particular
1	IUPAC Name	1-Amino-1-[2-(tetrazol-5-ylidene)hydrazinyl]guanidine
2	Chemical and other names	Tetrazene explosive, 1-(5-Tetrazolyl)-4-guanyl tetrazene
3	Molecular formula	$C_2H_8N_{10}O$
4	Molecular weight	188.1 gm/mole
5	Physical description	Pale yellow to colorless crystal plate
6	Solubility	Insoluble in water, alcohol, ether and benzene
7	Melting point	180°C
8	Boiling point	160°C
9	Flash point	N.A.
10	Density	1.7 g/cm³ at 20°C
11	Decomposition temperature	150–170°C

11.11.2 *Manufacturing process*

Hoffmann and Roth created it for the first time in 1910 by reacting aminoguanidine salts with a neutral sodium nitrite solution. Rathsburg proposed tetrazene's usage in explosive mixtures in 1921.

11.11.2.1 *Raw materials*

Aminoguanidine sulfate, Acetic acid, Sodium nitrite and Water

11.11.2.2 *Chemical reactions*

As per Fig. 11.13, 2 moles of aminoguanidine sulfate and 1 mole of sodium nitrite is undergoe a reaction in the presence of acetic acid and water to get tetrazene.

11.11.2.3 *Flow sheet diagram*

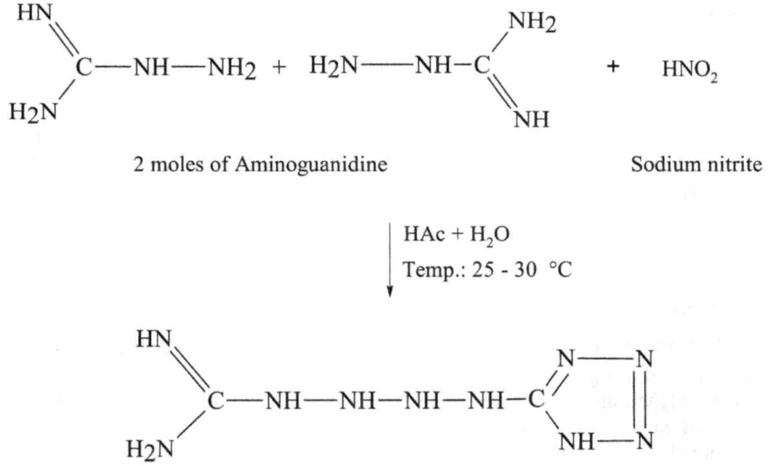

2 moles of Aminoguanidine   Sodium nitrite

HAc + H$_2$O
Temp.: 25 - 30 °C

Tetrazene

Figure 11.13 Synthesis of Tetrazene

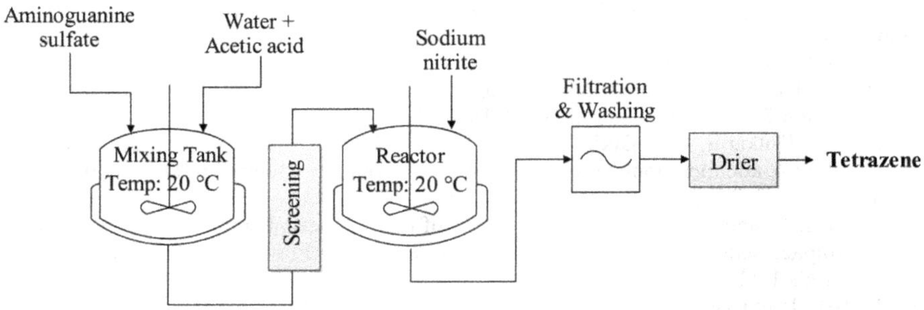

Figure 11.14 Flow sheet diagram of production Tetrazene

11.11.2.4 Process description

As per Fig. 11.14, aminoguanidine sulfate is taken in a jacketed reactor with a stirring facility. Also, the mixture of water and acetic acid is charged, stirred and warmedin order to dissolve the aminoguanidine sulfate in it. The solution is then cooled down to 30°C by circulating cooling water in it. The cooled solution is collected from the bottom of the reactor and screened to remove impurities. The screened solution is taken in another reactor having a stirrer. Also, solid sodium nitrite is charged with stirring. The stirring is continued till the sodium nitrite in it dissolves. After dissolution, the stirring is stopped and it is allowed to stand for three or four hours. To encourage precipitation, the mixture is agitated just enough, then left to stand for an additional 20 hours. A constant temperature of 25-30°C is maintained throughout the entire procedure. Tetrazene precipitate is filtered and thoroughly rinsed with water to remove any remaining acid.

11.11.3 Uses

Compared to mercury fulminate, tetrazene is slightly more impact-sensitive. Its sensitivity is diminished or destroyed when sufficiently pressed; this is known as dead pressing. Additionally, it disintegrates in hot water. It rapidly ignites on contact with fire and emits copious volumes of black smoke. To increase the explosive's sensitivity to percussion and friction, a little amount of tetrazene is added to the explosive mixture in ignition caps (Li et al., 2020).

References

Agrawal, J.P. 2010. High Energy Materials: Propellants. Explosives and Pyrotechnics, Wiley-VCH. https://doi.org/10.1002/anie.201003666

Alfaraj, W.A., McMillan, B., Ducatman, A. and Werntz, C. 2016. Tetryl exposure: Forgotten hazards of antique munitions. Annals of Occupational and Environmental Medicine 28(1). doi: 10.1186/s40557-016-0102-7

Dehkharghani, A.A. 2005. Explosive Classification in Mining Based upon Power Index, Iranian Mining Engineering Conference, Tehran.

Jin, S., Park, J.W., Baek, H., Jeon, S., Park, S.W. and Hwang, S. et al. 2017. Evaluation of nitroglycerin and cyclosporine A sorption to polyvinylchloride-and non-polyvinylchloride-based tubes in administration sets. Journal of Pharmaceutical Investigation 48: 665–672. https://doi.org/10.1007/s40005-017-0364-2

Kubota, N. 2007. Propellants and Explosives: Thermochemical Aspects of Combustion. Wiley-VCH. DOI: 10.1002/9783527693481

Kurzer, F. 2000. Fulminic Acid in the History of Organic Chemistry. Journal of Chemical Education 77(7). https://doi.org/10.1021/ed077p851

Lavanya, D., Kulkarni, P.K., Dixit, M., Raavi, P.K. and Krishna, L.N.V. et al. 2011. Sources of cellulose and their applications—A review. International Journal of Drug Formulation and Research 2(6): 19–38.

Li, J., Chang, C. and Lu, K. 2020. Optimization of the Synthesis Parameters and Analysis of the Impact Sensitivity for Tetrazene Explosive. Central European Journal of Energetic Materials 17(1): 5–19.

Liu, J. 2019. Pentaerythritol Tetranitrate, In book: Nitrate Esters Chemistry and Technology, 341–375.

Martel, R., Bellavance-Godin, A., Levesque, R. and Cote, S. 2010. Determination of Nitroglycerin and Its Degradation Products by SolidPhase Extraction and LC-UV, Chromatographia 71(3): 285–289. DOI:10.1365/S10337-009-1415-2

Millar, R.W., Arber, A.W., Endsor, R.M. and Hamid, J. 2011. Clean Manufacture of 2,4,6-Trinitrotoluene (TNT) via Improved Regioselectivity in the Nitration of Toluene. Journal of Energetic Materials 29(2): 88–114. https://doi.org/10.1080/07370652.2010.484411

Saleem, A., Rafi, N., Qasim, S., Ashraf, U. and Virk, N.H. et al. 2019. Synthesis of Picric Acid at Domestic Scales. International Journal of Innovations in Science and Technology 1(2): 62–78. DOI: 10.33411/IJIST/2019010202

Tarver, C., Tran, T.D. and Whipple, R.E. 2003. Thermal Decomposition of Pentaerythritol Tetranitrate. Propellants Explosives Pyrotechnics 28(4): 189–193. https://doi.org/10.1002/prep.200300004

Wilson, E. 2007. Mercury Fulminate Revealed, Chemical & Engineering News 85(36): 10.

Zapata, F. and Garcia-Ruiz, C. 2021. Chemical Classification of Explosives, Critical Reviews in Analytical Chemistry 51(7): 656–673. https://doi.org/10.1080/10408347.2020.1760783

CHAPTER 12

Polymer Industries

12.1 Introduction

Polymers are the most common versatile materials, used by mankind as well as industry. It is difficult to seek out a facet of our lives that's not affected by polymers. In fact, our body is formed by a lot of polymers, e.g., Proteins, enzymes and DNA. They have a wide range of applications in fibers, plastics, adhesives, paints, medical instruments, bottles, rubber and scientific instruments. Polymer is derived from the Greek words poly (many) and mer (parts or units). Additionally, they are known as macromolecules. This means that a polymer is described as "a big molecule with a high molecular weight, formed by the repetitive joining of small molecular units by chemical bonds". The repetitive units are called monomers. Polymerization is the name given to the chemical process that produces polymers from the corresponding monomers. The amount of monomers needed to make a polymer is referred to as the degree of polymerization. High polymerization oligopolymers are referred to as such, while oligopolymers are those with a degree of polymerization below 25 (short chain polymer or oligomer).

12.1.1 Classification of polymers

Polymers are classified according to their source, structure, molecular forces.

12.1.1.1 Classification of polymers according to their source

(A) **Natural polymers:** Such polymers are found in plants and animals. Proteins, cellulose, starch, some resins and rubber are examples of natural polymers.

(B) **Semi-synthetic polymers:** These are polymers formed by chemical reactions with natural polymers. Cellulose derivatives like cellulose acetate (Rayon) and cellulose nitrate are the usual examples of these polymers. Rayon is prepared by acetylation reaction of cellulose (natural polymer) with acetic anhydride in the presence of an acidic medium. Cellulose nitrate is manufactured by the nitration of cellulose with nitric acid and sulfuric acid.

(C) Synthetic polymers: These are totally man-made polymers. There are different synthetic polymers available like fibres (Nylon 6,6) and artificial rubbers (Buna-S, Buna-N, Neoprene) are examples of synthetic polymers, which are comprehensively utilized in day to day life and also in industries.

(D) Cross linked polymers: These are often formed from monomers with two or three functions, and they consist of robust covalent connections between different linear polymer chains, such as Melamine and Bakelite.

12.1.1.2 Classification of polymers according to molecular forces

Numerous uses of polymers in various industries rely on their distinctive mechanical characteristics, such as durability, elasticity and toughness. Intermolecular interactions, such as van der Waals forces and hydrogen bonds present in the polymer, are responsible for these mechanical qualities. The polymer chains are likewise bound by these forces. In this category, polymers are divided into the following three categories based on the strength of the intermolecular forces that they include.

(A) Rubbers: These solids have elastic characteristics. These elastomeric polymers are very amorphous because the polymer chains are randomly coiled and kept together by the smallest intermolecular pressures. The modest binding forces allow for stretching of these polymers. These polymers can revert to their original configuration because of a few "crosslinks" between the chains. Neoprene, Buna-S, and Buna-N are a few examples of rubber.

(B) Fibers: They are the thread-forming solids with a high modulus and lastingness. The strong intermolecular interactions like hydrogen bonds are frequently attributed to these properties. These powerful interactions also result in tightly packed chains which give crystals their crystalline form. Examples include polyesters (Terylene), polyamides (Nylon 6, 6), and others.

(C) Plastics: By applying heat and pressure, a polymer can be molded into sturdy utility items; this process is known as "plasticizing." They are partially crystalline because the intermolecular force between polymeric chains is halfway between that of elastomers and that of fibres. The two sub-classes of plastics are (a) thermoplastic plastics and (b) thermosetting plastics.

(a) Thermoplastic plastics: Some polymers soften when heated can be molded into any shape while retaining their shapes when cooled. These polymers, which soften on heating and stiffen on cooling, are known as "thermoplastics." The process of heating, reshaping, and retaining an equivalent on cooling is sometimes done numerous times. These are long-chain molecules that are linear or slightly branched and can repeatedly soften upon heating and solidify upon chilling. Between fibres and elastomers, these polymers offer intermediate intermolecular forces of attraction. Examples of thermoplastic polymers include polyethylene, PVC, nylon and seal.

(b) Thermosetting plastic: On heating, some polymers go through a chemical shift and transform into an infusible mass. They become a bulk upon heating and cannot be altered once set. Any polymer that solidifies into an insoluble mass when heated is referred to as a "thermosetting polymer." This type of polymer has cross-linked or thick branched chains in its structure. Therefore, heating results in the disintegration and infusibility of these chains. Therefore, it cannot be used again. Examples of thermosetting polymers are bakelite and other urea-formaldehyde resins.

12.1.1.3 Classification of polymers according to Tacticity (Configuration)

Based on the relative geometric arrangement of the functional (side) groups, polymers can be divided into one of the three groups listed below.

(A) Isotactic polymer: The polymer chain's functional groups are all on the same side.

(B) Syndiotactic polymer: All the functional groups are placed in a consistent pattern on the opposite sides of the polymeric chain.

(C) Atactic polymer: All the functional groups are distributed at random on both sides of the polymeric chain.

12.1.2 Types of polymerization (mechanism of polymerization)

There are two types of polymerization reactions or we can say that polymerization mechanisms can be explained in the following two ways.

12.1.2.1 Condensation or step growth polymerization

When the same or different monomers undergon polymerization, which involves intermolecular reactions it is called condensation. This polymerization occurs when monomers contain at least one functional group. High molecular mass condensation polymers may be created as a result of these polycondensation reactions, which may also result in the loss of some simple molecules like water and alcohol. This kind of polymerization typically entails a cycle of condensation reactions between two bi-functional monomers (two molecules linked together, losing tiny molecules as a result). Every step's end result is a bi-functional type once more, so the condensation process keeps on. This method is also known as step growth polymerization since each stage produces a distinct functionalized species and is independent of the others. The number of functional end groups in the monomer that may react determines the type of end polymer that is produced during condensation polymerization. Monomers with a single reactive group stop a chain from developing and produce products with a lower relative molecular mass as a result. Monomers with two reactive end groups are used to make linear polymers, and monomers with just two end groups resulting in cross-linked three-dimensional polymers. Condensation polymerization causes a gradual formation of polymer chains, often taking several hours to several days. The rapid oligomerization of each monomer results in a high concentration of developing chains. The polymerization mixture is frequently heated to high temperatures since the majority of the chemical processes used have rather large activation energies. Step-reaction polymerizations often produce polymers with molecular weights of 104–105 kg/mole, which is considered to be moderate.

12.1.2.2 Addition or chain growth polymerization

In this polymerization, the same monomer having double bonds undergo polymerization, which involves opening out of the double bonds of the monomers and their linking together by chain reactions. So, it is called chain polymerization. Moreover, it does not eliminate the small molecules during the polymerization. This polymerization is usually induced by light, heat or catalyst for opening up the double bonds and creating reactive sites. This polymerization's mechanism entails three sequential phases, the first of which is initiation. An initiator molecule is thermally broken down or allowed to interact chemically with a monomer to create an "active species" in the initiation stage. An anion, cation, or free radicals are examples of this active species. The carbon-carbon double bond of the monomer is then augmented by this active species, which then starts the polymerization process. In the growing polymer chain, the first monomer turns into the first repeat unit. Similar to how it did in the initiation step, the newly created active species adds to another monomer in the propagation step. This process is carried out repeatedly until the termination stage—the process's last step-takes place. The expanding chain terminates at this stage either by an interaction with another growing chain, a reaction with another species present in the polymerization mixture, or a spontaneous breakdown of the active site. According to the nature of active sites, the chain polymerization mechanism is divided into three types: (A) Free radical, (B) Ionic and (C) Co-ordination or stereo regular.

12.1.2.3 Co-ordination polymerization

Two scientists, Karl Zeigler (Germany) and Giulio Natta (Italy) discovered a new technique of polymerization, in which vinyl monomer undergoes polymerization in the presence of an organo-metallic catalyst (combination of Titanium tetra-chloride and triethyl aluminum), known as Ziegler-Natta catalyst. This polymerization is of prime importance, because it gives isotactic polymers (orderly arrangement of groupson the same side of the carbon chain) having high molecular weights, high densities and melting points. So, Ziegler and Natta got a Nobel Prize in 1963 (Lodge, Hiemenz, 2020; Hu et al., 2020; Carraher, 2018; Kotzenburg et al., 2017).

12.2 Polyvinyl Chloride (PVC)

PVC was unintentionally synthesized by German chemist Eugen Baumann in 1872. There after, the polymerization process of PVC was patented in 1913 by German inventor Friedrich Klatte and it was converted into a commercial product in 1933. The most adaptable and second-largest thermoplastic (after polyethylene) is polyvinyl chloride.

12.2.1 Properties

A colorless, brittle material, pure polyvinyl chloride is insoluble in alcohol but somewhat soluble in tetrahydrofuran. PVC has experienced a rapid increase in demand due to its special qualities, which include durability, ease of manufacturing, and cost effectiveness, despite several environmental concerns that have occasionally been raised. Chemically speaking, PVC is resistant to acids, salts, bases, fats, and alcohols. It is resistant to shock, abrasion, chemical degradation, corrosion, and the elements. PVC is easily cut, shaped, welded, and linked in a number of ways. Its low weight makes manual handling easier.

12.2.2 Manufacturing process

About 80% of the production involves suspension free radical polymerization because its products are the most versatile and suitable for a wide range of applications. Further, it gives polymers having average diameters of 100–180 μm. Two other methods are emulsion and bulk polymerizations.

12.2.2.1 Raw materials

Vinyl Chloride Monomer (VCM), Suspending Agent, Organic Peroxide Initiator, Buffer and Water

12.2.2.2 Chemical reactions

VCM has undergone additional suspension polymerization in the presence of an initiator and suspending agent to give PVC.

12.2.2.3 Quantitative requirements

For the production of 1000 kgs of PVC, we require 1100 kgs of VCM, 1 kg of suspending agent, 1 kg of organic peroxide initiator, 1 kg of buffer and 2400 kgs of water.

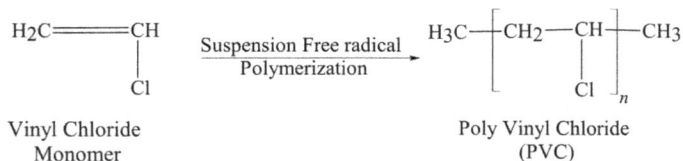

Figure 12.1 Synthesis of Polyvinyl chloride (PVC)

12.2.2.4 Flow sheet diagram

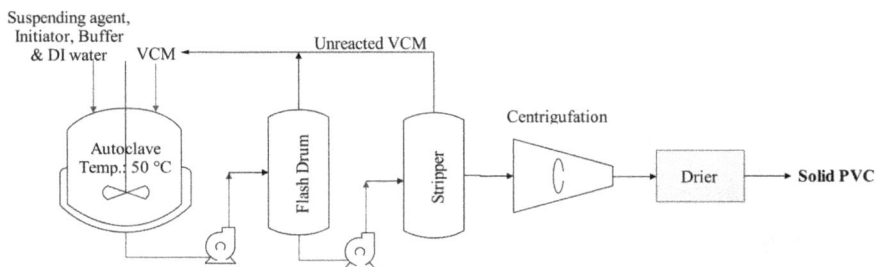

Figure 12.2 Flow sheet diagram of production of Polyvinyl chloride (PVC)

12.2.2.5 Process description

In suspending polymerization, VCM, suspending agent, organic peroxide initiator, buffer and water are fed into an autoclave. The suspending agents include clays, gelatin, inorganic salts, cellulose ether derivatives like methyl cellulose and hydroxypropyl methylcellulose, and several strongly hydrolyzed polyvinyl alcohols (PVA). In the autoclave, a temperature of 50°C is maintained. In most cases 90% of the monomer is transformed into PVC in 3–4 hours. PVC slurry is sent to the stripper and flash drum. In this instance, unconverted VCM is taken out of the PVC slurry. The residual VCM dissolved in the polymer is removed by steam in the stripper after the VCM gas is evacuated from the flash drum to the gasholder. The VCM liquid is put back into storage to be used again. The VCM that was not used in the PVC batch is almost entirely recovered and repurposed. Using a centrifuge and fluid bed drier, PVC was isolated from the slurry. The centrifuge removes the majority of the water, creating a wet cake of polymer. By running a stream of hot air bubbles over the cake, any remaining water is evaporated.

12.2.3 Uses

1. PVC is a significant plastic that is widely used in industries related to construction, transportation, packaging, electrical/electronics, and healthcare.
2. PVC is a strong, long-lasting building material that can be utilized in a variety of applications. It comes in a wide spectrum of colors, from white to black and everything in between.
3. Due to its inherent properties, PVC is widely employed in a variety of industries and produces a wide range of commonly used essential goods.
4. The government's massive infrastructure spending and emphasis on expanding irrigational land are what are primarily causing the rise in PVC consumption (Patrick, 2005; Li et al., 2015).

12.3 Low Density Polyethylene (LDPE)

The thermoplastic known as low-density polyethylene (LDPE) is made from the ethylene monomer. Free radical polymerization at high pressure was used by Imperial Chemical Industries (ICI) to prepare it.

12.3.1 Properties

The density range for LDPE is 0.900–0.940 g/cm^3. With the exception of powerful oxidizing agents and a few solvents, it is not reactive at room temperature. It can endure temperatures of 80°C continuously and 95°C briefly. It is robust and fairly flexible.

12.3.2 Manufacturing process

It is prepared by the addition polymerization of ethylene.

12.3.2.1 Raw materials

Ethylene, Initiator, Water and Solvent

12.3.2.2 Chemical reactions

Ethylene undergoes free radical polymerization at a temperature of 150–300°C and an extremely elevated pressure, i.e., 1000–3000 atmospheres using an initiator, a small amount of oxygen, and/or an organic peroxide to form LDPE.

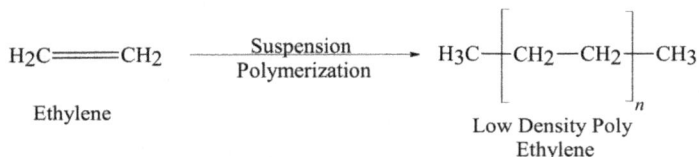

Ethylene   Low Density Poly Ethylene

Figure 12.3 Synthesis of Low Density Polyethylene (LDPE)

12.3.2.3 Quantitative requirements

For the production of 1000 kgs of LDPE, we require 1200 kgs of ethylene, 1 kg of initiator, 500 kgs of solvent and 2400 kgs of water.

12.3.2.4 Flow sheet diagram

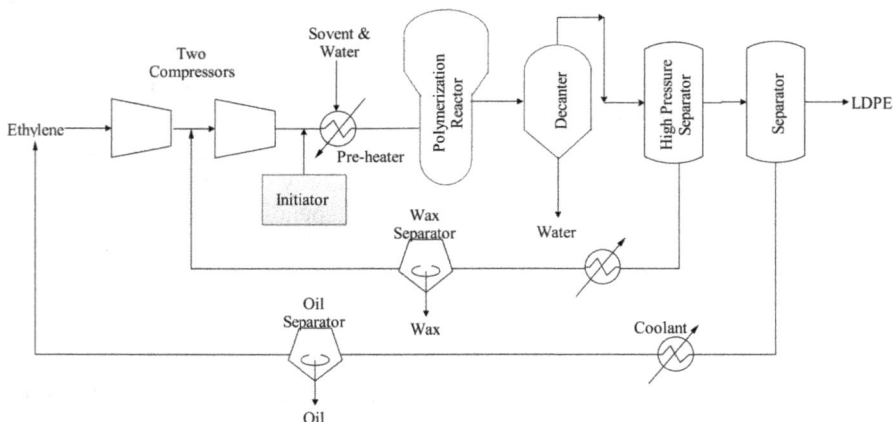

Figure 12.4 Flow sheet diagram of production of Low Density Polyethylene (LDPE)

12.3.2.5 Process description

A part of the gaseous ethylene and melted un-reacted gas from the process is inserted into the first compressor for compression. This new compressed gas is melted again with unreacted gas and compressed in the second compressor. An initiator (organic peroxide) is added after second compressor. Thereafter, materials are heated and mixed properly in a heater. Also, the solvent and water are added. Now, a mixture of ethylene, initiator, solvent and water are inserted into a polymerization reactor. This reactor may be tubular or an autoclave type. High pressure upto 1000–3000 atmospheres and moderate temperatures (150–300°C) are maintained in the polymerization reactor. After polymerization, the reactor mixture is transferred into a decanter, where water is separated from the bottom. Moisture free LDPE is collected from the top of the decanter. Then, it undergoes two levels of separation to isolate wax and oil respectively.

12.3.2.6 Major engineering problems

1. Jacketed valves make it possible to maintain the fluid temperature and to prevent polyethylene solidification.
2. Sealants (gaskets, seals, etc.) must be chosen to stop polyethylene from becoming contaminated.

12.3.3 Uses

1. LDPE is primarily used for low-cost packaging films.
2. Additionally, it is used to make toys, carrying bags, high temperature insulations, chemical tank linings, heavy duty sacks, general packaging, gas and water pipelines, squeeze bottles, and toys.
3. As well as carrying soft and pliable parts, plastic wraps, juice and milk cartons, it is used to prepare corrosion-resistant work surfaces, parts that need to be weldable, machinable, and flexible.

12.4 Linear Low Density Polyethylene (LLDPE)

DuPont developed LLDPE in the latter half of the 1950s. LLDPE is often created by copolymerizing ethylene with short-chain alpha-olefins. It may be a generally linear polymer with a significant number of short branches (for example, 1-butene, 1-hexene and 1-octene).

12.4.1 Properties

It is incredibly elastic and lengthens under pressure. It is frequently used in thinner films and has superior resilience to cracking from external stress. It is chemically resistant. It has excellent electrical qualities.

12.4.2 Manufacturing process

It is made by copolymerizing ethylene in coordination with a tiny amount of -olefins such as 1-butene, 1-hexene, or 1-octene. The reason for choosing these -olefins is that they have similar structural characteristics, functionalities, and applications to branched compounds. LLDPE is produced by the slurry-suspension process or gas phase fluidized bed polymerization in the presence of catalyst. Here, we discuss slurry polymerization.

12.4.2.1 Raw materials

Ethylene, α-Olefins, Catalyst, Cyclohexane and Deactivator

12.4.2.2 Quantitative requirements

For the production of 1000 kgs of LLDPE, we require 1150 kgs of ethylene, 5 kgs of α-olefins, 1 kg of catalyst, 1500 kgs of cyclohexane and 1 kg of deactivator.

12.4.2.3 Flow sheet diagram

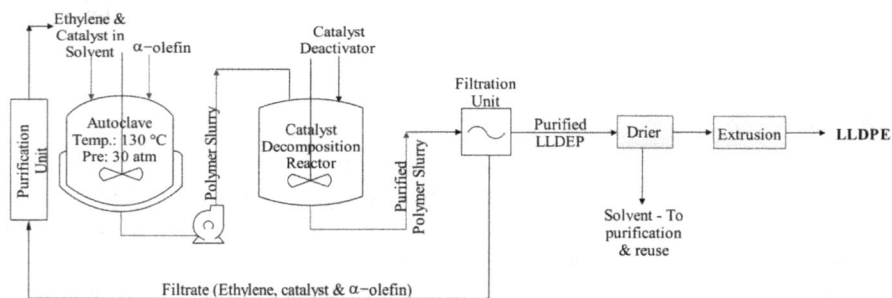

Figure 12.5 Flow sheet diagram of production of Linear Low Density Polyethylene (LLDPE)

12.4.2.4 Process description

As per Fig. 12.5, ethylene, α-olefins, catalyst and cyclohexane as the catalyst solvent are added to the polymerization reactor. It uses coordination catalysts such metallocene or Ziegler-Natta catalysts as well as supported Cr or Cr organometallic catalysts. This heat transfer medium, an alkene or cyclohexane solvent, suspends the catalyst in suspension. The reactor is kept at a temperature between 100–150°C and a pressure between 20–35 atmospheres. Following polymerization, the reaction's bottom slurry solution is collected and sent to a reactor for the breakdown or deactivation of any remaining catalyst. Here, CO, CO_2 or S is used as a decomposer. Purified polymer slurry is filtered to remove unreacted ethylene, catalyst and α-olefins as a filtrate. The residue is a wet cake of LLDPE, which is further dried to remove the catalyst solvent. This solvent is purified and re-used. The dry polymer undergoes extrusion.

12.4.3 Uses

LLDPE is commonly referred to as polyethylene and utilized in a variety of products, including geo-membranes, plastic bags and sheets (where it permits utilizing a thinner thickness than related LDPE), toys, covers, buckets, and containers, pipes, wrapping, stretch wrap, pouches, and toys.

12.5 High Density Polyethylene (HDPE)

In the year 1953, Karl Ziegler and Erhard Holzkamp developed high-density polyethylene (HDPE). Because of its versatile qualities, a low pressure was used in the process together with catalysts, which were used in a range of applications. This HDPE was created to prepare a pipe in 1955, two years afterwards. In 1963, Ziegler won the Chemistry Nobel Prize for developing HDPE.

12.5.1 Properties

It has good low temperature toughness (as low as –60°C), is flexible, transparent or waxy, weatherproof, easy to produce using most methods, is inexpensive, and has good chemical resistance.

12.5.2 Manufacturing process

Two types of catalysts are primarily used in the production of HDPE:

1. Organometallic Ziegler-Natta catalyst (titanium compounds with an aluminum alkyl group).
2. Phillips-type catalyst made of an inorganic substance (chromium (VI) oxide on silica).

Three different processes are used to make HDPE. The typical temperature is between 120 and 150°C. It is manufactured at relatively low pressures (10–80 atmospheres) in the presence of an inorganic or Ziegler-Natta catalyst. In each of the three procedures, ethylene and hydrogen are combined to regulate the polymer's chain length.

(i) Slurry process using Ziegler-Natta catalyst.
(ii) Solution process using Ziegler-Natta catalyst.
(iii) Gaseous phase process using Phillips-type catalyst.

About 65% of HDPE is prepared through slurry processes, 25% using gas-phase processes, and 10% using solution processes. So, we are discussing the slurry process.

12.5.2.1 Raw materials

Ethylene, Hydrogen and Ziegler-Natta catalyst

12.5.2.2 Chemical reactions

It is divided into two parts: (1) Preparation of Ziegler-Natta catalyst and (2) Preparation of HDPE, in which polymerization is conducted at a vacant position in the catalyst.

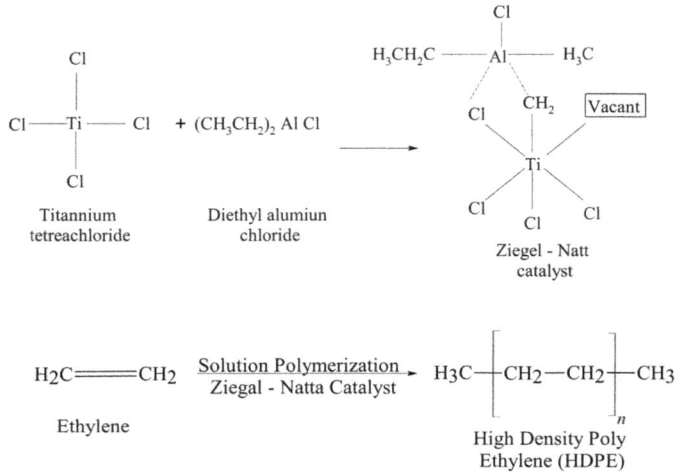

Titannium tetreachloride   Diethyl alumiun chloride   Ziegel - Natt catalyst

H_2C══CH_2 $\dfrac{\text{Solution Polymerization}}{\text{Ziegal - Natta Catalyst}}$ ⟶ $H3C$─$\left[CH_2\!-\!CH_2\right]_n$─$CH3$

Ethylene

High Density Poly Ethylene (HDPE)

Figure 12.6 Synthesis of High Density Polyethylene (HDPE)

12.5.2.3 Flow sheet diagram

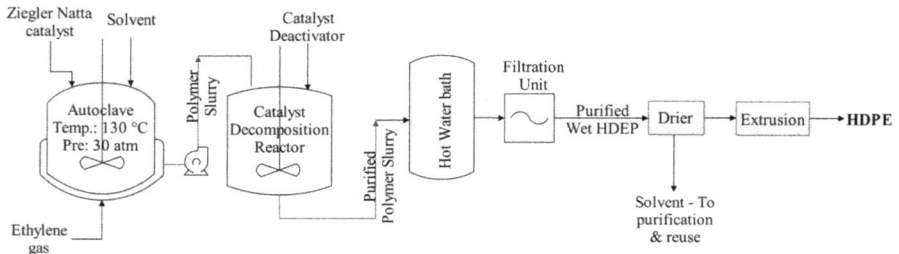

Figure 12.7 Flow sheet diagram of production of High Density Polyethylene (HDPE)

12.5.2.4 Process description

The first step involves reacting an organometallic compound, such as titanium tetrachloride, with a metal alkyl, such as diethyl aluminum chloride, in a reaction vessel at a temperature between 100 and 130°C and a pressure of about 20 atmospheres. In the reactor, Ziegler-Natta catalyst is created utilizing the afore mentioned raw ingredients. At this point, ethylene is injected into the reactor vessel at the reactor's bottom in the gas phase. Ethylene has a boiling point of about 100°C. HDLE is created when the catalyst's active site interacts with the ethylene. Heat is distributed through the solvent. High-density polyethylene has a melting point of roughly 130°C. As a result, the polyethylene that is created is solid. The unused catalyst is deactivated or degraded in a catalyst breakdown reactor after being added to the slurry solution. Here, CO, CO_2 or S is added as a decomposer. Purified polymer slurry is filtered to remove the unreacted ethylene, catalyst and α-olefins as a filtrate. Wet cake of HDPE is washed in a hot water bath to remove solvent. Wet HDPE is further dried to remove traces of the catalyst solvent. This solvent is purified and re-used. The dry polymer undergoes extrusion. HDPE is then extracted using a filtration and drying process. The Ziegler method yields polyethylene with a molecular weight between 20,000 and 1.5 million. The molecular weight is controlled during polymerization in a variety of ways, including chain transfer reagents, pressure of the reactor vessel (greater pressure, fewer branches), temperature during catalyst synthesis, and the ratio of Al/Ti catalyst introduced to the reactor.

12.5.2.5 Major engineering problems

In the polymerization reaction, an inert solvent is used in the reactor, because, the solvent must not vaporize or react with any of the compounds in the reactor.

12.5.3 Uses

High-density polyethylene (HDPE) is mainly employed in a high-performance pipe, blow-molded household and industrial bottles, oil bottles, injection-molded food containers, consumer durables and disposable goods; and film goods such as grocery sacks and merchandise bags.

HDPE is also used in raffia, blow molding, films, pipes and injection molding (Vasile and Pascu et al., 2005; Peacock, 2000; Malpass, 2010; Zhong, et al., 2018; Kesti and Sharana, 2019; Luing et al., 2014; Jordan et al., 2016).

12.6 Polypropylene (PP)

It was originally polymerized in 1951 by Paul Hogan and Robert Banks of Phillips Petroleum, and later by Italian and German scientists Natta and Rehn. It rose to prominence quite quickly because commercial production only started three years after Professor Giulio Natta of Italian chemistry first did so.

12.6.1 Properties

A thermoplastic polymer called polypropylene offers exceptional resistance to organic solvents, degreasers, and electrolytic attacks. While it is less impact-resistant than low- or high-density polyethylene, it performs better in terms of working temperatures and tensile strength. It has a low rate of moisture absorption, is lightweight, and is stain-resistant. This problematic material is heat-resistant and semi-rigid, making it perfect for moving hot liquids or gases. Its use is suggested for vacuum systems and locations with high heat and pressure. It is quite resistant to acids and alkalis, but less resistant to solvents that are aromatic, aliphatic, or chlorinated.

12.6.2 *Manufacturing process*

Various companies use different processes for manufacturing polypropylene (PP) according to their further applications like slurry process, shperipol process, fluidized bed reactor process, horizontal reactor process and vertical reactor process. In this process, a solvent is needed as a carrier for the catalyst or initiator feeds or as a diluent for the reactor solution and slurry suspension processes, while a co-monomer is used to control the polymer density of the final product. A volatile hydrocarbon solvent and co-monomer are used so that they can be easily separated from the polymer. Most commonly a vertical reactor is used.

12.6.2.1 *Raw materials*

Propylene, Catalyst, Hydrogen as carrier and Ethylene as co-monomer

12.6.2.2 *Chemical reactions*

It is prepared by the addition polymerization of propylene.

12.6.2.3 *Quantitative requirements*

For the production of 1000 kgs of PP, we require 1300 kgs of propylene, 1 kg of catalyst, 1 kg of hydrogen as a carrier and 1 kg of ethylene as a co-monomer.

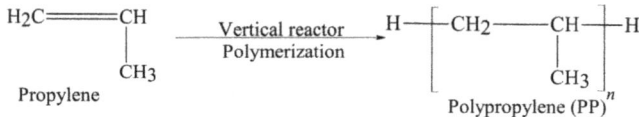

Figure 12.8 Synthesis of Polypropylene (PP)

12.6.2.4 *Flow sheet diagram*

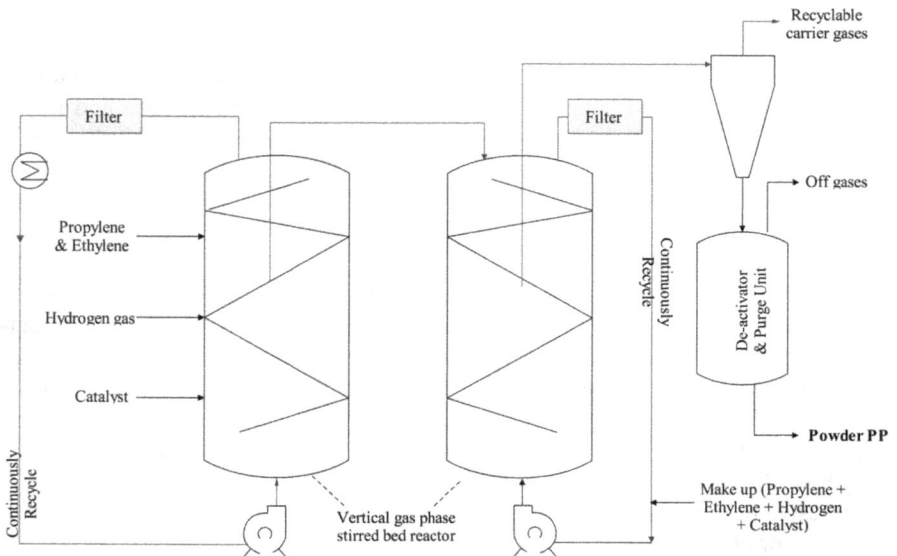

Figure 12.9 Flow sheet diagram of production of Polypropylene (PP)

12.6.2.5 Process description

In this two identical vertical gas-phase stirred bed reactors are taken for the polymerization process. All raw materials, i.e., propylene, ethylene, catalyst and hydrogen are inserted in the first reactor. The mixture is filtered, condensed and recycled continuously to the first reactor using a pump. The remaining polymerization is conducted in the second reactor. Here, raw materials are made-up to complete the polymerization. The polymer grade being produced determines the polymerization conditions (temperature, pressure, and reactant concentrations). The exothermic nature of the reaction necessitates the use of a flash heat exchanger (condenser), in which the liquefied reactor gas (mostly propylene) is combined with a new feed before being introduced into the reactor. Maximum heat exchange is guaranteed by the evaporating liquid in the polymer bed flash. After the polymerization reaction of propylene, the reaction mixture is undergoes the cycloneseparation process to remove the carrier gas. This carrier gas is compressed, condensed and recycled. The carrier gas is further purified and deactivated in a purging unit. Here, the off-gasses are removed from the top of unit. The polypropylene powder is collected from the bottom.

12.6.3 Uses

Among the many uses for polypropylene are packaging and labelling, textiles (such as ropes, thermal underwear, and carpets), stationery, plastic parts and reusable containers of various types, laboratory equipment, loudspeakers, automobile components, transvaginal mesh, and polymer banknotes. An addition polymer composed of the propylene monomer, is tough and remarkably resistant to numerous chemical solvents, bases, and acids (Karger-Kocsis and Barany, 2019; Visakh and Poletto, 2018).

12.7 Polystyrene (PS)

An apothecary from Berlin, Eduard Simon invented in 1839.

12.7.1 Properties

An amorphous, glassy polymer that is typically hard and reasonably priced is polystyrene. Unfilled polystyrene often goes by the names crystal polystyrene or general-purpose polystyrene (GPPS) because of its sparky appearance. By affixing rubber or butadiene copolymer, the polymer's toughness and impact strength improveand high impact polystyrene grades (HIPS) are created as a result. At temperatures securely below their breakdown ranges, polystyrene has strong flow characteristics and is easily extruded, injected, or compression molded. A sizeable amount of polystyrene is prepared as heat-expandable beads with the right blowing agent, which finally produces well-known foamed polystyrene products.

12.7.2 Manufacturing process

Today, the bulk polymerization method is used to generate a majority of all polystyrene produced. Batch and continuous mass polymerization are the two types, with the latter being far more popular. A polymerization section with agitated vessels polymerizes up to 80% conversion in a batch process using batch mass polymerization. The styrene monomer itself and ethyl benzene are the usual solvents used in this method.

12.7.2.1 Raw materials

Styrene, Solvent, Additive and Catalyst

12.7.2.2 Chemical reactions

Styrene undergoes addition polymerization to give polystyrene.

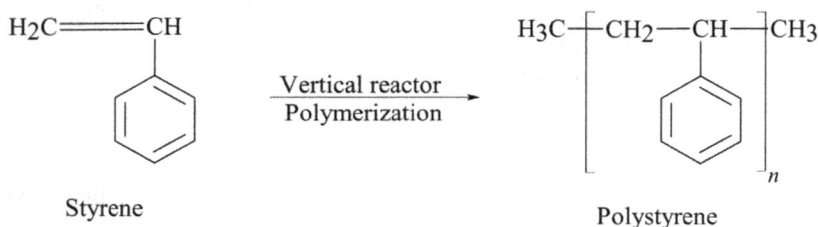

$$H_2C=CH$$

Vertical reactor
Polymerization

$$H_3C-[CH_2-CH-]CH_3$$

Styrene   Polystyrene

Figure 12.10 Synthesis of Polystyrene

12.7.2.3 Flow sheet diagram

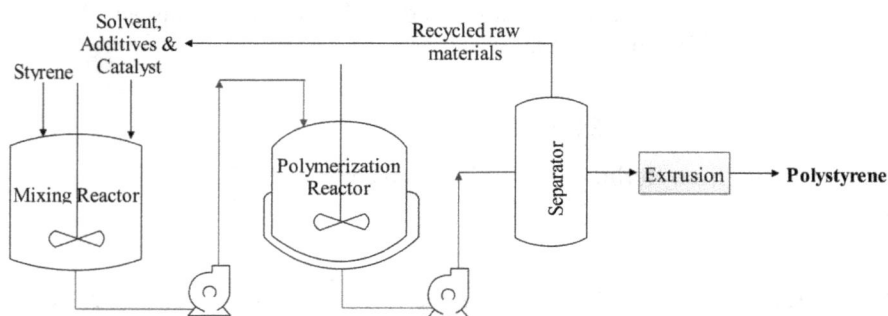

Figure 12.11 Flow sheet diagram of production of Polystyrene

12.7.2.4 Process description

In a batch reactor, styrene, additive, catalyst and solvent are mixed properly with stirring in a mixing tank. This mixture is transferred into a polymerization reactor. The polystyrene and unreacted raw materials are collected from the bottom of the reactor and fed into a separator column. Here, all the unreacted raw materials are separated and recycled. The remaining polystyrene is extruded.

12.7.3 Uses

1. Polystyrene (PS) is used for making CDs, dinnerware, disposable plastic cutlery, jewelery cases, smoke detector housings, license plate frames, plastic model assembly kits, and many other objects where a rigid, economical plastic is desired. Polystyrene Petri dishes and other laboratory containers such as test tubes and microplates play an important role in biomedical research and science.

2. Polystyrene is employed in building insulation materials, such as in insulating concrete forms and structural insulated panel building systems. Graphite incorporated with grey polystyrene foam has better insulation characteristics.

3. Expanded polystyrene (EPS) is a rigid and tough, closed-cell foam. It is usually white and made from pre-expanded polystyrene beads. EPS is employed for several applications, e.g., trays, plates, bowls and fish boxes. Other uses include molded sheets for building insulations and packings ("peanuts") for cushioning fragile items inside boxes.

4. Extruded polystyrene is also used in model-making and crafts, particularly for architectural models. Due to the extrusion manufacturing process, XPS doesn't require facers to take care of its thermal or property performance. Thus, it makes a more uniform substitute for corrugated board (Lynwood, 2014; Fry, 1999; Aitchison et al., 2011).

12.8 Styrene Butadiene Rubber (SBR or Buna-S)

Styrene butadiene rubber (SBR) is the largest produced synthetic rubber.

12.8.1 Properties

SBR performs better during processing, heat ageing, and abrasion resistance when compared to natural rubber, but it performs worse when it comes to elongation, hot tear strength, hysteresis, resilience, and tensile strength.

12.8.2 Manufacturing process

It is prepared by the polymerization of styrene and butadiene.

12.8.2.1 Raw materials

Styrene, Butadiene, Emulsifier, Modifier, DI water, Initiator, Antioxidant, Sodium hydroxide and Sulfuric acid

12.8.2.2 Chemical reactions

As per Fig. 12.12, styrene and butadiene undergo addition polymerization to obtain BUNA – S.

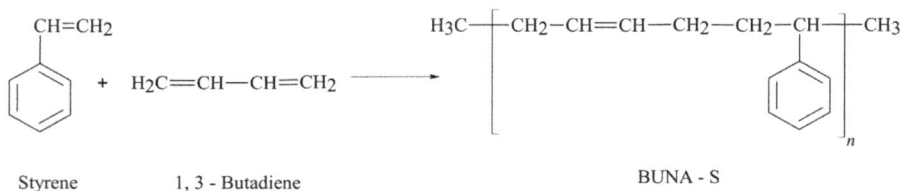

Figure 12.12 Synthesis of BUNA-S

12.8.2.3 Flow sheet diagram

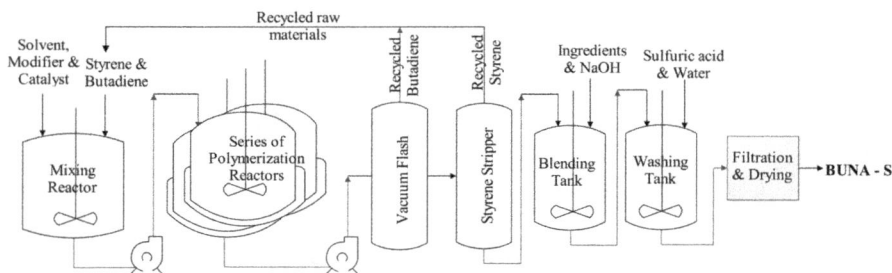

Figure 12.13 Flow sheet diagram of production of BUNA-S

12.8.2.4 Process description

In the first step, an emulsifier, modifier, initiator and DI water are mixed properly in a mixing tank, filtered and fed into a polymerization reactor. Also, styrene and butadiene are inserted at the top of the tank. Typically, a series of 10 to 15 glass or stainless steel reactors containing stirrers are used. A temperature of 50°C is maintained during the polymerization. After completion of polymerization (about 5–6 hours), the reaction mass is transferred into a finishing reactor. Thereafter, it is sent into a vacuum flash, in which butadiene is separated and recycled. The reaction mass is then transferred into a styrene stripper. Here, styrene is separated and recycled. The monomer-free emulsion is then added with certain compounding ingredients in a blending tank. The latex is coagulated to rubber using NaOH. Rubber is then sent to a washing tank, where it is washed with sulfuric acid and water to remove the catalyst, emulsifiers and other soluble impurities. Buna-S rubber is further filtered and dried.

12.8.3 Uses

It is mainly divided into solution SBR (S-SBR) and emulsion SBR (E-SBR) with respect to its application.

1. S-SBR is used by speaker driver manufacturers as the material for Low Damping Rubber Surrounds. It is used in some rubber cutting boards. It is also used as a binder in lithium-ion battery electrodes, in combination with carboxymethyl cellulose as a water-based alternative for, e.g., Polyvinylidene fluoride.

2. E-SBR is predominantly used for the production of car and light truck tires and truck tire retreading compounds. It also utilized in manufacturing house-ware mats, drain board trays, shoe soles, heels, chewing gums, food container sealants, tires, conveyor belts, sponge articles, adhesives and caulks, automobile mats, brakes and clutch pads, hoses, V-belts, floorings, military tank pads, hard rubber battery box cases, extruded gasketsand rubber toys.

3. It is also used in molded rubber goods, shoe soling, cable insulation and jacketing, pharmaceutical, surgical, and sanitary products and food packaging (Berki et al., 2017).

12.9 Acrylonitrile-Butadiene Rubber (NBR or Buna-N)

It is also known as Nitrile rubber and Perbunan. It is sold under numerous trade names including iNipol, Krynac and Europrene.

12.9.1 Properties

Generally, this rubber is resistant to chemicals, oil, and fuel. It can endure temperatures between -40 and $+125°C$. Its resistance increases with an increase in acrylonitrile content. It possesses good abrasion and compression set resistance, as well as a tensile strength greater than $10N/mm^2$. In comparison to natural rubber, it is weaker and less flexible. Due to the presence of a double bond in the butadiene portion of the chemical backbone, it exhibits poor resistance to ozone, UV, and weathering.

12.9.2 Manufacturing process

The emulsion polymerization reaction of 76% ACN and 34% 1,3-butadiene is carried out to form Buna-N rubber at 30–40°C and 2–5 hours.

12.9.2.1 Raw materials

Acrylonitrile and 1,3-Butadiene

12.9.2.2 Chemical reactions

76% ACN and 34% 1,3-butadiene are undergoe emulsion polymerization to form Buna-N rubber at 30–40°C and 2–5 hours.

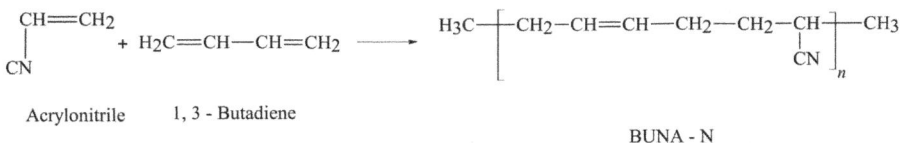

Figure 12.14 Synthesis of BUNA-N

12.9.2.4 Flow sheet diagram

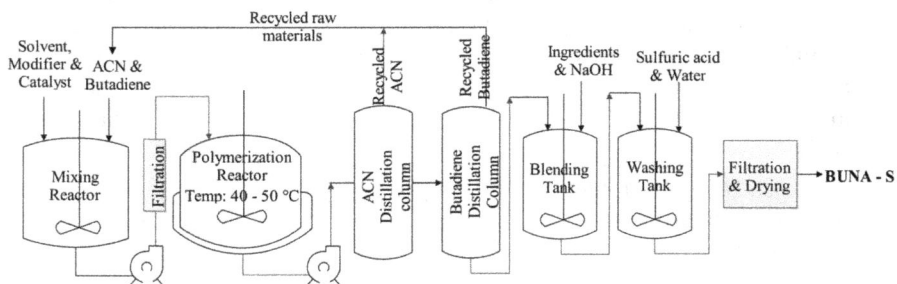

Figure 12.15 Flow sheet diagram of production of BUNA-N

12.9.2.5 Process description

In the first step, stoichiometric quantities of ACN and Butadiene; emulsifiers and additives are mixed properly in a mixing tank. Thereafter, it is filtered and transferred into a polymerization reactor having stirring and heating facilities. Heat the reaction mass upto 30–40°C with stirring. Maintain this temperature for 2–5 hours with stirring. About 70–75% polymerization occurs. After polymerization, the mass is collected from the bottom of the reactor and transferred into distillation columns. Here, monomers are separated into individual columns and recycled. Traces of monomers are removed into a flash column. In blend tanks, the monomer-free emulsion is next added along with some compounding materials. NaOH is used to coagulate the latex into rubber. Rubber is then washed with sulfuric acid and water to remove additives and other soluble impurities. Buna-N rubber is further filtered and dried.

12.9.3 Uses

Acrylonitrile Butadiene Rubber (NBR) or Nitrile rubber is one among the foremost popular compounds for automotive applications. It is used to create fuel and oil handling hoses, seals, and grommets in the automotive and aerospace industries. It is also used in the nuclear business to create protective gloves. Additionally, it is a good material for temporary lab cleaning and examination gloves. The upgraded form of Buna-N, also known as XNBR, is used for hoses, rubber belts, sealing parts, special purpose articles in oilers, reciprocating oil seals, rubber seals, gaskets, roll covers, shoe soles,molded parts for shoe heels, and O-rings. Buna-N latex is used in the preparation of adhesives and as a pigment binder (Berki et al., 2017).

12.10 Neoprene

It is also referred to as PC-rubber or Polychloroprene. Neoprene was created in 1930 by DuPont scientists.

12.10.1 Properties

It maintains flexibility over a good temperature range and has strong chemical stability. Neoprene is sold in the market in two different forms: solid rubber and latex. It resists the sun, wind or water, hot and cold temperatures, various oils and chemicals.

12.10.2 Manufacturing process

It is manufactured by the free-radical polymerization of chloroprene. It is prepared by batch or continuous polymerization.

12.10.2.1 Raw materials

Chloroprene, Ammonia, Emulsifier and Additives

12.10.2.2 Chemical reactions

Industrially, free radical emulsion polymerization of chloroprene at 0–5°C is carried out to form neoprene.

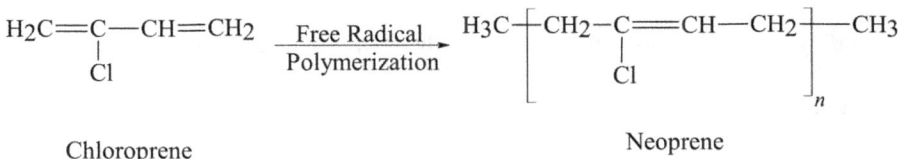

Figure 12.16 Synthesis of Neoprene

12.10.2.3 Flow sheet diagram

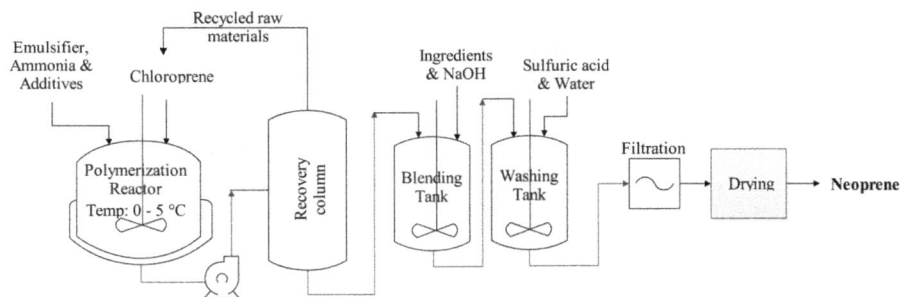

Figure 12.17 Flow sheet diagram of production of Neoprene

12.10.2.4 Process description

As per the figure, chloroprene, ammonia, emulsifiersand additives are taken in a series of polymerization reactors having stirring and jacketed facilities. Here, ammonia is used as a coolant to maintain a temperature of 0–5°C in the reactor. About 98–99% of the polymerization is conducted at this temperature. After polymerization, the reaction mass is taken into a recovery column, where the unreacted monomer is separated and recycled. Traces of the monomer are separated from the flash drum. After that, specific compounding materials are added to the monomer-free emulsion in blend tanks. NaOH is used to coagulate the latex into rubber. Rubber is then washed with sulfuric acid and water to remove additives and other soluble impurities. Neoprene rubber is further filtered and dried.

12.10.3 Uses

Due to its versatile properties, it can be used in a variety of products. It is used to prepare wet suits, fly fishing waders, diving suits, life jackets, knee and elbow pads. Additionally, it is utilized in hoses, electrical insulations, automotive fan belts, gaskets, and weather stripping materials. The use of neoprene rubber in industrial items is rising due to its incredible versatility (Pop and Gheorghe, 2010).

12.11 Nylon-6

Nylon-6, also known as PA6 or Polycaprolactam, is a semi-crystalline polyamide that is frequently used in a variety of industries. It goes by several trade names, including Perlon in Germany, Dederon in the former Soviet Union and its satellite states, Nylatron, Capron, Ultramid, Akulon, and Kapron in the United States. It also goes by the brand name Durethan.

12.11.1 Properties

It is strong, has a high tensile strength, is elastic, and has a glossy finish. It is particularly resistant to abrasion, wrinkles, and chemicals like acids and alkalis. Its fiber can absorb up to 2.4% water, however this reduces durability. 47°C is the glass transition temperature. Although nylon 6 is an artificial fiber, it is typically white. However, before manufacture, it is frequently dyed to provide a variety of colors. It has a density of 1.14 gm/cc and a tenacity that ranges from 6 to 8.5 gm/den. It has a melting point of 215°C and, on average, can withstand heat up to 150°C.

12.11.2 Manufacturing process

Ring splits and polymerization occurs when caprolactam is heated with water at around 350–370°C for about 4–5 hours in an inert environment of nitrogen.

12.11.2.1 Raw materials

Caprolactam, Catalyst and Additives

12.11.2.2 Chemical reactions

It is prepared by ring-opening polymerization of caprolactam.

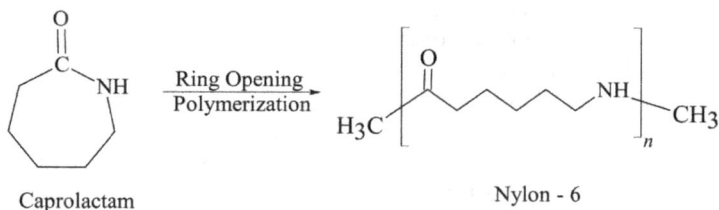

Figure 12.18 Synthesis of Nylon-6

12.3.2.3 Quantitative requirements

For manufacturing 1000 kgs of nylon-6, we require 1050 kgs of caprolactam and 50 kgs of (catalyst + additives).

12.11.2.4 Flow sheet diagram

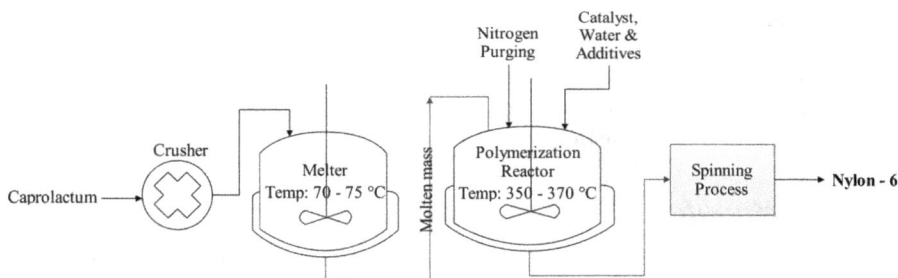

Figure 12.19 Flow sheet diagram of production of Nylon-6

12.11.2.5 Process description

As per the flow-sheet, caprolactam is crushed in a crusher. The crushed mass is sent to a steam jacketed reactor called melter. Here, the mass melts at 70–75°C. The catalyst, water and other additives are added into molten caprolactam. This molten mass is transferred into a pre-heated polymerization reactor having stirring and heating facilities. The reaction mass is heated upto 350–370°C with stirring under pressure. Maintain this temperature for 4–5 hours under pressure. During polymerization, nitrogen is purged into the reactor. Here, caprolactam molecule is broken and polymerized. Thereafter, the desired viscosity is achieved in a VK-tube, a vertical jacketed vessel. About 90–92% polymerization occurs under these conditions. After polymerization, the molten mass is passed through spinnerets to form fibers of nylon 6. These fibers are sent to a washing tower, where they are washed with hot water at 90–100°C. Here, unreacted caprolactam is separated and recycled. The fiber is dried further.

12.11.3 Uses

Nylon 6 may be a versatile synthetic material to form fibers, sheets, filaments or bristles. It is extensively used in cloth, yarn and cordage. Hosiery, knit garments, and parachutes are all made with fine filament. In the textile business, it is commonly used to create non-woven fabrics. Fabric made of nylon 6 is vibrant, lightweight, and robust. In contrast to many other fabrics, it is commonly simply dyed under air pressure in brighter and deeper hues. It is used to create plastic films, particularly for packaging, that are widely utilized in the food business. Nylon 6 bristles are used in tooth and bristle brushes. Nylon 6 can be used to produce molded goods like toy cars, skateboard wheels, and gun frames by combining it with other polymers to create intricate structures (Bansal and Raichurkar et al., 2016; Farina et al., 2019).

12.12 Nylon 6, 6

Nylon 6, 6 (nylon 6-6, nylon 6/6 or nylon 6,6) is a type of polyamide or nylon.

12.12.1 Properties

Nylon 6,6, semi crystalline polyamide, has a high melting point to give resistance to heat and friction. It is resistant to variuos solvents like water, alcohol, acids and alkalis. It is highly elastic.

12.12.2 Manufacturing process

The starting materials for nylon 6, 6 are adipic acid and hexamethylene diamine (HMDA).

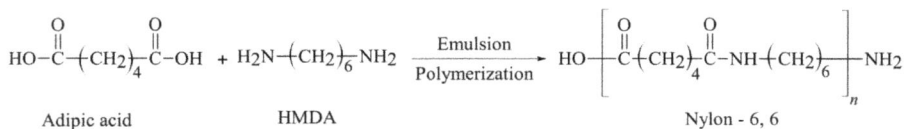

Figure 12.20 Synthesis of Nylon-6,6

12.12.2.1 Raw materials

Adipic acid, Hexamethylene diamine, Acetic acid and Water

12.12.2.2 Chemical reactions

It is made by the emulsion polymerization of adipic acid and hexamethylene diamine (HMDA) at temperatures between 250 and 270°C and pressures between 20 and 23 atmospheres. Adipic acid's carboxylic acid group and HMDA's amine group combine to form a segment of nylon 6,6.

12.12.2.4 Flow sheet diagram

12.12.2.5 Process description

First, a solution of HMDA and adipic acid is charged into individual reactors. These reactors are equipped with a stirrer and jacket. Sufficient water and acetic acid are also charged with stirring to form a solution. Here, acetic acid is added into the reaction mass to stabilize the chain length. These solutions are heated upto 250–270°C. Pre-heated solutions of HMDA and Adipic acid are charged into the polymerization reactor. Water and acetic acid are charged into it, if required. A temperature of 250–270°C is maintained for 4–5 hours under pressure. The monomers undergo polymerization at this temperature to form Nylon 6, 6 salt. This salt is concentrated by a steam evaporation process in an evaporator to remove water. Concentrated purified molten polymer mass is transferred into the reactor. Here, molten viscous polymer is forced out of the bottom onto a casting wheel by purified nitrogen under pressure for the spinning process to make polymer fiber.

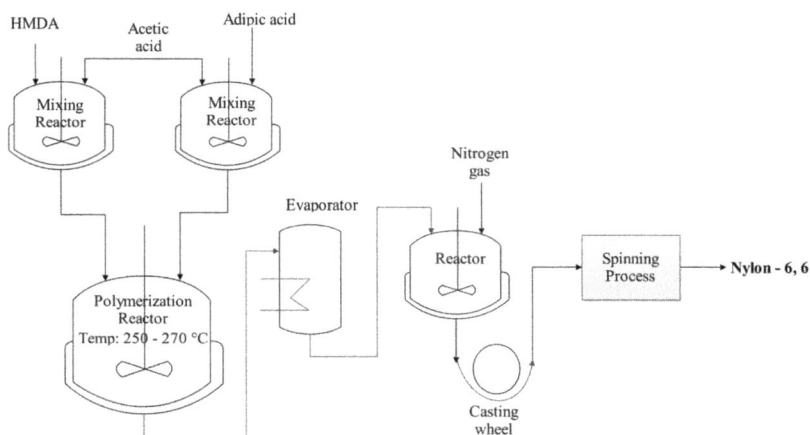

Figure 12.21 Flow sheet diagram of production of Nylon-6,6

12.12.3 Uses

It is used to prepare hosiery materials, weaving and wrap knitting, coated fabrics, tires, conveyor belts, gaskets and floor covers (Kanu, et al., 2019).

12.13 Teflon

Roy Plunkett, working for DuPont, unintentionally discovered Teflon (Polytetrafluoroethylene-PTFE) in 1938 while trying to create a new chlorofluorocarbon refrigerant. Then, PTFE was developed by a number of scientists.

12.13.1 Properties

PTFE is a solid, white material that is highly resilient, adaptable, and self-lubricating. It is water-resistant, dielectrically strong, has a high density, and has a low coefficient of friction. Also, it can withstand higher temperatures upto 260°C and cryogenic temperaturesas low as –240°C. It is also a non-stick low friction food grade compliant compound.

12.13.2 *Manufacturing process*

First, tetrafluoroethylene is prepared using chloroform and hydrogen fluoride, and then it undergoes polymerization to get Teflon.

12.13.2.1 *Raw materials*

Chloroform, Hydrogen fluoride, Ammonium persulphate, Borax and Catalyst (Disuccinic acid peroxide)

12.3.2.2 *Chemical reactions*

The manufacturing process of PTFE is conducted by three different chemical reactions. In the first, chloroform is fluorinated using hydrogen fluoride to give chlorodifluoromethane, which on pyrolysis at 550–750°C yields monomer tetrafluoroethylene. The obtained monomer undergoes free radical emulsion polymerization at 85–90°C to give Teflon.

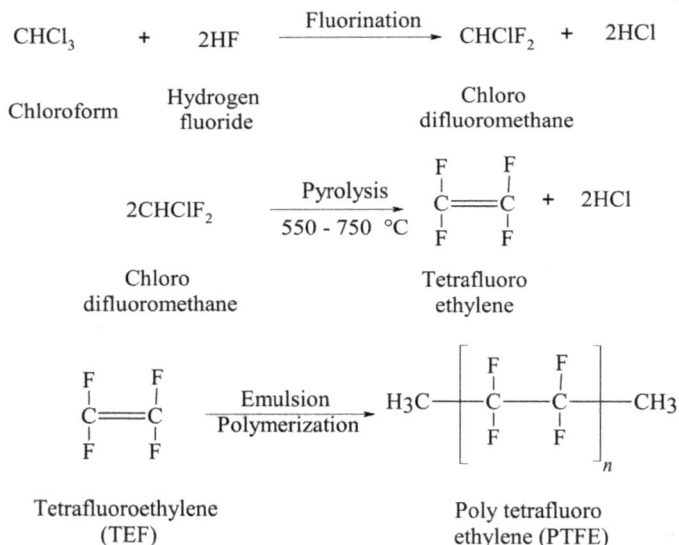

$$CHCl_3 \quad + \quad 2HF \xrightarrow{\text{Fluorination}} CHClF_2 \quad + \quad 2HCl$$

Chloroform · Hydrogen fluoride · Chloro difluoromethane

$$2CHClF_2 \xrightarrow[\text{550 - 750 °C}]{\text{Pyrolysis}} CF_2{=}CF_2 \quad + \quad 2HCl$$

Chloro difluoromethane · Tetrafluoro ethylene

$$CF_2{=}CF_2 \xrightarrow[\text{Polymerization}]{\text{Emulsion}} H_3C{-}[{-}CF_2{-}CF_2{-}]_n{-}CH_3$$

Tetrafluoroethylene (TEF) · Poly tetrafluoro ethylene (PTFE)

Figure 12.22 Synthesis of PTFE

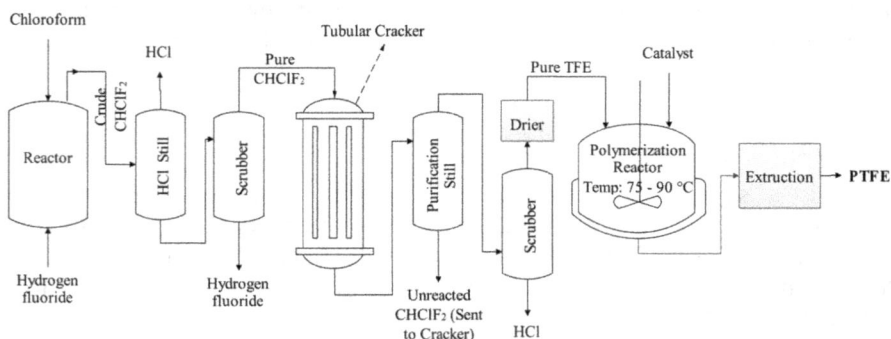

Figure 12.23 Flow sheet diagram of production of PTFE

12.13.2.3 Flow sheet diagram

12.13.2.4 Process description

As per the flow sheet diagram, the required quantity of chloroform is inserted into a reactor. Also, gaseous hydrogen fluoride is added into the bottom of the reactor. Here, interaction with the volatile liquid and gas occurs on its own, so, an agitator is not required to form crude gaseous chlorodifluoromethane. The obtained gaseous reaction mass is collected from the top of the reactor, and undergoes purification. This mass is taken into a HCl still for HCl removal from the bottom and recycling thereafter. The HCl free reaction mass is scrubbed using water to remove HF gas and recycled. After removing HCl and HF, pure chlorodifluoromethane is injected into a tubular cracker having a temperature of 550–750°C. For the synthesis of PTFE, monomer tetrafluoroethylene is required in very pure form. Impurities will have an impact on the finished product. Furthermore, even at the initial temperatures, lower than room temperature, tetrafluoroethylene can polymerize violently. Therefore, crude tetrafluoroethylene is purified to remove unreacted chlorodifluoromethane and recycled. Traces of HCl are scrubbed using water and dried. Now, a silver-plated polymerization reactor with stirring and agitation facilities is initially filled with ammonium persulfate, Borax, catalyst and water, and a pH of 9.0 to 9.2 is maintained. Pure tetrafluoroethylene is also added slowly into the reactor. The reaction mass is heated upto 85–90°C. The reactor is agitated for one hour at 85–90°C to complete the polymerization reaction upto 85–87%. Then the reaction mass is cooled and extruded.

12.13.3 Uses

1. PTFE is utilized for wiring in aerospace and computer applications, such as hookup wires, coaxial cables, carbon fiber composites, and fiber-glass composites, accounting for 50% of the total production.
2. It is frequently used to coat kitchen appliances, especially non-stick frying pans, tawas and cookers.
3. Additionally, it is used to prepare well-drilled components, industrial pipe lines, pump interiors, washers, expansion joints, hose assemblies, rings, seals, and spacers.
4. It is frequently used to separate conducting surfaces in capacitors as well as in wire and cable wrappings.
5. Numerous laboratories employ PTFE polymer piping, tubing, and vessels.
6. To improve specific qualities, a variety of fillers, including glass fibers, glass beads, carbon, graphite, molybdenum disulfideand bronze, can be mixed with a PTFE based resin.
7. Additionally, it is utilized in the production of chips, and encapsulators for quartz heaters.
8. In addition, PTFE is used to produce printed circuit boards, hookup wires, coaxial cables, outdoor clothing, school uniforms, footwear, insoles, and orthotics (Teng, 2012; Sobey, 2010).

12.14 Terylene

Terylene, commonly known as Polyethylene Terephthalate, PET, PETE, or PET-P, is the most widely used and fourth-most produced polyester thermoplastic polymer, behind PE, PP, and PCV. James Tennant Dickson, John Rex Whinfield, and the Manchester, England-based Calico Printers' Association use them. The 1941 patent for E. I. DuPont de Nemours in Delaware, United States. The Mylar trademark was first used in June 1951, and it was established for registration in 1952. Terylene is another name for it, as is Lavsan in Russia and the former Soviet Union, Dacron in the US, and other brand names.

12.14.1 Properties

PET is a transparent transplant resin that ranges in crystallinity from amorphous to semi crystalline. It is a robust, stiff, strong, and dimensionally stable material that either doesn't absorb any water or absorbs very little. It possesses strong chemical resistance, with the exception of alkalis, and gas barrier qualities. Excellent wear characteristics, electrical qualities, and dimensional stability. Because polyesters are semi-crystalline, they have great chemical resistance and high environmental stress crack resistance, especially when compared to polycarbonates. Even at high temperatures, the creep rate is relatively low. It may frequently be made quickly in a variety of hues.

12.14.2 Manufacturing process

It is prepared using two processes with either dimethyl terephthalate (DMT) or terephthalic acid, in which DMT is the most preferred product, because it gives a higher purity product.

12.14.2.1 Raw materials

Dimethyl terephthalate (DMT), Monoethylene glycol (MEG) and Zinc acetate-antimony trioxide

12.14.2.2 Chemical reactions

As per Fig. 12.24, DMT and MEG undergone copolymerization at 150–200°C in presence of zinc acetate-antimony trioxide catalyst, followed by trans-esterification step proceeds at 270–280°C to yield PET.

12.14.2.3 Flow sheet diagram

12.14.2.4 Process description

First, MEG/catalyst and DMT are inserted into a paste preparation reactor having stirring and heating facilities. The reaction mass is heated upto 150–160°C with stirring. Here, DMT is melted and mixed properly with the MEG/catalyst. The molten reaction mass is collected from the reactor and pumped into an esterification reactor. The reaction mass is stirred properly under atmospheric pressure to get the esterified product. This esterified mass undergoes crystallization and passes

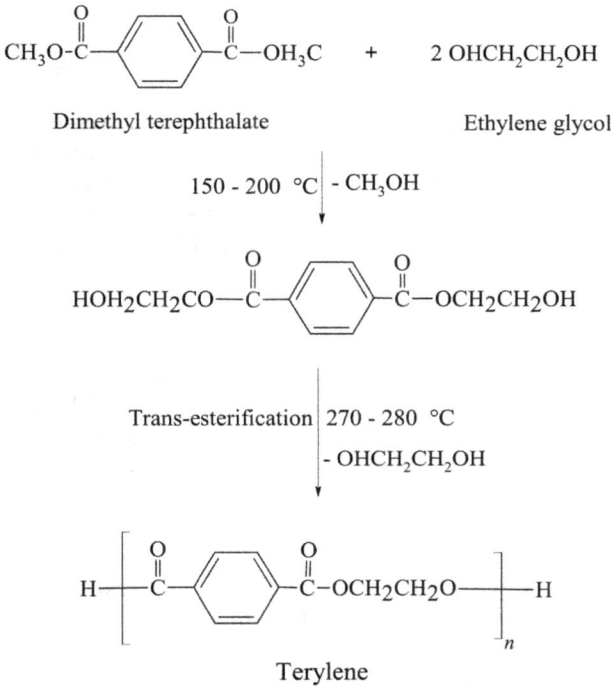

Figure 12.24 Synthesis of Terylene

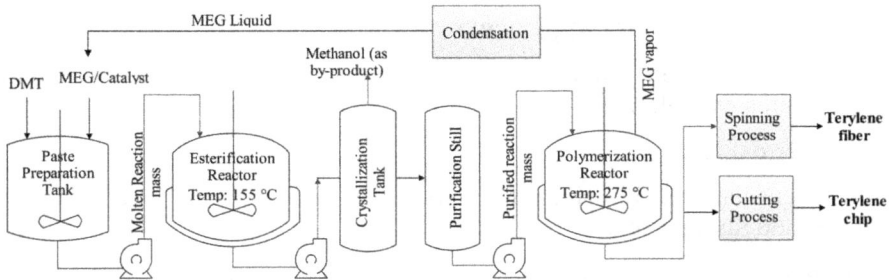

Figure 12.25 Flow sheet diagram of production of Terylene

through the methanol still to get methanol as a by-product. This methanol free reaction mass is inserted into a purification still, in which the reaction mass is purified. The purified reaction mass is pumped into a polymerization reactor. In this reactor, the temperature is increased upto 270–280°C with stirring under vacuum to get a molten mass of PET. Here, MEG vapor is collected from the top of the polymerization reactor, condensed and reused. The molten mass of PET is undergoes either a spinning process to make fibers or cutting and crystallization to get chips.

12.14.3 Uses

PET is mainly used for bottles, jars, containers and packaging applications. Under the brand names Dacron®, Trevira®, and Terylene®, it is also utilized to create fibers for a very broad range of industrial and textile applications. Its films are used for capacitors, graphics, film bases, and recording tapes, among other things. It is widely utilized in the textile industry to create durable clothing such as sarees, tapestries, and dress materials. It serves as an automatic clothing vacuum packing machine for laundry equipment (Gries et al., 2008).

12.15 Kevlar

In 1964, the DuPont group began exploring for novel, strong, lightweight material that is both light and durable for utilization in tyres. In 1971, Polish-American chemist Stephanie Kwolek introduced a fibre with a substantially higher tenacity and strength under the brand name Kevlar. This fibre is a polyamide fiber.

12.15.1 Properties

This fiber offers great chemical and ultraviolet radiation resistance, high strength, excellent absorption resistance, no melting point, low flammability, excellent heat resistance, and cut resistance. It can be carbonized at and over 425°C and has high glass transition temperatures close to 370°C. It does not melt or burn readily.

12.15.2 Manufacturing process

It is manufactured from 1,4-phenylene-diamine (*p*-phenylenediamine) and terephthaloyl chloride.

12.15.2.1 Raw materials

p-Phenylenediamine, Terephthaloyl chloride, N-methyl-pyrrolidone, Calcium chloride and Sulfuric acid

12.15.2.2 Chemical reactions

As per Fig. 12.26, A *p*-Phenylenediamine and terephthaloyl chloride solution is undergoes condensation polymerization in the presence of solvents, N-methyl-pyrrolidone and calcium chloride to give Kelar with HCl as a by-product.

12.15.2.4 Flow sheet diagram

12.15.2.5 Process description

In a single step polymerization, solvents (N-methyl-pyrrolidone and calcium chloride) are charged into a reactor having stirring facilities. With stirring, add *p*-phenylenediamine and terephthaloyl chloride. After 2–3 hours, solution polymerization is completed. Thereafter, it undergoes drying to give a dry

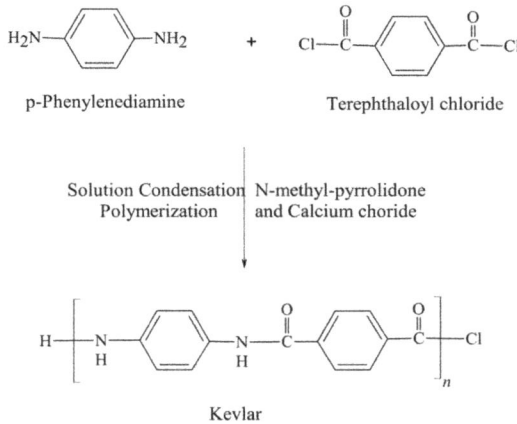

Figure 12.26 Synthesis of Kevlar

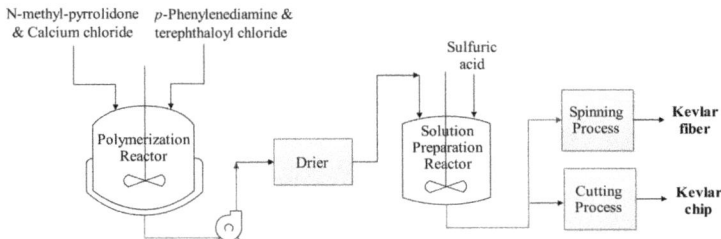

Figure 12.27 Flow sheet diagram of production of Kevlar

product. The polymer solution is prepared using sulfuric acid as a solvent. Kevlar solution undergoes either a spinning process to make fibers or cutting and crystallization to get chips.

12.15.3 Uses

This fiber belongs to the Aramid class, which is used in aerospace and military applications, as well as in bicycle tires, marine cordage, marine hull reinforcement, body armor fabric and composites, combat helmets, ballistic face masks, and ballistic vests. It also serves as an asbestos substitute. Because of its low thermal conductivity and great strength in comparison to other materials for suspension, Kevlar is commonly used in the field of cryogenics. It is also used to make motorcycle safety apparel, particularly in regions with padding like the shoulders and elbows. It is used for chaps, coats, sleeves, gloves, and other garment items. It is used in a variety of products, including tires, bowed string instruments, drumheads, woodwind reeds, frying pans, shoes, audio equipment, smart phone accessories, ropes, cables, sheaths, expansion joints, and wind and marine current turbines. Additionally, it is used to make plastrons, breeches, protective coats, and bibs for masks. The monocoque bodies of F1 racing cars, helicopter rotor blades, tennis, table tennis, badminton, and squash rackets, kayaks, cricket bats, field hockey, ice hockey, and lacrosse sticks are all made from this fiber as well as other materials like carbon and glass fibers (Joven and Maranon, 2008).

12.16 Bakelite

Known as phenol-formaldehyde resin, bakelite was the first plastic created from synthetic materials. Leo Baekeland, a Belgian-American chemist, invented it in Yonkers, New York, in 1907, and it was patented on December 7, 1909. It was used to create electrical insulators, radio and telephone casings, as well as a variety of products like jewellery, cooking utensils, toys for kids, and weapons. The American Chemical Society recognized the significance of Bakelite as the first synthetic plastic on November 9, 1993, when it named it a National Historic Chemical Landmark. Its chemical name is polyoxybenzyl methylene glycol anhydride. The trade name or brand name for its alkaline form is ALPHABOND-1612.

12.16.1 Properties

It is a thermoplastic polymer that can be swiftly molded, is smooth, keeps its shape, and is impervious to heat, scratches, and harmful chemicals. It has a low conductivity and great resistance to chemical, electrical, and thermal activity.

12.16.2 Manufacturing process

It is manufactured from phenol and formaldehyde.

12.16.2.1 aw materials

Phenol, Formaldehyde, Sodium hydroxide and Methanol

12.16.2.2 Chemical reactions

Phenol and formaldehyde undergo step-growth polymerization in the presence of sodium hydroxide at 55–60°C to get bakelite.

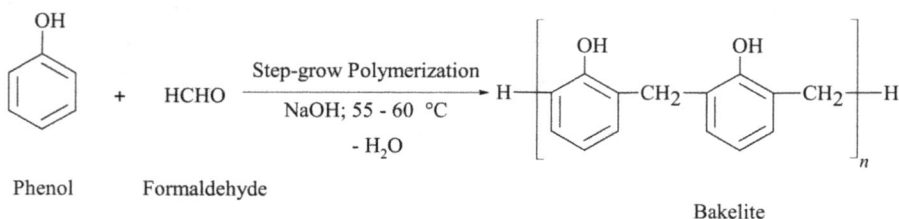

Figure 12.28 Synthesis of Bakelite

12.16.2.3 Quantitative requirements

For manufacturing 1000 kgs of bakelite, we require 490 kgs of phenol, 600 kgs of 37% formaldehyde, 200 kgs of NaOH and 10 kgs of methanol.

12.16.2.4 Flow sheet diagram

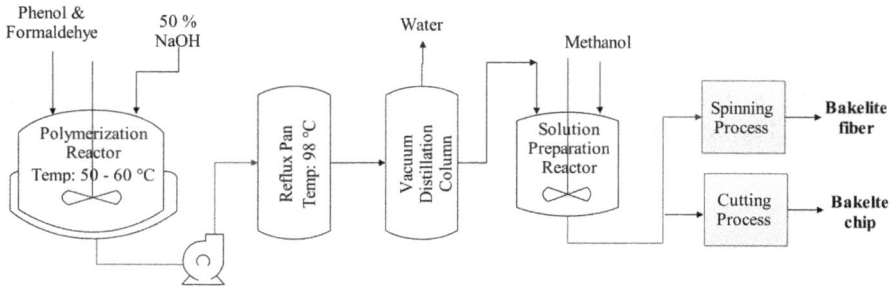

Figure 12.29 Flow sheet diagram of production of Bakelite

12.16.2.5 Process description

First, phenol and formaldehyde (37%) are inserted into a jacketed reactor having a stirring facility. The mixture is heated upto 55–60°C with agitation. The 50% NaOH solution is charged into it with stirring. The reaction mass is collected from the bottom of threactor and pumped into a reflux pan. Here, the reaction mass is refluxed for 30 minutes at 98°C. Thereafter, vacuum distillation is carried out to remove water molecules. The reaction mass is cooled and methanol is added to it for dilution. Resin solution either undergoes spinning to make fiber or cutting to form a laminate sheet.

12.16.3 Uses

Saxophones, whistles, cameras, electric guitars, telephone handsets, appliance casings, pipe stems, and buttons are made from bakelite. It is also utilized in inexpensive board and tabletop games, jewellery items, and toys, as well as small precision-shaped components, molded disc brake cylinders, saucepan handles, electrical plugs, switches, and parts for electrical irons (Sasmitha and Uma, 2018).

12.17 Melamine Formaldehyde Resin

This resin is also known as melamine or MF resin. It was discovered by William F. Talbot and patented on 21st October, 1941. Melit, Cellobond, Melmex, Isomin, Epok, Plenco, Melsir, Melopas, and Melolam are some brand names of this resin.

12.17.1 Properties

It is a thermoplastic polymer, having good resistance to stains and strong solvents including water.

12.17.2 Manufacturing process

It is prepared from melamine and formaldehyde.

12.17.2.1 Raw materials

Melamine, Formaldehyde, n-Butanol and Xylene

12.17.2.2 Chemical reactions

Melamine and formaldehyde undergo condensation polymerization in the presence of solvent, n-butanol at 100–105°C to obtain MF resin.

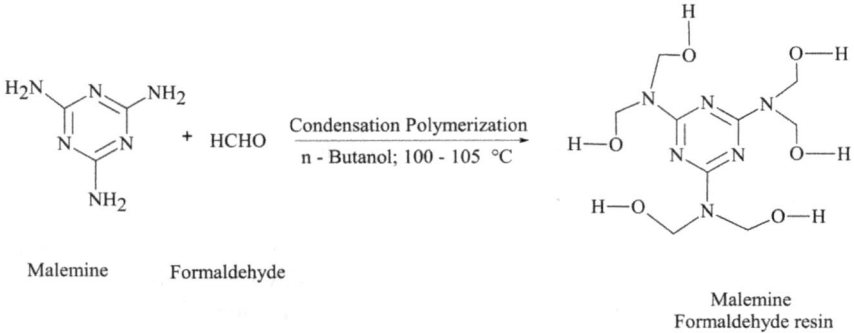

Figure 12.30 Synthesis of Melamine Formaldehyde (MF) resin

12.17.2.3 Quantitative requirements

For manufacturing 1000 kgs of bakelite, we require 250 kgs of melamine, 600 kgs of 37% formaldehyde, 500 kgs of n-butanol and 20 kgs of xylene.

12.17.2.4 Flow sheet diagram

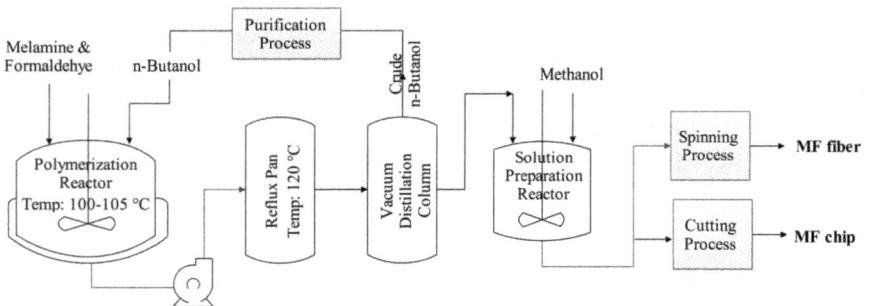

Figure 12.31 Flow sheet diagram of production of Melamine Formaldehyde (MF) resin

12.17.2.5 Process description

First, n-butanol is inserted into a jacketed reactor having a stirring facility. The mixture is heated upto 100–105°C with agitation. Charge the melamine and formaldehyde (37%) with stirring into it. Collect the reaction mass from the bottom of the reactor and pump it into a reflux pan. Here, reflux the reaction mass for 3–4 hours at 120°C. Measure the acidic pH of the solution and the viscosity of the polymer solution. Thereafter, vacuum distillation is carried out to remove n-butanol from the reaction mixture. n-Butanol is purified and reused. Cool the reaction mass and add xylene for dilution. Resin solution is either undergoes spinning to make fibers or cutting to form laminate sheets.

12.17.2.6 Major engineering problems

For polymerization, ideal conditions of temperature and acidic pH are continuously measured to obtain hexa-hydroxymethyl derivative, which on further condensation and cross linking giving mono-, di-, and polyethers having hetero Oxygen atoms in them. The cross-linkage and density of melamine resin can be controlled by co-condensation with bifunctional analogues of melamine, such as benzoguanamine and acetoguanamine.

12.17.3 Uses

For more upscale applications that call for a binder with good aesthetic qualities, such as laminates for countertops and cabinets and surface coatings for automobiles, MF resin is used. It is utilized as a plywood and particleboard adhesive. Additionally, it is commonly used in polyester appliance coatings, epoxy coatings, and automotive coatings. Melamine formaldehyde laminate sheet is used to manufacture decorative laminated panels for use in public transportation as well as surface walls, cabinets, and counters. Its foam has recently been utilized as a cleaning abrasive in addition to its primary uses as an insulating and soundproofing substance (Binder and Dunky, 2004; Kohlmayr et al., 2011).

12.18 Epoxy Resin

Various types of epoxy resins are available in market, like Bisphenol A, Bisphenol F, Novolac, Cycloaliphatic, Glycidylamine epoxy resins, in which Bisphenol A is considered as the most common epoxy resin. It may be available in either liquid or solid state. These epoxy resins were first patented by Pierre Castan in 1943, in which three major epoxy resins were produced in his patent.

12.18.1 Properties

This type of resin has excellent properties of biocompatibility, environmentally friendly, flame resistant and resistance to chemicals. It has very good mechanical strength. It has better gap-filling capabilities. An epoxy resin is heat, radiation, and steam resistant.

12.18.2 Manufacturing process

It is prepared from Bis phenol A and Epichlorohydrin.

12.18.2.1 Raw materials

Bisphenol A, Epichlorohydrin, Sodium Hydroxide, Methylisobutyl ketone and Xylene

12.18.2.2 Chemical reactions

Bisphenol A and epichlorohydrin undergo condensation polymerization using sodium hydroxide solution as a solvent at 140–160°C to give epoxy resin.

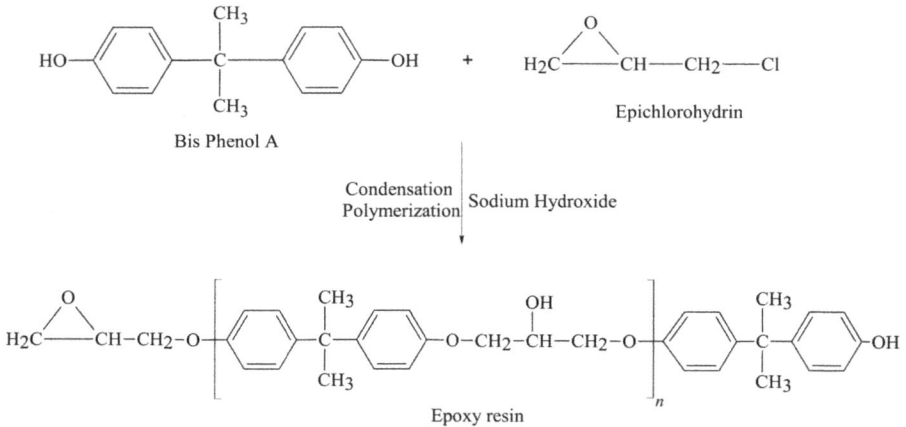

Figure 12.32 Synthesis of Epoxy resin

12.18.2.3 Flow sheet diagram

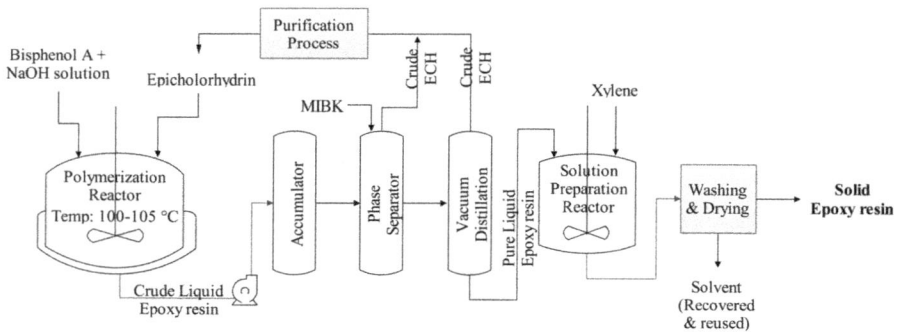

Figure 12.33 Flow sheet diagram of production of Epoxy resin

12.18.2.4 Process description

First, Bisphenol A and ECH are charged into a reactor having stirring and jacket facilities. Thereafter, 20–30% NaOH solution is added into a polymerization reactor. The reaction mixture is heated to 140–150°C to get epoxy resin. This liquid crude reaction mass undergoes accumulation and phase separation. Here, accumulators are usually installed in hydraulic systems to store energy and to smooth out pressure development in the system. Thereafter, liquid and gaseous phases are separated. The gaseous phase containing ECH is purified and reused. The liquid phase is collected and traces of ECH are also removed. Now, pure epoxy resin liquid is ready to use. Pure epoxy resin liquid undergoes solution preparation in a reactor, in which solvent, xylene is added. The solution is washed to remove impurities. It is thereafter dried to obtain solid epoxy resin.

12.18.3 Uses

Epoxy resin is used as an adhesive in the construction of vehicles, snowboards, aircrafts and bicycles. Epoxy resin is employed in the production of insulators, motors, transformers and generators. It is used in painting, coating and sealing. It also provides effective corrosion protection in chemical, pharmaceutical, energy, pulp industry, desulphurization plants, and water treatment plants (Prasoetsopha et al., 2009; Xie et al., 2018).

References

Aitchison, D.R., Brooks, H., Bain, J.D. and Pons, D. 2011. Rapid manufacturing facilitation through optimal machining prediction of polystyrene foam. March Virtual and Physical Prototyping 6(1): 41–46.https://doi.org/10.1080/17452759.2010.533961

Bansal, S. and Raichurkar, P. 2016. Review on the Manufacturing Processes of Polyester-PET and Nylon-6 Filament Yarn. International Journal on Textile Engineering and Processes 2(3): 23–28.

Berki, P., Gobl, R. and Karger-Kocsis, J. 2017. Structure and properties of styrene-butadiene rubber (SBR) with pyrolytic and industrial carbon black, Polymer Testing, 61.https://doi.org/10.1016/j.polymertesting.2017.05.039

Binder, W.H. and Dunky, M. 2004. Melamine-Formaldehyde Resins, In book: Encyclopedia of Polymer Science and Technology. https://doi.org/10.1002/0471440264.pst498

Carraher, C.E. 2018. Carraher's polymer chemistry, CRC Press.

Farina, I., Singh, N., Colangelo, F. and Luciano, R. 2019. High-Performance Nylon-6 Sustainable Filaments for Additive Manufacturing, Materials 12(23): 3955. https://doi.org/10.3390/ma12233955

Fry, B. 1999. Working with Polystyrene, In Speaking of Plastics Manufacturing Series, Speaking of Plastics Manufacturing Series.

Gries, T., Chennoth, A., Mahlmann, I. and Fischer H. 2008. Polyester staple fibre production and processing, 29th International Cotton Conference Bremen Proceedings, Germany, 95–103.

Hu, R., Qin, A., Tang, Ben Z. and Zhao, Z. 2020. Synthetic polymer chemistry: innovations and outlook, In Polymer chemistry series No. 32, Royal Society of Chemistry. DOI https://doi.org/10.1039/9781788016469

Jordan, J.L., Casem, D.T., Bradley, J.M., Dwivedi, A.K., Brown E.N. et al 2016. Mechanical Properties of Low Density Polyethylene, Journal of Dynamic Behavior of Materials 2: 411–420. https://doi.org/10.1007/s40870-016-0076-0

Joven, R. and Maranon, A. 2008. Manufacturing Kevlar Panels by Thermo-Curing Process, Proceedings of the IV conference of Electro-Mechanical Engineering, Colombia.

Kanu, N.J., Chavan, S., Jadhav, G. and Hude, N.S. 2019. Manufacturing of Nylon 6, 6 Nanofibers using Electrospinning, International Journal of Analytical Experimental and Finite Element Analysis (IJAEFEA) 6(2). https://doi.org/10.26706/ijaefea.2.6.20190404

Karger-Kocsis, J. and Barany, T. 2019. Polypropylene Handbook: Morphology, Blends and Composites, Springer International Publishing. https://doi.org/10.1007/978-3-030-12903-3

Kesti, S. and Sharana, S. 2019. Physical and Chemical Characterization of Low Density Polyethylene and High Density Polyethylene. Journal of Advanced Scientific Research, 10(3): 30–34.

Kohlmayr, M., Zuckerstatter, G. and Kandelbauer, A. 2011. Modification of melamine-formaldehyde resins by substances from renewable resources, Journal of Applied Polymer Science 124(6): 4416–4423. https://doi.org/10.1002/app.35438

Koltzenburg, S., Maskos, M. and Nuyken O. 2017. Polymer Chemistry. Springer-Verlag Berlin Heidelberg.

Li, W., Belmont, B. and Shih, A. 2015. Design and Manufacture of Polyvinyl Chloride (PVC) Tissue Mimicking Material for Needle Insertion, Procedia Manufacturing. https://doi. org/10.1016/j.promfg.2015.09.078

Lodge T.P. and Hiemenz, P.C. 2020. Polymer Chemistry, CRC Press. https://doi. org/10.1201/9780429190810

Luing, W.S., Ngadi, N. and Abdullah, T.A. 2014. Study on Dissolution of Low Density Polyethylene (LDPE). Applied Mechanics and Materials 695. https://doi.org/10.4028/ www.scientific.net/AMM.695.170

Lynwood, C. 2014. Polystyrene: Synthesis, Characteristics and Applications. In Chemistry Research and Applications, Nova Science Pub Inc.

Malpass, D. 1010. Introduction to Industrial Polyethylene, Wiley. DOI: 10.1002/9780470900468

Patrick, S. 2005. Practical Guide to Polyvinyl Chloride, Rapra Publishing.

Peacock, A. 2000. Handbook of Polyethylene, In Plastics Engineering, CRC Press.

Pop, P.A. and Gheorghe, B.M. 2010. Manufacturing process and applications of composite materials, Fascicle of Management and Technological Engineering 2(2).

Prasoetsopha, N. 2009. Studies of modified natural rubber/epoxy resin blend. Suranaree University of Technology.

Sasmitha, S.S. and Uma R.N. 2018. A Critical Review on the Application of Bakelite as a Partial Replacement of Fine and Coarse Aggregate. International Journal for Science and Advance Research in Technology 4(11): 174–178.

Shafik, E., Saleh, N., Younan A.F. and El-Messieh, S.L.A. 2020. Novel plasticizer for acrylonitrile butadiene rubber (NBR) and its effect on physico-mechanical and electrical properties of the vulcanizates. Bulletin of Materials Science 43(1). https://doi. org/10.1007/s12034-020-02196-2

Sobey, E. 2010. The Way Kitchens Work: The Science Behind the Microwave, Teflon Pan, Garbage Disposal, and More, Chicago Review Press.

Teng, H. 2012. Overview of the Development of the Fluoropolymer Industry. Applied Sciences 2(4): 496–512. https://doi.org/10.3390/app2020496

Vasile, C. and Pascu, M. 2005. Practical Guide to Polyethylene, Rapra Publishing.

Visakh, P.M. and Poletto, M. 2018. Polypropylene-Based Biocomposites and Bionanocomposites. In Thermoplastic Bionanocomposites Series, Wiley-Scrivener.

Xie, X., Zhang, X., Jin, Y. and Tian, W. 2018. Research Progress of Epoxy Resin Concrete, IOP Conference Series Earth and Environmental Science 186(2): 012038. doi :10.1088/1755-1315/186/2/012038

Zhong, X., Zhao, X., Qian, Y. and Zou, Y. 2018. Polyethylene plastic production process. Insight - Material Science 1(1). doi: 10.18282/ims.v1i1.104

CHAPTER 13

Period – II Chemical Industries

13.1 Introduction

The periodic table, often called the periodic table of elements, is a chart that displays chemical elements in the order of their atomic numbers, atomic weights, electron configurations, and common chemical properties. This type of arrangement was first provided by Dmitri Mendeleev to assista further understanding of established elements. The Modern Periodic Table is separated into 7 different periods (rows) and 18 groups (columns). Periodic trends are visible in the table's structure. The seven rows of the table known as periods, consist of 18 elements, typically having metals on the left and nonmetals on the right. The elements in the columns, referred to as groups, have comparable chemical behaviors. Also, this table is divided into s, p, d and f - blocks according to the entry of the last electron. In this chapter we discuss period - II chemicals. Lithium, beryllium, boron, carbon, nitrogen, oxygen, fluorine, and neon are all elements found in the second period. The main properties of period - II elements are discussed as per Table 13.1.

13.2 Lithium

Lithium having symbol 'Li' and atomic number 3, is a soft, silverywhite alkali metal. It is highly reactive, so, it is not freely available. It is found in ionic compounds usually. It must be kept in mineral oil because it is highly combustible. When sliced, it exhibits a metallic brilliance. Most lithium produced today is taken from salars, which are brine reserves found in high-altitude regions of Bolivia, Argentina, and Chile. It contains 170–700 ppm lithium. Besides salar sources, lithium is extracted from various ores such as spodumene ($LiAlSi_2O_6$), lepidolite [$K(Li,Al)_3(Si,Al)_4O_{10}(F,OH)_2$], zinnwaldite [$KLiFeAl(AlSi_3)O_{10}(F,OH)_2$], amblygonite [$(Li,Na)Al(PO_4)(F,OH)$]and petalite [$LiAlSi_4O$], clays. Lithium ores are found in Chile, Australia, China, Argentina

Table 13.1 Atomic and physical properties of element of period – II

Properties	Element of period – II							
	Li	Be	B	C	N	O	F	Ne
Atomic number	3	4	5	6	7	8	9	10
Electron configuration	[He]2s1	[He]2s2	[He]2s2p1	[He]2s2p2	[He]2s2p3	[He]2s2p4	[He]2s2p5	[He]2s2p6
Chemical Series	Alkali Earth Metal	Metalloid	Reactive nonmetal	Reactive nonmetal	Reactive nonmetal	Reactive nonmetal	Reactive nonmetal	Noble gas
Atomic Radius (pm)	182	153	180	170	155	152	147	154
Ionization Energy (kJ/mol)	520	899	801	1086	1400	1314	1680	2081
Electron Affinity (kJ/mol)	59.6	1.57	26.7	153.9	7.0	141.0	328	0
Electro-negativity (Pauling scale)	0.98	1.57	2.04	2.55	3.04	3.44	3.98	-
Melting Point (°C)	180.5	1560	2349	3800	63.2	54.3	53.5	24.6
Boiling Point (°C)	1560	2742	4200	4300	77.3	90.5	85.2	27.0
Density (g/cm³)	0.534	1.85	2.34	2.267	0.00125	0.00143	0.001696	0.0008999

and Zimbabwe. It has a larger lithium content than brine, but the mining process and extraction are very lengthy, requiring high energy and materials. Thus, brine extraction is more feasible than ore extraction. There are lots of methods by which lithium is recovered from brine. These methods include precipitation-evaporation, ion-exchange method, liquid-liquid extraction, ionic liquid method, membrane process and adsorption-desorption process.

It has only one valence electron, which is quickly lost to create a cation, making it a highly reactive element and a good conductor of heat and electricity. It reacts with many organic and inorganic reactants actively. It reacts with oxygen to yield its monoxide and peroxide. It has a very low viscosity and density. Lithium metal is mainly used in the glass and ceramic industries and in the production of aluminum. Lithium compounds are mainly used in the preparation of lithium-iron-phosphate batteries. They are also used in ceramics, glasses, rocket propellants, vitamin A synthesis, silver solders, underwater buoyancy devices, lubricants and grease. There are different lithium compounds such as lithium oxide, lithium acetate, lithium aluminate, lithium borate, lithium chloride and lithium cyanide (Garrett, 2004; Gershon and Shopsin, 1973).

13.3 Beryllium

Beryllium having symbol 'Be' and atomic number 4 was formerly known as glucinium until 1957. It was first discovered by Nicolas-Louis Vauquelin in 1798 and around 1828, Friedrich Wohler and Antoine A.B. Bussy separated it as the metal. It is the lightest alkaline-earth metal in Group 2 (IIa) of the periodic table. A metallic element called beryllium can be found in rocks, coal, soil, and volcanic ash. There are about 30–35 ores containing beryllium such as beryl ($Al_2Be_3Si_6O_{18}$-beryllium aluminum silicate), bertrandite ($Be_4Si_2O_7$ $(OH)_2$-beryllium silicate), phenakite (Be_2SiO_4), and chrysoberyl ($BeAl_2O_4$), helbertrandite [$Be_4(Si_2O_7)(OH)_2$], genthelvite [$Zn_4(BeSiO_4)_3$], euclase ($Be_2Al_3Si_2O_7$), helvite ($Be_3Fe_3Zn_3Mn_4SSi_5O1_6$) and epididymite ($Be_2Na_2Si_6O_{15}$). Off these ores, beryl and bertrandite have industrial importance and about 90–95% of beryllium is extracted in the form of beryllium hydroxide or beryllium oxide. These ores are found in Brazil, USA, South Africa, Southern Rhodesia, Mozambique, Argentina, Morocco, Mexico, Russia, India, China and England. These ores undergo various processes such as roll crushing, screening, ball milling, blundering with lime, conditioning using fatty acid, cleaning and scavenging to increase the concentration of beryllium in the form of its halide, hydroxideor oxide, among others. Thereafter, pure beryllium is converted using different processes such as reduction of beryllium oxide and beryllium halide, hydrogen reduction, thermal decomposition of BeI_2, electrolytic reduction and distillation among others.

It is a bivalent element having a steel gray color. Among light metals, it has the highest melting point. It has excellent thermal conductivity. When it is exposed to air, at normal temperatures and pressures, beryllium withstands nitric acid assaults and is resistant to oxidation. It is a very hazardous element because it can damage the lungs, heart and other organs. Beryllium-10 (^{10}Be) is a radioative isotope of bryllium, mostly created by the cosmic ray spallation of oxygen and nitrogen in the Earth's atmosphere. Beryllium has special characteristics that are visible to X-rays. As a result, the X-ray device's foil is used as the window material. In nuclear reactors, beryllium metal and beryllium oxide are employed to regulate fission processes. Beryllium is employed in the trigger mechanisms for nuclear weapons. It is also used in radar systems, and military infrared counter measure devices. Beryllium is used to prepare various alloys with aluminum, copper, antimony, transition metals, iron and titanium to enhance the thermal conductivity, high strength and hardness, non-magnetic properties, resistance and dimensional stability. These alloys are used to manufacture connectors, springs, IR target acquisitions, heat sink constraining cores, switches, and other components of electronic and electrical devices for aerospace, automobile, computer, defense, medical, telecommunications, and other products (Dyer, 2014; Walsh, 2009; Grew, 2002).

13.4 Boron

Boron, having symbol "B" and atomic number 5, is found in the Solar System and the Earth's crust. It is an electron deficient low-abundance element with an oxidation state of +3. It was first segregated by Joseph-Louis Gay-Lussac, Louis-Jacques Thenard and Sir Humphry Davy by boiling boron oxide (B_2O_3) with potassium metal in 1808. Due to its water solubility, it occurs in more common naturally occurring compounds, the borate minerals. It is obtained in combination with oxygen and other natural elements in nature. It is available in the oceans, sedimentary rocks, coal, shale and some soils. There are a number of ores of boron, but the most common ores are ulexite ($NaCaB_5O_9$), borax [$Na_2B_4O_5(OH)_4$], colemanite ($Ca_2B_6O_{11}$) and kernite [$Na_2B_4O_6(OH)_2$]. These ores are found in Turkey, USA, Argentina, Chile, Russia, China, and Peru. These ores are crushed, mixed with sulfuric acid, filtered, evaporated for increasing concentration, cooled and dried to get born in the form of boric acid. It is also extracted from brine from natural or industrial salt mines. This brine undergoes different processes such as treatment with hydrochloric acid, washing, drying, crystallization and solvent extraction. Further, pure boron is prepared by the reduction of its halide with hydrogen on an electrically heated tantalum filament.

Pure boron is a dark brownish black amorphous powder. It has the highest melting as well as boiling point among the metalloids. It has one isotope, i.e., boron-10, which is utilized as a neutron absorber in nuclear reactors and is part of emergency shutdown systems. Due to its outstanding properties such as heat and wear resistance, high strength, and rigidity, boron and its compounds are widely used in different industries. A very small amount of boron is required for the growth of various land plants and indirectly essential for animal life in the form of boric acid or borate. Food, the environment, and supplements containing boron are all used. In medicine, boron is used to strengthen bones, treat osteoarthritis, aid in muscle growth, raise testosterone levels, and enhance cognitive function and motor coordination. Boron fiber is used as a part of light composite materials for airships. Boric acid is used as a food preservative, especially for fish and margarine. This acid is also used as an astringent or to prevent skin infections. Boric acid is also used as an eye wash. Alloys of boron and iron are used in microphones, magnetic switches, loudspeakers, headphones, particle accelerators, and many technical applications (Zhang et al., 2017; Zhu et al., 2022; Soriano-Ursua et al., 2014).

13.5 Nitrogen

In 1772, Scottish physician Daniel Rutherford made the first discovery and isolation of nitrogen, which has the chemical symbol N and atomic number 7. French chemist Jean-Antoine-Claude Chaptal proposed the name nitrogen in 1790 since it is present in nitric acid and nitrates. Various chemicals including nitrogen gas were being prepared and utilized in different fields. Most nitrogenous compounds are used in fertilizers such as NPK fertilizers, urea, ammonium nitrate and ammonium sulfate (Follett and Hatfield, 2001).

13.5.1 Nitrogen gas

At ordinary temperatures, elemental nitrogen, a colorless, odorless, tasteless, and essentially inert diatomic gas, makes up 78% of the Earth's atmosphere by volume. All organisms that are alive have nitrogen. It is a component that all amino acids, proteins, and nucleic acids must have (DNA and RNA). Almost all neurotransmitters have nitrogen in their chemical makeup, and alkaloids may also contain nitrogen as a significant component.

13.5.1.1 Properties

Its main properties are as follows.

Table 13.2 Physical & Chemical Properties of Nitrogen

Sr. No.	Properties	Particular
1	IUPAC name	Nitrogen molecule
2	Chemical and other names	Nitrogen-14, Dinitrogen
3	Molecular formula	N_2
4	Molecular weight	28.01 gm/mole
5	Physical description	Odorless Colorless Extremely Cold Liquid
6	Solubility	Soluble in water and insoluble in alcohol, ether, benzene
7	Melting point	$-210.2°C$
8	Boiling point	$-196°C$
9	Flash point	-
10	Density	1.251 at 0°C

13.5.1.2 Manufacturing process

It is extracted from air using three processes: (1) Polymeric Membrane, in which air (mixture of nitrogen and oxygen) is passed through a special type of polymeric membrane. This membrane separates nitrogen and oxygen based on the solution diffusion mechanism. (2) Pressure Swing Adsorption (PSA), in which compressed air passes into a carbon molecular sieve (CMS) from bottom-to-top using a pump and separates nitrogen from oxygen and other gases. (3) Cryogenic distillation, in which air is cooled cryogenically and then, nitrogen is distilled out. This technology has been practiced for over 100 years. Cryogenic distillation of air is one of the most popular and efficient separation techniques for making nitrogen, and it is used in this process. We discuss cryogenic distillation in this chapter.

Here, we discuss cryogenic distillation.

(A) Raw materials: Air (mixture of nitrogen and oxygen)

(B) Flow chart

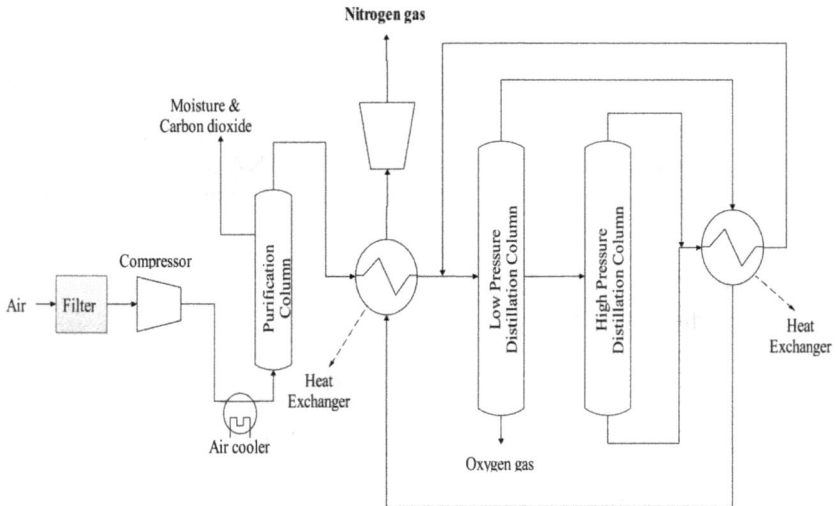

Figure 13.1 Flow sheet diagram of production of Nitrogen gas

(C) Process description: Air is filtered, compressedand cooled down. This cooled air is purified in a column to remove moisture and carbon dioxide. Purified air is cooled down in a heat exchanger at a temperature of −170 to −180°C (cryogenically). This air enters two fractional distillation, medium-pressure and low-pressure columns. Pure oxygen and nitrogen are continuously created at the top and bottom of the low-pressure column, respectively, by the exchange of matter and heat between the rising steam and the falling liquid.

13.5.1.3 Uses

1. Solids and liquids that are combustible or explosive are shielded from contact with the air by nitrogen blanketing.
2. Nitrogen is used as a shield gas during the heat treatment of iron, steel, and other metals as well as for treating the melt in the production of steel and other metals. It serves as a process gas for the reduction of the carbonization and nitrating processes by interactions with other gases.
3. It is used in the rubber and plastics industry to lower temperatures to make hard and brittle material.
4. It is also used in foods and drinks for extensive freeze-down, which reduces ice crystal damage to cells and improves the look, taste, and texture.
5. To guard against oxidation or moisture adsorption, some medicines are packed with nitrogen as a shield gas.
6. Many goods are directly tested in harsh environments using nitrogen as a coolant, or it is employed as a refrigeration source to cool circulating dry air.
7. Blood and livestock sperms are frozen with nitrogen and stored for a number of months to years. It is also used in vaccinations against viruses (Yinchange et al., 2012; Khalil, 2018).

13.5.2 Nitric acid

Aqua fortis, which means "strong water" in Latin, and spirit of niter are other names for nitric acid, a mineral acid that is extremely corrosive.

13.5.2.1 Properties

Its main properties are as follows.

Table 13.3 PhySical & Chemical Properties of Nitric acid

Sr. No.	Properties	Particular
1	IUPAC name	Nitric acid
2	Chemical and other names	Hydrogen nitrate, Azotic acid
3	Molecular formula	HNO_3
4	Molecular weight	63.01 gm/mole
5	Physical description	Colorless, yellow, or red, fuming liquid with an acrid, suffocating odor.
6	Solubility	Miscible in water and alcohol
7	Melting point	–41.6°C
8	Boiling point	83.1°C
9	Flash point	-
10	Density	1.092 at 0°C

13.5.2.2 *Manufacturing process*

There are several methods available in the market like the Ostwald process (Ammonia oxidation process), Chilli Salt Peter's method ($NaNO_3$ + H_2SO_4 method), Wisconsin process and Nitrogen fixation by nuclear fission fragments. Out of these, Nitric acid is most commonly manufactured by the Ostwald process. Nitric acid was produced and patented by Wilhelm Ostwald in 1902. The primary raw material for manufacturing the most popular type of fertiliser is provided by the Ostwald process, which is the foundation of the contemporary chemical industry.

(A) Raw materials: Air, Ammonia and Alloy (90% Platinum + 10% Rhodium)

(B) Chemical reactions: Air-oxidation of ammonia is conducted into two steps to form nitric oxide.

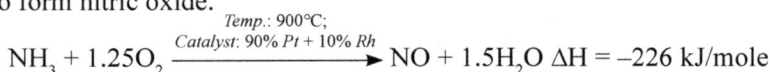

$$NH_3 + 1.25O_2 \xrightarrow{\substack{Temp.:\ 900°C;\\ Catalyst:\ 90\%\ Pt\ +\ 10\%\ Rh}} NO + 1.5H_2O \ \Delta H = -226 \text{ kJ/mole}$$

$$2NO + O_2 \xrightarrow{Temp.:\ 40°C} 2NO_2 \ \Delta H = -114 \text{ kJ/mole}$$

Thereafter, nitric oxide is absorbed in water to yield nitric acid.

$$3NO_2 + H_2O \longrightarrow 2HNO_3 + NO; \ \Delta H = -135.6 \text{ kJ/mole}$$

(C) Flow chart

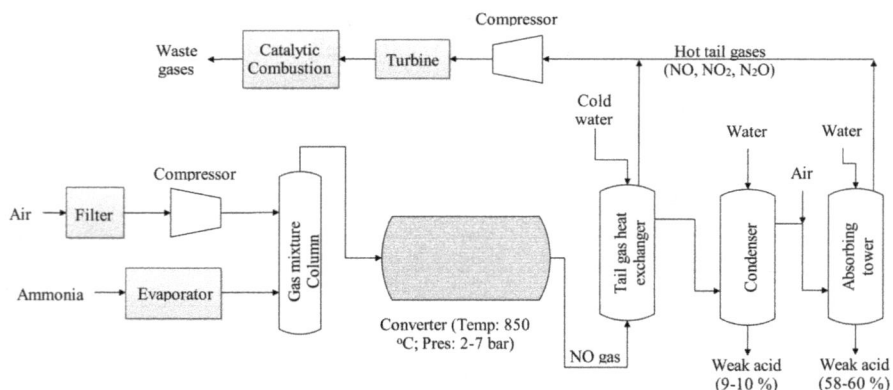

Figure 13.2 Flow sheet diagram of production of Nitric acid using Oswald process

(D) Process description: As per the flow-sheet, air is compressed by the compressor and heated upto 250°C into a pre-heater. Anhydrous Ammonia goes into an evaporator to form superheated ammonia. These gases, air and ammonia, are mixed in a gas mixer at a concentration of 90% air and 10% ammonia. This gaseous mixture is fed into a converter at 2–7 bar pressure. Since nitrogen oxide breaks down into N_2 and O_2 at temperatures above 900°C and forms at temperatures below 800°C, this does not result in nitric acid formation when dissolved in water, and the temperature of oxidation has to be be closely controlled within the range of 800–900°C. A higher temperature and catalyst increase the rate of the oxidation reaction. The product gas containing

95% nitrogen monoxide at 900°C, is inserted into a tail gas heat exchanger through the top. Also, cool water is charged from the top of the exchanger. Here, nitrogen monoxide gas is cooled down to 40°C from 900°C and further oxidation of nitrogen monoxide occurrs to form NO_2 (nitrogen dioxide). A lower temperature increases the rate of this oxidation reaction. This nitrogen dioxide is inserted into a condenser from the top. Water is charged to cool it down. Also, some part of nitrogen dioxide is reacted to form weak nitric acid (9–10%). A major part of the nitrogen dioxide is mixed with air and inserted into an absorption tower. Also, cooling water is charged from the top of the column. The nitric acid (57–60%) formed is collected from the bottom of the column. A 93–96% yield is obtained by this process.

Treatment of Tail-gases: The hot tail-gas from the tail gas heat exchanger and absorption tower contains nitric oxide (NO), nitrogen dioxide (NO_2) and nitrous oxide (N_2O). These gases are compressed and sent into a turbine. Here, steam and energy are generated and used for further processes. These cool gases undergo catalytic combustion and are converted to elemental nitrogen and thereafter, pass into the atmosphere.

(E) Major engineering problems

(a) Thermodynamic and kinetic considerations: All the important oxidation reactions are exothermic. This reaction has an extremely favorable rate constant in order that a 1 step heat converter design can also be used. Operation at low pressure has a little equilibrium advantage (1 atmosphere).

Reaction kinetics within the ammonia oxidation stage is often summarized as:

(i) By raising the temperature until an optimum is obtained, which rises with increased gas velocities, the reaction to produce NO is encouraged. This happens because back diffusion of NO into the area with greater NH_3 concentrations is prevented.

(ii) Rhodium alloying with platinum increases the yield under specific circumstances.

(iii) The diffusional transport of ammonia molecules to the catalyst surface roughly matches the rate of NO production.

(b) Process design modifications: Most plants use intermediate (3–4 atmospheres) or high (8 atmospheres) pressure processes instead of an entire atmospheric process or at a combination pressure of 1 atmospheric pressure oxidation and high pressure absorption. The maximum pressure is restricted by the pressure vessel costs.

The drawbacks of higher pressure include, however, reduced oxidation yields, higher catalyst losses, absence of effective filtering, and high power requirements.

13.5.2.3 Uses

1. Nitric acid is primarily employed as a nitrating agent in the creation of several products, including fertilizers like calcium nitrate., explosives like trinitrotoluene (TNT), nitroglycerine, cotton, colors and fragrances.

2. It is also used to clean silver, gold, platinum, and other metals.
3. Nitric acid is used to etch designs onto objects made of copper, brass and bronze.
4. It serves as a laboratory reagent (Singh et al., 2020; Minivadia et al., 2019).

13.5.3 Ammonia

Due to its numerous applications, one of the inorganic substances that is most commonly produced is ammonia. Large-scale nitrogen production can be done using a variety of techniques. In 2010, 131 million tons of nitrogen (or 159 million tons of ammonia) were produced industrially.

13.5.3.1 Properties

Its main properties are as follows.

Table 13.4 Physical & Chemical Properties of Ammonia

Sr. No.	Properties	Particular
1	IUPAC name	Azane
2	Chemical and other names	Ammonia
3	Molecular formula	NH_3
4	Molecular weight	17.03 gm/mole
5	Physical description	Colorless gas with a pungent, suffocating odor
6	Solubility	Soluble in water and alcohol
7	Melting point	−77.7°C
8	Boiling point	−33.35°C
9	Flash point	132.0°C
10	Density	0.688 at 0°C

13.5.3.2 Manufacturing process

In this process, gaseous hydrogen is produced by the steam reforming of natural gas (i.e., methane), LPG (liquefied petroleum gases, such as propane and butane), or petroleum naphtha. The Haber-Bosch or Kellogg process is then used to mix the hydrogen with nitrogen (obtained from the air) to create ammonia. Fritz Haber and Carl Bosch developed the Haber-Bosch process to produce ammonia from nitrogen in the air in 1909, and it was granted a patent in the year of 1910. Factors that affect this process are the cost, pressure, temperature, catalyst, raw material purity, and—most crucially—heat recycling and recovery. Modifications to the Haber and Bosch processes have been started in order to produce excellent materials at a lower cost and replaced by the modern Kellogg process. Here, we discuss the Kellogg and Haber-Bosch processes.

Manufacturing by Kellogg process

(A) Raw materials: Natural gas and Air (Oxygen)

(B) Flow chart

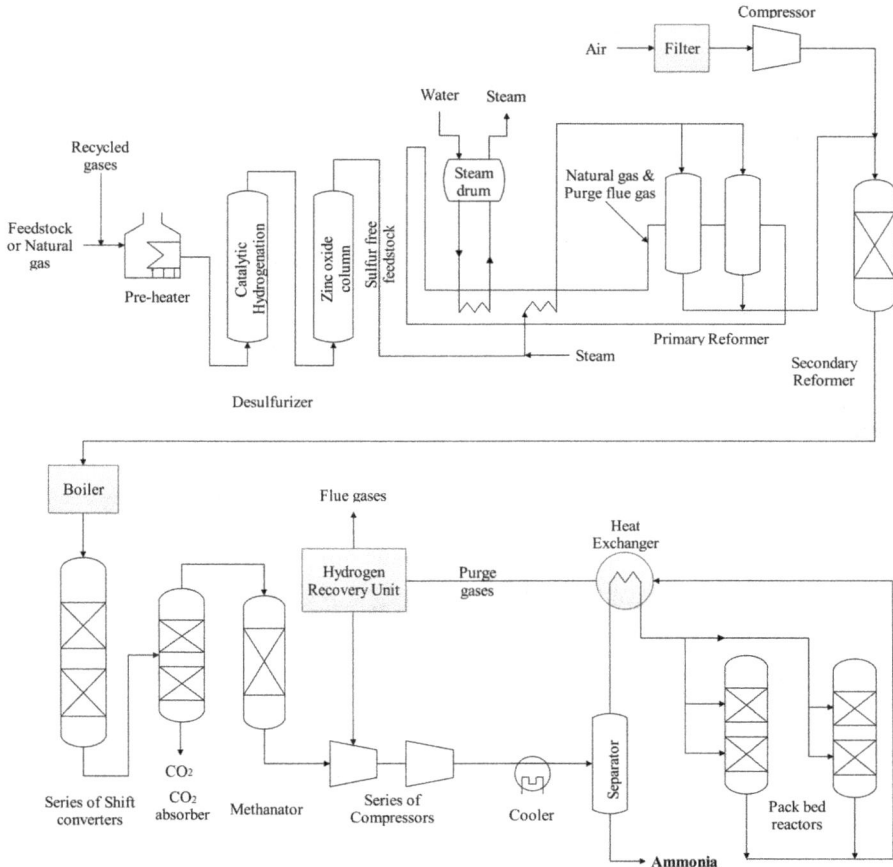

Figure 13.3 Flow sheet diagram of production of Ammonia using Kellogg process

(C) Process description: As per the flow-chart, the natural gas feedstock is heated in a heater. Heated feedstock is transferred into a catalytic hydrogenation column, where sulfur compound is converted into gaseous hydrogen sulfide. The gaseous mixture is collected from the bottom of the column.

$$H_2 + RSH \rightarrow RH + H_2S \text{ (gas)}$$

The gaseous mixture is passed through a zinc oxide bed, where it is converted to solid zinc sulfide. The remaining mixture is collected from the bottom of the column.

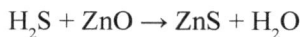

$$H_2S + ZnO \rightarrow ZnS + H_2O$$

Now, sulfur free feedstock is transferred into a primary catalytic steam reformer column from the its top. Here, methane is converted into carbon monoxide and hydrogen gas at a temperature of 1000°C and 20 atmospherespressure. This is an endothermic reaction. Steam is inserted into the tower to maintain the pressure.

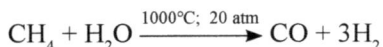

$$CH_4 + H_2O \xrightarrow{\text{1000°C; 20 atm}} CO + 3H_2$$

This gaseous reaction mixture is inserted into a secondary reformer. Here, nitrogen in form of air is also inserted into the reformer. The remaining methane is converted. The reaction is exothermic, so, heat is liberated during the reaction. Here, steam is utilized inthe primary steam reformer. The gaseous mixture is transferred into a shift converter first at a higher temperature (400°C) and thereafter, at a lower temperature (200°C) respectively. Here, carbon monoxide is converted carbon dioxide and more hydrogen, and even traces of carbon monoxide are converted at the lower temperature.

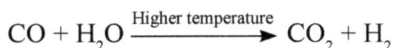

$$CO + H_2O \xrightarrow{\text{Higher temperature}} CO_2 + H_2$$

This reaction mixture is inserted into a carbon dioxide adsorption column from its top. Here, the column is filled with monoethanolamine solution (MEA), sulfinol, propylene carbonate and others, to absorb the carbon dioxide. Thereafter, a gaseous mixture is inserted into a catalytic methanator, which converts traces of CO and CO_2 with hydrogen in the presence of a catalyst into methane. The amount of methane is insignificant compared to both these oxides. The remaining gaseous mixture is pure hydrogen and nitrogen.

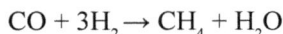

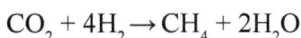

$$CO + 3H_2 \rightarrow CH_4 + H_2O$$
$$CO_2 + 4H_2 \rightarrow CH_4 + 2H_2O$$

Pure hydrogen and nitrogen are transferred into a series of compressors, where they are compressed upto 300 atmospheres to form ammonia. The entering gas stream is combined with a combination of ammonia and unreacted gases that have already been around the loop and chilled to 5°C. The gases are heated to 400°C at a pressure of 300 atmospheres, the ammonia is removed, and they are then passed over an iron catalyst. Ammonia is created when 26% of the hydrogen and nitrogen are combined under these circumstances. The discharge gas from the ammonia converter is cooled from 220°C to 30°C. More over half of the ammonia is condensed during this cooling step, which is followed by separation. With the use of a hydrogen recovery device, the residual gas is cooled and blended with the stream.

(D) Major engineering problems

(a) Thermodynamic and kinetic considerations:

$$N_{2(g)} + 3H_{2(g)} \rightarrow 2NH_{3(g)} \quad \Delta H = -22.0 \text{ kcals}$$

The maximum product yield from the above reaction can be achieved at a high pressure and low temperature which can be stated as follows:

$$Y_{NH_3} = \sqrt{K_P \cdot Y_{N_2} \cdot Y_{H_2}{}^3 \cdot P_T}$$

where, the equilibrium constant is measured at a particular absolute temperature.

$\Delta F = -RT \ln K_p = -19000 + 9.92\ T \ln T + 1.15 \times 10^{-3} T^2 - 1.63 \times 10^{-6} T^3 - 18.32\ T$

This exothermic reaction, similar to the oxidation of SO_2, is best carried out at low temperatures from an equilibrium perspective, but the reaction kinetics need a compromised temperature of 500–550°C in a single stage convertor.

Multistage operations, as employed with SO_2 oxidation, are not economically feasible for ammonia production since the cost of a high pressure reaction system is high.

Therefore, based on the following factors, the design challenge for space velocity optimization is eliminated. The equation shows that as the flow rate or space velocity increases, the fraction of $NH_3(x)$ in the exit gas drops.

$x = fV^{-n}$

where,

$n < 1$ if the bed is at the correct temperature and mass transfer rates are improved

$n > 1$ when the bed temperature is too low because of a high cooling gas velocity
The space time yield (Y) is

$$Y = V \cdot x = \frac{Cum\ of\ product}{(hr)(Cum\ of\ catalyst)}$$

$$Y = V.\ V^{-n} = V^{(1-n)}$$

In addition to a very high space velocity, the cooling bed increases the cost of NH_3 recovery because x is lower and also increases the pumping cost hence based on these considerations an optimized cost is calculated.

(b) Catalyst development: Alkali-promoted iron oxide or nonferrous metal oxides like K_2O (1–2%) and Al_2O_3 (2–4%) are frequently employed as catalysts. In an electric furnace, iron oxide is first melted before promoters are added. The bulk is ground and sieved until the desired particle size is obtained. Iron oxide is transformed into porous iron in the synthesis reactor's early phases of operation. The catalyst fuses above its maximum functioning temperature of roughly 620°C. Recently, a promoted iron catalyst (Mont Cenis method) was created in Europe that enables functioning at a very low temperature (400°C) and pressure (100 atmosphere). The catalyst's lifespan is not deterministically predicted.

Some process design modifications are required, which are mentioned below.

Conversion, recirculation rates, and refrigeration all depend on process parameters. The various processes used with different process parameters are as follows

1. Very high pressure (900–1000 atmospheres, 500–600°C, 40–80% conversion)
2. High pressure (600 atmospheres, 5000°C, 15–25% conversion)
3. Moderate pressure (200–300 atmospheres, 500–550°C, 10–30% conversion)
4. Low pressure (100 atmospheres, 400–425°C, 8–20% conversion)
5. Mont Cenis: Uses a new type of iron catalyst, iron cyanide.

Due to the relatively high cost of pressure vessels, the present tendency is towards lower pressure and higher recirculation loads. The huge single train plants with capacities as high as 1000 tonne/day from a single reactor and low production costs, utilizing centrifugal compressors, are commonly used.

Manufacturing by the Haber-Bosch process

This method employs a metal catalyst in a reaction with hydrogen while at high temperatures and pressures, turning atmospheric nitrogen into ammonia.

(A) Raw materials: Nitrogen, Hydrogen and Catalyst (Magnetite-Fe_3O_4)

(B) Chemical reactions: Atmospheric nitrogen gas is reacted with hydrogen gas at a temperature of 400–500°C and pressure of 150–250 atmospheres in the presence of iron catalyst (Magnetite-Fe_3O_4).

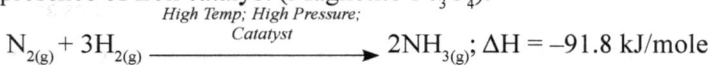

$$N_{2(g)} + 3H_{2(g)} \xrightarrow{\substack{\text{High Temp; High Pressure;} \\ \text{Catalyst}}} 2NH_{3(g)}; \Delta H = -91.8 \text{ kJ/mole}$$

(C) Flow chart

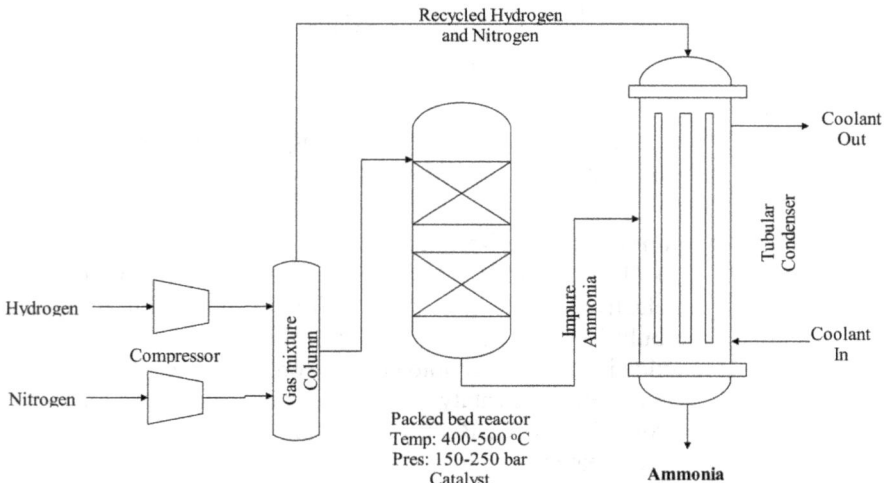

Figure 13.4 Flow sheet diagram of production of Ammonia using Haber-Bosch process

(D) **Process description:** A simple process diagram for manufacturing ammonia using the Haber-Bosch process is shown in Fig. 13.4, in which nitrogen and hydrogen gas are individually compressed and mixed in gas mixture column. This gaseous mixture is fed into a packed bed reactor with iron oxide (magnetite), in which a temperature of 400–500°C and a pressure of 150–250 atmospheres is maintained to prepare ammonia. Impure ammonia is collected from the bottom of the column and sent to a tubular condenser. The gaseous mixture is cooled down and liquid ammonia is separated; and collected from the bottom. Un-reacted nitrogen and hydrogen are collected as overhead products and recycled in the mixture column.

(E) **Major engineering problems:** The following conditions are to be maintained during the process.

(a) **Low Temperature:** A low working temperature of 500°C is preferred for an an optimum product yield as, the reaction liberate sheat energy (heat).

(b) **High Pressure:** Four mols of reactants yield two mols of product, meaning that the volume is halved. As a result, ammonia production is carried out at its highest between 200 and 900 atmospheres.

(c) **Use of Catalyst:** Catalyst use speeds up the reaction in the direction of product synthsis. Therefore, for an optimal product output, catalysts like molybdenum and promoters like iron are required.

(d) **Purity of Reactants:** The quality of the reactants, hydrogen (H_2) and nitrogen (N_2), contribute to a higher product yield.

(e) **Concentration of Reactants:** The production increases as the reactant concentration increases. Therefore, hydrogen and nitrogen are employed in high concentrations.

13.5.3.4 Uses

1. It is mostly (about 80% of ammonia) used in the agricultural and fertilizer industries for manufacturing ammonium phosphate, ammonium nitrate, ammonium sulfate, livestock feeds for cattle, sheep and goats.
2. It is used to manufacture TNT-Trinitro toluene.
3. To balance the acid components in crude oil and minimize equipment corrosion, the petroleu industry employs ammonia.
4. It is also used in rubber, pulp, textile and plastics industries (Bhavam et al., 2021, El-Moneim et al., 2018; Appl, 1999; Liu, 2013)

13.5.4 Sodium nitrate

Sodium nitrate is mainly found by mining in the form of nitratine, but it can also be manufactured industrially.

13.5.4.1 Properties

Its main properties are as follows.

Table 13.5 Physical & Chemical Properties of Sodium nitrate

Sr. No.	Properties	Particular
1	IUPAC name	Sodium nitrate
2	Chemical and other names	Nitrate of soda, Chile saltpeter, Cubic niter, Sodium saltpeter
3	Molecular formula	$NNaO_3$
4	Molecular weight	84.99 gm/mole
5	Physical description	Colourless or White Hygroscopic Crystals
6	Solubility	Slightly soluble in ethanol and methanol
7	Melting point	–308 °C
8	Boiling point	-
9	Flash point	-
10	Density	2.26 gm/cm³ at 20°C

13.5.4.2 Manufacturing process

There are various processes by which sodium nitrate is prepared. In the production of sodium nitrate, nitric acid is neutralized with soda ash to produce a nitrate and carbonic acid. Additionally, sodium nitrate and aluminum hydroxide are created by combining aqueous solutions of aluminum nitrate and sodium hydroxide.

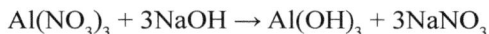

$$Al(NO_3)_3 + 3NaOH \rightarrow Al(OH)_3 + 3NaNO_3$$

Aqueous sodium nitrate solution and white aluminum hydroxide solid are produced when aluminum nitrate and sodium hydroxide react.

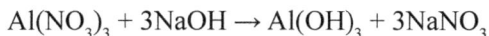

$$Al(NO_3)_3 + 3NaOH \rightarrow Al(OH)_3 + 3NaNO_3$$

It is also prepared by mixing solutions of iron nitrate and sodium hydroxide to give aqueous solutions of sodium nitrate and solid iron hydroxide.

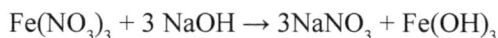

$$Fe(NO_3)_3 + 3\ NaOH \rightarrow 3NaNO_3 + Fe(OH)_3$$

Off these methods, we discuss using nitric acid with soda ash.

(A) Raw materials: Nitric acid and Soda ash

(B) Chemical reactions: Nitric acid is neutralized with soda ash to yield sodium nitrate and carbonic acid.

$$Na_2CO_3 + 2HNO_3 \longrightarrow 2NaNO_3 + H_2CO_3$$

(C) Flow chart

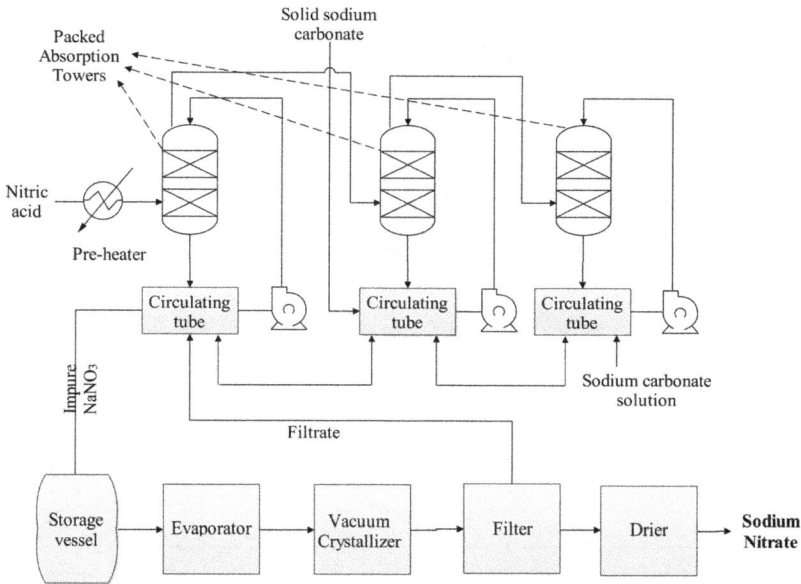

Figure 13.5 Flow sheet diagram of production of Sodium nitrite

(D) Process description: Three packed absorption towers with individual circulating tubes at the bottom of the tower were used to manufacture sodium nitrate. Also, the solution is conducted through an arrangement of interchangeable circulation tubes and pumped into respective towers. First, liquid nitric acid is converted to a gas using a pre-heater. Liquid sodium carbonate is fed into a circulating tube. The sodium carbonate solution is introduced into the top of the tower and flows downwards countercurrently with nitric acid gas which is introduced into the bottom of the tower. Also, solid sodium carbonate is added into the circulating tube to increase the concentration. Impure sodium nitrate is collected in a storage vessel from the circulating tube. It is evaporated, vacuum crystallized and filtered to collect solid pure sodium nitrate. The filtrate is recycled into a circulating tube. The sodium nitrate is further dried.

13.5.4.3 Uses

It is utilized for the preparation of potassium nitrate, fertilizers, explosives, high-strength glass and some pharmaceuticals. Sodium nitrate is also utilized to preserve meats (Yadav et al., 2020; Barnum, 2003).

13.5.5 Diammonium phosphate (DAP)

Ammonium phosphate was the first fertilizer made available in the 1960s, and after that, DAP quickly replaced it as the most popular fertilizer in this category.

13.5.5.1 Properties

Its main properties are as follows.

Table 13.6 Physical & Chemical Properties of Diammonium phosphate

Sr. No.	Properties	Particular
1	IUPAC name	Diammonium phosphate
2	Chemical and other names	DAP, Ammonium hydrogen phosphate, Ammonium orthophosphate, Ammonium phosphate
3	Molecular formula	$H_9N_2O_4P$
4	Molecular weight	132.06 gm/mole
5	Physical description	Odourless White Crystals or Powder
6	Solubility	Soluble in water and insoluble in alcohol, acetone, benzene
7	Melting point	155°C
8	Boiling point	-
9	Flash point	-
10	Density	1.678 gm/cm³ at 27°C

13.5.5.2 Manufacturing process

The starting materials for manufacturing DAP are ammonia and phosphoric acid.

(A) Raw materials: Ammonia and Phosphoric acid

(B) Chemical reactions: Phosphoric acid undergoes ammoniation (or neutralization) with ammonia to yield DAP.

$$2NH_3 + H_3PO_4 \longrightarrow (NH_4)_2HPO_4$$

(C) Flow chart

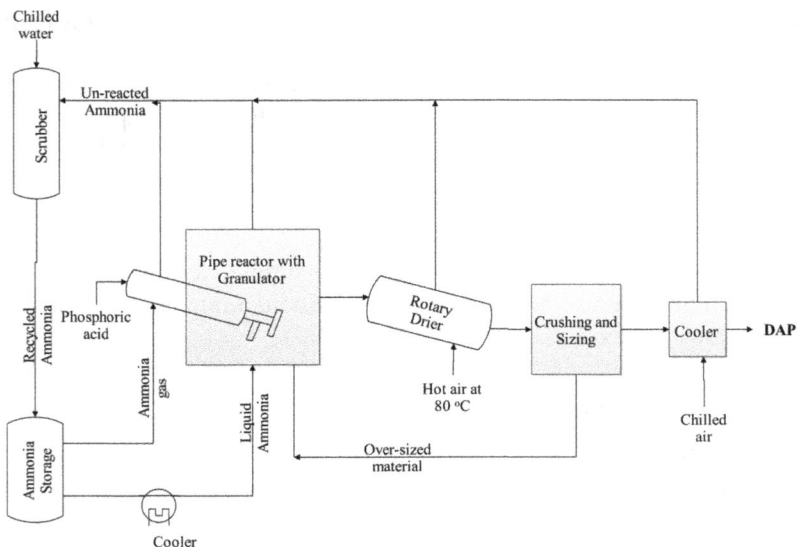

Figure 13.6 Flow sheet diagram of production of Diammonium phosphate (DAP)

(D) Process description: A two stage ammoniation of phosphoric acid is conducted in this process. First, phosphoric acid and ammonia gas are fed into a pipe reactor or a pre-neutralization downstream reactor. The pipe reactor is horizontal and mounted into a drum called, granulator. In this reactor, mono ammonium phosphate slurry is formed to reduce the amount of vaporized water during granulation. This slurry is sent to the granulator, in which liquid ammonia is inserted from the bottom. Here, the slurry is reacts with additional liquid ammonia to form DAP granules. The DAP granules are dried in a rotary drier using hot air at 85°C. Then they undergo crushing and sizing to maintain a proper size. Over-sized materials are further recycled into the granulator. The sized DAP granulesare cooled using chilled air and packed. Un-reacted ammonia from the granulator, drier and cooler is collected from the top, scrubbed with water in a scrubber and recycled into ammonia storage.

13.5.5.3 Uses

The most widely used phosphorus (P) fertilizer in the world is diammonium phosphate (DAP). It is composed of two components that are frequently used in the fertilizer industry, and it is popular because of both its outstanding physical characteristics and relatively high nutrient content. It is also used in fire retardants, metal finishings, wines and the chess manufacturing process (Gargouri et al., 2011; Manjare et al., 2011).

13.6 Carbon

Chemically speaking, carbon is an element with the symbol C and atomic number 6. Tetravalent and non-metallic, it offers four accessible electrons for covalent chemical bonding. It is a member of group 14 in the periodic table. Carbon is an extremely common element carbon. Like diamond and graphite, it can be found in pure or almost pure form, but it can also interact with other elements to form molecules. Carbon has three separate isotopes: stable isotopes 12C and 13C that are found in nature, and the radioactive isotope 14C that has a half-life of about 5,730 years. After hydrogen, helium, and oxygen, carbon is the fourth most prevalent element in the cosmos by mass and the fifteenth most abundant element in the earth's crust. Because of its abundance, distinctive variety of organic compounds, and unusual capacity to form polymers at typical Earth temperatures, carbon serves as a fundamental component of all known life. After oxygen, it makes up around 18.5% of the mass of the physical body and is the second most prevalent element there. The fundamental components of people, animals, plants, trees, and soils are these carbon-based molecules. Like fossil fuels, which are mostly made of hydrocarbons, some greenhouse gases, such as CO_2 and methane, also contain molecules with a carbon base. All petrochemicals such as olefins, aromatics, and a third group that includes synthesis gas contain carbon as the main atom. The manufacturing processes of each petrochemical were mentioned earlier (Silva, 2003).

13.6.1 Carbon dioxide

Colorless carbon dioxide, also known as a greenhouse gas, has a density that is roughly 60% more than that of dry air. It is naturally present as a trace gas in the Earth's atmosphere. Volcanoes, hot springs, and geysers are examples of natural sources. By dissolving in water and acids, carbonate rocks are removed by dissolution in water and acids. CO_2 is a gas that naturally occurs in groundwater, rivers, lakes, ice caps, glaciers, and oceans due to its solubility in water. It can be found in gas and oil reserves.

13.6.1.1 Properties

Its main properties are as follows.

Table 13.7 Physical & Chemical Properties of Carbon dioxide

Sr. No.	Properties	Particular
1	IUPAC name	Carbon dioxide
2	Chemical and other names	Carbonic acid gas, carbonic anhydride, dry solid ice
3	Molecular formula	CO_2
4	Molecular weight	44.01 gm/mole
5	Physical description	Colorless, odorless gas
6	Solubility	Miscible with water, hydrocarbons and most organic liquids
7	Melting point	−56.5°C
8	Boiling point	−78.48°C
9	Flash point	-
10	Density	1.58 gm/cm³ at 27 °C

13.6.1.2 Manufacturing process

As this gas is naturally available, it is recovered from various natural sources. The following are important commercial methods by which carbon dioxide is produced.

1) By burning of carbonaceous materials

$$C + O_2 \rightarrow CO_2$$

2) From synthesis gas in ammonia production

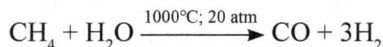

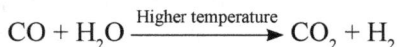

$$CH_4 + H_2O \xrightarrow{\text{1000°C; 20 atm}} CO + 3H_2$$

$$CO + H_2O \xrightarrow{\text{Higher temperature}} CO_2 + H_2$$

3) In the production of H_2 by steam water gas

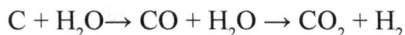

$$C + H_2O \rightarrow CO + H_2O \rightarrow CO_2 + H_2$$

4) From the production of ethanol by fermentation

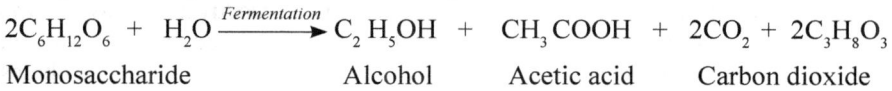

$$2C_6H_{12}O_6 + H_2O \xrightarrow{Fermentation} C_2H_5OH + CH_3COOH + 2CO_2 + 2C_3H_8O_3$$

| Monosaccharide | Alcohol | Acetic acid | Carbon dioxide |

But, the most important method for commercial method is the burning of carbonaceous materials like oil, natural gas or coke.

(A) Raw materials

Any organic fuel such as natural gas, kerosene, diesel and Ethanolamines

(B) Chemical reactions

$$C_nH_{2n} + 3n/2 O_2 \xrightarrow{Burning} nCO_2 + nH_2O$$

(C) Flow chart

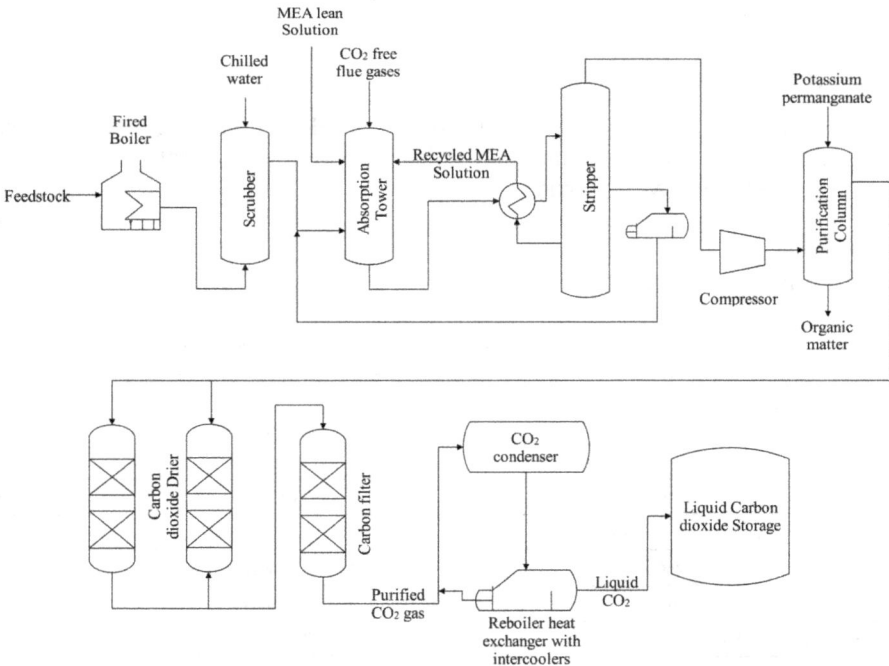

Figure 13.7 Flow sheet diagram of production of carbon dioxide

(D) Process description:

Fuel is burned to produce 200–250 psig of steam in a fire boiler as per Fig. 13.7. 10–18% CO_2 and flue gases are removed at 345°C from the boiler and sent via cooling towers where chilled water is used to cool and clean the gases. Flue gases that have been cooled are fed into the base of the absorption tower, where CO_2 is selectively absorbed by a monoethanolamine (MEA) solution flowing against the gas stream. From the tower's summit, CO_2 free flue gases are released into the environment. The solution containing CO_2 was sprayed from the top of the stripper after being collected at the bottom of the absorption tower. Heat is used to remove

CO_2 from the amine solution, and the MEA solution is then returned to the absorption tower through colder machinery. The reactivation tower's top gas cooler, where steam condenses and refluxes back down to the tower, was pumped with steam and CO_2. Gases are compressed to 5 atmospheres pressure and pass through a purified tower, in which organic matter is removed from the bottom using $KMnO_4$. The gas is then purified using two carbon dioxide driers and carbon filters to get purified carbon dioxide. It is liquidized using the CO_2 condenser and reboiler heat exchanger and stored.

13.6.1.3 Uses

It is mainly used as a fire extinguisher. Gaseous CO_2 is used as a neutralizing agent. It is used to create an inert atmosphere in aerobic digestion. Solid CO_2 is used in the refrigeration process. It is the primary raw material for the manufacturing of Na_2CO_3 and $NaHCO_3$ (Brahmbhatt et al., 2015; Sankaranarayanan and Srinivasan, 2012).

13.6.2 Carbon monoxide

Carbon monoxide is a combustible gas that is tasteless, colorless, and odorless. It is also somewhat lighter than air. It is poisonous for animals that employ hemoglobin as an oxygen carrier.

13.6.2.1 Properties

Its main properties are as follows.

Table 13.8 Physical & Chemical Properties of Carbon monoxide

Sr. No.	Properties	Particular
1	IUPAC name	Carbon monoxide
2	Chemical and other names	Carbonic oxide, carbon(II) oxide
3	Molecular formula	CO
4	Molecular weight	28.01 gm/mole
5	Physical description	Colorless, odorless gas
6	Solubility	Miscible with water, hydrocarbons and most organic liquids
7	Melting point	−205.2°C
8	Boiling point	−191.1°C
9	Flash point	-
10	Density	1.250 gm/cm³ at 0°C

13.6.2.2 Manufacturing process

(A) **Raw materials:** Natural gas and Air (Oxygen)

(B) **Flow chart**

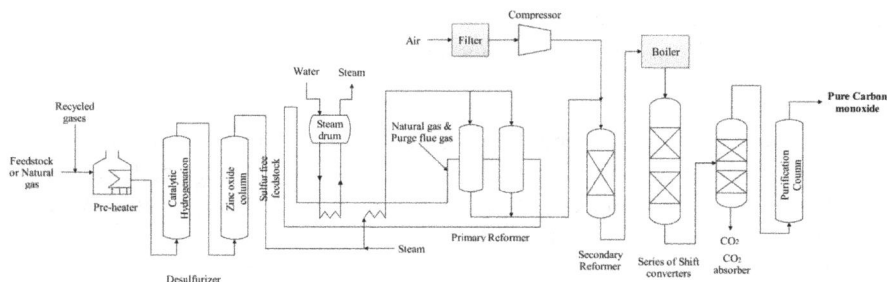

Figure 13.8 Flow sheet diagram of production of carbon monoxide

(C) Process description: As per the flow-chart, natural gas feedstock is taken and heated in a heater. Heated feedstock is transferred into a catalytic hydrogenation column, where sulfur compoundsare converted into gaseous hydrogen sulfide. The gaseous mixture is collected from the bottom of the column.

$$H_2 + RSH \rightarrow RH + H_2S \text{ (gas)}$$

The gaseous mixture passes through a zinc oxide bed, where it is converted to solid zinc sulfide. The remaining mixture is collected from the bottom of the column.

$$H_2S + ZnO \rightarrow ZnS + H_2O$$

Now, sulfur free feedstock is transferred into a primary catalytic steam reformer column from its top. Here, methane is converted into carbon monoxide and hydrogen gas at a temperature of 1000°C and 20 atmospheres pressure. This is an endothermic reaction. Steam is inserted into the tower to maintain the pressure.

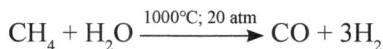

$$CH_4 + H_2O \xrightarrow{1000°C; \ 20 \ atm} CO + 3H_2$$

This gaseous reaction mixture is inserted into a secondary reformer. Here, nitrogen in the form of air is also inserted. The remaining methane is converted. The reaction is exothermic, so, heat is liberated during the reaction. Steam for this reaction is utilized in the primary steam reformer. The gaseous mixture is transferred into a shift converter, first at a higher temperature (400°C) and thereafter, at a lower temperature (200°C). Here, carbon monoxide is converted to carbon dioxide and more hydrogen; even traces of carbon monoxide are converted at a lower temperature.

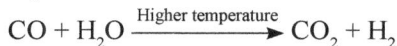

$$CO + H_2O \xrightarrow{Higher \ temperature} CO_2 + H_2$$

This reaction mixture is inserted into a carbon dioxide adsorption column from its top. Here, the column is filled with monoethanolamine solution (MEA), sulfinol, propylene carbonate and others, to absorb the carbon dioxide. Thereafter, it undergoes purification in a column to get pure carbon monoxide.

13.6.2.3 Uses

1. Carbon monoxide is employed extensively in the chemical industry (benzaldehyde and citric acid), inorganic chemicals (metal carbonyls, titanium dioxide), and chemical intermediates (toluene and diisocyanates, used to produce polyurethane).
2. Syngas, a fuel gas used as a natural gas substitute, is created when carbon monoxide is mixed with various other gases (hydrogen, nitrogen, methane, and carbon dioxide).
3. When refining metals, carbon monoxide is also used as a reducing agent.
4. In the food packaging industries, it is incorporated with meat, poultry and fish to preserve freshness and deliver suppleness in delivery (Harriet et al., 2022).

13.6.3 Carbon disulfide

Carbon disulfide is a highly volatile and flammable liquid having a low ignition temperature. Small amounts of this compound are released by volcanic eruptions and marshes.

13.6.3.1 Properties

Its main properties are as follows.

Table 13.9 Physical & Chemical Properties of Carbon disulfide

Sr. No.	Properties	Particular
1	IUPAC name	Carbon disulfide
2	Chemical and other names	Carbon disulphide, Carbon bisulfide, Dithiocarbonic anhydride, Carbon bisulphide
3	Molecular formula	CS_2
4	Molecular weight	76.15 gm/mole
5	Physical description	Colorless to faint-yellow liquid with sweet ether-like odor
6	Solubility	Soluble in water, alcohol, acetone, chloroform
7	Melting point	−111.7°C
8	Boiling point	46°C
9	Flash point	−30°C
10	Density	1.263 gm/cm³ at 27°C

13.6.3.2 *Manufacturing process*

The conventional process for producing carbon disulfide involves a sulfur and charcoal reaction that occurs at a temperature of about 850°C. Since sulfur reacts slowly with coke at all practical temperatures, charcoal is employed. The reaction involving methane, ethane, and ethylene from natural gas has largely supplanted this coal-burning process.

(A) **Raw materials:** Methane, Sulfur and Catalyst

(B) **Chemical reactions:** Natural gas (methane) is reacted with molten sulfur at 600–700°C in the presence of silica gel or alumina catalyst to yield carbon disulfide.

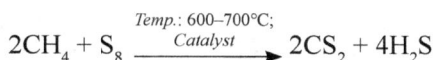

$$2CH_4 + S_8 \xrightarrow[\text{Catalyst}]{\text{Temp.: 600–700°C;}} 2CS_2 + 4H_2S$$

(C) **Flow chart**

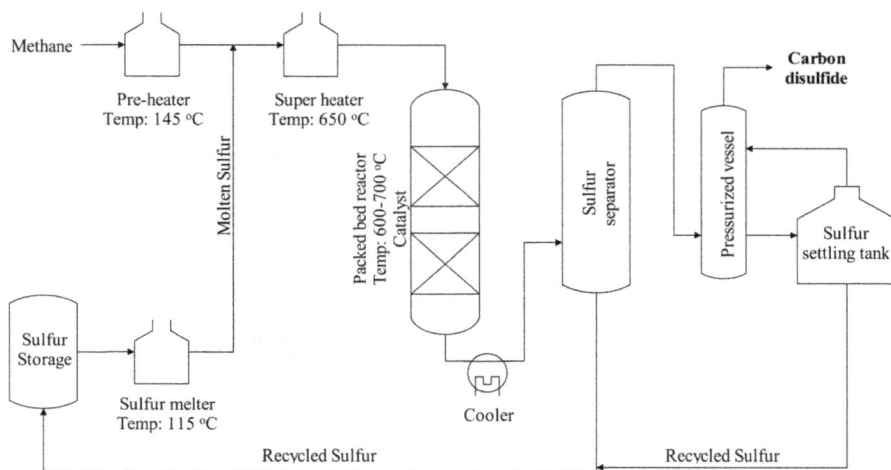

Figure 13.9 Flow sheet diagram of production of carbon disulfide

(D) **Process description:** As per simple flow chart in Fig. 13.9, methane is heated upto 140–150°C in a pre-heater. Sulfur is taken from sulfur storage and heated in a heater at 110–115°C to molten sulfur. The pre-heated methane and molten sulfur are then heated upto 600–700°C; and fed into a packed bed reactor from the top. This reactor is already impregnated with catalyst. After completion of the reaction, the mass is collected from the bottom and cooled. The cooled reaction mass is transferred to a sulfur gas separator. Here, sulfur is separated from the bottom and recycled. The remaining mass is fed into a pressurized vessel, in which traces of sulfur are transferred from the bottom and carbon disulfide is separated as the overhead product.

13.6.3.3 Uses

Carbon disulfide is employed principally in the manufacture of rayon, cellophane, and carbon tetrachloride. Additionally it is used to prepare rubber chemicals and pesticides. It is used as a solvent for iodine and phosphorous. It is also used in generating petroleum catalysts (Clark, 2018).

13.6.4 Phosgene

Phosgene is a very toxic gas that is created when chlorinated hydrocarbons are burned or when these substances are exposed to ultraviolet light. There is no phosgene present in the environment naturally. Since, phosphorus gas is heavier than air, it may be occurring more in low-lying locations. It is exceedingly dangerous when inhaled acutely (shortly). During World War I, phosphorus was frequently employed as a choking (pulmonary) agent. Phosgene was the poison used in the war that caused a vast majority of casualties.

13.6.4.1 Properties

Its main properties are as follows.

Table 13.10 Physical & Chemical Properties of Phosgene

Sr. No.	Properties	Particular
1	IUPAC name	Carbonyl dichloride
2	Chemical and other names	Phosgen, Chloroformyl chloride, Carbon oxychloride, Carbonic chloride
3	Molecular formula	$COCl_2$
4	Molecular weight	98.91 gm/mole
5	Physical description	Colorless gas with a suffocating odor like musty hay
6	Solubility	Slightly soluble in water and freely soluble in alcohol, benzene, toluene, glacial acetic acid
7	Melting point	−118.5°C
8	Boiling point	8.2°C
9	Flash point	-
10	Density	1.432 gm/cm³ at 27°C

13.6.4.2 Manufacturing process

Starting materials of phosgene are carbon monoxide and chlorine gas.

(A) Raw materials

  Carbon monoxide, Chlorine gas and Catalyst

(B) Chemical reactions

Industrially, phosgene is manufactured by the reaction of purified carbon monoxide and chlorine gas at a temperature of 220–230°C and a pressure of 7 atmospheres in the presence of a catalyst like porous activated carbon.

$$CO + Cl_2 \xrightarrow[\text{Pres.:7 atm; Catalyst}]{\text{Temp.: 220–230°C;}} COCl_2; \Delta H = -107.6 \text{ kJ/mole}$$

The reaction is exothermic.

(C) Flow chart

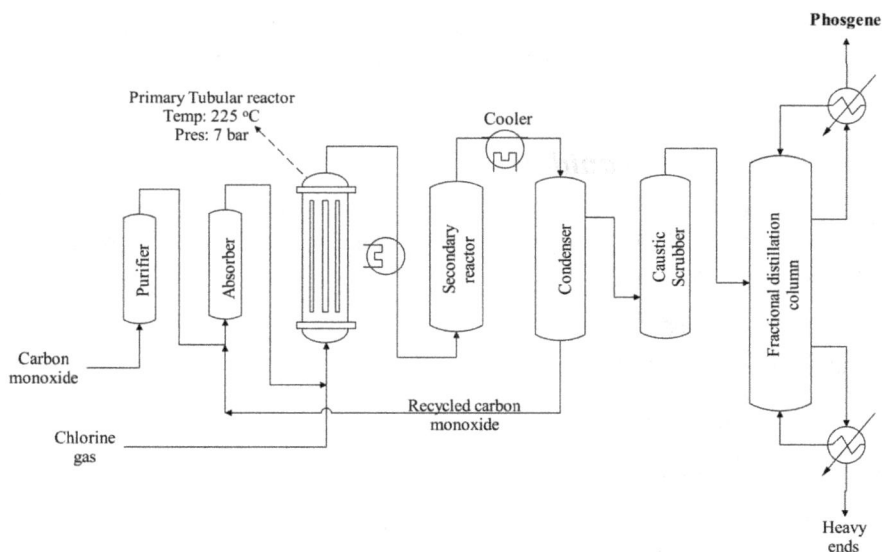

Figure 13.10 Flow sheet diagram of production of Phosgene gas

(D) Process description

First, carbon monoxide is purified in an absorber. Chlorine gas is mixed with purified carbon monoxide and transferred into the bottom of a primary tubular reactor. This reactor is impregnated with porous activated carbon to create a porous bed. The reaction is exothermic, so, cooling water is fed into the jacket to maintain a temperature of 220–230°C and a pressure of 7 atmospheres. About 90–92% of the reaction is completed. The reaction mass is cooled and fed to a secondary reactor, where the remaining raw materials are reacted to yield phosgene. It is cooled down and transferred into a condenser. Here, the un-reacted carbon monoxide and chlorine; and by-product carbon tetrachloride are separated from phosgene. This stream is passed through an absorber to yield purified carbon monoxide and recycled. Raw phosgene is purified in a caustic scrubber and a fractional distillation column to get pure phosgene as an over-headed product.

(E) Major engineering problems

Here, highly purified carbon monoxide is utilized to reduce the carbon tetrachloride by product. As the manufacturing reaction of phosgene is exothermic, the reactor must be cooled. Two stage reactors are utilized to get better conversion.

13.6.4.3 Uses

Phosgene is mainly used in the herbicide-pesticide industry as a chlorinating agent. It is used as an intermediate in the production of isocyanates, carbamates, organic carbonatesand chloroformates. It is also used in the manufacture of dyes, pharmaceuticals, synthetic foams, resins and polymers (Senet, 2004; Vobnacker et al., 2021).

13.6.5 Hydrogen cyanide

Hydrogen cyanide is an extremely hazardous conjugate acid of a cyanide, which is used as a chemical weapon agent. It is a flammable liquid and boils marginally above room temperature.

13.6.5.1 Properties

Its main properties are as follows.

Table 13.11 Physical & Chemical Properties of Hydrogen cyanide

Sr. No.	Properties	Particular
1	IUPAC name	Formonitrile
2	Chemical and other names	hydrogen cyanide, hydrocyanic acid, Prussic acid
3	Molecular formula	HCN
4	Molecular weight	27.02 gm/mole
5	Physical description	Colorless or pale-blue liquid or gas with a bitter, almond-like odor
6	Solubility	Miscible with water, ethanol, ethyl ether
7	Melting point	$-13.4°C$
8	Boiling point	$26°C$
9	Flash point	$-18.1°C$
10	Density	0.689 gm/cm^3 at 27°C

13.6.5.2 Manufacturing process

It is prepared by the BMA process or Andrussow process. Methane and ammonia are reacted in the presence of a platinum catalyst in the Degussa process, also known as the BMA process or Degussa process, which was created by the German chemical company Degussa. The term is shortened from the German words Blausäure (hydrogen cyanide), Methan (methane), and Ammoniak (ammonia).

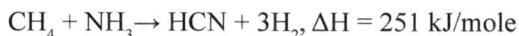

$$CH_4 + NH_3 \rightarrow HCN + 3H_2, \Delta H = 251 \text{ kJ/mole}$$

The reaction is extremely endothermic and requires an approximate temperature of 1400°C. So, this process is not preferable.

Another process, Andrussow process was developed by Leonid Andrussow in 1927, in which a reaction of ammonia, natural gas and air (oxygen) in the presence of a platinum catalyst is performed to give HCN. The process is conducted at room temperature, and hence, does not requirea higher temperature

Here, we discuss the Andrussow process.

(A) Raw materials

Ammonia, Natural gas (Methane) and Air (Oxygen)

(B) Chemical reactions

Ammonia, natural gas and air are reacted over a platinum catalyst to form hydrogen cyanide.

$$2CH_4 + 2NH_3 + 3O_2 \longrightarrow 2HCN + 6H_2O; \Delta H = -939.62 \text{ kJ/mole}$$

(C) Flow chart

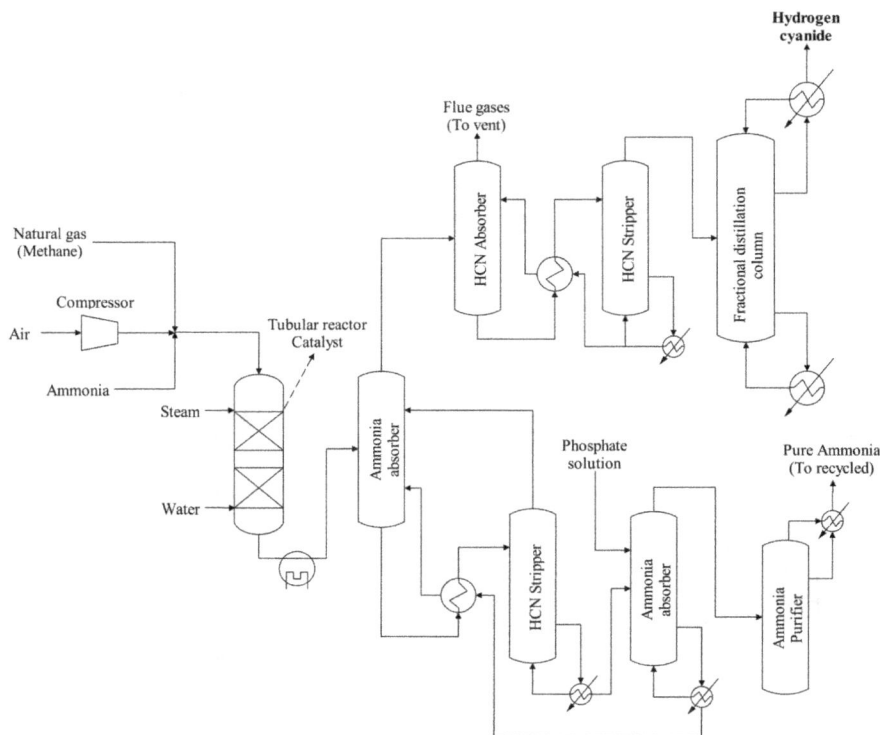

Figure 13.11 Flow sheet diagram of production of Hydrogen cyanide

(D) Process description: Air is compressed, mixed with natural gas and air; and fed to tubular reactor, having a catalyst. Here, the reaction is exothermic, so, the liberated heat is absorbed using water to generate steam for the boiler. After the reaction, the mass is cooled down using a cooler. The cooled mass is fed to an ammonia absorber, where two streams are differentiated. The first stream containing ammonia solution is transferred into a HCN stripper, where HCN is recovered from ammonia and transferred into the ammonia absorber again. The solution containing ammonia is goes to the ammonia absorber and ammonia purifier to get ammonia, which is recycled into the feed. The second stream containing HCN solution is fed to the HCN absorber to remove flue gases. Further, it goes to the HCN stripper and the fraction distillation column to get hydrogen cyanide as an over-head product.

13.6.5.3 Uses

It is used to manufacture adiponitrile, which is then used to produce nylon. Additionally, it is used to make acrylonitrile, which is subsequently used to make acrylic fibres, synthetic rubber, and plastics. Hydrogen cyanide is used as a fumigant rodenticide, and as a fertilizer. It is used to manufacture nitrites and polyurethane foam. It is used to prepare sodium and potassium cyanide, which are widely used in metal processing including electroplating and hardening (Eisner et al., 1975).

13.6.6 Cyanogen chloride

First cyanogen chloride was prepared using chlorine and hydrocyanic acid (aka prussic acid); and thus it was entitled 'oxidized prussic acid" in 1787. In 1815, the exact formulation for cyanogen chloride was discovered. It was applied in 1916 during World War I.

13.6.6.1 Properties

Its main properties are as follows.

Table 13.12 Physical & Chemical Properties of Cyanogen chloride

Sr. No.	Properties	Particular
1	IUPAC name	Carbononitridic chloride
2	Chemical and other names	Cyanogen chloride, Chlorine cyanide, Chlorocyanide, Chlorocyanogen, Chlorcyan
3	Molecular formula	CClN
4	Molecular weight	61.47 gm/mole
5	Physical description	Colorless gas or liquid with an irritating odor
6	Solubility	Soluble in water, ethanol, ethyl ether and other organic solvent
7	Melting point	−6.5°C
8	Boiling point	13°C
9	Flash point	51.1°C
10	Density	1.22 gm/cm^3 at 27°C

13.6.6.2 Manufacturing process

The starting materials of cyanogen chloride are hydrogen cyanide and chlorine gas.

(A) Raw materials

Hydrogen cyanide, Chlorine gas and Catalyst (Cupric and Ferric ions)

(B) Chemical reactions: Cyanogen chloride is prepared by reacting hydrogen cyanide and chlorine gas in gaseous phase in the presence of a catalyst (preferably low activity carbon) at a temperature of 280–300°C and high pressure.

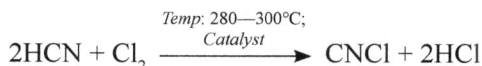

$$2HCN + Cl_2 \xrightarrow[\text{Catalyst}]{\text{Temp: 280—300°C;}} CNCl + 2HCl$$

(C) Flow chart

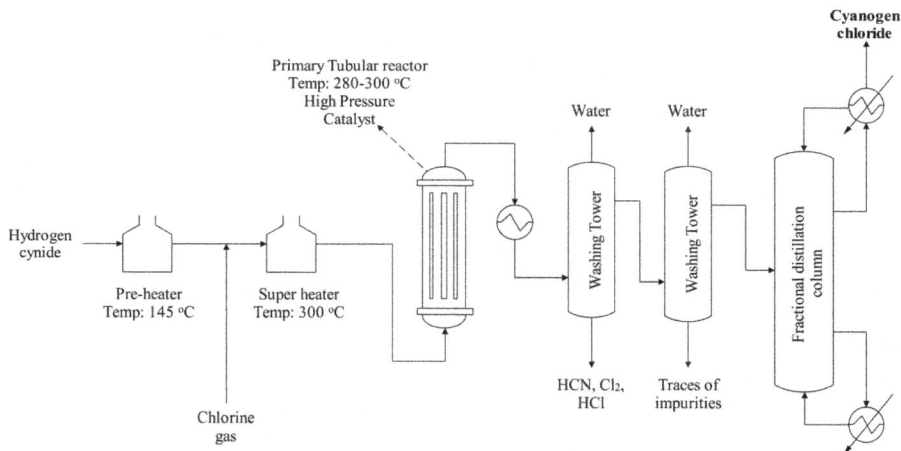

Figure 13.12 Flow sheet diagram of production of Cyanogen Chloride

(D) Process description: First hydrogen cyanide is heated upto 145°C in a pre-heater to convert it to vapor cyanide. This vapor is mixed with chlorine gas and heated to 280–300°C. The hot gaseous mixture is inserted at the bottom of a tubular reactor, having a catalyst. Hydrogen cyanide and chlorine gas react within a few seconds. The reaction mixture is collected from the top of the reactor and cooled down. The cooled reaction mixture goes through two stages in a washing tower, where it is washed with water to remove un-reacted hydrogen cyanide and chlorine gas, by-product hydrochloric acid and other impurities. Finally, fractional distillation is carried out to get cyanogen chloride as an over-head product.

13.6.6.3 Uses

About 65–70% cyanogen chloride is used in the preparation of the triazine-class pesticides, especially atrazine, It is also used as an intermediate in the manufacture of dyes, pharmaceuticals, brightening agents, synthetic resins, rubbers, explosives and plastics. It is moreover utilized in the Lonza process in the preparation of extremely pure malononitrile (Zheng et al., 2004).

13.7 Oxygen

The non-metallic element oxygen, which has the chemical symbol "O" and atomic number 8, is very reactive. Carl Wilhelm Scheele discovered oxygen in 1772. He did it by heating potassium nitrate, mercuric oxide, and a variety of other chemicals. It is a potent oxidizer that easily produces oxides when combined with other elements, such as carbon, sulfur, nitrogen, phosphorusand iron. After hydrogen and helium, it is the third most highly abundant element. Two atoms of the element combine to generate dioxygen, an odorless, tasteless, and colorless diatomic gas, at an ordinary temperature and pressure. It is a vital gas in the atmosphere, as all living organisms depend on it. Without oxygen, all living organisms would be dead. This gas is taken up by living organisms and; converted into carbon dioxide. During photosynthesis, all plants use carbon dioxide as a source of carbon and release oxygen back into the atmosphere. Oxygen makes up about 20.95% of the Earth's atmosphere. Additionally, various states of oxygen, including ores, rocks, ozone, gases, oxides, acids and alkalis make up half of the Earth's crust. Oxygen is combines with various metallic and non-metallic compounds to form oxides, which may be acidic (like sulfur, carbon, aluminum and phosphorus), basic (calcium, magnesium and iron) or salt-like compounds. Extraction of oxygen from these oxides is not practically possible, due a strong bond between oxygen and metal. Further, for complete combustion oxygen is required to generate heat and energy. All hydrocarbon fuels (oil, coal, petrol, natural gas) undergo combustion reaction only in the presence of oxygen to form carbondioxide. Gaseous oxygen is converted into liquid oxygen cryogenically and utilized widely as an oxidizing agent for use in missiles and rockets. It reacts with hydrogen to provide a remarkable thrust. It is used in the production and growth of bacteria in sewage as well as industrial wastewater. These bacteria remarkably reduce the various pollutants from wastewater (Decker and Holde, 2011).

13.8 Fluorine

Fluorine, having chemical symbol "F" and atomic number 9, is extremely reactive and the lightest halogen. It reacts with all is attract electrons and a small atomic size. Attraction of electrons is due to the fact that it has 5 electrons in its 2P shell, required one electron to complete its octet. It is a pale yellow toxic diatomic gas at STP (Standard Temperature and Pressure). It has various isotopes having a different number of neutrons; the most common isotope is F19, which has a natural abundance of 100%. It was first discovered by Georgius Agricola in 1530. He initially found it in the compound Fluorspar, which was used to encourage the fusion of metals. Thereafter, hydrofluoric acid (HF) was first isolated by heating fluorspar with concentrated sulfuric acid by Carl Wilhelm Scheele in 1771. The average person's body has 3 mgs of fluoride since it is essential for preserving the teeth and bones. It is added to the area's drinking water, which indicates a fluorine shortage. Toothpaste also contains fluoride. It is a mineral that naturally exists in a variety of foods and liquids. About 650 ppm of fluorine can be found in the Earth's crust. Various minerals of fluorine are available, in which major sources of fluorine are fluorspar (CaF_2), cryolite (Na_3AlF_6), fluorapatite ($Ca_5[PO_4]_3[F,Cl]$), topaz [$Al_2SiO_4(F,OH)_2$], and lepidolite [$K(Li,Al,Rb)_2(Al,Si)_4O_{10}(F,OH)_2$]. These minerals are found in USA, Germany, France, Russia, China and India. Most of the fluorine is derived in form of hydrogen fluoride (HF) from fluorspar. For the production of HF, fluorspar is first distilled with concentration sulfuric acid in a lead or cast-iron vessel. Liquid HF is formed with solid calcium sulfate, which is easy removed by filtration. Now, raw HF further undergoes fractional distillation to get pure HF. In other areas, the source of fluorine is drinking water in deep wells, containing 3–5 ppm fluorine. Also, HF is isolated along with the wastes from petroleum refining. This HF is purified and used as catalyst in gasoline production.

Industrially, Fluorine gas is produced by Moissan's method, in which a mixture of hydrogen fluoride and potassium fluoride undergoes electrolysis under 8–12 volts at a temperature of 70–130°C. Here, fluorine gas is obtained at the carbon block anode and hydrogen gas is obtained at the steel cathode. Fluorine is used to prepare uranium hexafluoride, which is required in the nuclear power industry to separate uranium isotopes. It is also utilized for the preparation of sulfur hexafluoride, which is an insulating gas for high-power electricity transformers. Fluorine gas is also used in rocket fuels, production of various polymers and plastics and air conditioning. It combines with oxygen and used as a refrigerator coolant. Initially it was used for the production of chlorofluorocarbons (CFCs), which were used in aerosols, refrigerators, air conditioners, foam food packaging, and fire extinguishers. CFCs are banned due to the fact that they are green-house gases and contribute to ozone depletion. Its most important constituent, Hydrofluoric acid is used in refrigerants, plastics, aluminum, weed killers, and semiconductors for computers. It is used for etching glass and cleaning alloys. Additionally, it is used for grinding titanium, rust treatment in commercial and household laundry goods, electroplating, petroleum exploration, refining, and oil fields, among other things (Poeppelmeier and Tressaud, 2016; Yamazki et al., 2009).

13.9 Neon

Neon, which has an atomic number of 10 and the chemical symbol "Ne", is a noble gas that does not interact with other substances due to its complete electron octet. Neon is derived from the Greek word neos, which means "new." After helium, neon is the second-lightest inert gas. The fifth most plentiful element is neon. At STP, it is a monatomic gas that is inert, colorless, and odorless. Along with argon and krypton, it was discovered in 1898 in London by the British chemists Sir William Ramsay and Morris W. Travers. It is present in trace amounts in both the atmosphere and the rocks that make up the planet's crust. It is extracted commercially via the fractional distillation of liquid air. Ne^+, $NeAr^+$, NeH^+, and $HeNe^+$ are a few neon ion forms that are used in optical and mass spectrometric studies. Additionally, an unstable hydrate occurs. 20Ne, 21Ne, and 22Ne are the three stable isotopes of neon. An essential cryogenic refrigerant is liquid neon. Compared to liquid helium and nearly three times that of liquid hydrogen, it has a refrigerating capacity that is over 40 times greater per unit volume. When it complies with refrigeration criteria, it is more affordable, inert, and smaller than helium. It is utilized in wave meter tubes, Geiger counters, lightning arrestors, and television tubes. Gas lasers are created using neon and helium (Aguilar et al., 2020).

References

Aguilar, Cavasonza, M.L.A. and Ambrosi, G. 2020. Properties of Neon, Magnesium, and Silicon Primary Cosmic Rays Results from the Alpha Magnetic Spectrometer. Physical Review Letters 124(21). DOI: 10.1103/PhysRevLett.124.211102

Appl, M. 1999. Ammonia: principles and industrial practice, Willey-VCH.

Barnum, D.W. 2003. Some History of Nitrates. Journal of Chemical Education 80(12). https://doi.org/10.1021/ed080p1393

Brahmbhatt, S.R. 2015. CO_2-an Important and Valuable Chemical for the Manufacturing Industry, Carbon Management Technology Conference, Texas.

Clark, F.F. 2018. The Production of Carbon Disulfide by the Reaction between Natural Gas and Hydrogen Sulfide, ProQuest LLC.

Decker, H. and Holde K.E. 2011. Oxygen, Its Nature and Chemistry: What Is so Special About This Element? In book: Oxygen and the Evolution of Life.

Dyer, P. 2014. Beryllium: Physicochemical Properties, Applications and Safety Concerns, In Chemistry Research and Applications, Nova Science Pub Inc.

Eisner, H.E., Wood, W.F. and Eisner, T. 1975. Hydrogen Cyanide Production in North American and African PolydesmoidMillipeds, Psyche: A Journal of Entomology 82(1): https://doi.org/10.1155/1975/35601

El-Moneim, N.A., Ismail, I. and Nasser M. 2018. Simulation of Ammonia Production from Synthesis Gas, The International Conference on Chemical and Environmental Engineering 9(6): 85–95. DOI: 10.21608/ICCEE.2018.34649

Follett, R.F. and Hatfield, J.L. 2001. Nitrogen in the Environment Sources Problems and Management, Elsevier Science. https://doi.org/10.1016/B978-0-444-50486-9.X5000-6

Gargouri, M., Chtara, C., Charrock, P., Nzihou, A. and El-Feki, H. et al. 2011. Synthesis and Physicochemical Characterization of Pure Diammonium Phosphate from Industrial Fertilizer. Industrial & Engineering Chemistry Research 50(11): 6580–6584. https://doi.org/10.1021/ie102237n

Garrett, D.E. 2004. Handbook of Lithium and Natural Calcium Chloride, Academic Press.

Gershon, S. and Shopsin, B. 1973. Lithium: Its Role in Psychiatric Research and Treatment, Springer US.

Ghavam, S., Vahdati, M., Wilson, I.A.G. and Styring, W.P. 2021. Sustainable Ammonia Production Processes. Frontiers in Energy Research 9. https://doi.org/10.3389/fenrg.2021.580808

Grew, E.S. 2002. Mineralogy, Petrology and Geochemistry of Beryllium: An Introduction and List of Beryllium Minerals. Reviews in Mineralogy and Geochemistry 50(1): 1–76. https://doi.org/10.2138/rmg.2202.50.01

Harriet, K., Wang, L., Tong, L., Cao, H. and Ding, Y. et al. 2022. Industrial Carbon Monoxide Production by Thermochemical CO_2 Splitting-a Techno-Economic Assessment. SSRN Electronic Journal. http://dx.doi.org/10.2139/ssrn.4075927

Heiden, Z.M., Mosquera, M.E.G. and Singh, H.B. 2019. Inorganic chemistry of the p-block elements, Dalton Transactions, 48 (20). https://doi.org/10.1039/C9DT90098E

Khalil, A., 2018. Production of Nitrogen from the Air During Cryogenic Process And Analyzing the Air feed In Raslanlf Utilities For Oil & Gas. International Conference and Exhibition of Oil and Gas, Benghazi.

Liu, H. 2013. Ammonia Synthesis Catalysts: Innovation and Practice. World Scientific Publishing Company. https://doi.org/10.1142/8199

Manjare, S.D. and Mohite, R. 2011. Application Life Cycle Assessment to Diammonium Phosphate Production. Advanced Materials Research 354–355: 256–265. https://doi.org/10.4028/www.scientific.net/AMR.354-355.256

Minivadia, R.S., Patel, N., Patel, N.J. and Patel, R. 2019. Process Analysis and Plant Design for Manufacturing of Nitric Acid. International Journal of Engineering Trends and Technology, 67 (4), 76-80. DOI: 10.14445/22315381/IJETT-V67I4P21

Poeppelmeier, K.R. and Tressaud, A. 2016. Photonic and electronic properties of Fluoride Materials: progress in Fluorine science series, Elsevier Ltd. https://doi.org/10.1016/C2013-0-18987-8

Sankaranarayanan, S. and Srinivasan, K. 2012. Carbon dioxide—A potential raw material for the production of fuel, fuel additives and bio-derived chemicals, Indian Journal of Chemistry-Section A 51: 1252–1262.

Scerri, E.R. 2011. A Review of Research on the History and Philosophy of the Periodic Table, Journal of Science Education 12(1): 4–7.

Senet, J.G. 2004. The Recent Advance in Phosgene Chemistry, National Company of Powders and Explosives, SA.

Silva, S.R.P. 2003. Properties of Amorphous Carbon (EMIS Datareviews). The Institution of Engineering and Technology.

Singh, B.P. and Malhotra, I. 2020. Concentrated Nitric Acid Production by Recycling and Rectification: Optimistic Approach for Better Yield. International Journal for Research in Applied Science & Engineering Technology 45: 98–105.

Soriano-Ursua, M.A., Das, B. and Trujillo-Ferrara, J. 2014. Boron-containing compounds: Chemico-biological properties and expanding medicinal potential in prevention, diagnosis and therapy. January Expert Opinion on Therapeutic Patents 24(5). DOI: 10.1517/13543776.2014.881472

Vobnacker, P., Wust, A. and Keilhack T. 2021. Novel synthetic pathway for the production of phosgene. Science Advances 7(40). DOI: 10.1126/sciadv.abj5186

Walsh, K.A. 2009. Beryllium Chemistry and Processing, ASM International.

Yadav, K., Sharma, M. and Jha, M. 2020. Extraction of nanostructured sodium nitrate from industrial effluent and their thermal properties. Water Environment Research 92(2). https://doi.org/10.1002/wer.1307

Yamazaki, T., Taguchi, T. and Ojima I. 2009. Unique Properties of Fluorine and Their Relevance to Medicinal Chemistry and Chemical Biology, In book: Fluorine in Medicinal Chemistry and Chemical Biology. DOI: 10.1002/9781444312096

Yinchang, L.I., Yang, D.U., Xinsheng, J., Bo, W., Dong, W. et al. 2012. Study on nitrogen preparation system based on simulation experiment of intrinsically safe operation of tank, International Symposium on Safety Science and Technology. Procedia Engineering 45: 496–50. doi: 10.1016/j.proeng.2012.08.192

Zhang, Z., Penev, E.S. and Yakobson, B. 2017. Two-dimensional boron: Structures, properties and applications, Chemical Society Reviews 46(22): 6746–6763. https://doi.org/10.1039/C7CS00261K

Zheng, A., Dzombak, D.A. and Luthy, R. 2004. Formation of Free Cyanide and Cyanogen Chloride from Chloramination of Publicly Owned Treatment Works Secondary Effluent: Laboratory Study with Model Compounds. Water Environment Research 76(2): 113–20. DOI: 10.2175/106143004x141636

Zhu, Y., Cai, J., Hosmane, N. and Zhang, Y. 2022. Introduction: basic concept of boron and its physical and chemical properties. In book: Fundamentals and Applications of Boron Chemistry. https://doi.org/10.1016/C2019-0-00779-0

CHAPTER 14

Period – III Chemical Industries

14.1 Introduction

A chemical element that belongs to the third row (or period) of the table of chemical elements is known as a period 3 element. As atomic number of elements increases, it is located with same chemical behavior with skipping itself. The eight elements that make up the playing time are argon, sodium, magnesium, aluminum, silicon, phosphorus, sulfur, and chlorine. While the others are found in the p-block of the periodic table, the first two, sodium and magnesium, are members of the s-block. These three elements and at least one stable isotope are both naturally occurring. Table 14.1 represented atomic and physical properties of these elements including atomic number, electron configuration, ionization energy electron affinity electro-negativity, etc.

14.2 Sodium

Sodium (atomic number 11, symbol "Na") is a delicate, silvery-white metal that is extremely reactive. The Latin word "natrium" is the source of the English word sodium. Humphry Davy first created sodium in 1807. It was isolated sodium element by electrolysis of fused sodium hydroxide. Sodium is element of group 1 in periodic table, because it has only one electron in its outer shell. This electron is rapidly donated and become Na^+ cation. 23Na is the only stable isotope of it. In the crust of the Earth, it is the sixth most prevalent element. Sodium reacted quickly with water to give sodium hydroxide, hydrogen and energy. Sodium is exposed to air, it produce sodium oxide coating having opaque grey color on its surface. Nitrogen and sodium do not interact, not even at very high temperatures. Common ores of sodium are chile saltpeter ($NaNO_3$), trone ($Na_2CO_3 \cdot 2NaHCO_3 \cdot 3H_2O$), borax ($Na_2B_4O_7 \cdot 10H_2O$) and common salt ($NaCl$).

Table 14.1 Atomic and physical properties of element of period – III

Properties	Element of period – III							
	Na	**Mg**	**Al**	**Si**	**P**	**S**	**Cl**	**Ar**
Atomic number	11	12	13	14	15	16	17	18
Electron configuration	[Ne]3s1	[Ne]3s2	[Ne]3s13p1	[Ne]3s13p2	[Ne]3s13p3	[Ne]3s13p4	[Ne]3s13p5	[Ne]3s13p6
Chemical Series	Alkali Metal	A l k a l i E a r t h Metal	Post-transition nonmetal	Metalloid	R e a c t i v e nonmetal	R e a c t i v e nonmetal	R e a c t i v e nonmetal	Noble gas
Atomic Radius (pm)	186	160	143	118	110	102	99	94
Ionization Energy (kJ/mol)	502	744	584	793	1017	1006	1257	1526
Electron Affinity (kJ/mol)	52.8	0.0	42.5	133.6	200.0	141.0	349	0
Electro-negativity (Pauling scale)	0.93	1.31	1.61	1.9	2.19	2.58	3.16	-
Melting Point (°C)	98	639	660	1410	44	113	–101	–189
Boiling Point (°C)	883	1090	2467	2680	280	445	–35	–186
Density (g/cm³)	0.534	1.85	2.34	2.267	0.00125	0.00143	0.001696	0.0008999

In addition, it plays a significant role in the formation of feldspar and mica, two silicate minerals. Massive rock salt deposits can be found all over the world, while sodium nitrate deposits can be found in Chile and Peru. Additionally, it appears naturally in some foods.

Industrially, sodium is isolated from rock salt having mixed with insoluble impurities of sand and bits of rock. It dissolved in solvent, filtered and dried. In an electric furnace, this dried powder is melted at 800–1000°C combined with some calcium chloride to lower the melting point of sodium chloride. The molten salt with calcium chloride is undergone reductive electrolysis in special electrochemical cell called the downs cell. In this downs cell, the interior lined is made of graphite, which is acted as the anode. Additionally, a cylindrical iron cathode encloses the cell. Chlorine gas produced at the anode is prevented from contacting sodium metal produced at the cathode by using an iron mesh screen. Now, high current is passed in the mixture to provide enough heat to differentiate sodium and chlorine gas. Chlorine gas is collected from graphite anode, while molten sodium metal along with trace of calcium is obtained at cathode. Due to much higher density of calcium compared to sodium, they are mixed and separated easily. Sodium has a role in the control of bodily fluids and is essential for nerve and muscle function. Additionally, it affects how the body regulates blood volume and pressure.

Its chloride salt, sodium chloride is very important compound, found in the living environment. It's another salt, caustic soda is essential and important chemical. Chemical involved in caustic soda and related chemical is Chlor-Alkali industries. Around 74% of India's output of important chemicals is made up of the chlor-alkali sector, which is a significant part of the global basic chemicals industry. Chlorine, sodium hydroxide, soda ash, and chlorine, along with hydrogen and acid, make up the Chlor-alkali industry's primary components. They are used in a variety of sectors, including those that produce textiles, chemicals, paper, PVC, water treatment systems, alumina, soaps and detergents, glass, and chlorinated paraffin wax, among others. Over the previous five years, the demand for the two sub-segments, caustic soda and soda ash, has grown rapidly, with respective compound annual growth rates (CAGR) of 5.6% and 4.7%. The nation's Chlor-Alkali Industry primarily creates sodium hydroxide, chlorine, and sodium carbonate. (Rojas et al., 2015; Zhigach et al., 2012).

14.2.1 Sodium chloride

One of the most prevalent minerals on Earth is sodium chloride, also known as table salt, which is a crucial nutrient for many animals and plants. The extracellular fluid seen in many multicellular animals and the salinity of seawater are both specifically caused by salt.

14.2.1.1 Properties
14.2.1.2 Manufacturing process

Seawater or brine is a major source of this salt.

(A) Raw materials: Seawater or Brine

(B) Flow chart

Table 14.2 Physical & Chemical Properties of Sodium chloride

Sr. No.	Properties	Particular
1	IUPAC name	Sodium chloride
2	Chemical and other names	Halite, Table salt, Saline, Rock salt, Common salt
3	Molecular formula	NaCl
4	Molecular weight	58.44 gm/mole
5	Physical description	White crystalline odorless powder
6	Solubility	Soluble in water, methanol, ether and other solvents
7	Melting point	800°C
8	Boiling point	1465°C
9	Flash point	-
10	Density	2.17 gm/cm³ at 27°C

(C) Process description: Calcium sulfate (0.12), calcium chloride (0.003), and magnesium chloride (0.007%) are only minor components of brine, which is primarily composed of water (80–70%) and sodium chloride (30–20%). Sometime, concentration of sodium chloride is increased by electro-dialysis or reserve osmosis. First, brine is aerated to remove H_2S. Trace of H_2S is removed by adding the chlorine gas by displacement reaction. The sulfur free brine is transferred to settling tank, where solution of caustic soda and soda ash added. After settlement, upper layer containing pure sodium chloride is collected, where sludge containing calcium, magnesium and ferric ions is taken out of the settling tank's bottom. It is then undergone evaporation by five or three stages multi-effect evaporator. It is also evaporated by natural solar evaporation. After evaporator, concentrated solution of sodium chloride is sent into the washer, where fresh brine is used to wash the salt crystals. Purified solution is crystalized, centrifuged and dried using hot air to yield pure dried sodium chloride (purity: 99.8%).

14.2.1.3 Uses

It is also used to purify the molten metal and improve the structure of certain alloys. It is widely used in the food industry as a taste enhancer and as a food preservative. It is a crucial raw element in the production of many industrial

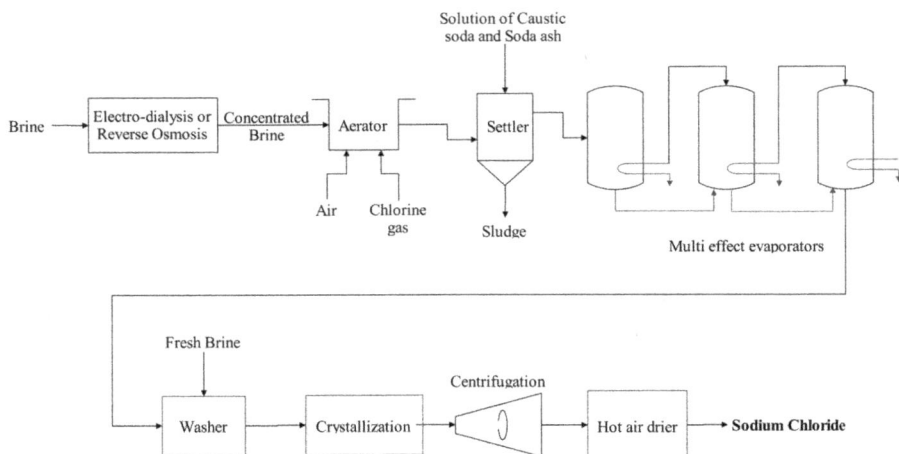

Figure 14.1 Flow sheet diagram of production of Sodium chloride from Brine

compounds, including sodium carbonate and sodium hydrogen carbonate. It has various types of compounds, which are used in different industries, such as glass, food, metal, etc. It is utilized to prepare various organic compounds and esters (Zsombor-Murray, 2021; Synowiec and Bunikowska, 2005).

14.2.2 Caustic Soda

Lye, commonly known as sodium hydroxide, is a white solid ionic compound made up of sodium cations and hydroxide anions. At normal room temperatures, sodium hydroxide, a base and alkali that is highly caustic and should result in serious chemical burns, breaks down proteins.

14.2.2.1 Properties

Table 14.3 Physical & Chemical Properties of Caustic Soda

Sr. No.	Properties	Particular
1	IUPAC name	Sodium hydroxide
2	Chemical and other names	Caustic Soda, Sodium hydrate, Soda lye, White caustic
3	Molecular formula	NaOH
4	Molecular weight	39.99 gm/mole
5	Physical description	White or colorless odorless powder or flask
6	Solubility	Soluble in water, methanol, ether and other solvents
7	Melting point	323°C
8	Boiling point	1388°C
9	Flash point	-
10	Density	2.31gm/cm^3 at 27°C

14.2.2.2 Manufacturing process

Two important chemicals, caustic soda and chlorine, are produced as co-products by the electrolysis of brine. During the electrolysis of brine, the chlorine is liberated at the anode, while caustic soda and hydrogen are produced at the cathode. Each procedure offers a unique way to keep the hydrogen and caustic soda created at the anode, as well as the chlorine produced there, apart. Mercury, diaphragm, and membrane cells are used for these operations, with membrane cells being used most frequently. As well as the soda-lime process, sodium carbonate and calcium hydroxide are combined to create caustic soda. However, electrolysis of brine in India yields more than 95% of chlorine and 80% of caustic soda. It is mainly manufactured by membrane process, in which electrolysis of pure saturated brine is electrolyzed to give caustic soda, chlorine and hydrogen gas. It is also prepared by diaphragm and mercury cell.

Manufacturing Process by Membrane process

(A) Raw materials: Brine (concentration: 35%), Sodium chloride, Hydrochloric acid, Water, 98% Sulfuric acid and Caustic soda

(B) Chemical reactions

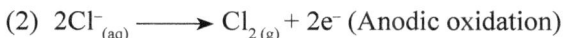

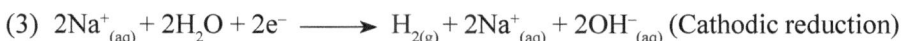

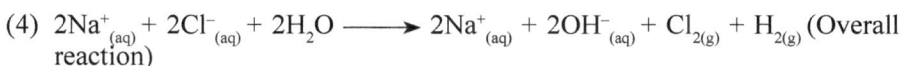

(1) $NaCl \longrightarrow Na^+ + Cl^-$ (Ionic conversion)

(2) $2Cl^-_{(aq)} \longrightarrow Cl_{2\,(g)} + 2e^-$ (Anodic oxidation)

(3) $2Na^+_{(aq)} + 2H_2O + 2e^- \longrightarrow H_{2(g)} + 2Na^+_{(aq)} + 2OH^-_{(aq)}$ (Cathodic reduction)

(4) $2Na^+_{(aq)} + 2Cl^-_{(aq)} + 2H_2O \longrightarrow 2Na^+_{(aq)} + 2OH^-_{(aq)} + Cl_{2(g)} + H_{2(g)}$ (Overall reaction)

(C) Flow chart

(D) Process description: For membrane cell electrolysis process, brine is first saturate with sodium chloride and thereafter, undergoes precipitation and filtration to remove unwanted precipitates. Brine is further purified by ion exchange process to remove calcium and magnesium salt from brine. Purified saturated brine (about concentration: 35%) is electrolyte using electric current in Membrane cell. Membrane cell consists of two big chambers separated by ion selective membrane. Each chamber contents one entrance and one exit point from side. Also, it contains one exit point from the top of each chamber for collection of gas during electrolysis process. Anode is kept in first chamber and cathode is kept is second chamber i.e. after membrane. Also, this membrane is porous chemically active plastic sheet, which has allow to flow across only positives ions, and does not allow negative ions.

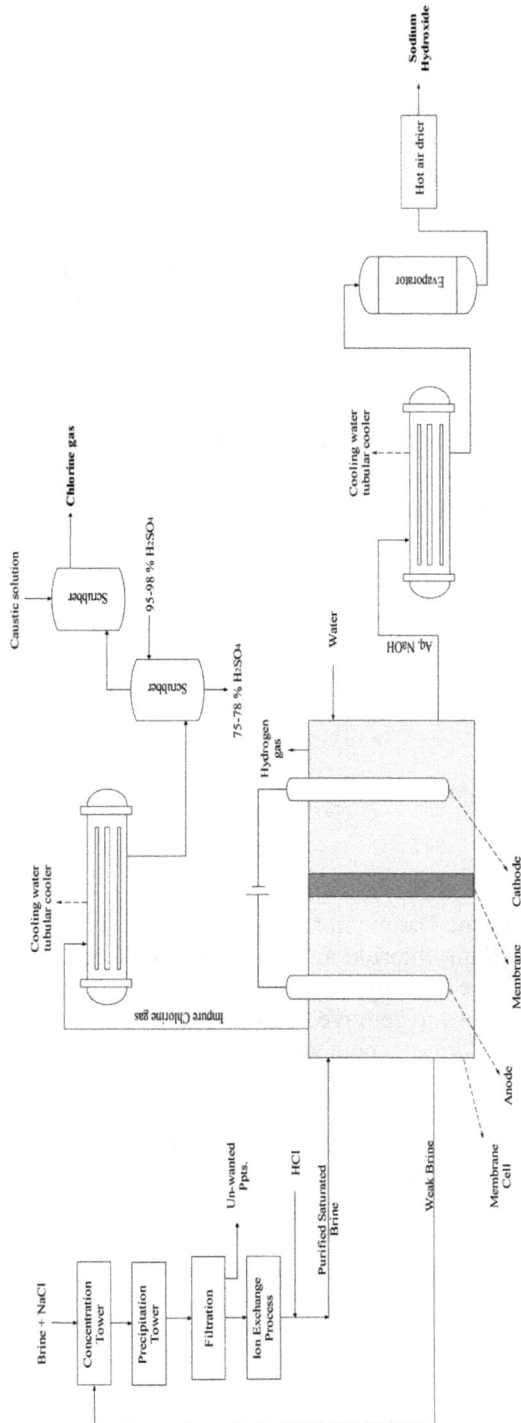

Figure 14.2 Flow sheet diagram of production of Sodium hydroxide from membrane process

For all processes the dissolving of salt, sodium chloride is

$$NaCl \rightarrow Na^+ + Cl^-$$

Now, electrical voltage about 3.0 to 6.0 V is applied at electrolytical cell. The anode is kept in the first chamber of the cell, which is then filled with these ions (saturated brine). At the anode, the chloride ions undergo oxidation, losing electrons and converting to chlorine gas.

$$2Cl^-_{(aq)} \rightarrow Cl_{2\,(g)} + 2e^-$$

This wet impure chlorine gas is collected from the top of first chamber. It is then cooled in tubular cooler with cooling water, dried and purified by scrubbing 98% sulfuric acid and caustic solution respectively to get purified chlorine gas. The counter ion Na^+ can now freely pass from the first chamber to the second chamber thanks to the ion selective membrane, but anions like hydroxide (OH^-) and chloride cannot passed. In second chamber, Na^+ is react with water at cathode to form individual Na^+ and OH^- ion. Then, these ions combine to form caustic soda. Here, hydrogen gas is liberated from cathode, which is carefully collected from the top of second chamber.

The cathode reaction is:

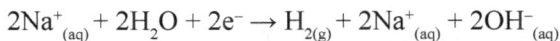

$$2Na^+_{(aq)} + 2H_2O + 2e^- \rightarrow H_{2(g)} + 2Na^+_{(aq)} + 2OH^-_{(aq)}$$

Now, caustic soda solution (purity: about 35%), collected from second chamber, is cooled using tubular cooler with cooling water, concentrated in evaporator using steam, dried and stored.

Here, chloride (Cl^-) is continuously used from brine in first anodic chamber, so, brine is become weaker after some times. Weak brine is then taken from chamber, concentrated with NaCl and re-used it. In second cathodic chamber, water is continuously feed up to increase the concentration of hydroxyl ion (OH^-). Also, pH of brine is adjusted to 3 by adding HCl prior to charging in to first chamber to lessen hydroxyl ion back migration that causes byproducts to develop at the anode.

(E) Major engineering problems: It required high purity brine. Also, cost of membrane is another affect on the process.

Manufacturing Process by Soda Lime Process

Soda Lime process is basically used for water softening process, but it is also manufactured caustic soda. In this process, reaction of soda lime with sodium carbonate is involved to get caustic soda.

(A) Raw materials: Soda lime and Sodium carbonate

(B) Chemical reactions: Soda lime is reacted with sodium carbonate at temperature is 80 to 90°C to get caustic soda and calcium carbonate.

$$Ca(OH)_2 + Na_2CO_3 \longrightarrow CaCO_3 + 2NaOH$$

(C) Flow chart

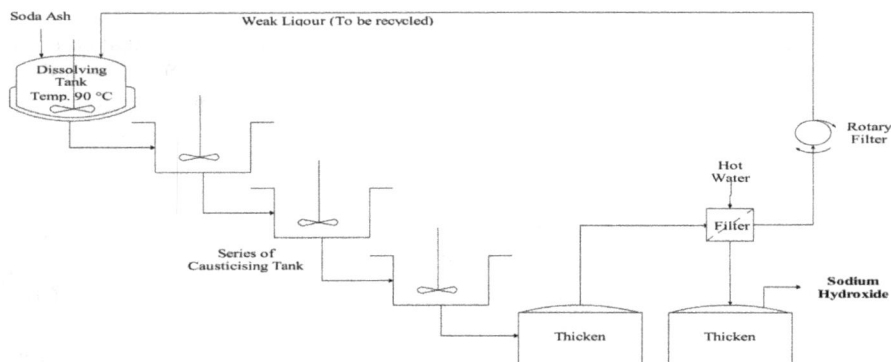

Figure 14.3 Flow sheet diagram of production of Sodium hydroxide from soda-lime process

(D) **Process description:** In the simple flow chart, soda ash having concentration 20% is added to the dissolving tank with stirring. Slake lime (sodium carbonate) is added with stirring. The reaction mixture is heated to 80–90°C with jacketed heater. The reaction mixture is undergone in series of causticising tank with stirring facility, in which bottom part of mixture is collected. Now, reaction mass is transferred in thickener, in which sludge having calcium carbonate is settled down at bottom, while NaOH is obtained from the overflow of the thickener. The sludge is filter in rotary filter and weak liquor is recycled into dissolving tank. NaOH is further dried.

14.2.2.3 Uses

1. It is used in pharmacy.
2. It is also utilized in pH control.
3. It is used as reagent in dehydrochlorination.
4. It is source of sodium in synthesis.
5. It is gas scrubbing in various industries.
6. It is used as cleaning agents (Kumar et al., 2012; Huebner and Wootton, 2008).

*The difference between membrane, diaphragm and mercury cell are mentioned below.

Table 14.4 Difference between membrane, diaphragm and mercury cell for preparation of Caustic soda

Membrane Cell	Diaphragm Cell	Mercury Cell
The cathode and anode reactions are separated in this procedure using a membrane. The membrane allows just sodium ions and a small amount of water to pass.	A porous diaphragm is used in a diaphragm cell to separate two compartments, enable brine to flow, and prevent chlorine and hydrogen gas from mixing.	A saturated brine solution floats on top of a thin layer of mercury in the Castner-Kellner process, sometimes referred to as the mercury cell process.
In the anode compartment, where sodium ions flow across the membrane and chlorine gas is produced, diluted brine is fed. Hydrogen is developed at the cathode in the cathode section, leaving behind hydroxyl ions that, when combined with sodium ions, produce caustic soda.	The anode compartment receives the brine, which then flows through the diaphragm and into the cathode compartment. While hydrogen gas and sodium hydroxide solution are created immediately at the cathode, chlorine gas is created at the anodes.	In this procedure, the sodium ions at the cathode are changed into sodium, which joins the mercury at the cathode to form an amalgam. Moreover, chlorine gas In the decomposer, this sodium is subsequently reacted with water to create NaOH. Also recycled in amalgam is mercury that is removed from the decomposer.
Anode: $2\,Cl^- \rightarrow Cl_2 + 2e^-$ Cathode: $2Na^+ + 2\,H_2O + 2e^- \rightarrow H_2 + 2Na^+ + 2OH^-$ Overall: $2Na^+ + 2Cl^- + 2H_2O \rightarrow 2NaOH + Cl_2 + H_2$	Anode $2Cl^- \rightarrow Cl_2 + 2e^-$ Cathode: $2H_2O + 2e^- \rightarrow 2OH^- + H_2$ Overall: $2NaCl + 2H_2O \rightarrow Cl_2 + 2NaOH + H_2$	Anode: $2Cl^- \rightarrow Cl_2 + 2e^-$ Cathode: $Na^+ + nHg + e^- \rightarrow NaHg_n$ Overall: $2Cl^- + 2Na^+ + 2nHg \rightarrow Cl_2 + 2NaHg_n$ Decomposer: $NaHg_n + H_2O \rightarrow NaOH + nHg + H_2$
Strength of NaOH is required upto 33–38%	Strength of NaOH is required upto 12%	Very high strength of NaOH is required, i.e., upto 50%
Cell voltage required in membrane cell is 3.0–6.0 V	Cell voltage required in diaphragm cell is 2.9–3.5 V	Cell voltage required in membrane cell is 3.9–6.2 V
Advantages: Low energy consumption Low capital investment High purity caustic Insensitivity to cell load variations and shutdowns	Advantages: Use of well brine Low electricity consumption	Advantages: 50% caustic direct from cell High purity chlorine and hydrogen Simple brine purification
Disadvantages: Cost of membrane Use of solid salt, high purity brine High oxygen content in chlorine	Disadvantages: Use of asbestos High steam consumption Low purity caustic Low chlorine quality	Disadvantages: Use of mercury Expensive cell operation Large floor space Costly environment protection

14.2.3 Sodium bicarbonate and Sodium carbonate

14.2.3.1 Properties

Both chemical have almost same properties, so, properties of sodium bicarbonate are as follows.

Table 14.5 Physical & Chemical Properties of sodium bicarbonate

Sr. No.	Properties	Particular
1	IUPAC name	Sodium hydrogen carbonate
2	Chemical and other names	Sodium bicarbonate, Baking soda, Carbonic acid monosodium salt, Sodium hydrocarbonate
3	Molecular formula	CHNaO$_3$
4	Molecular weight	84.07 gm/mole
5	Physical description	Colorless or white odorless crystalline
6	Solubility	Soluble in water, methanol, etc.
7	Melting point	50°C
8	Boiling point	851°C
9	Flash point	-
10	Density	2.107 gm/cm³ at 27°C

14.2.3.2 Manufacturing process

The Solvey procedure is used to produce both the chemicals, sodium bicarbonate and sodium carbonate. The ammonia-soda, or Solvay, process was developed in 1864 by a Belgian chemist called Ernest Solvay to produce sodium carbonate from sodium bicarbonate. In this procedure, ammonia is dissolved in brine and then treated with carbon dioxide to produce sodium bicarbonate precipitates. These precipitates are then calcined to produce high-purity sodium carbonate.

(A) Raw materials: Lime stone (Calcium carbonate), Brine, Ammonia and Water

(B) Chemical reactions: The following three equations represent the general chemical reactions that take place throughout the Solvay process.

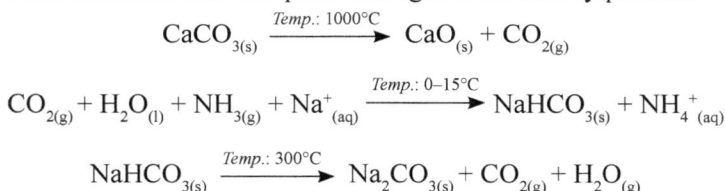

$$CaCO_{3(s)} \xrightarrow{\text{Temp.: } 1000°C} CaO_{(s)} + CO_{2(g)}$$

$$CO_{2(g)} + H_2O_{(l)} + NH_{3(g)} + Na^+_{(aq)} \xrightarrow{\text{Temp.: } 0-15°C} NaHCO_{3(s)} + NH_4^+{}_{(aq)}$$

$$NaHCO_{3(s)} \xrightarrow{\text{Temp.: } 300°C} Na_2CO_{3(s)} + CO_{2(g)} + H_2O_{(g)}$$

(C) Flow chart

(D) Process description

Following major steps are involved in Solvey process.

1. **Preparation of Ammoniated brine:** In first step, the brine is saturated with solid sodium chloride and purified to remove calcium and magnesium salt.

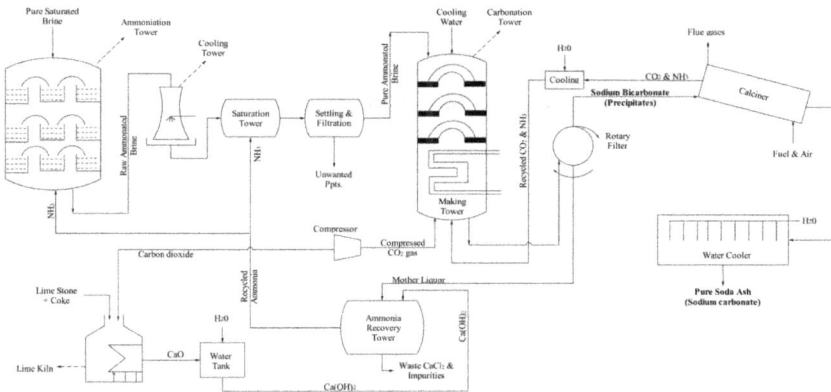

Figure 14.4 Flow sheet diagram of production of Sodium bicarbonate and Sodium carbonate by Solvey process

The purified brine is allowed to trickle down the ammonia tower as ammonia gas is forced counter clockwise through the bottom. As a result, the needed amount of ammonia and heat are removed from the brine solution. The brine that falls from the tower absorbs the gas that escapes from the solution in the tank. The carbon dioxide that remains is precipitated as insoluble carbonate after being absorbed by ammonia. The ammoniated brine cooled to about 30°C, allowed to settle down and filtered to remove unwanted precipitates.

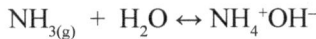

$$NH_{3(g)} + H_2O \leftrightarrow NH_4^+OH^-$$

$$CO_{2(g)} + OH^- \leftrightarrow HCO_3^-$$

2. **Preparation of Carbon dioxide:** For the preparation of CO_2 gas, the limestone (Calcium carbonate) and coke is heated upto 1000°C in lime kiln. It yields carbon dioxide and calcium oxide. Carbon dioxide is used whenever required. The calcium oxide is reacts with water to give lime (calcium hydroxide). This calcium hydroxide is further utilized to recover ammonia.

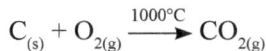

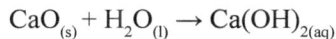

$$CaCO_{3(s)} \xrightarrow{1000°C} CaO_{(s)} + CO_{2(g)}$$

$$C_{(s)} + O_{2(g)} \xrightarrow{1000°C} CO_{2(g)}$$

$$CaO_{(s)} + H_2O_{(l)} \rightarrow Ca(OH)_{2(aq)}$$

3. **Carbonation of Ammoniated brine:** Ammoniated brine percolates down the bottom of the carbonating tower after being squeezed and pushed through the bottom by CO_2 from the lime kiln. Cast iron towers with a diameter ranging from 1.6 to 2.5 meters are used as carbonating towers in series with multiple precipitation towers. Recompressed ammoniated brine and

CO$_2$ gas (90–95%) from the bicarbonate calciner. During the precipitation cycle, cooling coils installed about 20 feet above the bottom are used to maintain temperatures of 20 to 25°C at both ends and 45 to 55°C in the center, respectively. As sodium bicarbonate cakes form on the cooling coils and shelves, the tower gradually floods. The fouling tower's cooling coils are turned off. The tower is then fed with hot, fresh ammoniated brine. To create ammonium carbonate solution, solid NaHCO$_3$ are here dissolved. The (NH$_4$)$_2$CO$_3$, unconverted NaHCO$_3$-containing solution is permitted to decelerate a second tower, known as the making tower. A sequence of boxes and slanted baffles make up the making towers. Cooling coils reduce the heat produced by exothermic reactions.

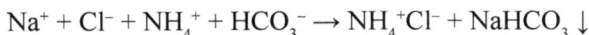

$$CO_{2(g)} + OH^- \leftrightarrow HCO_3^-$$

$$Na^+ + Cl^- + NH_4^+ + HCO_3^- \rightarrow NH_4^+Cl^- + NaHCO_3 \downarrow$$

4. **Filtration:** The carbonation tower's slurry is next filtered through a rotary vacuum filter, which aids in drying the bicarbonate and recovering the ammonia. The filter cake is rinsed with water after the salt and ammonium chloride have been removed, and is then either put to a centrifugal filter to remove the moisture or immediately calcined. 10% of the NaHCO$_3$ that is washed off also makes it into the filtrate. Lime from a lime kiln is used to treat the filtrate, which contains sodium chloride, ammonium chloride, sodium bicarbonate, and ammonium bicarbonate. Ammonia is recovered and recycled in this place. Calcium salt that is still present is separated and regarded as waste.

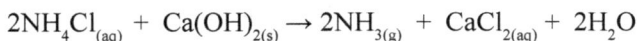

$$2NH_4Cl_{(aq)} + Ca(OH)_{2(s)} \rightarrow 2NH_{3(g)} + CaCl_{2(aq)} + 2H_2O$$

5. **Calcination:** The horizontal calciner, which is either ignited at the feed end by gas or steam heated equipment, calcines the NaHCO$_3$ from the drum filter at a temperature of around 200°C. The production of lumps of bicarbonate is prevented by the heating being applied through the shell parallel to the product. Hot soda ash from the calciner is transferred to a rotary chiller and bagged. In order to produce liquid ammonia, some gases, including CO$_2$, NH$_3$, steam, and others, are cooled and condensed. The rich CO$_2$ gas is then chilled and returned to the carbonating tower. We get light soda ash from the calciner. Light soda ash is ground with enough water to produce dense soda ash.

$$2NaHCO_{3(s)} \rightarrow Na_2CO_{3(s)} + CO_{2(g)} + H_2O_{(l)}$$

(E) Major engineering problems

1. **Absorption units:** The design of the absorption units should allow increasing sodium bicarbonate crystals to move downward. To achieve this, each unit is made to resemble a very large single bubble cap with downward sloping floors. The absorption is given in towers that are overflowing with liquid. CO$_2$ must therefore be compressed. At the peak of the carbonating cycle, the partial pressure and solubility of CO$_2$ rose as a result of compression.

2. **Making tower:** It is important to ensure that the precipitated bicarbonate is unquestionably filterable and effectively washable before the suspension of the bicarbonate of soda created inside the manufacturing tower is drawn off. It is managed by adjusting the manufacturing tower's temperature and concentration. Fine sodium bicarbonate crystals are allowed to develop while the temperature gradient is kept constant at 20°C at both end and 45°C in the center. The heat of reaction raises the temperature from 20°C to $45 \pm 55°C$, while the use of coils lowers it.

3. **Development of suitable calcining equipment:** Bicarbonate of soda that is too moist will pile up on the kiln's walls and block the shell from receiving heat effectively. To prevent caking, a hefty scraper chain must be installed within the kiln, and wet-filter cake must be combined with dry product. Utilizing fluidized bed calciners can frequently prevent these issues.

4. **Filtration unit:** Vacuum is used to achieve filtering on the drum filter. It aids in recovering ammonia and drying the bicarbonate.

5. **Ammonia recovery:** Because ammonia recovery is more expensive than Na_2CO_3 inventory, losses must be maintained to a minimum. With the right kit design and upkeep, losses are less than 0.2% of the recycle load, 0.5% per kg of product, or 1 kg per tonne of washing soda.

6. **Waste disposal:** During the procedure, a lot of $CaCl_2$-NaCl liquid is produced. These alcoholic beverages can either be used or discarded as waste.

14.2.3.3 Uses

In many fields, sodium carbonate is frequently used to neutralize acidic solutions. Body reactions or processes need sodium carbonate. Additionally, sodium carbonate is used in the production of glass, water softening, culinary additives, and cooking. Our bodies contain sodium bicarbonate, which is an essential substance. It aids in regulating and balancing the blood's excessive acidity. Additionally, sodium bicarbonate is utilized as a deodorizer, cleaning or exfoliating agent, and occasionally as a temporary extinguisher. Intravenously administered sodium bicarbonate is used for heart resuscitation, poor kidney function, cocaine toxicity, preventing kidney damage from certain allergy medications, reviving newborns, poisoning, preventing chemotherapy side effects, preventing muscle breakdown, and fluid buildup in the lungs brought on by a specific chemical. Additionally, it is employed as a component of baking soda (Sircus, 2010; Brescia et al., 1975; Bedekar et al., 2007).

*Difference between Solvey process and Modified Solvey [DUAL] process:

Chinese chemist Hou Debeng created this method in the 1930s. Hou's Process' initial stage is similar to the original procedure, except ammonium chloride is used in place of calcium chloride (NH_4Cl). Instead of adding lime to the leftover calcium chloride solution, carbon dioxide and ammonia are added to it until it reaches saturation at 40 degrees Celsius. The solution is then cooled to 10 degrees Celsius, leading the ammonia chloride to be filtered out. After that, the ammonium chloride is recycled and utilized again to make sodium carbonate. Due to the high expense of imported rock salt and the usage of ammonium chloride as a fertilizer, particularly in rice farming, the DUAL process is significant in Japan. The major steps of DUAL process are as follows.

Process description

1. Ammonia absorption

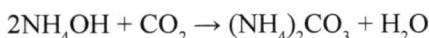

$$NH_3 + H_2O \rightarrow NH_4OH$$

$$2NH_4OH + CO_2 \rightarrow (NH_4)_2CO_3 + H_2O$$

2. Carbonation of the ammonia brine and production of sodium bicarbonate and ammonium chloride

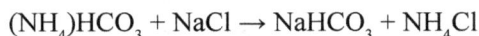

$$(NH_4)_2CO_3 + CO_2 + H_2O \rightarrow 2NH_4HCO_3$$

$$(NH_4)HCO_3 + NaCl \rightarrow NaHCO_3 + NH_4Cl$$

3. Filtration of sodium bicarbonate
4. Crystallization and separation of ammonium chloride
5. Decomposition of bicarbonate into soda ash and recovery of carbon dioxide

$$2NaHCO_3 \xrightarrow{\text{Heat}} Na_2CO_3 + H_2O + CO_2$$

14.2.4 Sodium sulfate

The anhydrous sodium salt version of sulfuric acid is sodium sulfate. In water, sodium ions and sulfate ions are produced via the dissociation of sodium sulfate anhydrous. It is referred to as Glauber's salt since Johann Rudolf Glauber discovered the sodium sulfate in Austrian spring water around 1625. Johann referred to it as sal mirabilis (miracle salt) because of its therapeutic qualities.

14.2.4.1 Properties

Table 14.6 Physical & Chemical Properties of sodium sulfate

Sr. No.	Properties	Particular
1	IUPAC name	Disodium sulfate
2	Chemical and other names	Sodium sulfate, Salt cake, Thenardite
3	Molecular formula	Na_2SO_4
4	Molecular weight	142.04gm/mole
5	Physical description	White orthorhombic odorless crystals or powder
6	Solubility	Soluble in water, methanol, etc.
7	Melting point	884°C
8	Boiling point	1430°C
9	Flash point	-
10	Density	2.684 gm/cm³ at 27°C

14.2.4.1.2 Manufacturing process

Industrially, it is manufacturing from sodium chloride. This process is called as Manheim Process. It is also produced by natural brine. It is made when magnesium sulfate and sodium chloride react in a solution, then crystallized. The Hargreaves method, which involves reacting sodium chloride with water, sulfur dioxide, and oxygen, can also be used to make it. In numerous processes, including waste acid neutralization, ion exchange regeneration, flue-gas desulfurization, and rayon production, sodium sulfate is also produced as a waste or byproduct.

Manufacturing process by Manheim Process

(A) Raw materials: Sodium chloride and Sulfuric acid

(B) Chemical reactions: Sodium chloride and sulfuric acid is reacted to yield sodium hydrogen sulfate, which is further heated with sodium chloride at 550–600°C to get sodium sulfate.

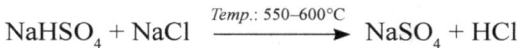

$$NaCl + H_2SO_4 \longrightarrow NaHSO_4 + HCl$$

$$NaHSO_4 + NaCl \xrightarrow{\textit{Temp.: 550–600°C}} NaSO_4 + HCl$$

Overall reaction

$$2NaCl + H_2SO_4 \longrightarrow Na_2SO_4 + 2HCl$$

(C) Flow chart

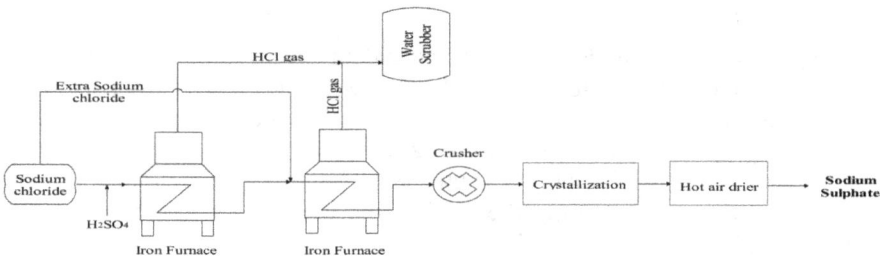

Figure 14.5 Flow sheet diagram of production of Sodium sulfate from Sodium chloride

(D) Process description: Calculated amount of sodium chloride is taken into cast iron furnace from its storage. Also, sulfuric acid is mixed with it and heated this mixture using burner. This mixture is transferred into another furnace, where more sodium chloride is added. The reaction mixture is heated upto 550–600°C to get sodium sulfate. Both furnaces are liberated hydrochloric acid, which are scrubbed in scrubber using water. The solid cake is crushed, crystallized and dried.

Manufacturing process by natural brine

Natural brines are a secondary source of sodium sulfate. About 60% of the natural sodium sulfate of the market is a by-product. The natural brines contain 7–11% Na_2SO_4 having traces of NaCl and $MgSO_4$.

(A) Raw materials: Natural brine

(B) Flow chart

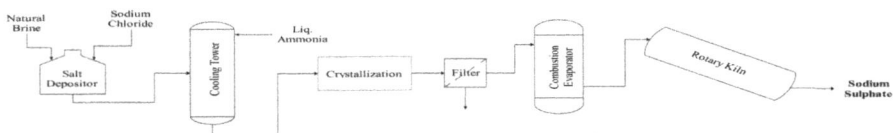

Figure 14.6 Flow sheet diagram of production of Sodium sulfate from natural brine

(C) Process description: The natural brine was charged into the salt depositor. Also, sodium chloride is added to lower the sodium sulfate solubility of the brine. The brine containing salt was collected from bottom of tank and chilled upto –10 to –60°C using ammonia to precipitate the sodium sulfate. It is then crystallized to concentrate it. The solid Glauber's salt is separated using filtration. The depositor is recycled with the mother liquor. The salt is sent to submerged combustion evaporator to melt it. It also removes most of water. Molten sodium sulfate is dried into rotary kiln dryer to get pure dried sodium sulfate.

14.2.4.3 Uses

It mainly serves to breakdown lignin and aid in the digestion of pulpwood. Additionally, the industry for soap and detergent uses it. It is used to make sodium salts, ceramic glazes, textile fibre processing, medicinal production, etc. It fluxes the glass while preventing the creation of scum by molten glass during the refining process. Additionally, it is utilized to remove minute air bubbles and flaws from molten glass during the casting and blowing procedures (Garrett, 2001; Young, 2007).

14.2.5 Sodium hypochlorite

It is inorganic compound having sodium cation (Na^+) and a hypochlorite anion (OCl^- or ClO^-). Its anhydrous form is unstable and ready decomposed, but its pentrahydrate form $NaOCl \cdot 5H_2O$ is stable. It is also called bleaching powder.

14.2.5.1 Properties

Table 14.7 Physical & Chemical Properties of Sodium hypochlorite

Sr. No.	Properties	Particular
1	IUPAC name	Sodium hypochlorite
2	Chemical and other names	Antiformin, Sodium oxychloride, Chlorox
3	Molecular formula	NaOCl
4	Molecular weight	63.0 gm/mole
5	Physical description	colorless or slightly yellow watery liquid with sweetish odor
6	Solubility	Soluble in water, methanol, etc.
7	Melting point	-
8	Boiling point	111°C
9	Flash point	-
10	Density	1.093 gm/cm³ at 27°C

14.2.5.2 Manufacturing process

Raw materials of sodium hypochlorite are caustic soda and chlorine.

(A) Raw materials: Caustic soda, Water and Chlorine gas.

(B) Chemical reactions: Chlorine and sodium hydroxide combine to form sodium chloride, sodium hypochlorite, and water.

$$2NaOH + Cl_2 \longrightarrow NaCl + NaOCl + H_2O$$

(C) Flow chart

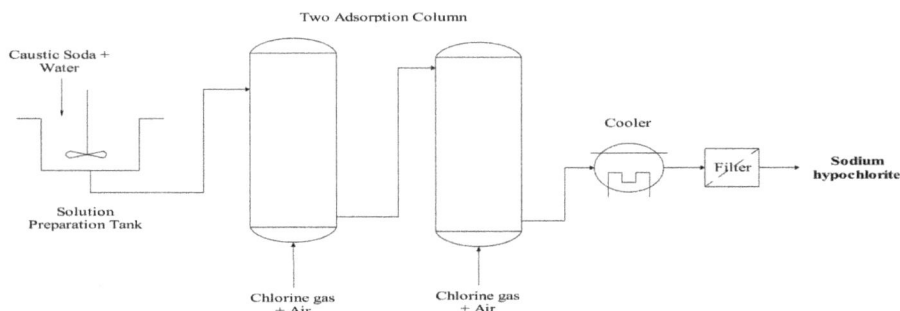

Figure 14.7 Flow sheet diagram of production of Sodium hypochlorite from natural brine

(D) Process description: In this process, reaction is carried out in two stage absorption columns. First, caustic soda is taken from its storage and added in solution preparation tank, where diluted caustic soda solution is prepared. Hot diluted caustic solution is cooled down upto room temperature and pumped from the top of first absorption column. Chlorine gas and air is bubbled out from the bottom of column, where caustic solution and chlorine is reacted to yield sodium hypochlorite. After first stage of reaction, the reaction mixture is transferred into second column, where, chlorine gas and air is bubbled out from the bottom of column. Remaining caustic solution is reacted to give solid precipitates of sodium hypochlorite. This reaction mixture is cooled down and filtered to yield pure sodium hypochlorite.

14.2.5.3 Uses

Its major component is sodium hypochlorite, which is also used in laundry bleach. In the textile, detergent, and paper & pulp sectors, it is widely employed as a bleach. Also used as an oxidant for organic compounds is sodium hypochlorite. Hypochlorite is used in the petrochemical sector for purifying petroleum products. Hypochlorite is used in the food processing industry to sterilize the food preparation equipment used in the manufacturing of fruit, vegetables, mushrooms, hog, beef, and poultry products, syrup, and fish. It is used to purify water in an efficient manner. It is extensively used for water disinfection, surface cleaning, bleaching, and odour elimination. It is frequently used to treat or prevent infections brought on by wounds, surgery, pressure ulcers, diabetic foot ulcers, or skin abrasions (Kamath, 2013; Ronco and Mishkin, 2007).

14.2.6 Sodium peroxide

It ignited in too much oxygen because it is a powerful base. So it can be found in many hydrates like $Na_2O_2 \cdot 2H_2O_2 \cdot 4H_2O$, $Na_2O_2 \cdot 2H_2O$, $Na_2O_2 \cdot 2H_2O_2$, and $Na_2O_2 \cdot 8H_2O$. Out of these hydrates, octahydrate form is easily synthesized.

14.2.6.1 Properties

Table 14.8 Physical & Chemical Properties of Sodium peroxide

Sr. No.	Properties	Particular
1	IUPAC name	Disodium peroxide
2	Chemical and other names	Sodium peroxide, Sodium dioxide
3	Molecular formula	Na_2O_2
4	Molecular weight	63.0 gm/mole
5	Physical description	Yellow-white to yellow granular solid
6	Solubility	Soluble in water
7	Melting point	465°C
8	Boiling point	654°C
9	Flash point	-
10	Density	2.81 gm/cm³ at 27°C

14.2.6.2 Manufacturing process

Industrially, it is prepared using sodium and oxygen in presence of catalyst. In a platinum or palladium tube, solid sodium iodide can also be created by exposing it to ozone gas. By treating sodium hydroxide with hydrogen peroxide, octahydrate sodium peroxide is formed.

(A) Raw materials: Sodium metal, Oxygen (Air) and Catalyst

(B) Chemical reactions: Sodium metal is reacted with oxygen at 130–200°C to get sodium peroxide.

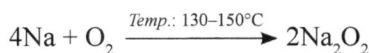

$$4Na + O_2 \xrightarrow{\textit{Temp.}: 130-150°C} 2Na_2O_2$$

(C) Flow chart

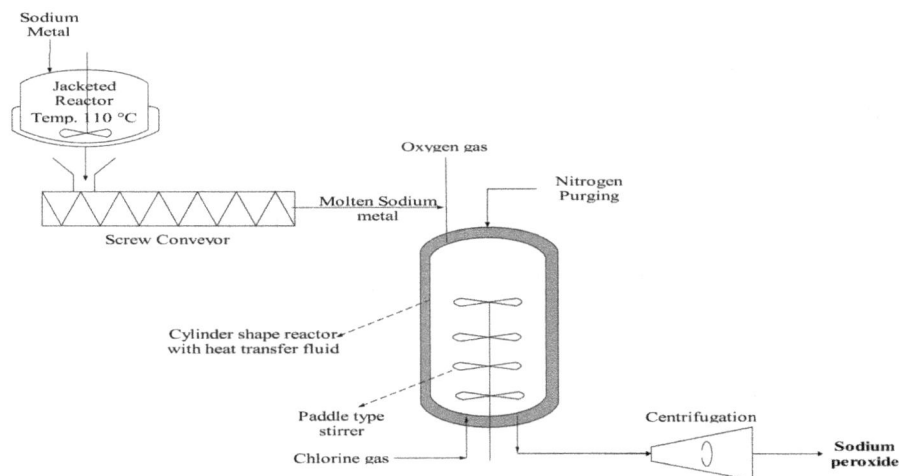

Figure 14.8 Flow sheet diagram of production of Sodium peroxide

(D) Process description: As per Fig. 14.8, sodium metal is taken from its storage and transferred into jacketed reactor having stirring facility. The metal heated at 100–110°C with stirring to melt it. This molten metal is taken from the bottom of reactor and transferred into special type reactor vessel via screw type conveyer and sodium inlet port. This conveyor has thermal jacket coating in order to prevent the solidification of the liquid sodium (Na). The reaction vessel has a cylindrical shape and thermal housing which is filled with a heat transfer fluid. This vessel is comprised of an oxygen inlet port, which is connected to oxygen nozzle set and located on top of the reactor inner vessel. This vessel has also paddle type stirrer at the top or bottom of vessel, preferably bottom of vessel, for proper mixing liquid sodium with air and preventing humidity ingression. This vessel has nitrogen purging facility to form inert layer during reaction to prevent the ingression of atmospheric air, oxygen and humidity. It has also discharged port at bottom of the reactor vessel to collect the product. After adding molten sodium, oxygen is inserted through nozzle with stirring. Also, nitrogen gas is inserted to create inert atmosphere. After completion of reaction, the product is collected from discharging port and centrifuged to get pure sodium peroxide.

14.2.6.3 Uses

By interacting with CO_2 to produce oxygen and washing soda, sodium peroxide serves as both an oxidant and an oxygen source. As a result, it is very helpful in submarines and scuba gear (Singh et al., 2022).

14.2.7 Sodium sulfite

An inorganic salt with the chemical formula Na_2SO_3 is sodium sulfite. Two sodium cations (Na^+) and one sulfite anion make up the ionic composition (SO_3^{2-}).

14.2.7.1 Properties

Table 14.9 Physical & Chemical Properties of Sodium sulfite

Sr. No.	Properties	Particular
1	IUPAC name	Disodium sulfite
2	Chemical and other names	Sodium sulphite
3	Molecular formula	Na_2SO_3
4	Molecular weight	126.51gm/mole
5	Physical description	White crystalline powder or colorless odorless crystals
6	Solubility	Soluble in water, methanol, etc.
7	Melting point	33.4°C
8	Boiling point	-
9	Flash point	-
10	Density	2.633 gm/cm³ at 27°C

14.2.7.2 Manufacturing process

Industrially, Sulfur dioxide gas and sodium carbonate solution are combined to form sodium sulfite. Additionally, sulfur dioxide gas and sodium hydroxide are combined to make it.

(A) Raw materials: Sulfur, Oxygen, Sodium carbonate and Sodium hydroxide

(B) Chemical reactions: Sodium carbonate solution is reacted with sulfur dioxide gas to get sodium bisulfate. It is further reacted with Sodium carbonate solution to yield sodium sulfite. Overall reaction is written as follows.

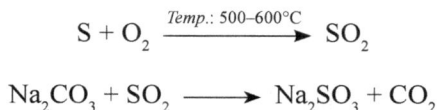

$$S + O_2 \xrightarrow{\textit{Temp.: 500–600°C}} SO_2$$

$$Na_2CO_3 + SO_2 \longrightarrow Na_2SO_3 + CO_2$$

(C) Flow chart

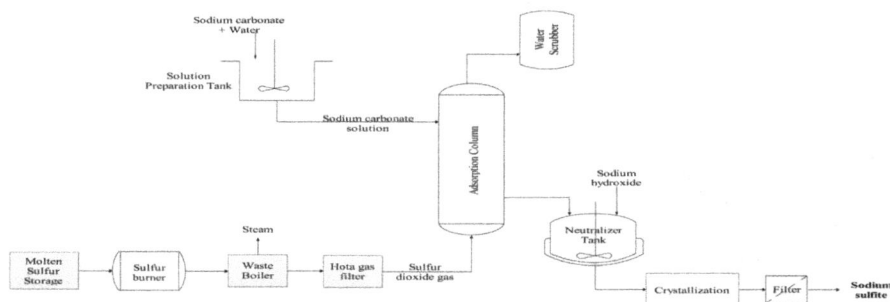

Figure 14.9 Flow sheet diagram of production of Sodium sulfite

(D) Process description: First, water is taken is sodium carbonate solution tank with having stirring facility. Proper amount of sodium carbonate is added in the tank to prepare diluted sodium carbonate solution. This carbonate solution is pumped from the top of tray absorption column. Solid sulfur is heated upto 110–120°C to convert liquid form. Thereafter, liquid sulfur is burnt in excess of air at temperature of 500–600°C to form sulfur dioxide gas. Hot sulfur dioxide gas filtered, cooled at 420°C and bubbled out from the bottom of absorption column to form sodium sulfite. Off-gases are scrubbed using water scrubber. The reaction mixture is transferred in neutralization vessel, where it is neutralized it with sodium hydroxide and forms yellow precipitates of sodium sulfite. It is crystallized and filtered.

14.2.7.3 Uses

Bleaching, desulfurization, and dechlorination are all processes that sodium sulfite exhibits. This chemical was used by the food sector to help keep food goods looking fresh. Many medications need it as a component to keep their efficacy and stability.

14.2.8 Sodium hydrosulfite

Sodium dithionite, also known as sodium hydrosulfite, can come in the form of a white, crystalline powder with a faint sulfur scent. Although it remains stable in the absence of oxygen, it disintegrates in harsh conditions and acidic solutions.

14.2.8.1 Properties

Its main properties are as follows.

Table 14.10 Physical & Chemical Properties of Sodium hydrosulfite

Sr. No.	Properties	Particular
1	IUPAC name	Sodium dithionite
2	Chemical and other names	Sodium hydrosulfite, Sodium hydrosulphite, Dithionous acid, Sodium sulfoxylate, Disodium dithionite
3	Molecular formula	$Na_2O_4S_2$
4	Molecular weight	63.0 gm/mole
5	Physical description	White to light yellow crystalline solid having a sulfur dioxide-like odor
6	Solubility	Soluble in water, alcohol, etc.
7	Melting point	52°C
8	Boiling point	-
9	Flash point	-
10		2.397 gm/cm³ at 27°C

14.2.8.2 Manufacturing process

It is prepared by reducing sulfur dioxide or sodium bisulfate using various processes such as zinc dust (sodium carbonate process), electrolytic reduction, and sodium borohydride and sodium amalgam process. Out of these processes, sodium-mercury amalgam is more preferable. Overall reaction is written as follows.

$$2Na + 2SO_2 \longrightarrow Na_2S_2O_4$$

(A) Raw materials: Sulfur dioxide, Sodium Amalgam (Na-Hg) and Buffer Sodium bisulfate

(B) Chemical reactions: In sodium amalgam process, sodium bisulfate is reacted with sodium metal to get sodium hydrosulfite.

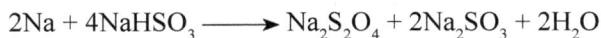

$$2Na + 4NaHSO_3 \longrightarrow Na_2S_2O_4 + 2Na_2SO_3 + 2H_2O$$

Regeneration Process of amalgam

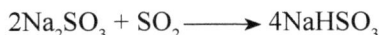

$$2Na_2SO_3 + SO_2 \longrightarrow 4NaHSO_3$$

(C) Flow chart

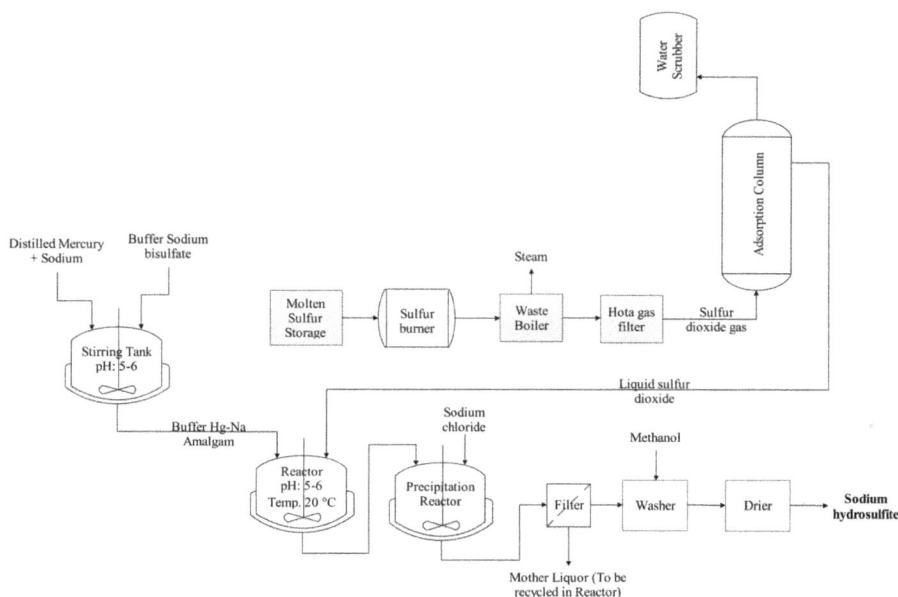

Figure 14.10 Flow sheet diagram of production of Sodium hydrosulfite

(D) Process description: First, distilled mercury is taken unto stainless steel reactor having stirring facility. Now, pure sodium is added into reactor with constant stirring to form sodium-amalgam. Also, buffer sodium bisulfate is added into amalgam solution to maintain pH 5-6 during the reaction. On another end, fresh pure sulfur dioxide is passed through absorption column to make liquid sulfur dioxide solution. This solution is kept under inert nitrogen gas to avoid oxidation. Now, sodium amalgam with buffer sodium bisulfate is charged into continuous flow stirrer-tank reactor. Also, liquid sulfur dioxide is added into reduce the quantity of mercury in amalgam. The reaction is conducted at pH 5–6 and temperature of 15–20°C to yield concentrated solution of sodium dithionite. The reaction mass is collected from bottom of reactor and transferred into precipitation reactor having stirrer. Now, sodium chloride is added into reactor with continuous stirring to precipitate the sodium dithionite. The reaction mass is undergone filtration to get solid sodium dithionite; and mother liquor is recycled into reactor. The product is washed with methanol and thereafter, dried under vacuum to get fine crystal of sodium dithionite.

14.2.8.3 Uses

It is mainly used as a reducing, sulfonating, chelating, fluorescent quenching and decolorizing agent in different organic reactions. It is also use to estimate the iron content in soil chemistry. It is also utilized in water treatment, gas purification, cleaning, leather, polymers, photography, etc. (Bahrle-Rapp, 2007).

14.2.9 Sodium silicate

With the chemical formula $Na_{2x}Si_yO_{2y+x}$ or $(Na_2O)_x \cdot (SiO_2)_y$, sodium silicate is also referred to as water glass or liquid glass. Examples of sodium silicate include sodium metasilicate Na_2SiO_3, sodium orthosilicate Na_4SiO_4, and sodium pyrosilicate $Na_6Si_2O_7$.

14.2.9.1 Properties

Its main properties are as follows.

Table 14.11 Physical & Chemical Properties of Sodium silicate

Sr. No.	Properties	Particular
1	IUPAC name	Dioxidodisodium silane
2	Chemical and other names	Sodium silicate, Sodium metasilicate, Waterglass, Sodium siliconate
3	Molecular formula	Na_2O_3Si
4	Molecular weight	122.03gm/mole
5	Physical description	Powdered or flaked solid substance with suffocating odor
6	Solubility	Soluble in water, acid,salt, methanol, etc.
7	Melting point	1890°C
8	Boiling point	2345°C
9	Flash point	-
10	Density	2.615gm/cm³ at 27°C

14.2.9.2 Manufacturing process

It can be manufactured using various processes as follows.

Silica is reacted with caustic soda at high temperature to yield sodium silicate solution and water. Also, sodium silicate is prepared by dissolving silica (SiO_2) in molten sodium carbonate at 1100–1200°C. Sodium sulfate, carbon and silica are heated upto 900–1000°C to get sodium silicate with liberating sulfur dioxide and carbon dioxide. Here, carbon is used as a reducing agent.

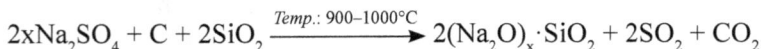

$$2xNa_2SO_4 + C + 2SiO_2 \xrightarrow{\text{Temp.: 900–1000°C}} 2(Na_2O)_x \cdot SiO_2 + 2SO_2 + CO_2$$

Industrially, mixture of sodium carbonate and caustic soda are reacted with silica to get sodium silicate at high temperature and pressure.

(A) Raw materials: Sodium carbonate, caustic soda and silica.

(B) Chemical reactions: Sodium silicate is prepared by dissolving silica in sodium carbonate and caustic soda at 1100–1200°C.

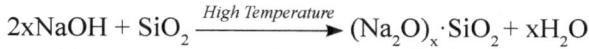

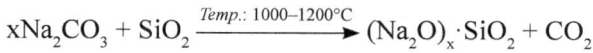

$$2x\text{NaOH} + \text{SiO}_2 \xrightarrow{\textit{High Temperature}} (\text{Na}_2\text{O})_x \cdot \text{SiO}_2 + x\text{H}_2\text{O}$$

$$x\text{Na}_2\text{CO}_3 + \text{SiO}_2 \xrightarrow{\textit{Temp.: } 1000-1200°C} (\text{Na}_2\text{O})_x \cdot \text{SiO}_2 + \text{CO}_2$$

(C) Flow chart

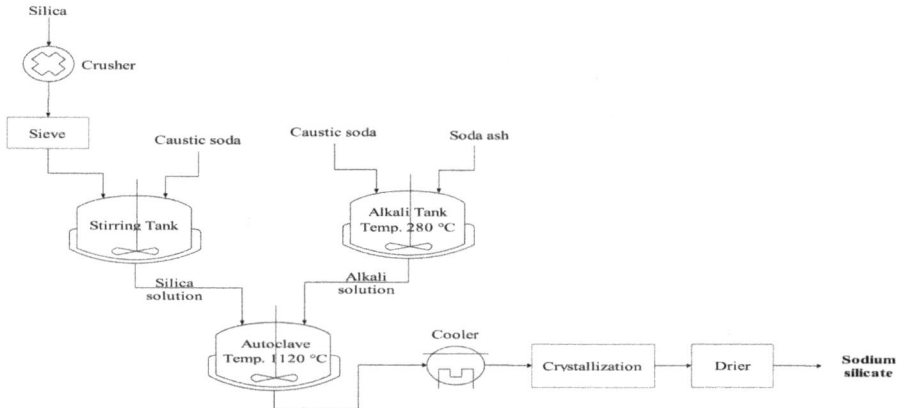

Figure 14.11 Flow sheet diagram of production of Sodium silicate

(D) Process description: In this process, quartz or sand containing silica is grinded, sieved and sent to silica vessel. Now, small quantity of caustic soda solution is charge into vessel added to maintain proper flow of silica. Now, sodium carbonate and caustic soda is added into another alkali vessel having stirring and heating facility. This mixture is heated upto 250–280°C with stirring. Silica- caustic soda solution; and sodium carbonate and caustic soda are charged into autoclave. The reaction mass is heated upto temperature and pressure of 1100–1200°C and 27–32 atmospheres respectively for 20–35 minutes. Then melt sodium silicate is undergone to controlled cooling upto room temperature. It is crystallized and filtered.

14.2.9.3 Uses

Detergents, paper, water treatment, and construction materials are the principal uses for sodium silicates. Additionally, it is used in sand casting, glue, drilling fluid, and passive fire prevention (Ismail et al., 2009).

14.2.10 Sodium cyanide

A poisonous substance with the formula NaCN is sodium cyanide.

14.2.10.1 Properties

Its main properties are as follows.

Table 14.12 Physical & Chemical Properties of Sodium cyanide

Sr. No.	Properties	Particular
1	IUPAC name	Sodium cyanide
2	Chemical and other names	Cyanide of sodium, Cyanide salts
3	Molecular formula	CNNa
4	Molecular weight	49.01 gm/mole
5	Physical description	White, granular or crystalline odorless solid
6	Solubility	Soluble in water, methanol, etc.
7	Melting point	563°C
8	Boiling point	1498°C
9	Flash point	83°C
10	Density	1.598 gm/cm³ at 27°C

14.2.10.2 Manufacturing process

Previously, sodium cyanide was manufactured the Castner process, in which sodium amide with and carbon were reacted at higher temperature.

$$NaNH_2 + C \longrightarrow NaCN + H_2$$

Now-a-days, it is prepared using hydrogen cyanide and sodium hydroxide

(A) Raw materials: Hydrogen cyanide and Sodium hydroxide

(B) Chemical reactions: Exothermic reaction is preformed between hydrogen cyanide and sodium hydroxide to form sodium cyanide.

$$HCN + NaOH \longrightarrow NaCN + H_2O$$

(C) Flow chart

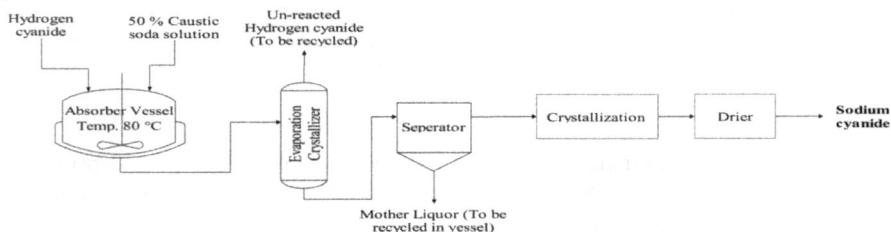

Figure 14.12 Flow sheet diagram of production of Sodium cyanide

(D) Process description: As per simple flow chart, hydrogen cyanide is inserted into absorber vessel. Also, 50% sodium hydroxide solution is also added into vessel. This exothermic reaction is controlled by circulating the reaction mixture of vessel through heat exchange. The temperature of mixture is maintained to 30–80°C. After completion of reaction, reaction mixture is collected from bottom and transferred to evaporation crystallizer, where un-reacted hydrogen cyanide is collected as over-headed product. Hydrogen cyanide is purified in HCN absorber column and recycled into vessel. While formed sodium cyanide is collected from bottom of evaporation crystallizer and transferred into solid liquid separator. Here, solid sodium cyanide is separated and mother liquor is collected from bottom; and recycled into vessel. It is crystallized and filtered.

14.2.10.3 Uses

It is also used economically for rodenticides, fumigation, electroplating, the extraction of gold and silver from ores, and the production of chemicals. It is also used in production of dye, hydrochloric acid, nitrites and chelating agents (Rogers and Porter, 1987; Acker, 1912).

14.2.11 Sodium ferrocyanide

Sodium ferrocyanide contains six coordination bond between a iron atom and six cyanide atoms. It is available in decahydrate form, i.e., $Na_4Fe(CN)_6 \cdot 10H_2O$. It is sometimes referred to as yellow soda prussiate.

14.2.11.1 (Introduction) properties

Its main properties are as follows.

Table 14.13 Physical & Chemical Properties of Sodium ferrocyanide

Sr. No.	Properties	Particular
1	IUPAC name	Iron(II)tetrasodiumhexacyanide
2	Chemical and other names	Tetrasodiumhexacyanoferrate, Sodium hexacyanoferrate(II), Sodium hexacyanoferrate(4-)
3	Molecular formula	$C_6FeN_6Na_4$
4	Molecular weight	303.91 gm/mole
5	Physical description	Yellow odorless solid
6	Solubility	Soluble in water
7	Melting point	435°C
8	Boiling point	-
9	Flash point	-
10	Density	1.481 gm/cm³ at 27°C

14.2.11.2 Manufacturing process

Starting materials for preparing of sodium ferrocyanide are ferrous chloride and hydrogen cyanide.

(A) **Raw materials:** Ferrous chloride, Hydrogen cyanide, Sodium hydroxide and Water

(B) Chemical reactions: First, hydrogen cyanide and sodium hydroxide is heated upto 80–90°C to give sodium cyanide. This sodium cyanide is reacted with ferrous chloride is give sodium ferrocyanide. Here, preparation of iron hydroxide is side-reaction.

$$HCN + NaOH \longrightarrow NaCN + H_2O$$

$$6NaCN + FeCl_2 \longrightarrow Na_4Fe(CN)_6 + 2NaCl$$

$$\text{Side Reaction: } FeCl_2 + 2H_2O \longrightarrow Fe(OH)_2 + 2HCl$$

(C) Flow chart

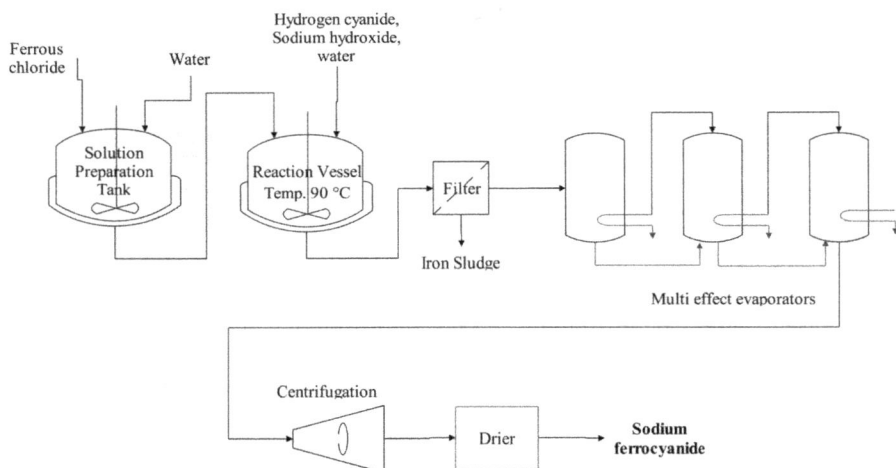

Figure 14.13 Flow sheet diagram of production of Sodium ferrocyanide

(D) Process description: As per simple flow sheet diagram 14.13, first ferrous chloride solution is prepared in tank and transferred into reaction having stirring and heating facilities. Also, water, hydrogen cyanide and sodium hydroxide are charged in reactor with stirring. Now, reaction mass is heated upto 80–90°C with steam. Maintain temperature of 80–90°C for 2 hours. Therefore, the reaction mass is filtered to remove iron hydroxide as sludge. The filtration get thicken in multi effect evaporator (MEE). It is then centrifuged to remove impurities and finally, dried to get pure dry sodium ferrocyanide.

14.2.11.3 Uses

It is majorly used as raw materials in different industries such as pharmaceuticals, pigment, tannage, metallurgy and chemical industries. It is also used for manufacturing iron blue and potassium ferricyanide, which is utilized as surface corrosion protecting agent for tannage and metal.

14.2.12 Sodium chlorite

It was first invented by in 1921 by E. Schmidt, when he was purified cellulose fibres with chlorine dioxide gas. Thereafter, it was first commercialized by Mathieson Chemical Corporation for bleaching function in 1960.

14.2.12.1 Properties

Its main properties are as follows.

Table 14.14 Physical & Chemical Properties of Sodium Chlorite

Sr. No.	Properties	Particular
1	IUPAC name	Sodium Chlorite
2	Chemical and other names	Textone, Chlorite sodium
3	Molecular formula	$NaClO_2$
4	Molecular weight	90.44 gm/mole
5	Physical description	White crystals or crystalline odorless powder
6	Solubility	Soluble in water, methanol, etc.
7	Melting point	-
8	Boiling point	-
9	Flash point	-
10	Density	2.468 gm/cm^3 at 27°C

14.2.12.2 Manufacturing process

It is manufacturing using sodium chlorate.

(A) Raw materials: Sodium chlorate, Air, Sulfuric acid, Sodium hydroxide and Sodium peroxide

(B) Chemical reactions: Sodium chlorate is undergone reaction with hydrogen peroxide and sulfuric acid to give chlorine dioxide gas, which on reaction with sodium hydroxide in presence of hydrogen peroxide at temperature below 35°C to give sodium chlorite.

$$2NaClO_3 + H_2O_2 + 2H_2SO_4 \longrightarrow 2ClO_2 + 2NaHSO_4 + O_2 + 2H_2O$$

$$2ClO_2 + 2NaOH + H_2O_2 \longrightarrow 2\ NaClO_2 + O_2 + 2H_2O$$

(C) Flow chart

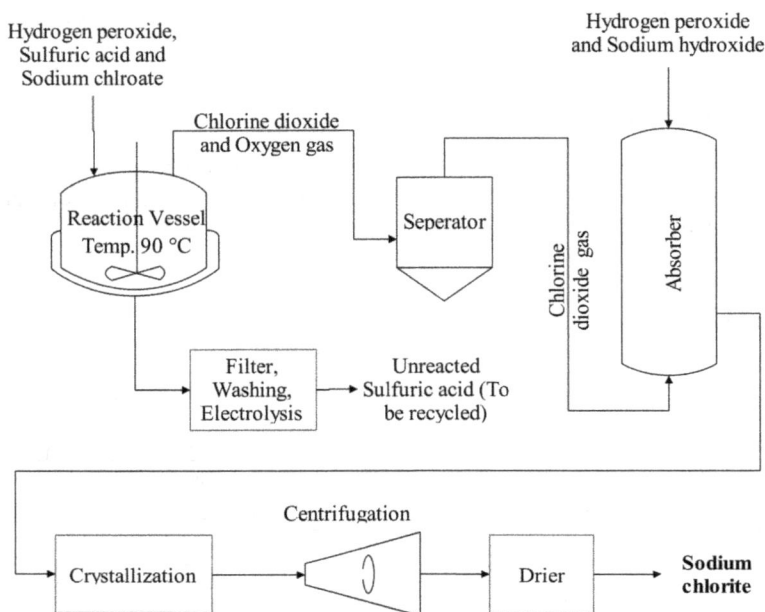

Figure 14.14 Flow sheet diagram of production of Sodium chlorite

(D) Process description: As per simple process diagram, first hydrogen peroxide and sulfuric acid is charged in reactor having stirring facilities. Now, sodium chlorate is charged into reactor with stirring. Sodium chlorate, hydrogen peroxide and sulfuric acid, chlorine dioxide is reacted to give sodium hydrogen sulfate and oxygen gas is formed. Here, oxygen is collected from top of reactor. While sodium hydrogen sulfate is collected from bottom, which is undergone filtration, washing and electrolysis to recovered sulfuric acid. Here, hydrogen is by-product. Now, collected chlorine dioxide along-with oxygen gas is undergone separator to separate chlorine dioxide from oxygen gas. This chlorine dioxide gas passed through carbon dioxide absorber from bottom. Also, hydrogen peroxide and sodium hydroxide solution is charged from tope of absorber. The aqueous sodium chlorate is collected from bottom, which is further purified by crystallized, centrifuged and dried.

14.2.12.3 Uses

As sodium chlorite is strong oxidizing agent, removes most of organic impurities, color and odor, it is used as disinfect in various applications, such as drinking water, industrial effluent, sewage water and swimming water. In the textile, pulp, and paper industries, it is utilized as a bleaching and stripping agent because it also literates carbon dioxide. It has properties of controlling bacterial growth, it is used in petroleum and food industries. It is also used as catalyst to recover copper from copper cyanide destruction. It is also used as an anti-mold agent in detergent, toothpaste and contact lens solution (Payuhamaytakul et al., 2018).

14.3 Magnesium

It is the fourth most abundant mineral after calcium, sodium and potassium in the human body and eighth abundant mineral in the Earth's crust. This mineral is extremely important for muscles and brain, as magnesium involves in several chemical reactions and retains good health. Magnesium was first discovered by Joseph Black and recognized as an element in 1755. Thereafter, it was isolated by Sir Humphry Davy in 1808 by electrolysis of mixture of magnesium oxide and mercuric oxide. Naturally, it is available in different forms such as oxide (magnesia - MgO), carbonate [magnesite - $MgCO_3$, and dolomite - $CaMg(CO_3)_2$], hydroxide [brucite – $Mg(OH)_2$], chloride (carnallite - $KMgCl_3·6H_2O$), and sulfate (kieserite - $MgSO_4·H_2O$). Industrially, it is produced by either electrolysis of its several forms or Pidgeon process of mixture of dolomite and ferrosilicon. It is silver shiny or gray colored metal having light in weight and strong. Due to its very low density, comparatively high strength and excellent machinability, 70% of magnesium is used to prepare alloy with elements such as aluminum, zinc, manganese or silicon. Its proportions are very with its applications. These alloys are utilized to create machinery, vehicles, portable tools, home appliances, aircraft, and spacecraft. It is also used as desulfuration agent from molten iron and steel. It has property of light-weight and high tensile strength, so, it is used to prepare car seats, luggage, laptops, cameras and power tools. As it is easily ignite in presence of air with bright light, it is used for making reworks, sparklers and ares. Its sulfate is utilized as a mordant for dye. Its oxide is used in heat-resistant bricks for fireplaces and furnaces. Some of its forms like hydroxide, sulfate, chloride and citrate are utilized for manufacturing medicines. It is also used for the treatment of skin-related diseases and deciencies. Chemically, it is very reactive metal, so it is used in Grignard reagent. It is used in production of explosive and pyrotechnic devices, as it is extremely flammable (Kaimer, 2004; Friedrich, 2004; Crerwinski, 2011).

14.4 Aluminum

As 8.1% of the Earth's crust was made of aluminum, it is the most common metallic element. History of aluminum is quiet old. Mesopotamia people (before 5000 BCE), Egyptians and Babylonians (4,000 years ago) were utilized aluminum compounds in different purposes like preparation of pottery, medicines and chemicals. Then, it was recognized that one of aluminum ore-alumina as the prospective origin of metal in 18th century. Attempted was made by Humphry Davy in 1807 to extract aluminum from its ore, but unsuccessful. Finally, in

1825, Danish chemist Hans-Christian was the first to extract aluminum in its pure form from its ore. In the Earth, it is always combined with other elements, as it is unexplored free in nature. There are several aluminum ores, but two most important and available ores are bauxite (alumina-$Al_2O_3 \cdot 2H_2O$) and feldspar ($KAlSi_3O_8$). Other ores are corundum (Al_2O_3), cryolite (Na_3AlF_6), Alunite [$K_2SO_4 \cdot Al_2(SO_4)_3 \cdot 4Al(OH)_3$] and kaoline ($2Al_2O_3 \cdot 6SiO_2 \cdot 2H_2O$), potassium aluminum sulfate [$KAl(SO_4)_2 \cdot 12H_2O$]. Industrially, pure aluminum is extracted by purification and electrolysis of bauxite. It is the sixth most ductile and second most malleable metal. It is corrosion resistant. So, pure metal is used in cans, foils, smartphone, tablets, laptops, computer monitors, utensils, window frames, ships and beer kegs. It is combined with other metals to increase its different properties and applications. Its alloy is useful in airplane part, transportation, building material, motor vehicles, spacecraft components, refrigerators, air conditioning, solar panels, etc. As it has high electrical conduction and also non-magnetic and non-sparking, it is widely used in electrical transmission lines (MacKenzie and Totten, 2005; Totten and MacKenzie, 2003; Kaufman and Rooy, 2004).

14.5 Silicon

In the crust of the Earth, silicon is the second most common element after oxygen and ranks seventh in the universe. It covers about 22.3% of crust. History of silicon was too early. In 1500 BCE, Egyptians were used silicon in glass, beads and small vases. Then, impure silicon was found in 1811. The discovery of pure silicon was made in 1824 by Swedish chemist Jons Jacob Berzelius by heating impure silicon with potassium. Pure Silicon is extremely reactive, so, it is found in rock, clay, soil and sand in different forms such as oxide (quartz-silicon oxide-SiO_2), silicon sulfide (SiS_2), hydrides (silane-SiH_4 and disilicon hexahydride-Si_2H_6), fluoride (silicon tetrafluoride-SiF_4), halogen (silicon tetrachloride-$SiCl_4$ and silicon tetraodide-SiI_4), but quartz ore is found highest amount from crust. It also available in some plants and body of certain animals. Ultrapure silicon is solid hard dark blue-grey with metallic luster. Majority of silicon is used in preparation of glass, cement, ceramic and semi-conductor devices. It is used as reducing agent in metallurgy. It is alloying with aluminum and iron, which is used in dynamo, transformer plate, engine block, cylinder head and machine tools. Silicon rubber is used in waterproofing systems in bathrooms, roofs, and pipes. It is also used in fire-brick and refractory materials. It is used in computer chips and solar cells (Searle, 1998; Singh et al., 2003).

14.6 Phosphorus

Phosphorus is second abundant element, after calcium in our body. Our bones are made up with phosphorus. Also, it helps stimulate our body in certain aspects. Phosphorus is also important part of plant, as it is responsible for plant growth. It was first invented by Hennig Brandt at Hamburg in 1669. Two phosphate ores are found in nature, i.e., phosphorite and apatite. Elemental phosphorus is prepared by heating phosphorite rock in electric arc furnace. It is a colorless, semitransparent, soft, waxy solid that glows in the dark. There are two types of phosphorus: (1) Red amorphous non-toxic solid phosphorus and (2) White waxy toxic solid phosphorus. Phosphorus compounds are widely used in fertilizers. Additionally, whereas red phosphorus is used in matchboxes, white phosphorus is employed in flares and other incendiary weapons (Alalq et al., 2019; Constant and Lacour, 2019). Two chemicals of phosphorus are discussed below.

14.6.1 Phosphoric acid

It was invented in bone by Johann Gottlieb Gahn and Karl Wilhelm Scheele in 1770. It is super-coolant liquid at room temperature.

14.6.1.1 (Introduction) Properties

Its main properties are as follows.

Table 14.15 Physical & Chemical Properties of Phosphoric acid

Sr. No.	Properties	Particular
1	IUPAC name	Phosphoric acid
2	Chemical and other names	Orthophosphoric acid, o-Phosphoric acid, Sonac, Phosphorsaeure
3	Molecular formula	H_3PO_4
4	Molecular weight	97.99 gm/mole
5	Physical description	Clear colorless liquid or transparent crystalline odorless solid
6	Solubility	Soluble in water, methanol, ether and other solvents
7	Melting point	42.2°C
8	Boiling point	407°C
9	Flash point	83°C
10	Density	1.874 gm/cm³ at 27°C

14.6.1.2 Manufacturing process

It is prepared using phosphorus ore, calcium phosphate as a raw material by three processes: (1) Wet process using sulfuric acid, (2) Wet process using hydrochloric acid (3) Electric furnace process. It is also manufactured by oxidation of liquid elemental phosphorus at temperatures of 1650–2760°C to yield phosphorus pentoxide, which on hydration with water to give strong liquid phosphoric acid.

(A) Wet process using sulfuric acid: Phosphoric acid is prepared by treating calcium phosphate with sulfuric acid and water.

(a) Raw materials: Calcium phosphate, Sulfuric acid and Water

(b) Chemical reactions: Strong sulfuric acid, extra water, and calcium phosphate are combined at temperatures between 80 and 90°C to produce phosphoric acid. The temperature is kept between 80 and 90°C because this reaction is exothermic.

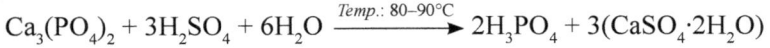

$$Ca_3(PO_4)_2 + 3H_2SO_4 + 6H_2O \xrightarrow{\textit{Temp.: 80-90°C}} 2H_3PO_4 + 3(CaSO_4 \cdot 2H_2O)$$

(c) Flow chart

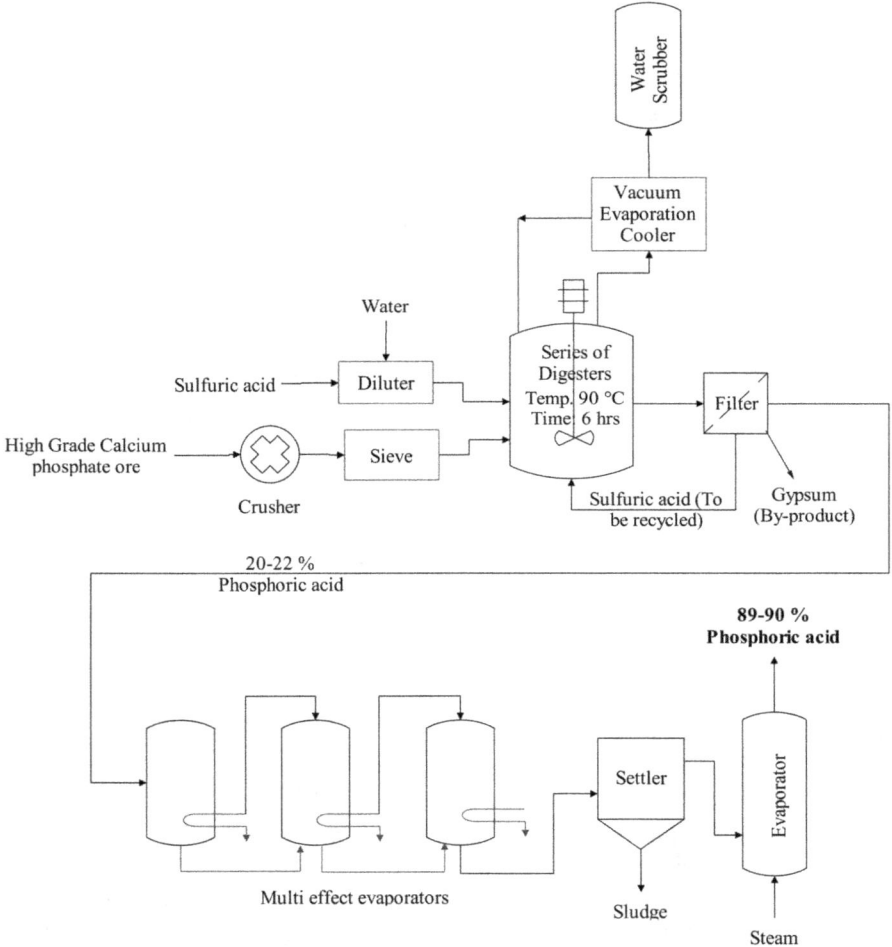

Figure 14.15 Flow sheet diagram of production of Phosphoric acid using Sulfuric acid

(d) Process description: As per Fig. 14.15, raw calcium phosphate is crushed, meshed and charged into digester. Also, fresh sulfuric acid with recycled sulfuric acid is fed into digester having stirring facilities. Here, acid proof brick-lined stainless steel digester is used. For 4–6 hours, the reaction mass is agitated. Since the reaction is exothermic, adding cool water helps to maintain the temperature between 80 and 90°C. The flue gases are collected from the top of reactor and scrubbed using water scrubber. After completion of reaction, the mass is filtered through travelling pan filter. First, raw phosphoric acid is separated from pan filter, thereafter, sulfuric acid is separated and recycled. And finally, gypsum is separated. Raw phosphoric acid (20–25% H_3PO_4) is undergone three stage multi-effect evaporator to concentrate the acid. Finally it is settled down to remove sludge and evaporated with steam to get pure dried phosphoric acid.

(e) Major Engineering Problems

1. If reaction temperature is above 100°C, then unwanted semihydrate ($CaSO_4 \cdot \frac{1}{2}H_2O$) and anhydrate ($CaSO_4$) is formed, which created problems in filtration process.
2. Recycled sulfuric acid contained certain amount of gypsum and pure phosphoric acid is not formed.
3. Also, this ore has impurities of calcium fluoride, so, it is also undergone reaction with sulfuric acid and water.

   Side reactions:

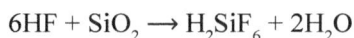

   $$CaF_3 + H_2SO_4 + 2H_2O \longrightarrow 2HF + CaSO_4 \cdot 2H_2O$$

   $$6HF + SiO_2 \longrightarrow H_2SiF_6 + 2H_2O$$

(B) Wet process using hydrochloric acid: Calcium phosphate is treated with sulfuric acid and water to produce phosphoric acid.

(a) Raw materials: Calcium phosphate, Hydrochloric acid and Water.

(b) Chemical reactions: Phosphoric acid is produced by treating calcium phosphate with strong hydrochloric acid and extra water at a temperature of 70 to 80°C.

$$Ca_3(PO_4)_2 + 6HCl + 6H_2O \xrightarrow{\text{\textit{Temp.: }70–80°C}} 2H_3PO_4 + 3CaCl_2$$

This reaction is exothermic, so, temperature of 70–80°C is maintained.

(c) Flow chart

(d) Process description: First, calcium phosphate is crushed, meshed and charged into digester. Also, fresh hydrochloric acid with recycled hydrochloric acid is fed into digester having stirring facilities. Here, rubber (or PVC)-lined mild steel digester is used. The reaction mass is stirred for

2–4 hours. The reaction is exothermic, so, temperature of 70–80°C is maintained by adding cool water. The flue gases such as carbon dioxide, hydrofluoric acid and hydrochloric acid are collected from the top of reactor and scrubbed using water scrubber. After completion of reaction, the reaction mass is collected from bottom of digester and charged to series of decantation unit. Arrangement of decantation is such that overflow of first decantation unit is moving down to the counter-current decantation for complete solvent (HCl) extraction process. This HCl is recycled in mixer. Mixture of calcium chloride and phosphoric acid is fed to two mixture-settler unit. Here, by-product calcium chloride is separated from upper part of settler and raw phosphoric acid is collected from bottom of settler. Finally, it is undergone distillation to get pure phosphoric acid as overhead product, while traces calcium chloride and HCl is separated from bottom of column.

(e) **Major Engineering Problems**

1. Solvent is recovery is major problem, associated with this process.
2. This method produces raw calcium chloride as a byproduct. It is either considered as solid hazardous waste or further purified and sold.
3. Also, this ore has impurities of calcium fluoride, so, it is also undergone reaction with hydrochloric acid and water.

Side reactions:

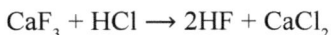

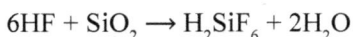

$$CaF_3 + HCl \rightarrow 2HF + CaCl_2$$

$$6HF + SiO_2 \rightarrow H_2SiF_6 + 2H_2O$$

(C) **Electric furnace process:** It is also prepared by heating calcium phosphate in electric furnace

(a) **Raw materials:** Calcium phosphate, Coke and Sand

(b) **Chemical reactions:** Calcium phosphate is heated with coke in electric furnace at 1800–2000°C to yield phosphoric acid having purity of 80–85%. Following series of chemical reactions are took place.

$$2Ca_3(PO_4)_2 + 6SiO_3 + 10C \longrightarrow P_4 + 10CO + 6CaSiO_3$$

$$P_4 + 10CO + O_2 \longrightarrow 2P_2O_5 + 10CO_2$$

$$P_2O_5 + 3H_2O \longrightarrow 2H_3PO_4$$

(c) **Flow chart**

(d) **Process description:** First, calcium phosphate is crushed, meshed, mixed coke and charged into electric furnace. This furnace is acid refractory brick-line stainless steel. Also, compressed oxygen is fed to furnace. The furnace is heated to 1800–2000°C to complete the combustion reaction. After complete reaction, the gaseous mixture containing phosphorus and other impurities is

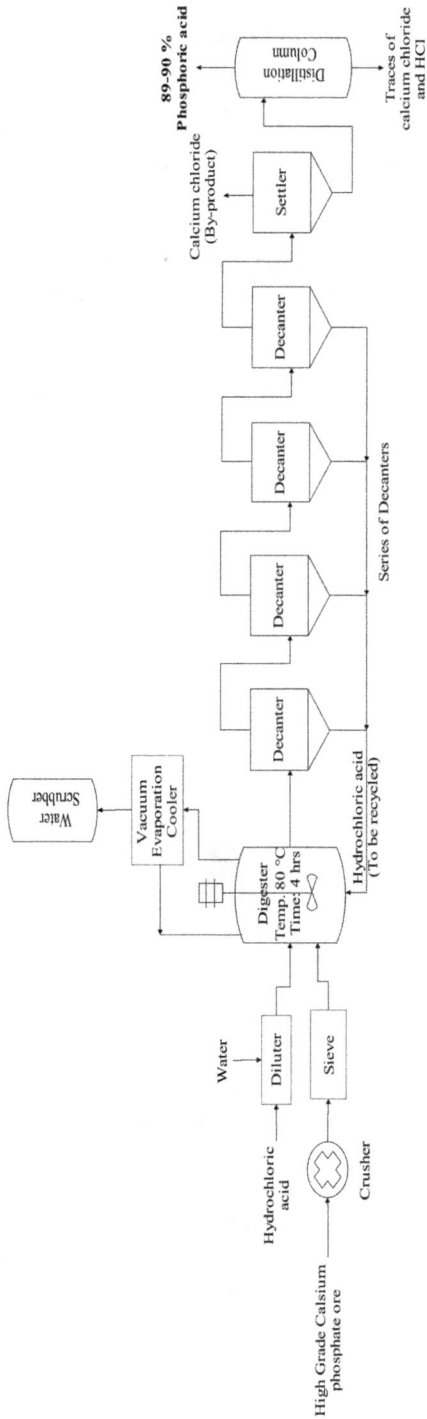

Figure 14.16 Flow sheet diagram of production of Phosphoric acid using Hydrochloric acid

collected from top of furnace. The gaseous mixture is passed through water sprayer to yield P_2O_5. Also, acid mist is formed, which scrubbed in water scrubber. Sulfuric acid and silica are used to eliminate calcium chloride and hydrogen fluoride, respectively, from a mist-free gaseous mixture. Arsenic is finally separated by H_2S being scrubbed against the stream during precipitating. The sludge containing calcium chloride, arsenic sulfide and hydrogen fluoride are removed using sand filter. Finally it is undergone distillation to get desired concentration of phosphoric acid.

(e) **Major Engineering Problems:** Sludge containing calcium chloride, arsenic sulfide and hydrogen fluoride is considered as solid hazardous waste. Removal of acid mist is complicated.

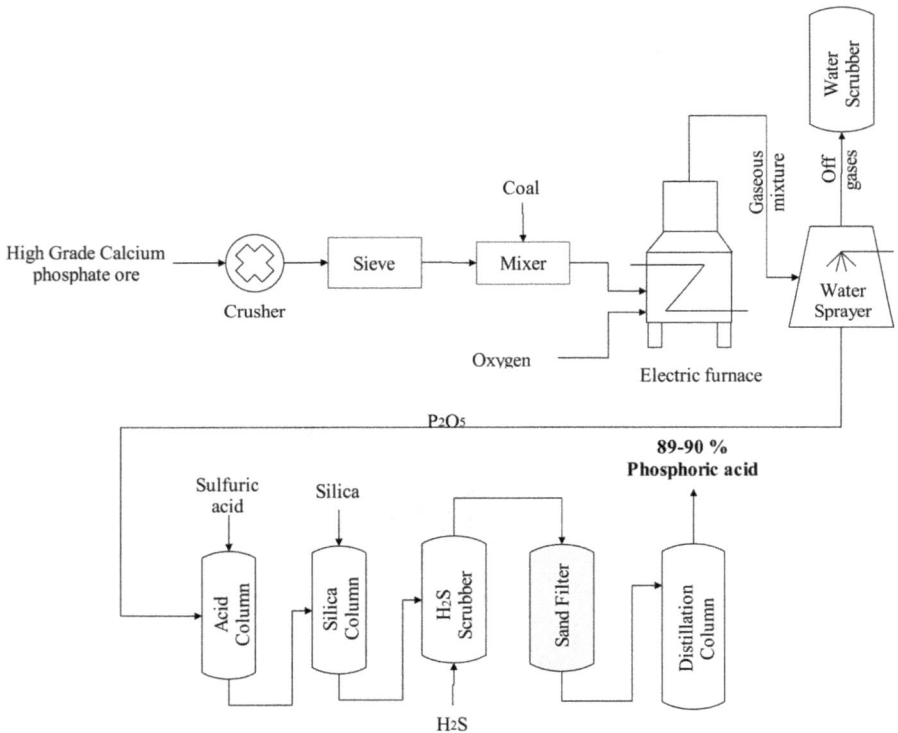

Figure 14.17 Flow sheet diagram of production of Phosphoric acid using Electric furnace process

14.6.1.3 Uses

About 80–90% of production, in the fertilizer industry, phosphoric acid is used. Detergent and cleaning goods also contain it. It is also used as rust removal from metals like iron, steel, etc. It is used as food additive in jam, cereal bars, cheese, and other bakery products. It is also used in personal care products and pharmaceutical industries (Gilmour, 2013; Salas et al., 2017).

14.7 Sulfur

A plentiful, multivalent non-metal, sulfur. Sulfur atoms normally combine to create cyclic octatomic molecules, which have the chemical formula S8. At normal temperature, elemental sulfur is a crystallized solid that is brilliant yellow. In terms of chemistry, all elements except for gold, platinum, iridium, nitrogen, tellurium, iodine, and noble gases react with sulfur. Sulfur has always played a significant role in global business. Due to the Industrial Revolution, there has been a sharp increase in the demand for sulfur, a key ingredient in the production of sulfuric acid, which suggests that overall industrial activity has increased as well. There are various ways to manufacture sulfur, but the most popular and practical approach is the Frasch sulfur mining process.

14.7.1 Frasch process

This mining method was developed by American chemist Herman Frasch, who is of German descent. This method involves boiling water to a temperature of about 170°C (340 °F) and forcing it into the deposit to melt the sulfur, which is then brought to the surface using compressed air. The sulfur has a melting point of about 115°C (240 °F). The 99 percent pure sulfur is then allowed to crystallize after a solution of sulfur and water is poured into bins. Three concentric tubes are concurrently inserted into the sulfur deposit during the Frasch process. Through the outermost tube, superheated water is fed into the deposit at a pressure of 25 atmospheres and a temperature of 160–180°C. Sulfur having melting point of 115°C is melts. As greater density of molten sulfur than water, it sinks and forms a pool around it. Due to the molten sulfur's lower viscosity, water pressure alone is unable to force the sulfur into the surface. Instead, hot air is introduced via the innermost tube to drool the sulfur, making it less thick, forcing it to the face, and collecting it from the middle tube. The sulfur obtained may actually be pure (99.7–99.8). In this form, it's light unheroic in color. However, it can be dark-multicolored; further sanctification isn't profitable, and generally gratuitous, if defiled by organic composites. Also, filtration is used to remove carbonaceous and mineral matter from sulfur. The data shows that United States in 1989 and Mexico in 1991 prepared 3.89 and 1.02 million tonne of sulfur respectively utilizing Frasch process. Deposits between 50 and 800 meters deep can be processed using this method. Every tonne of sulfur must be produced using 3–38 cubic meters of superheated water, which has a high energy cost. Also, sulfur can be prepared by heating hydrogen sulfide with air (oxygen). It is also recovered from pyrite ores using various processes such as Outokumupu flash-smelter process, Orkla process and Noranda process.

14.7.2 Sulfuric acid

The properties of sulfuric acid are mentioned below.

Table 14.16 Physical & Chemical Properties of Sulfuric acid

Sr. No.	Properties	Particular
1	Chemical Name	Sulphuric acid, Oil of vitriol, Dihydrogen sulfate, Battery acid
2	Molecular Formula	H_2SO_4
3	Molecular weight	98.07 gm/mole
4	Appearance	Colorless to dark brown, oily liquid
5	Odour	Odourless
6	Solubility	Alcohol and water are miscible together yet produce a lot of heat and experience volume contraction
7	Melting point	10°C
8	Boiling point	337°C
9	Density	1.84 g/ml at 20°C

14.7.2.1 Manufacturing of sulfuric acid

There are two processes available for manufacturing of sulfuric acid: (1) Chamber process and (2) Contact process. John Roebuck first developed the chamber process in 1746. More economical process, i.e., the contact process was patented by Peregrine Phillips in 1831. In this process, sulfur is burnt to form SO_2 and absorbed in lead-lined chambers. This process is used for dates back about 200 years, but it yield less than 80%. Now a days, contact process is widely used to manufacturing sulfuric acid in the high concentrations needed for industrial processes. In this process, sulfuric acid is produced by Single Absorption Contact Process and Double Contact Double Absorption (DCDA) method, but most of sulfuric acid is manufactured by DCDA method due to increase yield and reduce stack emission of unconverted SO_2.

Manufacturing process of Sulfuric acid by DCDA process:

The process can be performed using following three steps

1. Burning of sulfur in presence of excess of air
2. Conversion (oxidation) of SO_2 gas into SO_3 gas
3. Hydration of sulfur trioxide gas

(A) Raw materials: Following raw materials are used in DCDA process.

1) SO_2 gas is obtained from various sources like,
   a. Sulfur burning
   b. Pyrites roasting containing 40–45% sulfur
   c. Metal sulfide roasting and smelting

d. Sulfuric acid regeneration

e. Metal sulfate roasting

f. Combustion of H_2S or other sulfur-containing gases

g. Other processes

2) Catalyst: For oxidation of SO_2 into SO_3, vanadium pentoxide (V_2O_5) dispersed on s porous carrier in pellet form is used. Previously Platinum was widely used, but it has several disadvantages like less conversion ratio (SO_2 to SO_3), poisonous, fragile, rapid heat deactivation and high initial investment, arsenic impurities. But, vanadium pentoxide is used as catalyst, because it has higher conversation ratio, less poisonous and initial low investment. Vanadium pentoxide is regenerated by following process.

1. Oxidation of SO_2 into SO_3 by V^{5+}

$$2SO_2 + 4V^{5+} + 2O^{2-} \rightarrow 2SO_3 + 4V^{4+}$$

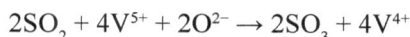

2. Oxidation of V^{4+} back into V^{5+} by oxygen (catalyst regeneration):

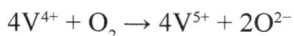

$$4V^{4+} + O_2 \rightarrow 4V^{5+} + 2O^{2-}$$

(B) Chemical reaction

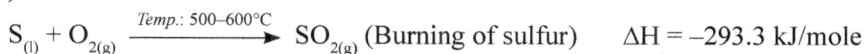

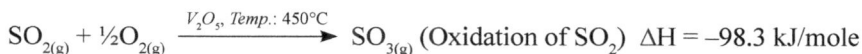

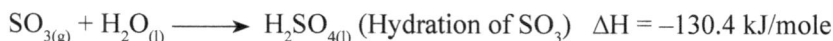

$$S_{(l)} + O_{2(g)} \xrightarrow{\text{Temp.: 500–600°C}} SO_{2(g)} \text{ (Burning of sulfur)} \quad \Delta H = -293.3 \text{ kJ/mole}$$

$$SO_{2(g)} + \tfrac{1}{2}O_{2(g)} \xrightarrow{V_2O_5, \text{ Temp.: 450°C}} SO_{3(g)} \text{ (Oxidation of } SO_2) \quad \Delta H = -98.3 \text{ kJ/mole}$$

$$SO_{3(g)} + H_2O_{(l)} \longrightarrow H_2SO_{4(l)} \text{ (Hydration of } SO_3) \quad \Delta H = -130.4 \text{ kJ/mole}$$

(C) Flow chart

(D) Process description

(1) Burning of sulfur in presence of excess of air: As per flow-chart, molten sulfur as a raw material is taken into storage tank and transferred in sulfur burner. Also, dry air is subjected to dry into drying tower using mixture of oleum and sulfuric acid. So, air contains 7–0% SO_2 gas is transferred in burner. Thereafter, liquid sulfur is burnt in excess of air at temperature of 500–600°C to form SO_2 gas. Hot sulfur dioxide gas filtered, cooled at 420°C and taken into converter. Heat is liberated during combustion reaction. It is necessary to thoroughly purify hot sulfur dioxide using techniques like cooling, dust collection, scrubbing with water, and sulfuric acid drying if source materials like sulfide ores (pyrites) and wasted or sludge acid are employed. Additionally, the reaction produces less heat than burning sulfur does. Sulfur is burned to produce heat that is put to use in waste heat boilers, steam generators, or power plants.

(2) Oxidation of SO_2 gas into SO_3 gas: Cleaned and dried sulfur dioxide at 430°C is passed with air over catalyst pellet beds called converters. The converter consists of four bed of vanadium (V) oxide as catalyst, with

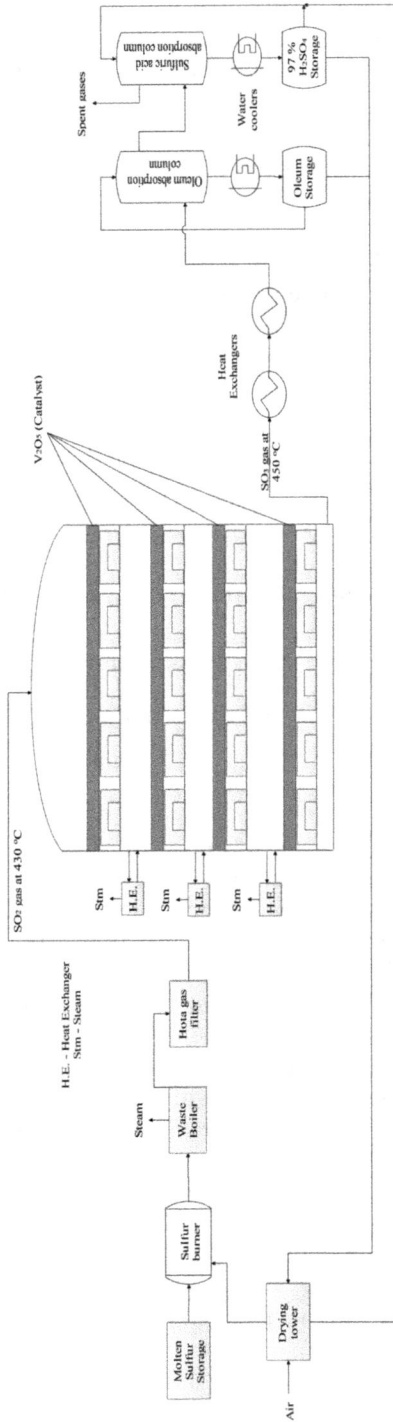

Figure 14.18 Flow sheet diagram of production of sulfuric acid using DCDA process

alkali metal sulfate promoters (mainly potassium or caesium sulfate) on a silica ceramic base in cylindrical type vessels. The catalyst bed may be 8 meter in diameter and 60 cm deep. The temperature of 500–600°C and pressure is normally 2 atmospheres is maintained. Conversion temperature is strictly maintained during the reaction 500–600°C, because a conversion is reversible above 650°C and catalyst is activated at 550°C. Continuous heat removal using heat exchanger during conversion is conducted, because of reaction is exothermic, so, heat is liberated.

At the beginning of the conversion process, SO_2 gas is converted to SO_3 while travelling through several catalyst beds. Both SO_2 and SO_3 are present in the exhaust gases from this step. These gases are routed via an intermediate absorption tower, where sulfuric acid trickles down packed columns and SO_3 combines with water to raise the concentration of sulfuric acid. Even while SO_2 also enters and exits the tower, it does so in an unreactive manner. After the required cooling, this stream of gas containing SO_2 is passed through the catalytic converter bed column once more, achieving up to 99.8% SO_2 to SO_3 conversion, and the gases are again passed through the final absorption column, resulting in not only high SO_2 conversion efficiency but also enabling the production of sulfuric acid with higher concentration.

(3) Hydration of Sulfur trioxide gas: Repeated heat exchanges allow hot sulfur trioxide gas to cool to 150°C. The cooled gas is then absorbed into an oleum-filled column before being followed by 97% sulfuric acid. The H_2SO_4 streams can be controlled to produce acid with concentrations ranging from 91 to 100% H_2SO_4 with varying additions of SO_3 and water. Over 99.5% of sulfur is converted to acid.

(E) Major engineering problems using DCDA process

(a) To maximize SO_3 generation, both stoichiometric and thermodynamic factors are taken into account. According to Le Chateier's Principle, the following variables are taken into account for SO_2/SO_3 systems to maximize SO_3 formation. According to Le Chatelier's Principle, an equilibrium system reacts to stress by regaining balance. Le Chatelier's theories are followed when sulfur dioxide and oxygen react to generate sulfur trioxide.

(i) Temperature: The stage-wise reaction with cooling in between catalyst beds uses the principle of removing heat (i.e., lower temperature) to achieve higher overall conversions. This is the same principle used in an 'isothermal' converter. For that, at each catalytical bed, heat exchange is provided to removing heat from converter.

(ii) Pressure: The impact of pressure is not significant for the standard contact process. However, a pressure sulfuric acid plant was developed and several plants were built using the principle that operating at higher pressures promotes the formation of SO_3, but very high pressure increases the rate of corrosion.

For selection of temperature and pressure for conversion, equilibrium constant for this reaction, calculated from partial pressures according to the law of mass, may be described as,

$$K\rho = \frac{\rho_{SO_3}}{\rho_{SO_2} X \rho_{O_2}^{\frac{1}{2}}}$$

Values of $K\rho$ is determined experimentally at constant pressure and found that maximum equilibrium constant (i.e., maximum conversion) is achieved at 400°C at 1 atmospheric pressure. But, at this temperature, the conversion is very slow. The rate at 500–550°C is 10 to 100 time faster than at 400°C. So, advisable temperature is 500–600°C for maximum conversion.

    (iii) Concentration of Products and Reactants: The removal of SO_3 from the gas mixture to promote higher overall conversions is the principle behind the double absorption plant. Also, some other factor that are also responsible for conversion like, increased oxygen concentration in the input side, removal of SO_3 (double contact double absorption process) from the reaction zone, selection of the catalyst to reduce the working temperature (equilibrium) and increase in pressure of converter.

(b) For a highly exothermic reaction, a multistage catalytic convertor was designed. To encourage better conversion, three or four stage converters are now used instead of the traditional two stage operation.

(c) To increase the space velocity in the catalyst chamber since it affects the reactor's fixed charges or pumping costs.

(d) To avoid the aforementioned issues, thin catalyst beds with a height of 30 to 50 cm are used. When there is insufficient convective gas velocity through the bed, longitudinal mixing might cause yield to decrease.

(e) Removing the heat from SO_3's acidic absorption, instead of pipe coolers with water pouring over the external surface, cast iron pipe with internal fins is designed for better heat transfer.

(f) The required minimal pressure drop necessitates the use of 8 cm stacked packing.

14.7.2.2 Uses

1. The fertilizer sector is the single largest user.
2. Primarily in the synthesis of phosphoric acid, which is used to create fertilizers such triple superphosphate, monoammonium phosphate, and diammonium phosphate.
3. It is used to make ammonium sulfate and superphosphate.
4. It is employed in oil refining for the refinement, alkylation, and purification of crude-oil distillates as well as in organic chemical and petrochemical processes requiring reactions including nitration, condensation, and dehydration.

5. It is also used in the production of hydrochloric acid, hydrofluoric acid, and TiO_2 pigments, among other inorganic chemicals.
6. It is employed in the non-ferrous metal purification and plating business, as well as in the pickling and descaling of steel, the leaching of copper, uranium, and vanadium ores, and the preparation of electrolytic baths.
7. Sulfuric acid is utilized in the production of chemical and textile fibres as well as leather tanning, is necessary for some wood pulping procedures in the paper sector.
8. It is utilized in the production of polymers, detergents, and explosives.
9. It is used in production of dyes, pharmaceuticals and other industries (Apodaca et al., 2017; Metzner and Thuiller, 1994, Kiss et al., 2006; Ashar and Golwalkar, 2013).

14.8 Chlorine

Chlorine gas is considered as second lightest among halogen, after fluorine. First, chlorine gas was produced by Swedish pharmacist, Carl Wilhelm Scheele in 1774 using hydrochloric acid and manganese dioxide in the laboratory. After several decades, English chemist Humphry Davy recognized chlorine as an element in 1810. This gas was first widely used as chemical weapon in World War I by Bruisers. Chlorine is toxic, corrosive, yellow-green gas at room temperature. As it has highest electron affinity, it is highly reactive and strong oxidizing agent. It has third highest electronegativity, after oxygen and fluorine according to Pauling scale. Naturally, chlorine gas is available about 2.0 and 0.017% of total volume of seawater and Earth's crust. Some chlorine minerals are sylvite (potassium chloride-KCl), bischofite ($MgCl_2 \cdot 6H_2O$), carnallite ($KCl \cdot MgCl_2 \cdot 6H_2O$), kainite ($KCl \cdot MgSO_4 \cdot 3H_2O$), sodalite ($Na_8Al_6Si_6O_{24}Cl_2$) and chlorapatite [$Ca_5(PO_4)_3Cl$]. Industrially, it is prepared by Chlor-Alkali process using brine along-with sodium chloride as per 14.2.2. As chlorine is toxic and fatal, it is largely employed as a disinfectant in drinking water, swimming pool water and wastewater. It is also used to disinfect the household area and textile industries. It is mixed with alkali such as caustic soda and prepare bleaching agent. It is used to manufacture variety of chemicals i.e. chloro-carbons (mainly chloroform and chlorofluorocarbon (CFC), carbon tetrachloride, etc.) and PVC. In organic chemistry, it is used oxidizing agent and in substitution reactions. As it is destroyed bacteria, it is used in food industries to prevent bacteria like E. coli, salmonella and a host of other foodborne germs from food. It is also used to prepare certain medicines for treatment of high obesity, arthritis pain and allergic personals. Blood bags, medical equipment, surgical sutures, contact lenses, safety glasses, and respiratory inhalers are all made with chlorine. Chlorine gas is utilized in the production of high-performance magnets, solar panel chips, hybrid car batteries, and home and commercial air conditioning refrigerants (Schmittinger, 2000; Tundo et al., 2016).

14.9 Argon

It is considered as most usable noble inert gas. This gas was first discovered by in 1785 by Henry Cavendish. By performing experiments, it was concluded that about 1% of air was not reacted even most extreme conditions. It was isolated by Lord Rayleigh and William Ramsay in 1894 from air. Argon is the third-most abundant gas, after nitrogen and oxygen. It constitutes about 0.934% and 0.00004% of the Earth's atmosphere and crust respectively. For industrial use, it is isolated by cryogenic fractional distillation of air. It is colorless odorless noncombustible gas. It is mildly soluble in water and roughly 1.4 times as heavy as air. Argon is used to develop inert blanket, when required. It is used in food and drink industry along-with nitrogen to prevent food and drink from souring and oxidation, especially fermented beverages like wine, liquor, beer, etc. It is used in neon and fluorescent tube in lighting. As argon isotope (^{40}Ar) is prepared by decay of ^{40}K by electron capture or positron emission, it is used for measurement the oldness of rocks using potassium-argon dating. It is used in production of titanium and other reactive metals. It is called as shield gas, as it is used during arc shielding of specially alloy. It is applied in welding of automobile frames, mufflers and other automotive parts. It is employed for preservation of old documentation to avoid deterioration effect. It is also used in 3-D printing and heat treating chemical processes (Xu et al., 2006; Sismanoglu and Pessoa, 2013).

References

Acker, C. 1912. The Manufacture of Sodium Cyanide. Industrial & Engineering Chemistry, 4(7): 552

Alalq, I., Gao, J. and Wang, B. 2019. Physical and Chemical Properties of Phosphorus, In book: Fundamentals and Applications of Phosphorus Nanomaterials. DOI: 10.1021/bk-2019–1333.

Apodaca, L.E., d'Aquin, G.E. and Fell, R.C. 2017. Sulfur and Sulfuric Acid, In book: Handbook of Industrial Chemistry and Biotechnology.

Ashar N.G. and Golwalkar, K.R. 2013. Cold Process of Manufacturing Sulfuric Acid and Sulfonating Agents, In book: A Practical Guide to the Manufacture of Sulfuric Acid, Oleums, and Sulfonating Agents. https://doi.org/10.1007/978-3-319-02042-6

Bahrle-Rapp, M. 2007. Sodium Sulfite and Sodium hydrosulfite, In book: Springer LexikonKosmetik und Korperpflege.

Brescia, F., Arents, J., Meislich, H., Turk A. and Weiner, E.R. 1975. Preparation and Reactions of Sodium Bicarbonate and Sodium Carbonate, In book: Fundamentals of Chemistry: Laboratory Studies.

Constant, S. and Lacour, J. 2005. New Aspects in Phosphorus Chemistry V, In Topics in Current Chemistry 250, Springer-Verlag Berlin Heidelberg. https://doi.org/10.1007/b98358

Czerwinski, F. 2011. Magnesium Alloys-Design, Processing and Properties, InTech Publication. DO: 10.5772/560

Friedrich, H. 2004. Magnesium Technology, Springer.

Garrett, D.E. 2001. Sodium Sulfate. Handbook of Deposits, Processing, Properties, and Use, Academic Press. https://doi.org/10.1016/B978-0-12-276151-5.X5000-1

Gilmour, R. 2013. Phosphoric Acid: Purification, Uses, Technology, and Economics, CRC Press. https://doi.org/10.1201/b16187

Huebner, J. and Wootton, E. 2008. The Action of Caustic Soda on Cotton. Journal of the Society of Dyers and Colourists 41(1): 10–18.

Ismail, G.A., Abd El-Salam, M.M. and Arafa, A.K. 2009. Wastewater reuse in liquid sodium silicate manufacturing in Alexandria, Egypt. Journal of the Egyptian Public Health Association 84(1–2): 33–49.

Kainer, K.U. 2004. Magnesium: Proceedings of the 6th International Conference Magnesium Alloys and their Applications, Wiley-VCH.

Kamath, A., Bijle, M.N.A and Patil, V. 2013. Sodium Hypochlorite-A Review, Journal of Dental & Oro-facial Research 9(2): 49–532.

Kaufman, J.G. and Rooy, E.L. 2004. Aluminum Alloy Castings: Properties, Processes And Applications, ASM International.

Kiss, A.A., Bildea, C.S. and Verheijen, P.J.T. 2006. Optimization studies in sulfuric acid production. Computer Aided Chemical Engineering 21: 737–742. https://doi.org/10.1016/S1570-7946(06)80133-1

Kumar, A., Du, F. and Lienhard, J.H. 2021. Caustic Soda Production, Energy Efficiency, and Electrolyzers. ACS Energy Letters 6(10): 3563–3566. https://doi.org/10.1021/acsenergylett.1c01827

MacKenzie, D.S. and Totten, G.E. 2005. Analytical Characterization of Aluminum, Steel, and Superalloys, CRC Press.

Metzner P. and Thuiller A. 1994. Sulfur reagents in organic synthesis, Hayka Publications.

Payuhamaytakul, K., Jitareerat, P., Uthairatanakij, A., Srilaong, V. and Renumarn, P. 2018. Effectiveness of sodium chlorite and acidified sodium chlorite to inhibit mesocarp browning of trimmed aromatic coconut, Acta Horticulturae. DOI: 10.17660/ActaHortic.2018.1210.11

Rogers, J.M. and Porter, H.F. 1987. US Patent No.: US4847062A, Process for production of sodium cyanide.

Rojas, R., Camara-Martos, M.F. and Lopez, M.A.A. 2015. Sodium: properties and determination, In book: The Encyclopedia of Food and Health, 1st Edition.

Ronco, C. and Mishkin, G.J. 2007. Disinfection by Sodium Hypochlorite: Dialysis Applications, Karger Publishers. DOI: 10.1159/isbn.978-3-318-01409-9

Bedekar, S.G. 2007. Properties of sodium carbonate-sodium bicarbonate solutions. Journal of Applied Chemistry 5(2): 72–75. https://doi.org/10.1002/jctb.5010050206

Salas, B.V., Wiener, M.S. Martinez, J.R.S. 2017. Phosphoric Acid Industry: Problems and Solutions, Intechopen Publication. DOI: 10.5772/intechopen.70031

Schmittinger, P. 2000. Chlorine: Principles and Industrial Practice, Willey-VCH.

Searle, T.M. 1998. Properties of Amorphous Silicon and its Alloys, Institution of Engineering and Technology.

Singh, R., Oprysko, M.M. and Harame, D. 2003. Silicon Germanium: Technology, Modeling, and Design, Wiley-IEEE Press.

Singh, Y., Singh, N.K. and Ram, M. 2022. Advanced Manufacturing and Process Technology, 1st Edition, CRC Press.

Sircus, M. 2010. Sodium Bicarbonate, International Medical Veritas Association.

Sismanoglu, B. and Pessoa, R. 2013. Argon: production, characteristics and applications, Nova Science.

Synowiec P.M. and Bunikowska, B. 2005. Application of Crystallization with Chemical Reaction in the Process of Waste Brine Purifying in Evaporative Sodium Chloride Production. Industrial & Engineering Chemistry Research 44(7): 2273–2280. https://doi.org/10.1021/ie040118e

Totten, G.E. and MacKenzie, D.S. 2003. Handbook of Aluminum: Volume 2: Alloy Production and Materials Manufacturing, Marcel Dekker Publication.

Tundo, P., He, L., Lokteva, E. and Mota, C. 2016. Chemistry Beyond Chlorine, Springer International Publishing. https://doi.org/10.1007/978-3-319-30073-3

Xu, J., Malek, A. and Farooq, S. 2006. Production of Argon from an Oxygen–Argon Mixture by Pressure Swing Adsorption. Industrial & Engineering Chemistry Research 45(16): 5775–5787. https://doi.org/10.1021/ie060113c

Young, J.A. 2007. Sodium Sulfate. Journal of Chemical Education 84(8).

Zhigach, S.A., Arkhipov, D.G., Lezhnin, S.I. and Usov, E.V. 2012. Development of a unified library of sodium properties. Thermal Engineering 59: 321–324. https://doi.org/10.1134/S0040601512040118

Zsombor-Murray, E. 2021. Sodium Chloride Balance, In book: Critical Care. https://doi.org/10.1201/9781315140629

CHAPTER 15

Miscellaneous Industries

15.1 Sugar Industries

15.1.1 Introduction

The production, processing, and selling of sugars are all included in the sugar industry (mostly saccharose and fructose). Sugar is a key ingredient in soft drinks, sweetened beverages, fast food, convenience meals, candy, sweets, confectionery, baking goods, and other related products. India is the world's largest grower of sugarcane and the second largest producer of sugar after Brazil (14% of worldwide production). Brazil accounts for about 22% of the world's sugar production, and India for nearly 14%. Sweet, soluble carbohydrates, many of which are found in food, are collectively referred to as sugar. Sugar comes in a variety of forms and is sourced from numerous places. Simple sugars are monosaccharides that contain glucose (also known as dextrose), fructose, and galactose. The properties of D-Sucrose are as follows.

Table 15.1 Physical & Chemical Properties of D-Sucrose

Sr. No.	Properties	Particular
1	IUPAC name	Beta-D-arabino-hex-2-ulofuranosyl alpha-D-gluco-hexopyranoside
2	Chemical and other names	D(+)-Sucrose, Saccharose, Cane sugar, Table sugar, White sugar, Saccharum, Amerfand
3	Molecular formula	$C_{12}H_{22}O_{11}$
4	Molecular weight	342.297 g/mol
5	Physical description	White odorless crystalline or powdery solid
6	Solubility	Highly soluble in water, methanol; slightly soluble in ethanol; insoluble in ethyl ether.
7	Melting point	185.5°C
8	Boiling point	Decomposed and gives characteristic caramel odor when heating
9	Density	1.5805 kg/m³at 17°C

15.1.2 Manufacturing process

70% of the sugar produced worldwide comes from sugar cane, primarily grown in tropical regions, and 30% primarily grown in lower temperate regions like the U.S. or Europe.

15.1.2.1 From Sugar cane

The grass family includes sugarcane, which contains a variety of spices, such as *Saccharum officinarum, Saccharum spontaneum, Saccharum sinense, Saccharum barberi, Saccharum robustum*, etc. It grows in warm and tropical climates. It has a bamboo like stalk, grows from 3 to 5 m height and contains 11 to 15% sucrose by weight. The manufacturing process is as follows.

(A) **Raw materials:** Sugar cane, Lime, Phosphoric acid, Carbon dioxide, Sulfur dioxide, Bone char and Hot water

(B) **Quantitative requirement:** 1 tonne of raw sugar (97% sucrose-95% yield) required 6.0–10.5 tonne sugarcane having 10 to 16% sucrose, 12–17 kg lime, and 6–10 kg sulfur dioxide.

(C) **Flow sheet diagram**

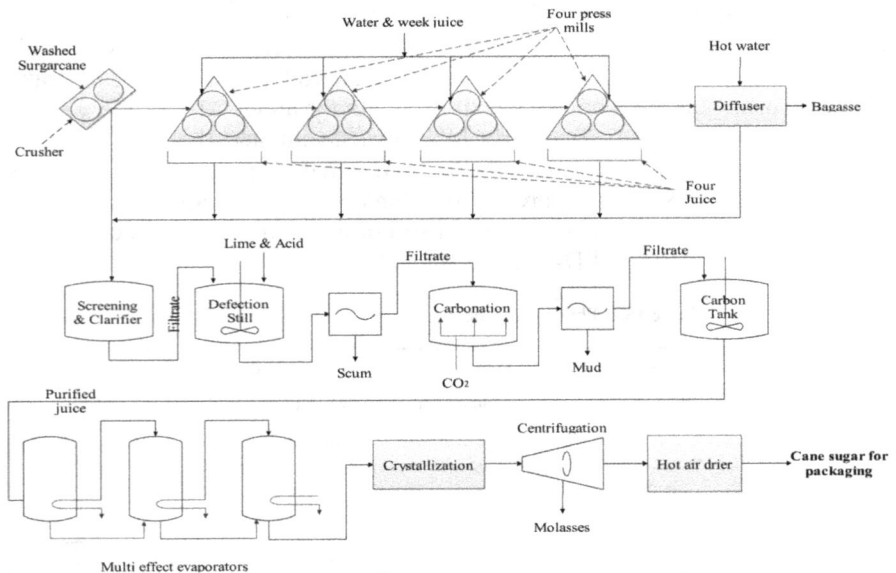

Figure 15.1 Flow sheet diagram of production of cane sugar from cane

(D) **Process description**

(a) **Juice Extraction:** The cane is first washed to remove mud and debris. Then, it is divided into little pieces by a cutter that rotates at a speed of 400 to 500 rpm while employing knives mounted on a cylindrical shaft. Knives are used to slice up the surgarcane. To extract the juice, canes are crushed and then shredded. Two rollers rolling in the opposite directions make up a crusher.

To extract the most juice, crushed canes are run through four pressure mills. Three cast iron rolls make up each pressure mill. The width of the rolls narrows as they go from the first to the last. To extract as much as possible, water and weak juices are employed. In cane, between 85 and 95 percent of the juice is extracted. Remaining juice is extracted using hot water in diffuser. Cellulosic material discharged from the diffuser is called "Bagasse" and is used a fuel in boilers.

(b) Juice Purification: The raw juice is a dark brown liquid that contains 15% sucrose, a minor amount of glucose, fructose, vegetable proteins, mineral salts, organic acid, coloring agents, gums, and fine bagasse particles suspended in the mixture. The juice is screened and clarified to remove floating impurities. The juice is collected from the bottom of clarified and transferred into steam jacketed defection steel tank with facility of agitator. Excess amount of lime solution (2–3%) as coagulant is charged into it and heated by means of high pressure stream in tank. In this process, vegetable proteins are coagulated and the organic acids are neutralized. These form precipitates of insoluble calcium salts called scum. The reaction mass is collected from the bottom of steel tank and subjected to filtration to remove scum. The filtration is collected and transferred into carbonation tank. In this tank, defecated juice is passed through carbon dioxide, in which excess lime is removed in form of calcium carbonate (called as mud) by means of filtration. The carbonated filtrate is decolorized with sulfur dioxide or bone charcoal, in which dark brown is bleached and produce with much lighter color. It completely removed lime to form calcium sulfite. Also, it coagulates hums and albuminoids. The juice is filtered to remove the precipitates.

(c) Concentration and Crystallization: The multi-effect evaporator receives this sugar solution. The 3–4 forward feed multi-effect evaporator receives the excess of the cleared liquor. Juice is concentrated at this location from 80–85% H_2O to 40% H_2O to prepare it for crystallization. The crystallizer is fed the clarified concentrated thick juice. It is once more cooked in vacuum pans here until a fine cloud of crystals is visible at a vapour temperature of 50–7°C. The masscuite, also known as the crystallizer, is a mixture of crystals and syrup. To separate mother liquor (molasses) from sugar crystals, it is then centrifuged in a basket type centrifuge at a speed of 800 to 1000 rpm. Drying is applied to the raw sugar to produce light brown sugar.

Also, by-product are very important, i.e., bagasse and ethanol from molasses. About 3 tonne of bagasse is produced from 10 tonne of raw materials, sugar cane. Bagasse is used to get energy by burning. Energy given by burning 1 tonne of bagasse is equivalent to 1 barrel crude oil. Further, 163 gallons of ethanol are theoretically produced for every tonne of sugar. The predicted actual recovery would be roughly 141 gallons per tonne of sugar when practical plant operations and the greatest yield possible are taken into account.

(E) Major Engineering Problems: Following major engineering problems are facing towards manufacturing sugar from sugarcane.

(a) **Extraction of juice from cane:** The optimization with design of roller with temperature and time are essential that the maximum juice is extracted from sugarcane and trace of sugar is remaining in baggage. Now a days, ultrasonic vibration is preferable.

(b) **Choice of flocculation agents:** For complete precipitation and fast filtration, there are more than 700 agents has been investigated. High magnesia lime is the one of the oldest choice because of it is readily available and economical. Also, carbon dioxide as carbonation is used in carbonator to reduce alkalinity, improve filtration process and decolonization process.

(c) **Evaporation and crystallization:** For evaporation, especially design Calandia-type evaporators are utilized to saturate the dilute juice. The reason behind that sucrose is not crystallize when saturated solution is obtained or supersaturate is reached. In the multi-effect evaporator, crystallization is done batch by batch, and supersaturate is measured by boiling-rise, vacuum measurement, and control. There are two crucial zones of supersaturation, and their resolution is as follows:

  (A) Transition region-At the greatest point of saturation, new nuclei are formed with the addition of powdered sugar seed.

  (B) Metastable region-By reducing the vacuum, this region is conquered, and the crystal will develop. At this stage, the total amount of crystals generated in one batch can be increased by adding feed liquid while evaporation is also occurring simultaneously.

(d) **Separation of crystals from syrup:** The older centrifugal machines are having 1000–1200 rpm is replaced by new designed automatic discharging machine having high speed 1800–2400 rpm. Viscosity and surface tension of syrup is considerable parameters for obtaining clean and rapid separation.

(e) **Inversion of sugar:** In handling sugar, some inversions take place according to the following reaction:-

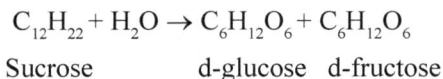

$$C_{12}H_{22} + H_2O \rightarrow C_6H_{12}O_6 + C_6H_{12}O_6$$

Sucrose   d-glucose   d-fructose

The product is called "Inverted sugar"

During the washing, the degree of inversion is measured by a polarimeter. If the value is $+ 97\,°$, i.e., no inversion and if it is $-20\,°$, completely invert sugar. The inversion is reduced by shading cutting and making quick delivery to the sugar cane process (Singh, 2015; Solomon, 2016).

15.1.1.2 From Sugar beet

The flat crown of the sugar beet is topped by a conical, white, sugar root. Sugar beets are processed in factories close to the farms and contain 16–18% sucrose. Because sugar beets are cultivated and harvested according to the seasons, factories typically run for four to seven months at a time. Sugar beet can be cultivated in salty soils where other crops cannot since it is a salt-tolerant crop.

In addition to being a source of sugar, sugar beets also produce ethanol, animal feed, and vitamin B10.

(A) Raw materials: Sugar beet, Slaked lime, Carbon dioxide, Sulfur dioxide, Bone char and Hot water

(B) Quantitative requirement: 1 tonne of raw sugar (97% sucrose–95% yield) required 8 tonne sugar beet having 16 to 18% sucrose, 0.35 tonne slaked lime, 0.8 kg sulfur dioxide, 0.9 tonne water and 1,100 kHh of electricity.

(C) Flow sheet diagram

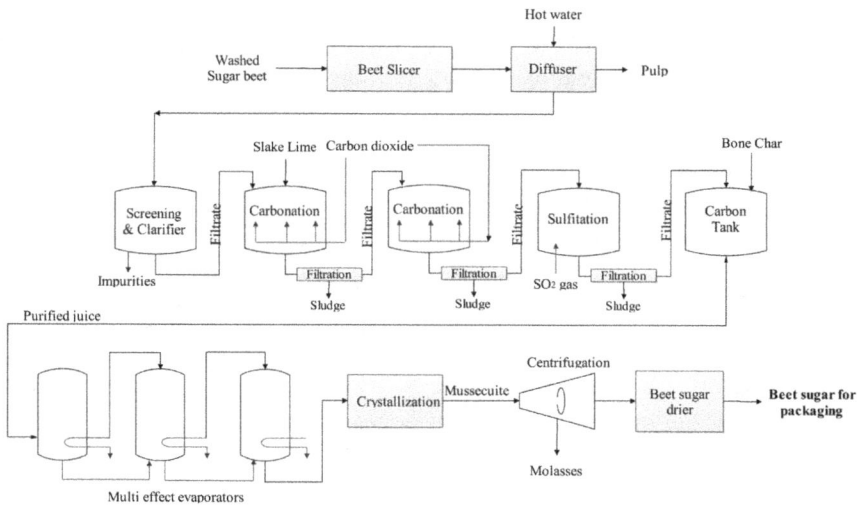

Figure 15.2 Flow sheet diagram of production of beet sugar from beet

(D) Process description

(a) Juice Extraction and Purification: The process involves washing and slicing sugar beets into cossettes, which are thin strips. The raw juice is taken from the cassettes while they are being processed by a diffuser, a specially constructed continuous countercurrent tank. The cossettes are gently lifted from the bottom to the top of the diffuser. Hot water at 70 to 80°C washes over them to absorb the sugar from beet-root. The beet pulp is what remains after sucrose-containing, uncooked, black-blue juice has been removed. Separately, this pulp is processed to create pellets for use in cow feed. The amount of sucrose in the raw juice is between 10 and 12 percent, and there are also trace amounts of glucose, fructose, vegetable proteins, mineral salts, organic acids, coloring agents, gums, and beetroot dispersed in fine particles. The juice is screened and clarified to remove floating impurities. The filtrate is collected from the bottom of clarifier and transferred into carbonation tank, in which slaked lime suspension is charged in it. Also, carbon dioxide gas is passed through the bottom of the tank. In this tank, all impurities are precipitated as sludge. The juice is collected from the bottom and subjected to filtration to remove sludge. The process of passing lime and gas; and filtration is repeated again. The resulting filtrate is collected in sulfitation, where sulfur dioxide as bleaching agent is passed thought from the bottom of tank. Here, lime and some impurities are precipitated and juice is partially bleached. The juice is collected from the bottom and subjected to filtration to remove precipitates. The filtrate is completely decolorized with bone charcoal, in which dark brown is bleached and produce with much lighter color.

(b) Concentration and Crystallization: Concentration of sugar in juice is increased by passed through multiple-effect evaporators and filtered. After feeding the resulting white, viscous fluid into the vacuum pan, crystals start to form. Massecuite is the name for the ensuing syrup and sugar crystal mixture. To extract the molasses, the mixture is centrifuged. It is then dried and packaged according to specifications. Molasses is utilized in this process to create ethanol.

Average yields from a tonne of sugar beet are 160 kg of sugar, 500 kg of wet pulp, and 38 kg of molasses.

(E) Uses: In addition to being used as a sweetener in food and soft drinks, syrup production, invert sugar, confections, jams and preserves, demulcents, and pharmaceutical items, sucrose is also used to make caramel. Detergents, emulsifying agents, and other sucrose derivatives can all be produced using sucrose as a chemical intermediary. Sucrose is a common component of plants' seeds, leaves, fruits, flowers, and roots, where it serves as a carbon source for biosynthesis as well as a source of energy for metabolism. Sucrose is used in food items as a sweetener, preservative, antioxidant, moisture-controlling agent, stabilizer, and thickening agent (Mall et al., 2012; Duraisam et al., 2017; Lambourne and Strivens, 1999).

15.2 Coating Industries

15.2.1 Introduction

The surface of an object used for coating is normally known as substrate. Coatings may be applied for practical, aesthetic, or both reasons. The actual coating may entirely cover the substrate or it may merely cover a portion of the substrate. A product label on many beverage bottles serves as an illustration of all of these coating kinds; on one side, the adhesive is entirely practical, and on the other, one or more ornamental coatings are printed in a suitable pattern to create the text and graphics. According to its intended use, coatings are divided into four categories.

15.2.1.1 Protective coating

Coating that guards against mechanical risks as well as the negative effects of the environment, primarily the atmosphere and chemicals. In addition to being anti-corrosion, these coatings might also be harder, more resistant to tribological wear, or have a more appealing appearance.

15.2.1.2 Decorative coating

The main purpose of decorative coatings, often known as ornaments, is to give the metal or non-metal object a beautiful external appearance. This is dependent on the coating's surface finish (hammer finish, webbling, crystal, crocodile skin), color, lustre, and resistance to tarnishing in addition to its shine attributes (fluorescence, phosphorescence, radioactivity).

15.2.1.3 Protective-decorative coatings

This kind of coating serves to both give the surface an appealing appearance and safeguard the object from corrosion and minor mechanical harm. Powder Coating Industries has a protective and adorning role in electrolytic coatings.

15.2.1.4 Technical coatings

Technical coatings are used to provide an object certain mechanical, electrical, and thermal qualities.

Simple classification of coating by material is metallic and non-metallic coating. Another classifications coating according to their manufacturing methods are galvanizing, immersion, spray, cladded coatings and crystallizing coatings. Most of coating technology involved the paint, powder and varnish as protective-decorative coating (Rafay and Singh, 2020).

15.2.2 Paint industries

Paint is a mixture of finely divided pigments dispersed in a liquid that is made up of a pigment suspended in an oil or water-based liquid or paste carrier. Paint is made chemically from a combination of pigment, resin, solvent, and a few additions. Using a brush, roller, or spray gun, paint is applied to a variety of surfaces, such as wood, metal, or stones. Although protecting the surface is its main function, it also serves as decoration.

15.2.2.1 Manufacturing process

It is manufactured as follows.

15.2.2.2 Raw materials

The preparation of the paints involves the use of three primary components: pigments, binders, and solvents. Additionally, additional compounds are added to give the product some extraordinary qualities. The following is a list of the primary paint ingredients and their uses.

(A) Pigment: The purpose of the pigment and filter (paddings) is to simply produce a multi-colored face that is aesthetically pleasant. In order to extend the life of the cosmetics and protect the essence from eroding, solid patches in makeup reflect light shafts. An inorganic compound that is comparable to titanium dioxide, worlds, lead, and zinc hues. A color is a wholly biological, irreversible hue. An inorganic carrier, such as aluminum hydroxide, barium sulfate, or complexion, was drenched in an organic color. One of the essential and crucial components of makeup is color. Colors should generally be opaque to ensure good covering power and chemically inert to ensure stability, leading to a long life. Colors should be harmless, or at the very least very low in toxicity, to both the painter and the people living in the space. Eventually, the chemicals that produce the film must moisten the hues, and they must also be inexpensive. The covering power per unit weight of various hues varies.

(B) Vehicles (Binders): The vehicle is the term used to describe the liquid element of the paint. They are either canvas or resin. Its purpose is to secure the pigment to the support. Its purpose is to dissolve the binders, adjust the makeup density, and give the carpeting face a uniform, consistent, and invariant consistence. Solvents and thinners such petroleum ether, toluene, and xylene are employed. Volatile and nonvolatile components make up vehicles: The use of nonvolatile vehicles varies depending on the type of paint.

(a) Oils, resins, driers, and additives are used in solvent-based paints.

(b) Lacquers paint: celluloses, resins, plasticizers, and additives (for coating wood).

(c) Styrene-butadiene, polyvinyl acetate, acrylic, various polymers and emulsions, copolymers, and additives are used in water-based paints.

(d) The most common volatile transporters include alcohol, aromatics, aliphatics, ketones, esters, and esters.

(C) Fillers: As a pigment extender, fillers serve to lower paint costs and regulate the viscosity (rheological qualities) of paints. Fillers include clay, talc, gypsum, and calcium carbonate.

(D) Driers: Dryers like cobalt, lead, zinc, zirconium, manganese, calcium, and barium are added to paint to speed up the drying process.

(E) Anti-skinning agents: In order to prevent the paint's surface from solidifying during storage, it is used next to the paints (unsaturated).

(F) Anti-settling agents: Anti-settling agent is used into paint to increase the efficiency with which pigments are dispersed into the vehicle and to stop pigments from settling while being store.

(G) Plasticizers: Specialized oils, phthalate esters, or chlorinated paraffins make up these substances to increase the flexibility of paint films and lessen their propensity to break.

(I) Other Raw Materials: For a specific purpose or application, dispersants, wetting agents, fire retardants, anti-floating, anti-foaming, etc. are also added to the paint. To increase the lifespan of water-based paints, preservatives are also utilized in their production. Between batches, use water-alkali solutions and solvents to wash and clean equipment. Antiseptics and detergents for cleaning floors are utilized.

15.2.2.3 Flow chart

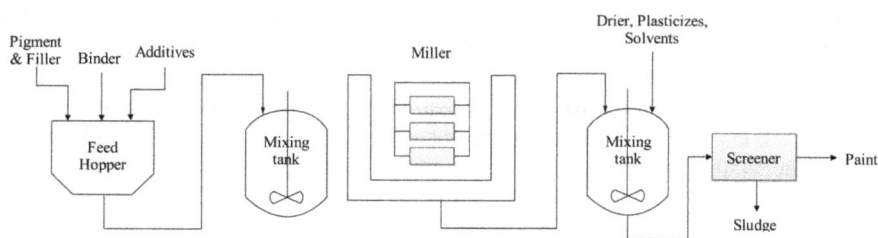

Figure 15.3 Flow sheet diagram of production of paint

15.2.2.4 Process description

In the process, resins, oils and pigment are taken in feed tank in desired proportion and thereafter it is transfer in mixing tank with an agitator. Vigorous mixing of all constituents is conducted in mixing tank. This batch paste is collected from the bottom of mixing tank and undergoes milling/ grinding process. Recently, high-speed agitator and high-speed stone mills, steel roller mills, ball-and-peddle mills, and sand mills have all been employed for proper grinding. After mixing and grinding the batch mass, it is transferred to another tank. Here plasticizers, drier and solvents are charged with stirring in this tank. Thereafter, non-dispersed pigments as sludge are removed using screening, centrifugal, or pressure filter technique. After being poured into cans or drums, the paint is tagged, packaged, and taken to storage.

15.2.2.5 Uses

It is used for interior and exterior home painting, as well as for protecting and enhancing furniture, boats, cars, planes, appliances, and a variety of other objects (Smith, 2010).

15.2.3 Powder coating industry

A type of coating known as powder coating is applied as a dry, free-flowing powder. Powder coating is chemically composed of resin, pigment, hardener, and filler. The main difference between a powder coating and a traditional liquid paint is that the powder coating doesn't need a solvent to keep the binder and filler components in a liquid suspension form. The coating is normally electrostatically deposited, and after being heated during curing, it flows and forms a "skin." Thermoplastic or thermoset polymers could make up the powder. It is typically employed to produce a hard finish that is more durable than regular paint.

15.2.3.1 Manufacturing process

The production of powder coatings involves several steps, namely batching, premixing, extrusion, and particle size reduction.

15.2.3.2 Raw materials

Raw materials and their applications are mentioned below.

(A) Resin: Two main categories of resin are used: thermoset and thermoplastic. The thermosetting resin incorporates a cross-linker into the formulation. When the powder is baked upto 150–200°C, it reacts with other chemical groups in the powder to polymerize. This coating is not become weak if they are exposed to high temperatures. Epoxy, polyester, polyester-urethane, and acrylate are some of the resins found in thermosetting powders. When applied, thermoplastic resin melts and flows into a single layer (the coating). If this coating is heated again, it will lose its strength. PVC and polyethylene are two examples.

(B) Hardener: This component of powder formulas is among the most crucial. The hardener cures the powder and, along with the resin, determines the unique qualities of the coating, including chemical resistance, thermal stability, physical and mechanical capabilities.

(C) Pigment: This substance adds color. Both organic and inorganic pigments exist. A decision is made between pigments that are light-fast, food-contact safe, or temperature stable depending on the application.

(D) Filler: Powder coatings use fillers to make a positive, technical contribution to the system. Filler can be used in a variety of ways, such as to improve abrasion resistance, resist sagging out of the film, and resistance to humidity.

(E) Additives: The powder coating process, which is described below, also uses some additives.
  (a) Accelerators-To hasten the curing system's reaction
  (b) Degassing additives-To allow trapped gases to escape via the coating's pores by widening them longer)
  (c) Thixotropic additives: viscosity regulator to prevent sagging outs.
  (d) A surface flow-improving flow-control agent
  (e) A matte-reducing ingredient, which lowers the gloss of powder coatings.

15.2.3.3 Flow sheet diagram

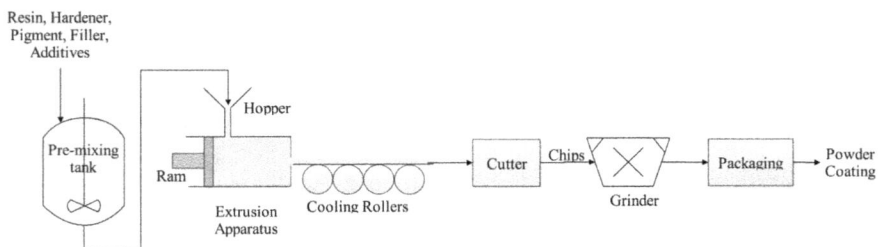

Figure 15.4 Flow sheet diagram of production of powder coating

15.2.3.4 Process description

As per simple diagram, all the raw materials, i.e., resin, hardener, pigment, fillers and additives are added in pre-mixing reactor at pre-required weight. After preparing homogenous mixture, the mass is collected from the bottom of reactor; and transferred into extrusion apparatus through hopper. Here, raw materials are heated to a melting point and combined in an extruder device while being compressed by a spiral driver. Then, the homogeneous melted mixture is come out of the extruder by pressuring the ram in controlled conditions. Thereafter, extruded material is cooled out upto room temperature using cooling belt. It is then cut to prepare chips. Finally, it is grinded and packed.

15.2.3.5 Uses

It is typically used to coat metals, including those found in household appliances, aluminum extrusions, drum hardware, and components for cars and bicycles. Other materials, such MDF (medium-density fiberboard), can now be powder coated in a variety of ways thanks to new technologies.

Advantages of powder coating

(A) Powder coatings release minimal to no volatile organic compounds (VOC) into the atmosphere and don't contain any solvents. As a result, finishers are no longer required to purchase expensive pollution control machinery. The American Environmental Protection Agency's laws can be complied with by businesses more easily and affordably.

(B) Compared to traditional liquid coatings, powder coatings may provide substantially thicker coatings without running or sagging.

(C) Compared to liquid coated goods, powder coated items often exhibit less visual variation between vertically and horizontally coated surfaces.

(D) Powder coatings make it simple to create a variety of unique effects that other coating techniques make impossible (Adigwe et al., 2022).

15.3 Starch Industries

15.3.1 Introduction

A polymeric carbohydrate known as starch or amylum has many glucose units connected by glycosidic linkages. Most green plants synthesize this polysaccharide as a source of stored energy. Potatoes, wheat, maize (corn), rice, and cassava are examples of common foods that contain substantial amounts of the most prevalent type of carbohydrate found in human diets. A white, flavorless, and odorless powder known as pure starch is insoluble in both cold water and alcohol. It is made up of two different kinds of molecules: branched amylopectin and linear and helical amylose. Depending on the plant, starch often has a weight ratio of 75 to 80% amylopectin and 20 to 25% amylose. Animals' stores of glucose are called glycogen, which is a more branched form of amylopectin.

15.3.2 Manufacturing process

Starch can be found in a variety of foods, including conventional maize, waxy maize, high-amylose maize, cassava, potatoes, wheat, rice, including waxy rice, peas, sago, oats, barley, rye, amaranth, sweet potatoes, and other exotics that are grown locally. In US and European countries, starch is prepared from maize and cassava, while in tropical zone like Asian countries, starch is prepared from potatoes, wheat and rice. Their wastes are also utilized to extract starch.

15.3.3 Flow sheet diagram

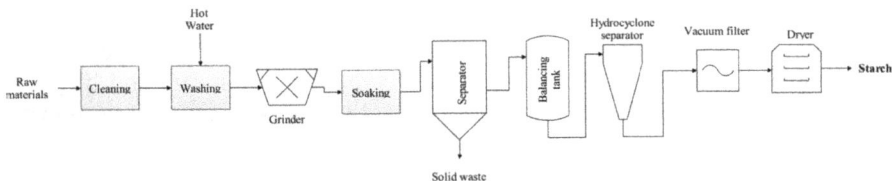

Figure 15.5 Flow sheet diagram of production of Starch

15.3.4 Process description

Raw materials such as maize, cassava or potatoes are cleaned, washed with hot water and grinded. It is then soaked for sometimes. Thereafter, it is separated in separator to remove solid waste. It is then undergone balancing tank to retention the sample. The solid starch is separated into hydrocyclone separator. Further, it is vacuum filter and dried to yield pure solid starch.

15.3.5 Uses

A wide variety of foods use starch for a number of functions, such as thickening, gelling, providing stability, and substituting for or extending more expensive ingredients. Because of their accessibility, relative affordability, and special qualities, starches are preferred. It is frequently used in meals including puddings, custards, soups, sauces, gravies, pie fillings, and salad dressings as a thickening and stabilizer. Additionally, it is utilized in the construction, textile, paperboard, cosmetic, and pharmaceutical industries.

15.4 Soap & Detergent Industries

Detergent and soap are compounds that dissolve in water and are used to clean surfaces like human skin, fabrics, and other substances.

15.4.1 Introduction

Chemically, soap is a long aliphatic chain-attached carboxylic acid that is sulfonate group with potassium or sodium. In contrast, detergent is an alkyl chain with a lengthy sulfonate group at the end that has been sulfonate group with potassium or sodium. As a surfactant, soap aids in the emulsification of oils in water by lowering the surface tension between a liquid and another substance. Water that is hard is soluble in detergent. The sulfonate group's inability to bind to the ions in hard water is thought to be the cause of its solubility. While detergent is made using synthetic derivatives, soap is made from natural sources like vegetable and animal oils. In addition, the chemical structures of soap and detergent are identical. Both a water-insoluble hydrophobic group and a water-soluble hydrophilic group must be present in their molecules. The hydrophilic component makes the molecule water soluble. The hydrophobic portion of the molecule typically attaches to the soil, solids, or fibres, whereas the hydrophilic portion typically attaches to water. Fatty acids or a carbon group with a relatively long chain, such as fatty alcohols or alkylbenzene, may be the water-insoluble hydrophobic group. The hydrophilic group that is water-soluble could be -COONa or a sulfo group like $-OSO_3Na$ or $-SO_3Na$ (such as in fatty alcohol sulfate or alkylbenzene sulfonate).

15.4.2 Soap

Long chain fatty acids are salts of sodium or potassium, and these are soaps. Triglyceride in fat or oil is hydrolyzed and transformed into soap and glycerol when it reacts with aqueous NaOH or KOH. This process is known as esters' alkaline hydrolysis.

15.4.2.1 Manufacturing process

Cold or hot processes are used to make soap. Typically, a heated process is used to make bath and laundry soaps. Cold method is used to create transparent soaps and other specialty varieties. Glycerol is a by-product that is separated off in both processes. Cold or hot processes are used to make soap.

(A) **Raw materials:** There are mainly raw materials for soap, i.e., fat and alkali (sodium hydroxide or potassium hydroxide). Previously, animal fat directly from slaughterhouse was used, but today, producing soap involves the use of numerous vegetable fats, such as olive oil, palm kernel oil, and coconut oil. Also, catalyst like zinc oxide is also used during saponification. Various additives such as color, texture, dye, fragrance, fillers, binding materials, perfumes and scent are also added after hydrolysis.

(B) **Chemical reactions:** Triglycerides in fat and oil are transformed into soap and glycerol when they interact with aqueous sodium hydroxide or potassium hydroxide. This process is known as esters' alkaline hydrolysis. This special type of hydrolysis is called Saponification process. Also, catalyst like zinc oxide is also used during saponification as per Fig. 15.6.

(C) **Flow chart**

(D) **Process description:** As per Fig. 15.7, the catalyst zinc oxide and the raw ingredients oil or/and fat are combined in a blending tank. This mixture is fed to hydrolysis still from the bottom of still. Also, hot aqueous sodium hydroxide solution and steam is added from top of still. This still contains steam coils to maintain temperature of 250°C and pressure of 40 atmospheres. The splitting of fat occurs continuously in a counter-flow fashion. The fat causes the aqueous phase to rise once more, dissolving glycerol in the process. Water containing glycerol is continuously discharged from bottom of still.

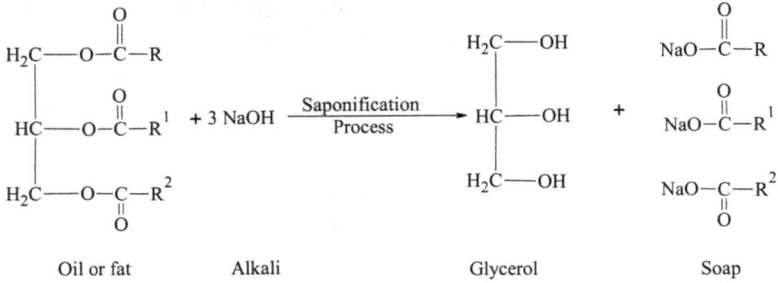

Figure 15.6 Chemical reactions for saponification process

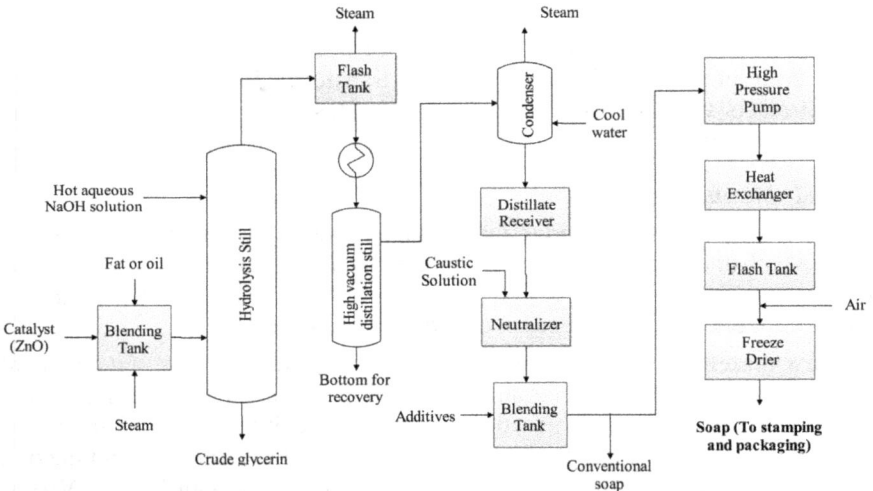

Figure 15.7 Flow sheet diagram of production of soap

The fatty acid is taken from top of still and fed to flash tank, where excess of water in form of steam is separated. This fatty acid is cooled down in heat exchanger and fed to high vacuum distillation still. Here, overhead product containing fatty acid is further undergone condenser, whereas bottom product is stored for recovery. The fatty acid further cooled down with cool water and collected distillate in distillation receiver. It is neutralized by caustic soda in a continuous neutralizer and blending with additives like color, texture, dye, fragrance, fillers, binding materials, perfumes, etc. as per required to form conventional soap. The conventional soap is eventually collected in a flash tank after being dried in a high pressure steam exchanger by heat and pressure. The soap is chilled in a freeze dryer after being combined with air. The prepared soap in further undergone various processes like bar conversion, stamping, packaging, etc.

15.4.1.2 Uses

Soaps are being used excellent for cleaning agent every day. It form gel, emulsifies the oil and lower the surface tension of water. It is also effective bactericide. It has good biodegradability (Isenberg, 1992).

15.4.2 Detergent

The sodium salts of a long chain of benzene sulfuric acid are detergents. It is used in both, powder and liquid and; sold as laundry powder.

15.4.2.1 Manufacturing process

For manufacturing detergent, surfactant is main raw materials, which is produced from petrochemicals, i.e., alkyl benzene.

(A) Raw materials: Alkyl benzene, Oleum and other additives like color, texture, dye, fragrance, fillers, binders, perfumes, etc.

(B) Chemical reactions: Alkylbenzene is reacted with oleum to give alklybenzene sulfonate. The reactions are as follows.

Main reaction:

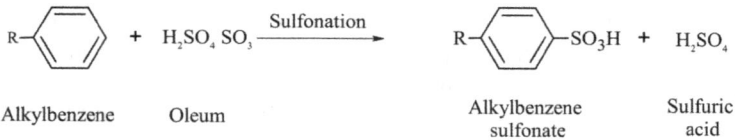

Side reaction:

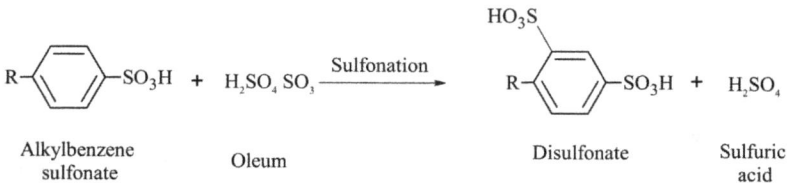

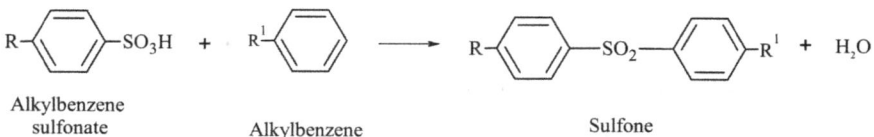

Figure 15.8 Synthesis of detergent

(C) Flow chart

(D) Process description: Initially, alkyl benzene and oleum are fed into sulfonation still, where alkyl benzene is prepared. It is stored in another storage tank and cooled down using tubular cooler. It is then neutralized with caustic solution to form surfactant, which store in separate tank. Now,

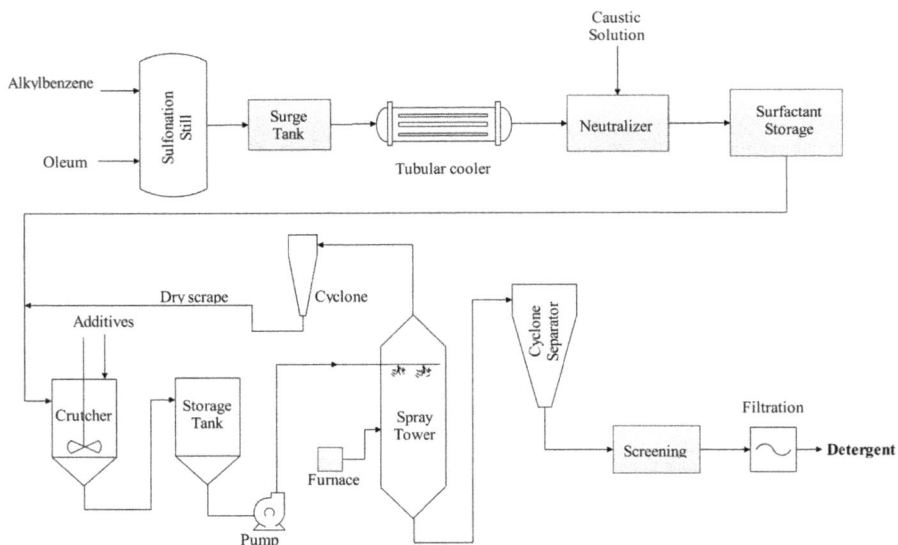

Figure 15.9 Flow sheet diagram of production of detergent

prepared surfactant and other additives are mixed into crutcher. It is store in storage tank. It is then pumped into spray tower, where surfactant mass is sprayed from the top of tower. Hot air is generated using furnace and passed through the bottom of tower to dry the surfactant mixture. Light particle is collected from top of column. It is undergone cyclone separator and heavy particles as dry scrape is recycled into crutcher. Dry surfactant mass is undergone cyclone separator, screening process and finally, filters to get detergent.

15.4.2.2 Uses

It is used to clean baseboards, doors, cabinets, carpet and molding. It is also used to clean tiles, floors, counters, tubs, dish, oil spills and toilets. It is also used to kill moss and weeds. It is used to prepare homemade bubble solution for bubble guns (Mashrabboyevna et al., 2022; Groot et al., 1991; Cheremisinoff and Rosenfeld, 2010).

15.5 Paper & Pulp Industries

15.5.1 Introduction

Trees are the primary source of cellulose fibre, which is used to make paper (or wood-pulp). Paper can also be created from resources like cotton, flax, esparto, straw, hemp, manilla, and jute in addition to wood pulp. The method used to extract the fibres from the timber affects some of the pulp's qualities. Chemical and mechanical processes are the two main ones. Papers are classified according to their different weights and sizes, and it is measured by GSM. GSM stands for gram per square meter, which is actual weight of paper. Various papers are available in market having their characteristic and application according to gsm varies from 80 to 400 gsm as per Table 15.2.

Table 15.2 Classification of paper

Type	GSM	Characteristics	Application
Layout paper	50	Thin transparent paper with a smoothsurface.	Used for concept development and sketching. It is typically utilized for tracing and is frequently used in the preparation of final concepts.
Tracing paper	60	Transparent, tough, and powerful. Due to its transparency, it is useful for tracing little details. You can use a craft knife to scrape off errors because to the resilience of the tracing paper.	Used for working drawings.
Photocopier paper	80	When purchased in quantity, it is quite affordable and comes in a variety of colors.	Typically used for drawing, although paint can also be used on it because of its excellent surface for pens, pencils, and markers. Additionally, it often is used for inkjet printing and photocopying.
Isometric paper	80– 100	Isometric paper is a ready-made printedsheet with isometric lines printed on it.	It is mainly used as an aid for isometricdrawings.
Grid paper	80– 100	A grid sheet with pre-printed vertical and horizontal lines. Working drawings with orthographic projection benefit from these lines. These grids have millimeter dimensions.	Mainly used for working drawings.
Sugar paper 100 gsm	100	Colored paper with contrasting colors is known as sugar paper. Tonal drawings are the major usage for it. For various needs, several weights and textures are offered.	It is mostly used for mounting drawings and exhibition work, however with time, sunshine will cause it to fade.
Cartridge paper	120–150	It costs more than copier paper. It is typically creamy white paper with a soft texture.	Typically used for drawing, although paint can also be used on it because of its excellent surface for pens, pencils, and markers. Watercolors, pastels, crayons, inks, and gouache can all be used with it.
Bleed proof paper	120–150	Similar to cartridge paper, bleed proof paper is especially effective at keeping water-based paints and inks from smearing on surfaces you don't want them to.	When quality is required for major presentations, it is mostly used.

Table 15.2 contd. ...

...Table 15.2 contd.

Type	GSM	Characteristics	Application
Ink jet card	120–280	All varieties of inkjet printers can use inkjet cards because of how they are processed.	It is only used with inkjet printers to provide high-quality print finishes.
Cardboard	125–300	Cardboard has a fantastic surface for printing and is an affordable, recyclable, rigid board.	Boxes and cartons, as well as packing, are its principal uses.
White board	200–400	The surface of a white board is ideal for printing and is made of a sturdy material thanks to bleaching.	It is mostly utilized for high-quality book covers and packaging.
Duplex board	230–420	A less expensive alternative to white board, duplex board offers several printing textures.	Due to the inability of recycled materials for this function, it is mostly utilized in food packaging.
Corrugated card	250+	Contains two or more layers of card with a fluted inner part that interlocks. The card's fluted inner section strengthens it without significantly increasing its weight.	It is primarily employed when shipping products that need to be protected.

15.5.2 Manufacturing of pulp

Basically, pulp is a commercially available fibrous material made from materials like bagasse (waste material), bamboo, wood, etc. Pulping is the process of reducing large, fibrous material to minute fibres. Mechanical, chemical, and semi-chemical methods are the three basic ways that pulp is produced. There are mainly two process by which pump is manufactured, i.e., (1) Sulfide process and (2) Sulfite process. Differences between these two processes are mentioned below.

The majority of chemical pulp is created using the colorte or Kraft method, which combines sodium colorte (Na_2S) and caustic soda ($NaOH$) in a chemical reaction (called white liquor). The resulting slurry has fibres that are dispersed yet still intact and strong. About half of the wood dissolves into "black liquor", which is mostly composed of the compounds lignin and hemicelluloses. The pulp, which is initially dark brown in color, can be bleached if necessary to achieve a high brightness. Prior to bleaching, the black liquid is removed from the pulp. The sulfite process is an alternate chemical pulping technique that works well with specialized pulp that has been bleached. The active ingredients in the pulping liquor used in the sulfite process, also known as the acid sulfite process, are bisulfite salt (combined SO_2) and physically dissolved SO_2. The bisulfite salt is created using a variety of bases, including ammonia, caustic, magnesium hydroxide, and lime. As a result, sulfite mills can be classed as ammonia, sodium, magnesium, or calcium based mills depending on the base they utilize. The requirements for chlorine-free goods for hygiene papers as well as for printing and writing papers are met by the sulfite pulp. Brown liquor is the common name for the liquor produced by sulfite pulping. The brown liquor can be burned in a recovery boiler, much like the Kraft pulping process, to produce steam and recover chemicals for the pulping process.

Due to its superior pulp strength properties compared to the sulfite process, its ability to be applied to all species of wood, and the effective chemical recovery systems that have been created and put into place, the Kraft process is currently the most widely used chemical pulping method worldwide. Large amounts of water are needed in the sulfite process to wash the used sulfite liquor, however this water can be recycled during the evaporation step. Also, sulfite process emits the polluted water as well as air. At a global scale, about 89% of the production of chemical pulp is manufactured using Kraft pulping, whereas about 5% is produced using sulfite pulping.

15.5.2.1 *Manufacturing of pulp using kraft process (sulfate process)*

In the Kraft process, the lignin that holds the cellulose fibres of the wood together is chemically dissolved under high temperature and pressure using white liquor, a water solution of sodium sulfide (Na_2S) and sodium hydroxide. After the wood chips have through this digestion, the wood pulp is either further dignified in an oxygen stage or bleached in a bleach plant, or it is washed, screened, and dried to produce unbleached pulp. Depending on the product's intended usage, a bleaching procedure may be added. The remaining Kraft processes are made to recover heat and chemicals.

(A) Raw materials: Chips (Cellulosic plant materials or other discard papers), Sodium sulfide, Sodium hydroxide, Sodium sulfite, Sulfur and Lime

(B) Flow chart

(C) Process Description

Following major processes are involved in manufacturing the pulp by Kraft process.

(1) Chipper bin: Chips are sent to this apparatus, where powerful knives on rotating discs reduce the wood to the size of 2–5 cm flat chips. In order to maximize the penetration of process chemicals, size reduction is carried out.

(2) Digester tower: Lignin and other non-cellulosic components are released after digestion. First chips are put from the top into a continuous digester tower that is 25 to 30 meters tall, warmed with volatilizing turpentine and non-condensable gases. Additionally, white liquor that has undergone chemical recovery is fed from the top. Cooking liquor is extracted as side streams and circulated through a heat exchanger to regulate the temperature of the digestive system. At 170°C, cooking takes around an hour and a half. Recycled black liquor is used to cool digested chips to prevent mechanical deterioration of fibres. Pressure is kept at around 10 atmospheres and the temperature is kept between 140 and 180°C. The temperature at the bottom is kept at 65°C. Distillation is used to separate the vapour collected from the digester's top into turpentine and water. The digester's bottom is used to collect the hot pulp slurry liquor that has been digested.

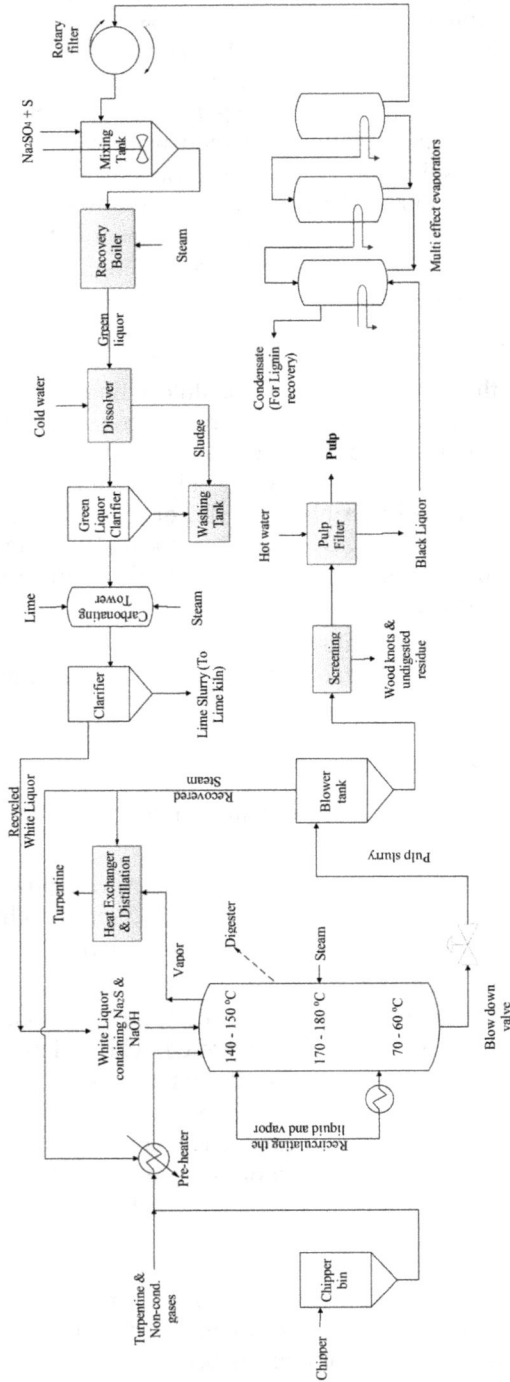

Figure 15.10 Flow sheet diagram of production of pulp using Kraft process (Sulfide process)

Black liquor undergone chemical treatment
for chemical recovery to make white liquor

(3) Blower: Through the blow down valve, heated, digested pulp slurry is transferred to the blow tank. Before entering the blow tank, this valve lowers the pressure of the stream from 80 to 1 atmosphere. In this instance, steam is produced by recovering heat. Chips are pre-heated with recovered steam, and distillation is also done with it. In the blow tank, which collects water from the tank bottom, there is a high concentration of pulp and a low concentration of water.

(4) Screener and Pulp filter: To get rid of wood knots and undigested leftovers, pulp is filtered. The pulp is then put into the digester tower after being filtered to separate the black liquor and put through a chemical recovery process to make the white liquor. For improved filtration, hot water is supplied to the second filter.

(5) Treatment of black liquor: The black liquor containing 15–18% solid is concentrated in multi effect-evaporator and disc evaporator. Here, black liquor is concentrated upto 60–65% solid. The product is then mixed using two agitators in the design. Make up ingredients, i.e., sodium sulfate and sulfur are added. After proper mixing, the mixture is heated into recovery furnace. Here, the black liquor's organic carbon is burned. Steam is created at high pressures of 28 to 30 atmospheres. The internal furnace reaction that results in molten slag (smelt). When combined with cold water, the molten chemical smelt dissolves instantly and produces green liquor (aqueous sodium carbonate) in the dissolver. After that, white liquor and calcium carbonate sludge are separated from green liquor by clarification. After clarification, impurities are removed in this tank by washing. The white liquor is put through a carbonating tank where lime is added to cause lime mud to precipitate. White liquor is ready to be added to the digester as the filtrate after the lime mud has been removed during clarification.

15.5.2.1 Manufacturing of pulp using sulfite process

In sulfite process, wood chips (hardwood and softwood) are digested into sulfurous acid and base materials to soften the chips.

(A) Raw materials: Sulfur, Air and Base materials (Ca, Na, Mg, NH_4, etc.)

(B) Chemical reactions

$$S + O_2 \longrightarrow SO_2$$

$$SO_2 + H_2O \longrightarrow H_2SO_3$$

$$SO_2 + H_2O + CaCO_3 \longrightarrow Ca(HSO_3)_2 + CO_2$$

$$SO_2 + H_2O + MgCO_3 \longrightarrow Mg(HSO_3)_2 + CO_2$$

$$2SO_2 + Mg(OH)_2 \longrightarrow Mg(HSO_3)_2$$

(C) Flow chart

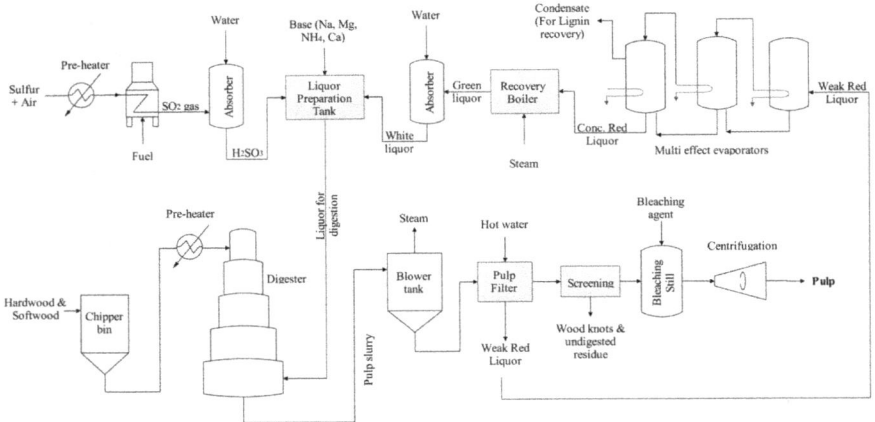

Figure 15.11 Flow sheet diagram of production of pulp using Sulfite process

(D) Process description: Following major processes are involved in manufacturing the pulp by sulfite process.

(1) Chipper bin: Both hardwood and softwood are fed into this apparatus, where rotary discs with powerful knives chop the wood into flat chips that range in size from 2 to 5 cm. Size reduction is conducted for maximize penetration of process chemicals.

(2) Preparation of liquor: Sulfur and air is pre-heated and burnt in furnace to get sulfur dioxide gas. This gas is passed through water in absorber to yield sulfurous acid. This acid is stored is liquor preparation tank. Base solution and recycled white liquor is also added in this tank.

(2) Digester tower: Lignin and other non-cellulosic components are released after digestion. First chips are put from the top into a continuous digester tower that is 25 to 30 meters tall, warmed with volatilizing turpentine and non-condensable gases. Also, liquor is also added from the bottom of digester tower. The digestion process is carried out for time duration of 6–12 hours, temperature of 120–150°C at pH 1.5–5. These parameters are depends upon type of chips that are to digester. The digested hot pulp slurry liquor is collected from the bottom of digester.

(3) Blower: The blow tank receives the heated pulp slurry after digestion. In this instance, steam is produced by recovering heat. Steam that has been recovered is used in another procedure. In the blow tank, which collects water from the tank bottom, there is a high concentration of pulp and a low concentration of water.

(4) Final processes: Red liquors separated from by filtration and undergo the chemical recovery process to make white liquor and inserted into digester tower. The screening of pulp is conducted to remove wood knots and undigested residues. It is then bleached using bleaching agent in bleaching tank. Finally, pulp is collected using centrifugation.

(5) Treatment of Red liquid: Weak red liquid contain 30–40% solid, which is undergone multi-effect evaporator (MEE) to increase solid contain. This MEE gives two products: (1) Condensate, which is used to extract lignin and alcohol, (2) Concentrated red liquor. This concentrated red liquor is further burnt in recovery boiler to collect base materials, called green liquor. This green liquor is adsorbed in absorber using water to form white liquor and recycled.

15.5.3 Paper making process

The beater, which consists of metal blades attached to a rotating drum, is used to process pulp that has either been made via the Kraft process or the sulfite process. The purpose of a beater is to mechanically separate the pulp fibres, resulting in paper that is more durable, even, dense, opaque, etc. To improve brightness, flexibility, softness, weight, and coloring agents, finely ground fillers are also added. The pulp slurry is inserted into Jordan. The Jordon is conical refiner or Jordan engine having metal bars and stones are set inside. Here, pulp is deformed, defibered and dispersed. Then, slurry is transferred into slurry storage. Now, slurry undergoes the process of web forming. In web forming, fiber slurry is made to run on an endless belt at a speed of 50 m/min to 500 m/min. The web-like arrangement of pulp fibres. Gravity drains the water out. To help the fibres on the mat better interlock, a shaking action is offered. White water, which contains 0.5% pulp, is collected and reused to save water, reduce the need for additives, and prevent pollution. The remaining free water is then pressed out using suction, water mark, and pressure rolls. 60–65% of the original water content remains. By using a sequence of metal drying rolls that are heated by steam and smoothing rolls, further water is removed. Water content is decreased from 60–65% to 5%. The paper is run through a succession of calendaring rolls in the finishing step to produce smooth paper. It has wounds on a significant winding roll (Rao et al., 1994; Rosenfeld and Feng, 2011; Henkel et al., 2007).

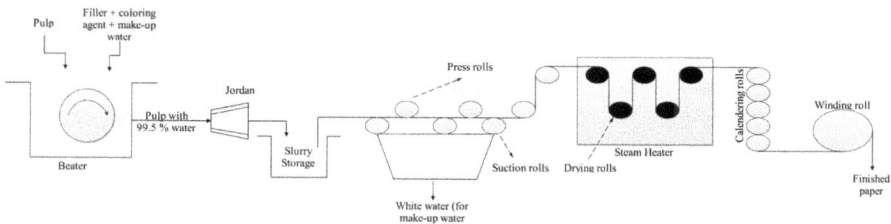

Figure 15.12 Flow sheet diagram of production of paper from pulp

15.6 Glass Industries

15.6.1 Introduction

Glass is described as a physically stiff, non-crystalline, amorphous solid that is frequently transparent and finds broad use in window panes, dinnerware, and optoelectronics as well as in practical, technological, and ornamental applications. Technically, the term "glass" refers to any solid that exhibits a glass transition when heated toward the liquid state and has an atomic structure that is non-crystalline, or amorphous. Alkali, alkaline earth, and more widely used metal silicates are fused together to form glass. Since it exhibits optical transparency, window panes are its main application. Glass can be cut and polished to produce optical lenses, prisms, fine glassware, and optical fibres for high-speed data transmission by light. Glass has the ability to transmit, reflect, and refract light. Traditional applications include bowls, vases, bottles, jars, and drinking glasses. It has also been used to make paperweights, marbles, and beads in its most solid forms. It transforms into a thermal insulating material when it is extruded as glass fibre and matted as glass wool in a way to trap air. These glass fibres are a crucial structural reinforcing component of the composite material fibreglass when they are implanted in an organic polymer plastic. Drinking glasses and reading glasses are examples of items that were previously created so frequently with silicate glass that they are simply referred to their constituent material. There are over 3000 types of glasses are available. But, in general commercial glass are classified as follows.

(A) **Fused silica:** Vitreous silica glass is silica (SiO_2) in glass or vitreous form, commonly known as fused quartz or fused silica (i.e., its molecules are disordered and random, without crystalline structure). Quartz glass is a remarkably versatile material utilized in a wide range of applications. Excellent optical transmission, good chemical resistance, great thermal qualities (1000–1500°C), low thermal expansion coefficient, superior electrical and corrosion performance. Additionally, it has the best resistance to weathering (caused in other glasses by alkali ions leaching out of the glass, while staining it). It is used to make mirror substrates, crucibles, trays, and boats as well as UV and IR transmitting optics (made of synthetic fused silica). For high temperature applications like furnace tubes, lighting tubes, etc., fused quartz is used.

(B) **Alkali silicate glass:** It is frequently referred to as "water glass" or "liquid glass," and it has numerous commercial and industrial uses. It primarily consists of an oxygen-silicon polymer backbone that holds water in the pores of the molecular matrix. Sodium carbonate, or Na_2CO_3, and silica sand, a common source of SiO_2, are often roasted in a furnace at temperatures between 1,000 and 1,400°C to create sodium silicate glass. This process releases carbon dioxide into the atmosphere. After cooling, it is powdered. Its chemical make-up ranges from Na_2O, SiO_2, to Na_2O_4, SiO_2. White sodium silicate powder dissolves easily in water to form an alkaline solution. In neutral and alkaline conditions, it is stable. In acidic liquids, the silicate ion and hydrogen ions combine to make silicic acid, which is then heated and roasted to create silica gel, a hard, glassy solid. Due to its versatile properties, it is used in metal repair, automobile repair leak, car engineer disablement, as adhesive, in aquaculture, in food preservation, as a passive fire protection, in concrete treatment, refractory uses.

(C) Soda-lime glass: The most widely used type of glass, glass contains both calcium and sodium. The raw ingredients for soda-lime glass are melted in a glass furnace at temperatures up to 1675°C, together with lime, dolomite, silicon dioxide (silica), aluminum oxide (alumina), and small amounts of fining agents (such as sodium sulfate, sodium chloride). This glass contains 72% silica, 14.2% sodium oxide (Na_2O), 10% lime (CaO), 2.5% magnesia (MgO), and 0.6% alumina as its constituents (Al_2O_3). The main properties of this glass are that good electrical insulator, very good transmission properties for visible light. Without being toughened, it can withstand temperatures of up to 80–90°C and can be honed or polished to the required finish. Toughening increases strength and lowers the maximum use temperature. Bottles, jars, windows in structures, tubes, rods, plates, and full glassware components can all be made using this glass at a reasonable cost. It is also employed for many different scientific and industrial applications, although it will not retain as high a temperature as borosilicate glass.

(D) Lead glass: Due its high refractive index and dispersion, it looks likes more brilliant, so it called "crystal". It is prepared by adding lead oxide in preheated calcium oxide in furnace at 1200 °C. Its composition is 59% silica, 25% lead oxide (PbO), 12% potassium oxide (K_2O), 2.0% soda (Na_2O), 1.5% zinc oxide (ZnO) and 0.4% alumina. It has numerous applications across a wide range of industries, from serving as tableware to protecting medical and scientific personnel from radiation. The most popular uses for lead crystal glass are as decorative items such drinking glasses, ornaments, decanters, jewellery, optical lenses, enamels and lacquers, glass sealants and solders, and as gamma and x-ray radiation shielding.

(E) Borosilicate glass: Silica (80–87%) and boric oxide (10–20%) make up the majority of borosilicate glass (also known as sodium-borosilicate glass), with aluminum oxide and other alkalis (sodium and potassium oxides) making up the remainder. Because sodium borosilicate glass has a low coefficient of expansion, superior shock resistance, chemical stability, and electrical resistance, it is widely used in the chemical industry, the pharmaceutical industry, for laboratory equipment, for ampoules and other pharmaceutical containers, for various high intensity lighting applications, as glass fibres for textile and plastic reinforcement, and for everyday cookware and ovens. Under the brand name "Pyrex," this glass is used to manufacture laboratory glassware.

(F) Special glass: The phrase "Special glass" describes a broad variety of technical glasses with radically diverse chemical compositions that are intended for radically different markets. Specialized furnaces and production techniques are used to make glasses. It comprises ceramic glasses that are colored, coated, opal, translucent, safe, and photochromic.

(G) Glass fiber: Fiberglass is another name for glass fibre. The method of making glass fibres after direct melting may also involve nozzle drawing, blowing, and rod drawing. 55% SiO_2, 10% B_2O_3, 14% Al_2O_3, 13% CaO, 5% MgO, and 0.5% Na_2O make up its composition. It is a substance created from incredibly fine glass fibres. Glass fibres that are incredibly fine are used to make fibreglass. It is a thin, incredibly strong, and robust substance. This glass' strength is less than that of carbon fibre, and it is also less stiff and naturally less brittle. However, compared to metals, its bulk strength and weight characteristics are also quite beneficial, and it can be easily manufactured utilizing molding procedures. The earliest and most well-known performance fibre is glass. Glass has been used to make fibres since the 1930s. Glass fibres are utilized in a variety of products, including fabrics for home furnishings, clothing, shoes, reinforced plastics, heat shields for aircraft, boat hulls and seats, fishing rods, wall paneling, and filament windings around rocket casings.

15.6.2 Manufacturing of glass

There are various types of glasses available and their raw materials with its particles sizes and preparation are different. But, common method is described as follows.

15.6.2.1 Raw materials

Glass usually contains between 70% and 90% silica (from sand) with most of the remainder consisting of a mixture of lime, soda ash and feldspar. Glass also contains small amounts of many other ingredients.

Sand: Over 99.5% of good glass sand is silica (quartz). The most popular glass-forming oxide and the foundation of most glasses is silica, sometimes known as silicon dioxide. The oxides of boron and phosphorus are the second most prevalent glass-forming oxides.

Lime: Depending on the kind of glass being prepared, either calcium oxide, magnesium oxide, or a combination of the two are used as limes. This lime is made from calcite or dolomite that has been mined. The most popular glass modifier, used to make glass easier to melt when heated, is lime.

Soda Ash: Sodium oxide, sometimes known as soda ash, is a typical flux used with glass. Glass is affected by fluxes at relatively low temperatures, which enhances the glass' forming properties. Insufficient stabilizers can actually make glass soluble in water when soda ash is used.

Feldspar: Alumina, one of the most significant stabilizers added to glass, is found in feldspar. The resistance of glass to chemicals, especially water, is increased by glass stabilizers. Glass also employs a wide variety of additional stabilizers. Feldspar's general formula is $R_2O \cdot Al_2O_3 \cdot 6SiO_2$, where R_2O can be either Na_2O, K_2O, or a combination of the two.

Other Ingredients: For various reasons, glass is enhanced with almost every stable element from the periodic table as well as a huge number of other compounds. Arsenic, antimony, barium, fluorine, iron, and lead are some of the substances that are frequently added to glass. Numerous oxides and sulfates are frequent additives to glass. Glass is enhanced using coke and slag from blast furnaces.

Glass can be processed and prepared more easily with some of these substances. Some enhance its consistency, strength, flexibility, opacity, optical qualities, heat resistance, chemical resistance, thermal shock resistance, and many other attributes. Glass is colored by metal oxides, such as selenium for red, chromium for green, and cobalt for blue.

15.6.2.2 Chemical reactions

$$Na_2CO_3 + CaCO_3 \longrightarrow Na_2Ca(CO_3)_2$$

$$Na_2Ca(CO_3)_2 + 2SiO_2 \longrightarrow Na_2SiO_3/CaSiO_3 + 2CO_2$$

$$Na_2CO_3 + 2SiO_2 \longrightarrow Na_2Si_2O_5 + CO_2$$

$$Na_2CO_3 + SiO_2 \longrightarrow Na_2SiO_3 + CO_2$$

15.6.2.3 Flow chart

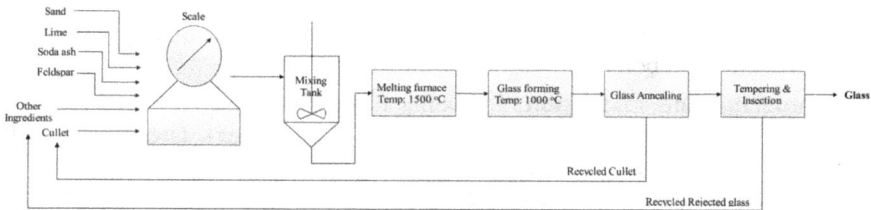

Figure 15.13 Flow sheet diagram of production of grass

15.6.2.4 Process description

The manufacturing process is divided into following parts.

(1) **Weighing and mixing:** All the raw materials, i.e., sand, lime, soda ash, feldspar and other ingredients with consistent properties are weighing in proper proportions. All raw materials must also have their particle sizes under tight control. The ratios of different particle sizes must be maintained constant, and particles must stay within certain size limits. This is done either by buying the right-sized particles or by screening and sorting the raw ingredients. All raw materials are appropriately blended after being weighed.

(2) **Melting:** The most crucial process for producing glass in the requisite amount and quality is melting. In refractory tanks, large-scale commercial melting occurs after high-temperature fusing of the raw materials in a furnace. The kind and volume of glass being made determines the furnace to use. Unit melters, recuperative melters, electric melters, and regenerative furnaces are among the several types of furnaces that are frequently employed. Usually, the entire furnace tank, including the cooling and melting tanks, comes into contact with the glass liquid before the moulding section. When the qualifying glass has finished being produced from the material melting and the glass liquid is in the convection condition, the temperatures fluctuate from 1200 to

1600°C during the entire process. Refractory blocks, which can function at temperatures above 1500°C, make up the majority of the furnaces. Bubbles emerge as a result of the carbon dioxide and water emissions that occur as the raw materials melt and react inside the furnace. High temperature and low viscosity are used to eliminate the gas bubble from the melt.

(3) **Glass forming:** A transitional step in the production of glass is glass shaping. Between glass melting and annealing, it occurs. After furnace process, melt glass is undergo controlled cooling from furnace temperatures to 1000–1100°C. Thereafter, it is shaping as per product requirement this temperature at like bottle, glass, window pane. In forming process, the broken glass, known as cullet, is further crushed and recycled into weighing.

(4) **Annealing:** This procedure is described as a method for reducing internal tensions produced during manufacturing by gently chilling hot glass products after they have been prepared. Glass products will break very easily if this strain is not relieved. After being drawn, rolled, or floated, flat glass is annealed before being cut. Glass containers are made, then annealed. The majority of annealing processes use fuel. These lessons come with the same risks as any fuel-fired device. The process of annealing involves recycling cullet. An essential step in maintaining the tension and quality of glass is annealing the labor.

(5) **Tempering and inspection:** By being heated in a tempering furnace and then carefully cooled, glass is significantly strengthened. The tension in the glass's core layers is balanced by the compressive forces introduced during the tempering process at the glass's surface. The glass is incredibly robust due to this internal force balance. Most tempering furnaces are fuel heated, just like annealing lehrs. Tempering furnaces have the same risks as any other fuel-fired equipment.

In whole process of glass manufacturing, cullet is a term for the crushed leftovers from the glassmaking process, such as broken or rejected glass. It has advantage that it is easily recyclable again and again without compromise its quantity and quality. About 20–30% of glass contains is recycled with reducing cost of new raw materials. It is also useful to our environment because it saves natural resources like raw materials, i.e., extend the plant life and also, saves energy.

15.6.2.5 Major engineering problems

1. Pollution is the considerable problems for glass industries. Fresh water is used to cool the furnace, compressor and unused molten glass. The amount of water used in factories varies greatly; for every tonne of molten glass, as little as one tonne of water may be used. About half of the one tonne evaporates to provide cooling, and the remaining portion becomes a wastewater stream.

2. Oil is included in the effluent because oil is used in gob-cutting shear blades and for cooling purposes. It is thought to be highly polluted wastewater. Oil traps are used to extract oil from effluent.

3. Gas-fired furnaces emit a lot of greenhouse gases, such as nitrogen and sulfur oxides, which have an adverse effect on the environment.

4. All of the raw ingredients used to make glass are dusty and come in either powder or fine-grained form. The dusty materials are challenging to manage. Additionally, recycled shattered glass, commonly known as cullet, is used in glass factories (Bourhis, 2007; Tackels, 2017; Brandt, 2009).

15.7 Cement Industries

15.7.1 Introduction

Cement is referred to as a binder, a substance used in building that hardens, sets, and can unite different materials. The two most crucial types of cement are employed in the creation of mortar for masonry work and concrete, which combines cement with aggregate to create a sturdy building material. Based on how they set and harden, cement is widely divided into hydraulic cements and non-hydraulic cements. Under water, hydraulic cements have the ability to set and solidify. In contrast, non-hydraulic cements are set and harden in air, and hence can't be used under water.

Further, cement is also classified into the following classes:

1) **Natural cement**: It is natural materials like limestone, having hydraulic properties. It is not useful due to properties ofquick-setting and have low strength.

2) **Pozzolane cement:** It is one of the oldest cementing materials, invented by Romans to prepare walls and domes. It is manufactured by simply mixing and grinding two materials: (1) slaked-lime and (2) natural pazzolana containing glassy materials, which is produced by rapid cooling of lava.

3) **High Aluminum Cement:** In this type of cement, the percentage aluminum is higher, which help to setting down very rapidly.

4) **Super Sulfate Cement (Slag cement):** It is prepared by furnace slag and hydrated lime, having very low hardening properties.

5) **Portland cement:** Beginning in the middle of the 18th century, natural cements were produced in Britain, from which Portland cement was formed. Because of its resemblance to Portland stone, a type of construction stone produced on the Isle of Portland in Dorset, England, it was given that name. The most significant and trustworthy cement in use today. Chemically, it is a mixture of complex silicates and aluminates of calcium containing less than 0.1% free lime and gypsum. It is prepared by calcining and clinkering the mixture of raw materials (calcareous, argillaceous material) and then adding gypsum. The Portland cement is further classified as follows.

1) **Regular Portland cement (Type-I):** This is general portland cement, normally used for common construction purpose. It is harden to full strength in 28 days. Other types of this cements are white cement, quick setting cement, oil well cement, etc. for special uses.

2) **Moderate heat of hardening cement (Type-II):** It is also called as sulfate resisting portland cement, used in general construction purpose, where moderate sulfur attack. It required moderate heat of hydration (Heat resulting from chemical reactions with water).

3) **High early strength cement (Type-III):** This type hardens up to three days sooner than regular Portland cement, as opposed to seven days. Though initially stronger, they become equal after two to three months. This cement is used to build roads because it is finely crushed, hardens quickly, and evaluates heat more quickly. Constructions made of bulk concrete do not employ it.

4) **Low heat Portland cement (Type-IV):** In comparison to regular Portland cement, it has lesser early strength (half the strength at 7 days and two thirds the strength at 28 days). Constructions made with bulk concrete employ it.

5) **Sulfate resisting portland cement (Type-V):** The heat of hydration that results from it is a little higher than what results from low heat cement. Its costs are higher than those of regular Portland cement, and its early strength is extremely poor.

15.7.2 Manufacturing of cement

Portland cement is prepared by mixing and burning the raw materials (calcareous and argillaceous materials) in rotary kiln. Pulverized coal is used as fuel and gypsum is added after kiln process.

Table 15.3 Details of raw materials used in manufacturing cement

Raw materials	Natural Sources	Chemical Constituent	Importance in cement
Calcareous materials	limestone, coral, marl	Lime (CaO)	It is the main component of cement. It has an impact on cement's strength. A surplus of lime will weaken cement because it causes the material to expand and break down. However, less lime will also weaken the bond since it changes cement's setting and hardening characteristics.
Argillaceous materials	Clay, shale, slate, ashes, blast furnace slag, cement rocks	Lime (CaO)	As per below
		Silica (SiO_2)	It imparts strength to cement.
		Alumina (Al_2O_3)	It is in charge of the cement setting. Alumina excess weakens the cement.
		Iron oxide (Fe_2O_3)	It provides color, strength and hardness to cement.
		Alkali	If it is present in excess, it can lead to cement efflorescence, a white coating of mineral salts on the face of concrete.
Gypsum ($CaSO_4$)	Gypsum	Calsuim sulphate	It enhances the initial time of cement.
		Sulphur trioxide	It imparts soundness property of cement.
Pulverized coal	-	-	It is used as a fuel. Sometime, peat coal is also utilized.

15.7.2.1 Raw materials

Raw materials used for manufacturing the cement are calcareous materials, argillaceous materials, gypsum and pulverized coal. Its chemical constituent and importance of each raw material are tabulated as per Table 15.3.

15.7.2.2 Chemical reactions

Various chemicals reactions are done in two different zones, which are mentioned below.

Calcination zone

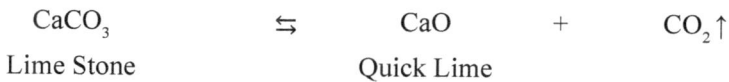

$$CaCO_3 \quad \leftrightarrows \quad CaO \quad + \quad CO_2\uparrow$$

Lime Stone   Quick Lime

Clinkering zone

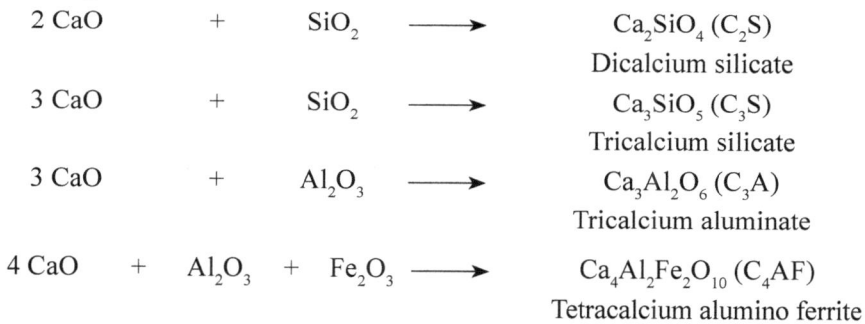

$$2\ CaO \quad + \quad SiO_2 \longrightarrow Ca_2SiO_4\ (C_2S)$$

Dicalcium silicate

$$3\ CaO \quad + \quad SiO_2 \longrightarrow Ca_3SiO_5\ (C_3S)$$

Tricalcium silicate

$$3\ CaO \quad + \quad Al_2O_3 \longrightarrow Ca_3Al_2O_6\ (C_3A)$$

Tricalcium aluminate

$$4\ CaO \quad + \quad Al_2O_3 + Fe_2O_3 \longrightarrow Ca_4Al_2Fe_2O_{10}\ (C_4AF)$$

Tetracalcium alumino ferrite

15.7.2.3 Flow sheet diagram

15.7.2.4 Process description

Manufacture of portland cement involved the following steps.

(A) Mixing of raw materials: It is conducted by two processes: (a) Wet process or (b) Dry process.

(a) Wet process: In this process, the raw materials are finely grounded in the required percentage. Argillaceous materials are washed with water thoroughly to remove organic matter. Powdered calcareous and washed argillaceous materials are added in proportioning tank. Thereafter, these raw materials are mixed properly in grinding mill to make paste or slurry. The slurry is fed into a rotary kiln.

(b) Dry process: To create small bits, the raw materials are crushed in gyratory crushers. They are ground in fine powder in ball mill and stored separately. Then, they are mixed properly and further, pulverized in tube mill. This is called a 'dry raw mix' and insert into rotary kiln.

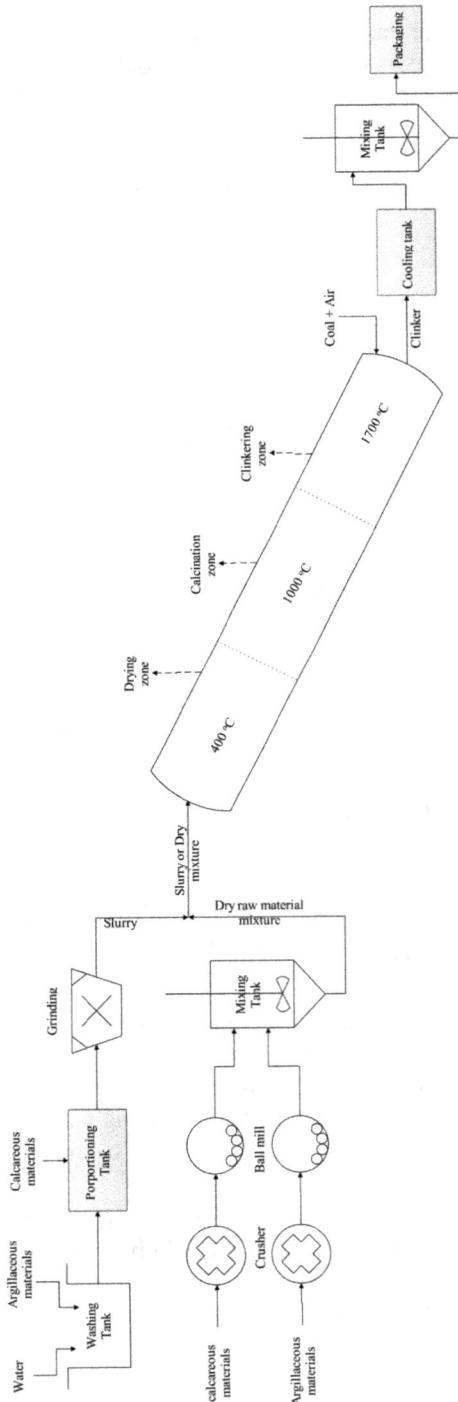

Figure 15.14 Flow sheet diagram of production of cement

(B) Burning of raw materials: This procedure is carried out in a steel tube-based rotary kiln that has been lined with refractory bricks. The length of kiln is about 90 to 120 meters, diameter about 2.5 to 3.0 meters and rotating at a speed of 0.5 to 2.0 rpm at longitudinal axis. The slurry or dry raw mix is inserted into upper part of kiln and burning fuel (Pulverized coal) and air is injected at the opposite lower end as per flow-sheet. The result is a lengthy, hot flame that raises the temperature within the kiln to a maximum of roughly 1750°C. Due to the slope and slow rotation of the kiln, the raw materials are continuously moved towards the hottest-end. Further, kiln is divided into three zones according to temperature and chemical process.

(a) Drying zone: This is upper part of kiln, where temperature is about 400°C. In this zone, lot of the water is get evaporated from slurry, therefore it is called drying zone.

(b) Calcination zone: This is the middle part of kiln, where temperature is about 1000°C. Quick lime and carbon dioxide, which is released later, breakdown limestone. Nodules are tiny lumps formed by the substances.

(c) Clinkering zone: Hottest zone of kiln having temperature of 1500 to 1700°C, is known as clinkering zone. Here, fusion of all raw materials is done to form aluminates and silicates of calcium.

(C) Grinding: The hot fused mass, also known as clinker, from the kiln is cooled to room temperature using rotary coolers. The cooled mass is then pulverized together with 2–6% gypsum in long tube mills.

(D) Packing: The ground cement is then packed and stored in moisture free area (Schafer and Hoenig, 2022)

References

Adigwe, O.P., John J.E. and Emeje, M.O. 2022. History, Evolution and Future of Starch Industry in Nigeria, In book: Starch, IntechOpen Publication. DOI: 10.5772/intechopen.94824

Bourhis, E.L. 2007. Glass: Mechanics and Technology, Wiley-VCH. DOI: 10.1002/9783527617029

Brandt A.M. 2009. Cement based composites: Materials Mechanical Properties and Performance, Taylor & Francis. https://doi.org/10.1201/9781482265866

Cheremisinoff, N.P. and Rosenfeld, P.E. 2010. Handbook of Pollution Prevention and Cleaner Production Vol. 2: Best Practices in the Wood and Paper Industries. https://doi.org/10.1016/C2009-0-20361-8

Duraisam, R., Salelgn, K. and Berekete, A.K. 2017. Production of Beet Sugar and Bio-ethanol from Sugar beet and it Bagasse: A Review. International Journal of Engineering Trends and Technology 43(4). DOI: 10.14445/22315381/IJETT-V43P237

Groot, W.H. 1991. Sulphonation Technology in the Detergent Industry, Springer Netherlands.

Henkel, M., Pleimling, M. and Sanctuary, R. 2007. Ageing and the Glass Transition, Springer. https://doi.org/10.1177/2277977919881399

Isenberg, C. 1992. The science of soap films and soap bubbles, Dover Publications.

Lambourne R. and Strivens, T.A. 1999. Paint and Surface Coatings Theory and Practice, A volume in Woodhead Publishing Series in Metals and Surface Engineering, Woodhead Publishing, 2nd Edition.

Mall, A.K., Misra, V., Santeshwari, Pathak A.D. and Srivastava, S. 2021. Sugar Beet Cultivation in India: Prospects for Bio-Ethanol Production and Value-Added Co-Products, Sugar Technology, 23, 1218-1234. https://doi.org/10.1007/s12355-021-01007-0

Mashrabboyevna, A.M., Voxidjon, M.G., Ugli, O.A.H. 2022. The composition of soaps and its effect on soap quality, ACADEMICIA-An International Multidisciplinary Research Journal 12(1): 398–402.DOI: 10.5958/2249-7137.2022.00084.2

Rafay, A. and Singh, N. 2020. Bright Paint Industries: Expansion Through Internationalization, South Asian Journal of Business and Management Cases 9(1): 40–53. https://doi.org/10.1177/2277977919881399

Rao, M., Xia, Q. and Ying, Y. 1994. Modeling and Advanced Control for Process Industries: Applications to Paper Making Processes, Advances in Industrial Control, Springer-Verlag London. https://doi.org/10.1007/978-1-4471-2094-0

Rosenfeld, P. and Feng, L.G.H. 2011. The Paper and Pulp Industry, In book: Risks of Hazardous Wastes. https://doi.org/10.1016/C2009-0-62341-2

Schafer, S. and Hoenig, V. 2022. Fuels of the future for the cement industry, Chemie Ingenieur Technik 94(9): 1293–1294. https://doi.org/10.1002/cite.202255294

Singh, R. 2015. Membrane Technology and Engineering for Water Purification: Application, Systems Design and Operation, 2nd Edition.

Smith, W.C. 2010. Smart Textile Coatings and Laminates, Woodhead Publishing.

Solomon, S. 2016. Sugarcane Production and Development of Sugar Industry in India, Sugar Technology 18: 588–602. https://doi.org/10.1007/s12355-016-0494-2

Tackels, G. 2017. Industrial Ecology and the Glass Industry, In book: Perspectives on Industrial Ecology, Routledge. https://doi.org/10.4324/9781351282086

Index

For Product Safety Concerns and Information please contact our EU
representative GPSR@taylorandfrancis.com
Taylor & Francis Verlag GmbH, Kaufingerstraße 24, 80331 München, Germany